Catalysts for Sustainable Hydrogen Production: Preparation, Applications and Process Integration

Catalysts for Sustainable Hydrogen Production: Preparation, Applications and Process Integration

Editors

Marco Martino
Concetta Ruocco

MDPI • Basel • Beijing • Wuhan • Barcelona • Belgrade • Manchester • Tokyo • Cluj • Tianjin

Editors
Marco Martino
Industrial Engineering
University of Salerno
Fisciano (SA)
Italy

Concetta Ruocco
Industrial Engineering
University of Salerno
Fisciano
Italy

Editorial Office
MDPI
St. Alban-Anlage 66
4052 Basel, Switzerland

This is a reprint of articles from the Special Issue published online in the open access journal *Catalysts* (ISSN 2073-4344) (available at: www.mdpi.com/journal/catalysts/special_issues/cshp).

For citation purposes, cite each article independently as indicated on the article page online and as indicated below:

LastName, A.A.; LastName, B.B.; LastName, C.C. Article Title. *Journal Name* **Year**, *Volume Number*, Page Range.

ISBN 978-3-0365-3672-9 (Hbk)
ISBN 978-3-0365-3671-2 (PDF)

Contents

About the Editors

Marco Martino

Dr. Marco Martino works in the ProCeed Laboratory of the Industrial Engineering Department of the University of Salerno, in the research group of Prof. V. Palma. Dr. Martino is currently involved in international research projects, in the field of hydrogen production (Pro.Me.Ca.) from CO-Water Gas Shift and propylene production from propane dehydrogenation (Macbeth). Dr. Martino studies the reforming of methane and bioethanol, and waste gasification processes. In 2019, he was named the Top Peer Reviewer 2019 powered by Publons, for ranking in the top 1% of reviewers in the global database of Cross-Field on Publons reviewers, determined by the number of peer review reports conducted during the award year 2018–2019. Since 2020, Dr. Martino has been a member of the Editorial Board of *Catalysts* (MDPI). In 2020, he was among the winners of the Outstanding Reviewer Awards 2019, powered by *Catalysts* (MDPI). Dr. Martino is a reviewer for several internationally prestigious journals published by ACS, Elsevier, MDPI, etc., and is the supervisor of several degree theses.

Concetta Ruocco

Dr. Concetta Ruocco is a Chemical Engineer and works in the ProCeed Laboratory of the Industrial Engineering Department of the University of Salerno, in the research group of Prof. V. Palma.

The research activity of Dr Ruocco is mainly devoted to the intensification of biomass-derived processes, involving the development of innovative catalysts for the production of H_2 from renewable sources. In recent years, Dr. Ruocco has served as a reviewer for several journals (including the *International Journal of Hydrogen Energy*, the *Chemical Engineering Journal*, *Catalysts* and *Energies*); moreover, she collaborated as a Guest Editor (together with Dr. Marco Martino) for the Special Issue "Catalysts for Sustainable Hydrogen Production: Preparation, Applications and Process Integration", which has been recently published in *Catalysts* by MDPI.

Editorial

Catalysts for Sustainable Hydrogen Production: Preparation, Applications and Process Integration

Concetta Ruocco * and **Marco Martino** *

Department of Industrial Engineering, University of Salerno, Via Giovanni Paolo II 132, 84084 Fisciano, SA, Italy
* Correspondence: cruocco@unisa.it (C.R.); mamartino@unisa.it (M.M.)

Citation: Ruocco, C.; Martino, M. Catalysts for Sustainable Hydrogen Production: Preparation, Applications and Process Integration. *Catalysts* **2022**, *12*, 322. https://doi.org/10.3390/catal12030322

Received: 9 March 2022
Accepted: 10 March 2022
Published: 11 March 2022

Publisher's Note: MDPI stays neutral with regard to jurisdictional claims in published maps and institutional affiliations.

The earth is experiencing a series of epochal emergencies, directly related to the overexploitation of natural resources. Climate change, economic and health conditions of poor countries as well as geopolitical tensions affect the availability of raw materials and, above all, the environmental and economic costs of energy. Researchers around the world have raised the alarm about the effect of excessive greenhouse gas emissions on the environment and on humans, advocating for the reduction of greenhouse gas emissions by at least 45% by 2030 [1] and carbon neutrality by 2050 to ensure a containment of global warming to +1.5 °C [2]. The world is hungry for cheap, clean energy; however, these two requirements seem difficult to reconcile. The current most practicable route involves the use of clean energy vectors, and, in this context, hydrogen appears to be extremely promising in supporting the electrification of processes, which requires long-term energy storage [1]. In this Special Issue, a series of articles tackle the issues associated with the production, storage, and purification of hydrogen.

Hydrogen production can be centralized or on-site, depending on the intended use. In both cases, different production processes are used; moreover, in the case of centralized production, the additional steps involved in the storage and distribution of hydrogen may present critical issues [3].

Hydrogen can be produced from a large variety of processes and raw materials; however, only hydrogen obtained from a renewable feedstock (by using renewable energy sources) can be defined as "green" [4] and hence clean. Most hydrogen is produced via the natural gas reforming processes; however, the biomass gasification and pyrolysis [5] to obtain bio-oil and followed by reforming is growing as a green alternative [6]. Model organic molecules derived from biomass, such as methanol, ethanol [7] or acetic acid [8] have been extensively studied in the last decades as hydrogen sources by pointing out the required improvement of the stability and selectivity of the involved catalytic systems. Water splitting is considered an ideal hydrogen production process as it can potentially substitute conventional fossil fuel-based processes with zero-carbon dioxide emissions [9]. However, the high cost of the most used Pt-based catalysts for the hydrogen evolution reaction is a major limitation to large-scale diffusion. Alternative catalytic systems, such as nickel sulfides on stainless steel, obtained by using electrodeposition and sulfurization techniques, may represent an effective solution [9]. Biological and photonic methods, such as fermentation, are extremely attractive but, despite meeting the demand for obtaining green hydrogen, are affected by the variability of the composition and high process costs, which represent a serious limitation to their diffusion [10].

Hydrogen storage materials, such as chemical hydride, are strategic hydrogen reserves that reconcile production with storage needs. An excellent example is the ammonia borane, which, with a high hydrogen capacity (19.6 wt%), low molecular mass, high solubility and stability in atmospheric pressure and ambient temperature, can be considered a candidate for the controlled storage and release of hydrogen [3]. On the other hand, on-site production allows the overcoming of issues related to distribution and storage, although applicable only to small scale productions. In the latter case, the size of the production plants plays a fundamental role.

When hydrogen is produced by reforming processes, further purification steps are required to obtain high purity hydrogen; in this regard, a critical issue is related to the reduction of the CO percentage [11]. To meet the specifications for fuel cell applications (ISO 14687:2019) the CO content must be less than 10 ppm, while for road vehicles and stationary appliances it must be less than 0.2 ppm [12]. CO water–gas shift is often used to reduce the amount of carbon monoxide and increase the amount of hydrogen in the reformate gas stream. Structured catalysts have been proposed as the best choice for the design of a single-stage process for small reactors [13]. However, the water–gas shift alone is not sufficient to reduce the CO content to levels of a few ppm; therefore, an integration with membranes is required. Pressure swing adsorption can be used to remove most of the contaminants, not only CO; however, single purification methods are limited. Therefore, in view of meeting the most stringent standards [14], the integration between two or even more purification technologies should be adopted.

The hydrogen economy is still a long way off, as a number of limitations and problems have not yet been resolved and massive investments in research are needed. A new generation of membranes could be the flywheel for integration with electrified hydrogen production processes. The articles presented in this Special Issue provide a comprehensive picture of what has already been done in this regard and what still needs to be done.

Author Contributions: Conceptualization, C.R. and M.M.; methodology, C.R. and M.M.; software, C.R. and M.M.; validation, C.R. and M.M.; formal analysis, C.R. and M.M.; investigation, C.R. and M.M.; resources, C.R. and M.M.; data curation, C.R. and M.M.; writing—original draft preparation, C.R. and M.M.; writing—review and editing, C.R. and M.M.; visualization, C.R. and M.M.; supervision, C.R. and M.M.; project administration, C.R. and M.M.; funding acquisition, C.R. and M.M. All authors have read and agreed to the published version of the manuscript.

Funding: This research received no external funding.

Acknowledgments: The Guest Editors would like to thank everyone who contributed to the success of this Special Issue. The Authors of the published articles for the quality of the submitted papers, and the Catalysts Editorial staff for the commitment and constant support.

Conflicts of Interest: The authors declare no conflict of interest.

References

1. Marnellos, G.E.; Klassen, T. Welcome to Hydrogen—A New International and Interdisciplinary Open Access Journal of Growing Interest in Our Society. *Hydrogen* **2020**, *1*, 6. [CrossRef]
2. What Is Carbon Neutrality and How can It Be Achieved by 2050? Available online: https://www.europarl.europa.eu/news/en/headlines/society/20190926STO62270/what-is-carbon-neutrality-and-how-can-it-be-achieved-by-2050 (accessed on 5 March 2022).
3. Liu, M.; Zhou, L.; Luo, X.; Wan, C.; Xu, L. Recent Advances in Noble Metal Catalysts for Hydrogen Production from Ammonia Borane. *Catalysts* **2020**, *10*, 788. [CrossRef]
4. Atilhan, S.; Park, S.; El-Halwagi, M.M.; Atilhan, M.; Moore, M.; Nielsen, R.B. Green hydrogen as an alternative fuel for the shipping industry. *Curr. Opin. Chem. Eng.* **2021**, *31*, 100668. [CrossRef]
5. Quiroga, E.; Moltó, J.; Conesa, J.A.; Valero, M.F.; Cobo, M. Kinetics of the Catalytic Thermal Degradation of Sugarcane Residual Biomass Over Rh-Pt/CeO_2-SiO_2 for Syngas Production. *Catalysts* **2020**, *10*, 508. [CrossRef]
6. Martino, M.; Ruocco, C.; Meloni, E.; Pullumbi, P.; Palma, V. Main Hydrogen Production Processes: An Overview. *Catalysts* **2021**, *11*, 547. [CrossRef]
7. Palma, V.; Ruocco, C.; Cortese, M.; Martino, M. Bioalcohol Reforming: An Overview of the Recent Advances for the Enhancement of Catalyst Stability. *Catalysts* **2020**, *10*, 665. [CrossRef]
8. Megía, P.J.; Carrero, A.; Calles, J.A.; Vizcaíno, A.J. Hydrogen Production from Steam Reforming of Acetic Acid as a Model Compound of the Aqueous Fraction of Microalgae HTL Using Co-M/SBA-15 (M: Cu, Ag, Ce, Cr) Catalysts. *Catalysts* **2019**, *9*, 1013. [CrossRef]
9. Youn, J.-S.; Jeong, S.; Oh, I.; Park, S.; Mai, H.D.; Jeon, K.-J. Enhanced Electrocatalytic Activity of Stainless Steel Substrate by Nickel Sulfides for Efficient Hydrogen Evolution. *Catalysts* **2020**, *10*, 1274. [CrossRef]
10. Kannah, R.Y.; Kavitha, S.; Preethi; Karthikeyan, O.P.; Kumar, G.; Dai-Viet, N.V.; Banu, J.R. Techno-economic assessment of various hydrogen production methods—A review. *Bioresour. Technol.* **2021**, *319*, 124175. [CrossRef] [PubMed]

11. Cifuentes, B.; Bustamante, F.; Cobo, M. Single and Dual Metal Oxides as Promising Supports for Carbon Monoxide Removal from an Actual Syngas: The Crucial Role of Support on the Selectivity of the Au–Cu System. *Catalysts* **2019**, *9*, 852. [CrossRef]
12. Bang, G.; Moon, D.-K.; Kang, J.-H.; Han, Y.-J.; Kim, K.-M.; Lee, C.-H. High-purity hydrogen production via a water-gas-shift reaction in a palladium-copper catalytic membrane reactor integrated with pressure swing adsorption. *Chem. Eng. J.* **2021**, *411*, 128473. [CrossRef]
13. Palma, V.; Gallucci, F.; Pullumbi, P.; Ruocco, C.; Meloni, E.; Martino, M. Pt/Re/CeO2 Based Catalysts for CO-Water–Gas Shift Reaction: From Powders to Structured Catalyst. *Catalysts* **2020**, *10*, 564. [CrossRef]
14. Du, Z.; Liu, C.; Zhai, J.; Guo, X.; Xiong, Y.; Su, W.; He, G. A Review of Hydrogen Purification Technologies for Fuel Cell Vehicles. *Catalysts* **2021**, *11*, 393. [CrossRef]

Review

Main Hydrogen Production Processes: An Overview

Marco Martino [1,*], Concetta Ruocco [1], Eugenio Meloni [1], Pluton Pullumbi [2] and Vincenzo Palma [1]

[1] Department of Industrial Engineering, University of Salerno, Via Giovanni Paolo II 132, 84084 Fisciano, SA, Italy; cruocco@unisa.it (C.R.); emeloni@unisa.it (E.M.); vpalma@unisa.it (V.P.)
[2] Air Liquide, Paris Innovation Campus. 1, Chemin de la Porte des Loges, 78350 Les Loges en Josas, France; pluton.pullumbi@airliquide.com
* Correspondence: mamartino@unisa.it; Tel.: +39-089-964027

Abstract: Due to its characteristics, hydrogen is considered the energy carrier of the future. Its use as a fuel generates reduced pollution, as if burned it almost exclusively produces water vapor. Hydrogen can be produced from numerous sources, both of fossil and renewable origin, and with as many production processes, which can use renewable or non-renewable energy sources. To achieve carbon neutrality, the sources must necessarily be renewable, and the production processes themselves must use renewable energy sources. In this review article the main characteristics of the most used hydrogen production methods are summarized, mainly focusing on renewable feedstocks, furthermore a series of relevant articles published in the last year, are reviewed. The production methods are grouped according to the type of energy they use; and at the end of each section the strengths and limitations of the processes are highlighted. The conclusions compare the main characteristics of the production processes studied and contextualize their possible use.

Keywords: hydrogen; reforming; gasification; water splitting; dark-fermentation; photo-fermentation; CO gas-fermentation; bio-photolysis; electrolysis

Citation: Martino, M.; Ruocco, C.; Meloni, E.; Pullumbi, P.; Palma, V. Main Hydrogen Production Processes: An Overview. *Catalysts* **2021**, *11*, 547. https://doi.org/10.3390/catal11050547

Academic Editor: Binlin Dou

Received: 7 April 2021
Accepted: 24 April 2021
Published: 25 April 2021

1. Introduction

In 2012 the UN Secretary-General stated that 'Energy is the Golden Thread', which connects economic growth, social equity, and environmental sustainability [1]. After more than 8 years, still today a significant percentage of the world population lives in conditions of energy poverty, and a large portion depends on highly polluting fuels and technologies [2]. The 2018 Intergovernmental Panel on Climate Change (IPCC) special report [3] on the impacts of global warming of 1.5 °C above pre-industrial levels and on the effects of global greenhouse gas emissions (GHG), has pointed out the need to drastically reduce these emissions. To effectively counteract the climate change, greenhouse gas emissions need to be reduced by at least 45% by 2030 [4], and carbon neutrality reached by 2050 [5] to stay below +1.5 °C of global warming. Therefore, we have two challenges ahead of us, we need more energy, but it has to be clean energy. Despite a wide variety of energy sources, fossil fuels (coal, oil and natural gas) still provide most of the energy needed to support human activities [6], but to achieve carbon neutrality the sources can only be renewable; therefore, it is necessary to reduce the consumption of fossil sources. Hydrogen is not a primary source but an energy vector [7], however some characteristics make it an extremely attractive candidate for making an energy transition to renewable sources. Hydrogen has been discovered for more than three hundred years [7,8], it is the most abundant element in the universe [9]; hydrogen can be produced from a wide range of technical processes and feedstocks, both fossil fuels, i.e., non-renewable source, and renewable sources such as biomass, moreover it is not toxic and it has heating values of 2.4, 2.8 and 4 times higher than those of methane, gasoline and coal, respectively [10]. Biomass is considered a promising source for hydrogen production (the hydrogen content in biomass is ~5–7 wt%), as a CO_2 neutral precursor, although the carbon footprint, in using biomass, is not effectively neutral [11]. It has been reported that 8.99×10^{-2} eqv·g·s^{-1} of CO_2 are emitted to produce 0.484 MJ·s^{-1} of hydrogen from 2.53×10^6 kg of biomass [11]. One of the critical aspects in the use of biomass is associated with land consumption, however alternatives like lignocellulosic, crops and organic waste sewage sludge, biooil, and biochar can be used as alternative. The nature and the availability of the feedstock is just one of the critical parameters, in the choice of the best process to be used, the energy required to sustain the process and the eventual use of catalysts, are equally crucial [12]. Different types and combinations of energy can be used, included thermal, electrical, bioenergy and photonic. The renewability of energy sources is another critical parameter, and plays a fundamental role, in the choice of the processes that will lead us to carbon neutrality.

This review aims to provide an overview of the major hydrogen production processes, mainly but not only focusing on those that make use of renewable sources, moreover for each process some relevant articles published in the last year are reviewed.

2. Thermal Methods

Currently, most hydrogen is produced by thermal processes, including reforming, gasification, and thermochemical methods. In the next sections a general overview of the main process will be provided.

2.1. *Steam Reforming*

The steam reforming is an endothermic equilibrium reaction, in which hydrogen is obtained through a catalyzed reaction between a hydrocarbon and steam (1):

$$C_nH_m + n\,H_2O \leftrightarrows n\,CO + (n+1/2m)\,H_2 \quad (1)$$

Among the reforming processes, methane steam reforming (MSR) is the most feasible route to convert methane into hydrogen [13].

2.1.1. Methane Steam Reforming

Methane, with an energy density of 55.5 MJ·kg^{-1}, is the simplest hydrocarbon molecule; in reforming it is reacted with steam at 700–1000 °C under a pressure range of 3–25 bar [14]. In addition to carbon monoxide and hydrogen, in the gas stream at the outlet of the reformer, unreacted methane and carbon dioxide are also present, therefore, further treatment/purification steps are necessary for obtaining pure hydrogen. In particular, one [15] or two stages of CO-water gas shift [16,17] followed by selective methanation [18], preferential oxidation [19] or a treatment with permselective membranes [20]. Although noble metal-based catalysts provide high activity and good stability, the high cost and the limited availability of the noble metals prevent their use, so Ni/Al_2O_3 is the most used commercial catalyst for MSR [21]. Unfortunately, these catalysts suffer of serious deactivation, due to the coke formation and Ni particle sintering. It has recently been shown that the application of a uniform positive electric field is able to modify the catalytic behavior of Ni-based catalysts during methane steam reforming by improving the methane activation at the surface while reducing the coke formation [22]. Both experimental and computational observations suggested that the positive field promotes the oxidation of the Ni catalyst, through the activation of water at the catalyst surface, in addition, it suppresses the polymeric carbon formation. Increased performance can be obtained by the addition of promoters, such as cerium oxide [23], which is able to enhance the methane conversion and provide a beneficial effect on coke resistance as well. Sintering-resistant $Ni@SiO_2$ catalyst has been recently reported in which the encapsulation in thermally stable SiO_2 nanospheres, prevent the Ni nanoparticles migration and thus avoided aggregation [24].

One of the major challenges in MSR process, is the managing of the high reaction temperatures; to reach 700–800 °C at the center of the catalytic bed, 1200 °C on the external reactor wall are needed. The use of structured catalysts, obtained by coating highly conductive carriers, with the catalytic formulation, provide a flattening of radial thermal profile on the catalytic bed, thus lowering the outside temperatures [13]. Alternatively, it has also been proposed to fill the voids of highly conductive open cell metal foams with small catalytic pellets, in order to exploit the radial heat transfer of the tubular reactor, due to the thermal conductivity of the interconnected solid matrix, while avoiding the washcoating of metal structures [25]. An inversion of the reactor configuration has been proposed, in which heat is provided from within the structure, so the reforming reaction is directly sustained by using the electrically driven SiC-based structured catalyst, obtained by washcoating a SiC heating element, without any external heat source [26]. Similar results have been obtained by using microwave heating for SiC-based structured catalysts, in which the dielectric properties of silicon carbide are used to transfer the heat directly to the catalyst, thus generating it directly inside the catalytic volume [27].

Intensification of the methane steam reforming process plays a key role, in order to achieve the objectives, set by the EU regulation. Indeed, electrification becomes competitive if the electricity used comes from renewable sources. The methane used in the reforming processes is mainly of fossil origin, however some studies have shown that the biomethane generated by the anaerobic digestion of sewage sludge can also be used [28]. A concrete alternative to methane comes from the use of biomass.

2.1.2. Bio-Oil Model Molecules Reforming

Concerning the thermal methods, two major routes to produce hydrogen from biomass are possible: the gasification to obtain syngas and the pyrolysis to obtain bio-oil, followed by reforming [29]. Bio-oil has larger energy density than biomass, composed by a mixture of organics including alcohols, carboxylic acids, aldehydes, ketones, furans, phenolics, etc. [30].

Several studies have recently been published in which the performance of bio-oil model molecules has been investigated. Comparative studies on the steam reforming of a series of organic molecules (methanol, formic acid, ethanol, acetic acid, acetaldehyde, acetone, furfural, guaiacol) derived from bio-oil, have demonstrated that the molecular structures drastically influence the reactivity and tendency to coking during steam reforming [29]. Methanol and formic acid could be reformed at low temperature, as they do not contain aliphatic carbons chain to be cracked, showing negligible coking (Figure 1). The steam reforming of ethanol, acetic acid, acetaldehyde, or acetone required much higher temperature, and generated remarkable amounts of coke deposits, especially acetone and acetaldehyde.

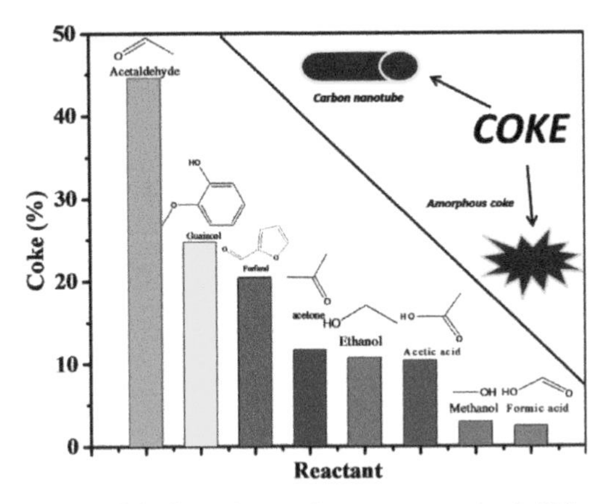

Figure 1. Coke formation trend per reactant molecule [29].

The characteristics of the catalyst play a fundamental role in the coke formation mechanisms. Comparative studies have showed that alumina support remarkably influences the catalytic stability of the catalyst, in methanol, acetic acid and acetone steam reforming [31].

The unsupported Cu showed lower stability than Cu/Al_2O_3, while the unsupported Ni showed higher stability than Ni/Al_2O_3, the unsupported Co was prone to coking.

In addition to the tendency to coke formation, a critical issue is the availability/renewability of the feedstock; bio-alcohols such as methanol, ethanol and glycerol can be easily obtained from renewable sources, therefore they seem to be a valuable alternative to natural gas.

2.1.3. Methanol Steam Reforming

Methanol can be obtained from biomass and CO_2, which makes it an attractive raw feedstock for reforming processes. The main advantage over ethanol is the lower tendency to coke formation, due to the high H/C ratio, moreover the absence of C-C bonds prevents the formation of a series of byproducts [32]. Cu-based catalysts are the most extensively studied catalysts for methanol steam reforming; although they show good activity at low temperature, the easily to sintering cause irreversible deactivation of the catalysts. To improve activity and stability of Cu-based catalytic systems, Cu-Al spinel oxide has been doped by adding MgO [33]. Mg^{2+} cations partially replaced Cu^{2+} incorporating into the spinel lattice, thus making it become hard to be reducible, consequently the doped catalysts showed a lower copper releasing rate and smaller copper particles. On the other hand, copper was successful used as promoter to increase the selectivity of Pd-based catalysts in methanol steam reforming [34]. The $CuPd/TiO_2$ bimetallic catalytic system showed improved performance, with respect the monometallic counterpart, by both thermo-photocatalytic and photocatalytic processes [35].

$CuFeO_2$-CeO_2 nanopowder catalyst, with a heterogeneous delafossite structure, prepared by the self-combustion glycine nitrate process, showed an improved H_2 generation rate of 2582.25 STP $cm^3 \cdot min^{-1} \cdot g_{cat}{}^{-1}$ at a flow rate of 30 sccm at 400 °C [36]. CuO/ZnO was successfully loaded onto a metal-organic framework material (Cu-BTC), thus improving the stability [37]. $(Ni_{0.2}Cu_{0.8})$/boron nitride nanohybrids have been studied, showing high catalytic stability and high CO_2 selectivity, moreover no carbon monoxide was detected during the full methanol conversion [38].

The effect of the support was also investigated in Cu/ZrAl-based catalysts, showing that the Zr/Al molar ratio of 0.4, in the support, improve the interaction between copper species and the support, resulting in a homogeneous distribution of highly dispersed Cu, with enhanced reducibility [39]. In a comparative study between different alumina supports, the commercial A520 MOF derived γ-Al_2O_3, in copper and palladium catalysts, has showed superior outcomes, attributed to the higher surface area, larger pore volume and possible defects in nanoscale alumina [40]. Different synthesis methods were also evaluated for CuZn/MCM-41 catalysts, revealing that the co-impregnation is the most effective method [41].

Alternative catalytic systems have been also proposed, for example $ZnCeZr_9Ox$ catalyst exhibited a full methanol conversion and an H_2 production rate of 0.31 $molh^{-1}g_{cat}{}^{-1}$ at 400 °C; the incorporation of Zn^{2+} into $CeZr_9Ox$ matrix, modulate the surface O_{Latt}/O_{Ads} ratio, and generate a Zn-O-Zr interfacial structure, corresponding to the lattice/bridge oxygen thus increasing the CO_2 selectivity [42]. Zn-modified Pt/MoC catalyst exhibited superior hydrogen production activity, with exceptionally low CO selectivity at low temperatures (120–200 °C), due to the formation of α-MoC_{1-x} phase and to the enhanced Pt dispersion [43]. In_xPd_y/In_2O_3 aerogels exhibited excellent CO_2 selectivity of 99% at 300 °C [44].

2.1.4. Ethanol Steam Reforming

Ethanol can be produced through fermentation of saccharides, and after distillation, bio-ethanol can be used in ethanol steam reforming (ESR); the post-ESR reformate has a high heating value of about 1450.9 $kJ \cdot mol^{-1}$ [45]. ESR has been investigated both by using simulated bio-ethanol feeding, and by using crude bio-ethanol or model mixture containing typical contaminants, however in the latter case, fast catalyst deactivation occurs [32]. Nickel-based are the most studied catalysts for ESR, due to the low cost, as alternative cobalt and platinum have also extensively investigated. The major issue in ESR process is the carbon formation, different strategies have been proposed to overcome this limitation, such as the use of promoters and method to control the metal particle size.

Potassium has been successfully used as promoter of Co/Al_2O_3–CaO catalysts. The K-promoted catalysts showed higher hydrogen yield and lower methane selectivity than the unpromoted catalysts, mainly due to the suppression of methanation reaction [46]. On the other hands, sodium doping decreases the catalytic activity, but significantly increases the CO_2 selectivity, improving the H_2 selectivity, in Pt/m-ZrO_2 catalysts [47]. The addition of sodium favors the decarboxylation route, the decomposition of acetate at lower temperatures, yields methane and adsorbed carbonate, which decompose to carbon dioxide. Moreover, sodium promotes C-C scission. Ga doping of ceria-based catalysts improves the H_2/CO_2 ratio in ESR reaction, by changing the product distribution and reducing the coke formation [48]. The stability of Co/CeO_2 and Ni/CeO_2 catalysts have been improved by the La_2O_3 promoter, the carbon formation rate has been reduced as results of an increased active phase dispersion and a strengthening of the metal-support interactions [49].

In a comparative study, the effect of a series of promoters, including Na, Mg, Zr, La, Ce and the elements from K to Zn in the periodic table of elements, on Co/Al_2O_3 catalyst was evaluated [50]. Na, K, Cu, Ni and Ce addition promoted the catalytic activity, while Mg, Ca, Sc, Ti, V, Cr, Mn suppressed it, Na, K, Ca, Fe, Zn and La helped to suppress coking while Cu or Zr enhanced. Na suppressed the formation of acetyl species, while Cu promoted the acetyl species, C=O and C=C formation, bringing Cu-Co/Al_2O_3 towards coking.

A dragon fruit-like Pt-Cu@$mSiO_2$ catalyst for low temperature ESR reaction has been successfully synthesized by encapsulation strategy, with Pt-Cu alloy nanoparticles, of about 50 nm, embedded in the $mSiO_2$ [51]. This catalyst showed better performance than Pt@$mSiO_2$, Cu@$mSiO_2$ and the supported Pt-Cu/$mSiO_2$ catalysts, in terms of activity, H_2 selectivity, and stability. The mesoporous SiO_2 shell prevents leaching and aggregation of active sites and spatially suppresses the carbon deposition on the active surface (Figure 2).

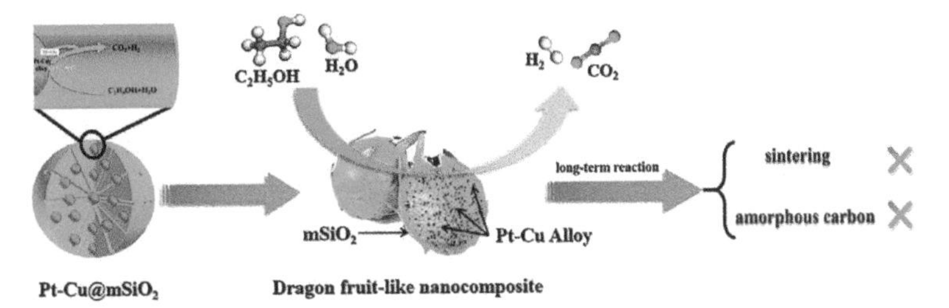

Figure 2. Schematic representation of the use in ESR of dragon fruit-like nanocomposite [51].

Graphene-encapsulated Ni nanoparticles (Ni@Gr), fabricated via in-situ growth method, showed good activity and durability in ESR at 550 °C [52]. Density functional theory calculations demonstrated that the presence of defects improved the adsorption energy of all reaction species. The primary reason for catalyst deactivation was the carbonaceous deposition. Spinel-type mixed manganese-chromium oxides $MnxCr_{3-x}O_4$, prepared by Pechini route have been used as support for the co-deposition of Ni and Ru; the strong metal-support interaction stabilized small clusters of metals/alloys, preventing carbon nucleation in ESR [53].

Biochar-supported Ni catalyst has also been used in ESR [54]. The results showed that biochar is a promising support as well as itself a reforming catalyst, in fact it contains alkali and alkaline earth metallic species and O-containing functional groups, which are factors affecting the catalytic performance.

2.1.5. Glycerol Steam Reforming

The use of glycerol in reforming processes is very promising; it can theoretically provide seven moles of hydrogen for every mole of $C_3H_8O_3$ [55], moreover glycerol is a by-product of biodiesel production, whose commercial value has been strongly affected by excessive supply. The use of glycerol for the production of value-added chemicals, such as hydrogen, appears to be the best way to exploit it and simultaneously increase the global biodiesel market [56]. Even in the case of glycerol steam reforming, the noble metal-based catalysts show excellent performance [57], in particular excellent activity has been found with Pt-based catalysts [55], moreover the presence of promoters suppresses the coke formation, however the high cost makes them uncompetitive compared to nickel-based catalysts [58].

The catalytic behavior of Ni catalyst supported on CaO-modified attapulgite in glycerol steam reforming, has shown that the addition of CaO promotes the dispersion of the active component, promoting the water gas shift reaction, thus leading to improved hydrogen yields [59]. Moreover, the addition of CaO enhances the inhibition of carbon deposition, prolonging the stability of the catalyst. The incorporation of CeO_2 into $NiAl_2O_4$ spinel is able to suppress the coke formation, through the formation of a $CeAlO_3$ phase which hinders the growth of filamentous carbon on nickel surface and enhances the gasification of carbon deposits by providing an oxidative environment around nickel active sites [60]. Recently, it has been showed that MgO is able to suppress the sintering of cobalt based catalysts in glycerol steam reforming, moreover the promotion with copper suppresses the coke formation [61].

The sintering of the active metal was suppressed in the bimetallic MNi/CNTs (M = Co, Cu and Fe, CNT = carbon nanotube) catalysts, in which Ni oxide was introduced into the cave and the other one was dispersed on the external wall of CNTs [62]. Flame spray pyrolysis has been successfully used to produce nano-sized Ni-based steam reforming catalysts for glycerol starting from $LaNiO_3$ and $CeNiO_3$ as base materials by varying the formulation, mixing them or incorporating varying amounts of ZrO_2 o SrO during synthesis [63]. The deactivation resistance was increased by improving the dispersion of nickel through the formation of Ni-La or Ni-Ce mixed oxide. ZrO_2 provided high thermal resistance, while a base promote /support, such as La_2O_3, downgraded the surface acidity of ZrO_2.

2.2. *Autothermal Reforming*

Autothermal reforming (ATR) is an interesting process which uses the heat of an exothermic reaction (partial oxidation, POX) to sustain the endothermic steam reforming reaction, by feeding air, steam, and the reacting feedstock, such as methane, methanol or ethanol to produce a H_2-rich stream. The main characteristic of ATR is the low energy requirements: by properly selecting the oxygen/fuel ratio, no external heat is required; moreover, oxygen availability may promote coke gasification reactions.

Ni-based catalysts are the most used in methane autothermal reaction, mainly due to the low cost; unfortunately, the sever reaction conditions can lead to the rapid deactivation of the catalyst. The use of promoters or bimetallic catalytic systems can improve the performance of these catalysts; series of 10Ni-M/$Ce_{0.5}Zr_{0.5}O_2/Al_2O_3$ catalysts (M = Pt, Pd, Re, Mo, Sn; molar ratio M/Ni = 0.003, 0.01, 0.03) have been tested in CH_4-ATR [64]. The catalysts with Pt, Pd, Re, or Mo, by contrast to the non-promoted sample, showed the ability to self-activate under the reaction conditions, making not necessary the catalyst pre-reduction, due to the improved reducibility. 10Ni-0.9Re/$Ce_{0.5}Zr_{0.5}O_2/Al_2O_3$ catalysts has shown high resistance to oxidation and sintering of the Ni active component as well as the resistance to coking.

The catalyst preparation method is one of the aspects that can improve the catalytic performance by enhancing the catalyst's physicochemical properties. These methods alter the metal-support interaction, thereby changing the kinetics of the catalyst which can result in enhanced productivity, reduced cost, and optimized energy requirements [65].

In a comparative study, bimetallic Cu-Ni catalysts, supported on binary oxides containing ZnO, ZrO_2, CeO_2 and Al_2O_3, were investigated, for the hydrogen production via the oxidative steam reforming of methanol [66]. At high temperature the most active catalyst was 30%Cu–10%Ni/$CeO_2Al_2O_3$, the good performance was attributed to the $Cu_{0.8}Ni_{0.2}$ alloy formation, as well as the high acidity and easy reducibility. At low temperatures, the best catalytic performance was obtained with 30%Cu–10%Ni/$ZrO_2Al_2O_3$. A series of CuO/$Ca_2Fe_2O_5$ catalysts, with different contents of copper were prepared as catalytic oxygen carrier (COC) which goes through the reduction → catalytic methanol conversion → re-oxidation route [67]. The results showed that the 40%Cu-loaded catalyst had the highest catalytic activity: the presence of $Ca_2Fe_2O_5$ tunes the redox activity and mobility of the lattice oxygen, obtaining a H_2 production rate of 37.6 $\mu mol \cdot H_2 \cdot g_{COC}{}^{-1} \cdot s^{-1}$ at 240 °C. Ni, Pt and a mixture of Ni and Pt supported on ZnO-rods were evaluated in methanol autothermal steam reforming, in the temperature range 200–500 °C [68]. The bimetallic catalysts showed the best catalytic activity, due to the formation of PtZn and NiZn alloys.

Pt-Ni/CeO_2-SiO_2 and Ru-Ni/CeO_2-SiO_2 catalysts were compared in an oxidative ethanol steam reforming reaction [69]. In both cases, the catalysts deactivated with time-on-stream, due to the severe reaction conditions; however, the Pt-based catalysts showed the highest ethanol conversion, hydrogen yield and the lowest carbon formation rate. A detailed kinetic mechanism has been also compared against experimental data and apparent kinetics, demonstrating that coke formation is associated with the 2-hydroxyethyl radical reaction path, explaining the effectiveness of the catalyst in coke suppression [70].

2.3. Gasification

Gasification is a process in which carbonaceous materials are converted in syngas, at high temperature, in presence of an oxidizing agent. Various types of biomass can be used as potential feedstocks for hydrogen production via gasification, including algae, food waste, municipal solid waste, and lignocellulosic biomass [71]. The hydrogen yield is strongly dependent on the process conditions of the biomass gasification, on the temperature of the steam flow but also on the type of raw material, which is a critical problem [72]. In a recent study, the optimization of biomass blending, which can be an effective way to overcome the problem of enormous feedstock variability, has been studied [73]. The simulation model used various biomass feedstock, including date pits, manure, and sewage sludge to correlate the biomass blending to the H_2/CO ratio in the obtained syngas, as alternative to the manipulation of process conditions.

One of the main problems of conventional gasification is the impossibility of using biomass with a high moisture content. Gasification in supercritical water (SCW) can directly use the biomass without drying process, since the reaction occurs in water phase. An experimental study on cornstalk gasification in SCW, carried out in the temperature range of 500–800 °C, a reaction time of 1–15 min and a feedstock concentration of 1–9%, has shown that the carbon gasification efficiency reaches 99% at the temperature of 700 °C, reaction time of 15 min and biomass concentration of 3% [74]. A parametric research on supercritical water gasification of food waste conducted with a micro quartz tube batch reactor (Figure 3), demonstrated that the complete gasification can be basically realized with a carbon gasification efficiency of 98% at a temperature of 850 °C, a concentration of 5 wt% and residence time of 10 min [75].

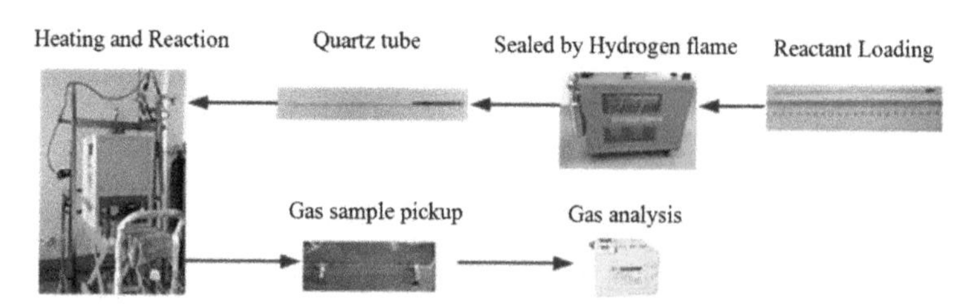

Figure 3. Supercritical water gasification experimental procedure with a micro quartz batch reactor [75].

2.4. *Thermochemical Water Splitting*

Water splitting is the reaction in which water is splitted into hydrogen and oxygen (2):

$$2\,H_2O + \text{heat} \rightarrow H_2 + O_2 \tag{2}$$

Complete decomposition in a single step can be obtained only at high temperatures (above 2000 °C), while thermochemical cycles, with multiple steps, and lower operating temperatures can supply the required heat [76]. The pure thermochemical cycles are driven either by only thermal energy (Figure 4a), while hybrid ones (Figure 4b) are driven by thermal and some other form of energy (e.g. electrical, photonic) [10].

In two-step thermochemical cycles, two separate reduction/oxidation steps utilizing a metal oxide as a reactive intermediate may be performed: (i) a higher valence metal oxide is reduced to the corresponding low valence metal oxide or metal and oxygen is produced (reduction step) and (ii) the lower valence metal oxide (or metal) reacts with water to form hydrogen and a higher-valence metal oxide (water splitting step), which is subsequently recycled in the first step [77]. The three-step processes can be obtained from the two-step one, in which the highest temperature reaction is replaced by a two-step reaction, thus achieving a reduction of the maximum temperature required. In most cases, the required temperature for water splitting decreases when more steps are employed, but the same decreasing trend also occurs for the efficiency potential due to the energy losses associated with heat transfer and products separation in each step. The sequence in the four-step cycles is: hydrolysis, evolution of hydrogen, evolution of oxygen and recycling of reagents [10].

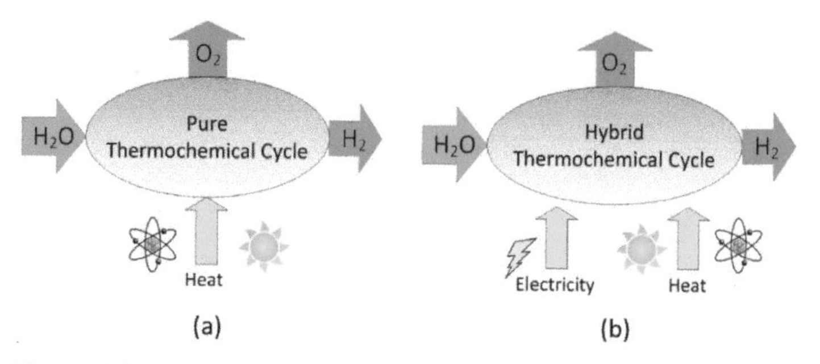

Figure 4. Schematic representation of pure (**a**) and hybrid (**b**) thermochemical cycles [10].

The four-step iron-chlorine (Fe-Cl) cycle was modelled by the Aspen Plus software package [78]. The results showed that the pressure does not significantly affect the reaction's production rates, while an increase in temperature favors oxygen production in reverse deacon reaction and magnetite production in hydrolysis and lowers hydrogen production in the hydrolysis step. Moreover, the steam/chlorine ratio is directly related to the HCl and oxygen production, in reverse deacon reaction and hydrogen production in hydrolysis.

Perovskite type mixed metal oxide compound, such as barium ferrite oxide, possesses good chemical and thermal stability under severe conditions, besides high oxygen storage capacity (OSC) and reducibility [79]. The addition of La^{3+} and Ga^{3+} to $BaFeO_{3-\delta}$ was investigated in water splitting, and the best performance was obtained with $Ba_{0.95}La_{0.05}FeO_{3-\delta}$, showing a H_2 production of about 1310 $\mu mol \cdot g^{-1}$ at 900 °C. Probably the smaller La^{3+} ionic radius compared to Ba^{2+} can lead to a larger lattice free volume for oxygen transport. The cubic perovskite $SrTi_{0.5}Mn_{0.5}O_{3-\delta}$ was tested in a water splitting cycle in which the material is thermally reduced at 1350 °C (pO_2, ~10−5 atm) and subsequently exposed to steam at 1100 °C (steam partial pressure of pH_2O = 0.4 atm), obtaining a hydrogen yield of 7.4 $cm^3 \cdot g^{-1}$ [80]. The cyclic operation did not show any degradation of the material, resulting in a constant 2: 1 yield of H_2/O_2.

$Ce_{0.9}Me_{0.1}O_y$ (Me = Fe, Co, Mn, and Zr) have been tested in thermochemical water splitting, both as powder and as mixed oxides reticulated porous ceramic (RPC) structures, obtained with sponge replica method [81]. The $Ce_{0.9}Fe_{0.1}O_y$ powder oxide showed the best hydrogen production (8.5 STP $cm^3 \cdot H_2 \cdot g_{material}{}^{-1} \cdot cycle^{-1}$) and stability during consecutive cycles, while the $Ce_{0.9}Fe_{0.1}O_y$ RPC sponge, showed an outstanding hydrogen production of 15 STP cm^3 $g_{material}{}^{-1} \cdot cycle^{-1}$ at a maximum temperature of 1300 °C, due to the to the open macroporosity of the reticulated porous ceramic structure, which enhanced both heat and mass transfer.

2.5. Remarks on Thermal Methods

The major drawback of thermochemical methods is the need to supply heat, which normally comes from fossil fuel combustion, typically by electric heating or by using a catalytic combustor [82]. To make these processes sustainable, not only must the feedstocks be renewable, but also the energy needed to the process should be sustainable too. For this reason, solar [83,84], wind and geothermal energy represent an interesting alternative to conventional methods, which, however, are feasible where cost savings are demonstrated. In a recent study a geothermal-assisted methanol reforming, incorporating a proton exchange membrane fuel cell, for hydrogen production, has been proposed [85]. Thermodynamic and economic assessment have showed that an annual cost-saving can be obtained of 20.9%, compared with the conventional system. Another issue is the significant production of CO_2 as byproduct, which can be captured by sorbents, such as CaO in the sorption enhanced steam reforming processes, thus improving the H_2 yield [86]. Membrane-assisted steam reforming is also an intriguing option for H_2 production from biomass, especially in refueling stations for automotive fuel cells [87]. The energy consumption required for H_2 compression can be significantly reduced if the process supplies H_2 at high pressures. Thermodynamic analyzes show that, in the case of ESR, reforming temperatures above 550 °C are required to obtain H_2 partial pressures that will allow operation without sweep gas, obtaining appreciable energy savings. Thermochemical water splitting presents some significance advantages, including no need of electricity in pure water splitting and no need of membranes for the hydrogen separation [10]. Biowaste-based biomethane as a feedstock for hydrogen production, in reforming processes, seems to be an intriguing option, as it can lead to negative life-cycle greenhouse gas emissions even without CO_2 capture and storage (CCS) [88]. Reforming-based hydrogen production processes with CCS can be considered a clean technology, as the life-cycle greenhouse gas emissions are lower than those of hydrogen from electrolysis, considering that the most of electricity supplied, is still largely based on fossil fuels [88].

3. Photocatalytic Methods

In the framework of photocatalytic methods, we basically refer to hydrogen production by water splitting using solar energy, through the generation of electron-hole pairs by photons and semiconductors [89].

Photoexcited electron-hole pairs can be separated efficiently using sacrificial agents, which allow the formation of hydrogen with reduced electron-hole pair recombination.

Nowadays, however, this process faces challenges in being implemented using visible light, given its low photon conversion efficiency [89]. To achieve photocatalytic water splitting, feasible photocatalysts must meet some fundamental criteria: they must absorb visible light (display suitable band gaps), they must be chemically stable under redox conditions, they must have a low cost, they must be recyclable, they must be chemically resistant and they must be adaptable for large-scale hydrogen production. In this sense, the design of efficient photocatalysts with high photo-conversion efficiency is the target for completing the photocatalytic hydrogen evolution [89].

In the last years different configurations have been proposed, including CuS- and NiS-based heterojunctions, titanium dioxide based core–shell structures and periodical structures with excellent adsorption ability, and imogolite hollow cylinders [90], as well as S-scheme heterostructure [91]. Among them, the titanium dioxide based ones are the most interesting, due to the chemical resistance properties, accessibility, and affordability [92]. Moreover, in recent years the scarce activation of titanium dioxide by visible sunlight has been mitigated; novel strategies of doping development of novel composites are allowing to obtain interesting results also in the field of visible light-activated photocatalysis [92]. The addition of different metals to titanium dioxide results in increasing the hydrogen production under visible light. In the case of palladium [93], a "series-parallel" reaction network has been proposed for describing the water splitting reaction using the mesoporous Pd-TiO_2 and ethanol as organic scavenger (Figure 5) [94].

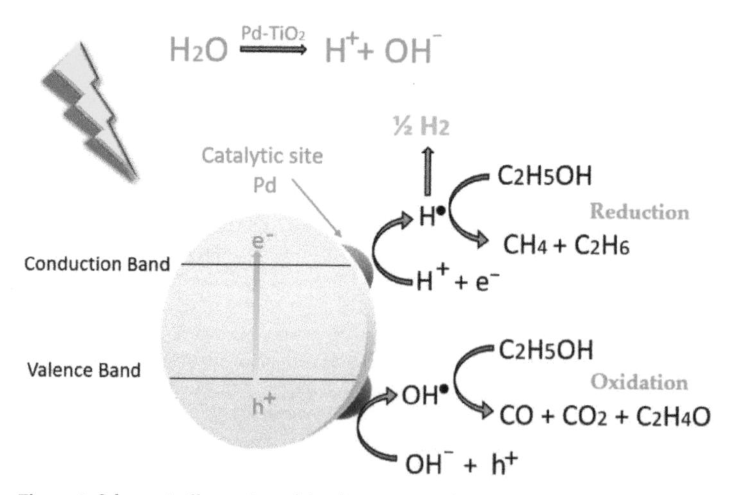

Figure 5. Schematic illustration of the three water splitting methods [94].

The addition of earth-abundant metals has a positive effect, in particular in the case of Ni [95] and Cu [96,97]. A high activity in hydrogen evolution under visible light has been demonstrated also for solid solutions of cadmium and manganese sulfides, due to their valence and conduction band position tuning, and for composite photocatalysts, CdS-β-Mn_3O_4-MnOOH, due to the ternary heterojunction formation [98]. Although reasonable research progress has been accomplished on the design of photocatalysts with high-conversion efficiency, there are still some issues to be addressed. For example, most metal chalcogenides-based heterojunctions can only split water in the presence of sacrificial agents.

Photocatalysis may also be used for hydrogen production from water, by using titania-containing gold nanoparticles [99], or wastewater pollutant removal, by using TiO_2-based catalysts [100,101] or metal halide perovskites [102]. In the latter processes, the contemporary targets of hydrogen production and the abatement of harmful pollutants may be reached, with an evident environmental benefit.

Hydrogen production may also be obtained through the so called "photoreforming processes" of organic substrates, such as methanol, ethanol, glycerol, sucrose, glucose, starch and wood [103], as well as of aromatic water pollutants [104]. In these processes, modified TiO_2-based catalysts are the most used materials for having higher hydrogen evolution.

The main problems of catalytic water splitting are the need for sacrificial reagents, typically organic, whose cost and environmental impact is not negligible, the development of more active visible light photocalysts and the design of low-cost reactors. These problems make these methods uncompetitive from an economic point of view.

4. Biological and Photonic Methods

Biological processes allow producing hydrogen from renewable resources such as biomass and solar energy; the main processes can be classified as direct/indirect photolysis, photo-fermentation, dark-fermentation, and CO gas-fermentation [105].

4.1. Dark-Fermentation

In dark-fermentation process, hydrogen is produced from organic materials, such as sugars, amino acids, waste materials, wastewaters and so on, without light, by using anaerobic organisms [106]. Dark-fermentation is considered a promising alternative to traditional hydrogen production methods, due to the low estimated production costs [107]. Hydrogen-producing bacteria can be classified as spore-forming obligate anaerobic bacteria, non-sporulating anaerobes and, facultative bacteria (Figure 6) [108]. The *Clostridium* bacteria, belonging the spore-forming obligate anaerobic microorganisms, are considered the most efficient bacteria in hydrogen production; the fermentation can be both acetate-type (3) and butyrate-type (4) [108].

$$C_6H_{12}O_6 + 2H_2O \rightarrow 2CH_3COOH + 4H_2 + 2CO_2 \ \Delta G^\circ = -206 \text{ kJ}\cdot\text{mol}^{-1} \quad (3)$$

$$C_6H_{12}O_6 \rightarrow CH_3CH_2CH_2COOH + 2H_2 + 2CO_2 \ \Delta G^\circ = -255 \text{ kJ}\cdot\text{mol}^{-1} \quad (4)$$

The yield of the process depends on a series of factors: pH, temperature, pressure, the hydraulic retention time [109], the type of organism, the composition of the substrate and the presence of metals, and several studies focus on the effects of changing these parameters. For example, the dark-fermentation can take place both with indigenous bacteria and by adding microbial inoculum; a recent study focused on determining if the indigenous bacteria associated with thermal pretreatment can impact on the performance [110]. The study has been carried out on seven organic substrates and has demonstrated that the indigenous bacteria are effective as the thermally pretreated exogenous bacteria in producing hydrogen. The highest hydrogen yield was obtained with Clostridiales and Enterobacteriales.

The invasive aquatic weed *A. philoxeroides* has been used to evaluate the effect of a steam-heated acid pretreatment and enzymolysis on the dark-fermentation, by using *Enterobacter aerogenes* ZJU1 mutagenized by ^{60}Co-γ irradiation as inoculum [111]. The study showed that the acid treatment significantly disrupts the fiber of the *A. philoxeroides*, contributing to the higher yield in reducing sugar and consequently a hydrogen yield increase by 59.9%. The effect of salinity and pH on dark-fermentation of swine wastewater pretreated with thermophilic bacteria, has showed that 1.5% of salinity and a pH = 6 are the optimal condition for hydrogen production, while 3.5% of salinity and pH = 5 are able to inhibit the production [112].

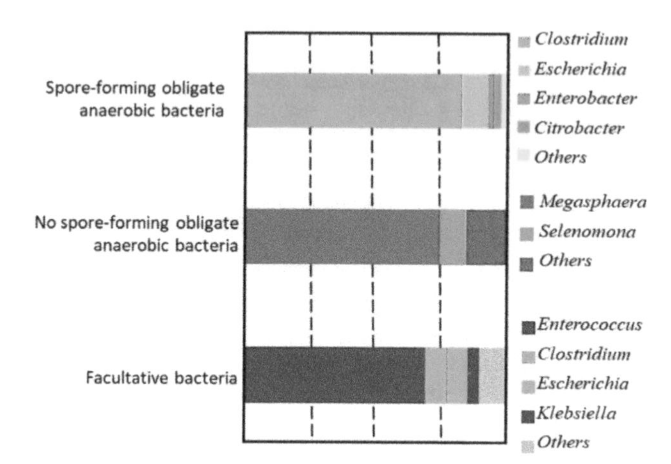

Figure 6. Three examples of the taxonomic composition of H_2-producing communities dominated by the different groups of H_2-producing microorganisms [108].

The use of lawsone and anthraquinone 2-sulphonate covalently immobilized on activated carbon, as redox mediators, has been evaluated in the dark fermentation of glucose by a pretreated anaerobic sludge [113]. The results showed that the use of lawsone increased the hydrogen production of 10%, while anthraquinone 2-sulphonate improved the hydrogen production rate of 11.4%. A remarkable increase in hydrogen production from waste activated sludge was obtained by freezing in the presence of nitrite [114]. The pretreatment accelerates the disintegration of sludge and promote the biodegradable organics released, thus providing more bio-available substrate for hydrogen production.

The iron hydroxide mineral ferrihydrite is able to promote the hydrogen production by Clostridium, redirecting the metabolic pathways and stimulating the bacterial growth, thus improving the carbon and electron conversion [115]. The addition of iron and nickel nanoparticles improve the fermentation process enhancing the hydrogen production by *Clostridium butyricum* [116]. Similar improvements were obtained by including NiO and CoO nanoparticles to dark fermentation of rice mill wastewater using *Clostridium beijerinckii* DSM 791 [117].

One of the main problems, related to this process is the homoacetogenesis (5) [108]:

$$4H_2 + 2CO_2 \rightarrow CH_3COOH + 2H_2O \ \Delta G^\circ = -104 \text{ kJ}\cdot\text{mol}^{-1} \quad (5)$$

Due to the negative impact on the hydrogen production, prevention, and control of homoacetogenesis are being studied. Currently there is no method to completely eliminate homoacetogenic bacteria, as their presence depends on the culture, substrate and process parameters. The strategy to minimize its impact is therefore to control CO_2 and H_2 concentrations during the process. For example, studies on the effect of shear velocity on hydrogen production, in a dynamic membrane bioreactor, containing a 50 μm polyester mesh as support material, have shown that the homoacetogenic pathway can minimized by choosing the optimal shear velocity, thus improving the hydrogen production [118].

4.2. Photofermentation

Photofermentation is a process in which hydrogen is produced from organic compounds through a nitrogenase-catalyzed reaction, in the presence of light energy, by photosynthetic or anaerobic bacteria, such as *Rhodobium*, *Rhodobacter*, *Rhodospirillum*, and Rh*odopseudomonas* [119]. The major limitations to the industrialization of photofermentation are the availability and distribution of light and to the need for specific substrates, i.e., small fatty acids, including acetate, propionate and butyrate [120]:

$$CH_3COOH + 2H_2O + Light \rightarrow 4H_2 + 2CO_2 \qquad \Delta G^\circ = +104\ kJ\cdot mol^{-1} \tag{6}$$

As in the case of dark-fermentation the influence of a series of process parameters have been recently studied, with the aim of optimizing the hydrogen production from photo-fermentation.

Lighting and mixing significantly affect the hydrogen production performance from agricultural waste (Figure 7); mixing enhance the mass transfer and shorten the lag phase, however the higher is the mixing speed the higher is the light intensity requirement [121]. Intermittent stirring has shown to increase the hydrogen production rate of 65.05% compared to continuous stirring, in corn stover hydrolysate photo-fermentation [122].

A combination of ultra-sonication and biosorption using banana peels waste pretreatment, for mixed effluents of 70% restaurant and 30% brewery, enhance the photo fermentative hydrogen production processes [123]. Addition of glycerol to *Arundo donax* L. can enhance the hydrogen production via photofermentation, due to redox potential [124]. The highest hydrogen yield (79.2 $cm^3\cdot g_{substrate}{}^{-1}$) was obtained with a glycerol: *Arundo donax* L. ratio 1:1.

Substrate concentration has been shown to be crucial in the process of photo-fermentation hydrogen production in a study using potato taken as a starch-rich agricultural leftover under fluctuating conditions [125].

The effects of NaOH and $Ca(OH)_2$ pretreatments on giant reed for the photofermentative hydrogen production, was evaluated [126]. The results showed that the 20% NaOH pretreated giant reed biomass reached the highest hydrogen yield (98.3 $cm^3\cdot g_{TS}{}^{-1}$; TS = total solids), which was 20% and 70% higher than the highest level obtained with Ca $(OH)_2$ pretreated (20% $Ca(OH)_2$) and untreated giant reed, respectively. The optimal substrate concentration of 25 $g\cdot dm^{-3}$ was found beneficial to hydrogen production, in cellulase and protease hydrolysis of *Chlorella* biomass [127].

The effect of enzymolysis time on the hydrogen production by photofermentation of energy grasses was also studied [128]. The results showed that the hydrogen production rate was depending on the kind of grass, however the highest hydrogen yield was obtained from *Medicago sativa* L. with an enzymolysis time of 60 h (147.64 $cm^3\cdot g_{TS}{}^{-1}$), the highest hydrogen production rate was obtained from *Arundo donax* with an enzymolysis time of 36 h (5.53 $cm^3\cdot h^{-1}\cdot g_{TS}{}^{-1}$) obtained, while the highest hydrogen production efficiency was obtained from *Miscanthus* with an enzymolysis time of 0 h (1.15 $cm^3\cdot h^{-1}\cdot g_{TS}{}^{-1}$).

Hydrogen production is strongly dependent on the pH and inoculation volume ratio [129]; it is well known that a decrease in pH, due to metabolic acid production, is a limiting factor in hydrogen production during the photofermentation from glucose [130].

A buffer solution Na_2HPO_4/NaH_2PO_4 has shown to be able to improve the buffer capacity of fermentation broth, thus improving the hydrogen production by photo-fermentation from corn stalk [131]. The highest energy conversion efficiency 9.84%, hydrogen yield 132.69 $cm^3\cdot g^{-1}$ of corn stalk, and hydrogen content 53.88% were achieved at pH value of 6. The initial pH value of phosphate buffer has a crucial role on the hydrogen production, in fact with the increase in pH values, the hydrogen production gradually delays, suggesting that the alkaline environment has a negative effect on the ability of photosynthetic bacteria [132]. A pH equal to 7 was found to be the optimal value for the hydrogen production from potato residue [133].

Response surface methodology was used to study the dependence of initial pH, substrate concentration, and cellulase loading on photo-fermentation hydrogen production by HAU-M1 from alfalfa, and to find the optimal conditions [134]. The highest hydrogen production yield of 55.81 $cm^3\cdot g^{-1}$ was observed at initial pH of 6.90, substrate concentration of 31.23 $g\cdot cm^{-3}$, and cellulase loading of 0.13 $g\cdot g^{-1}$.

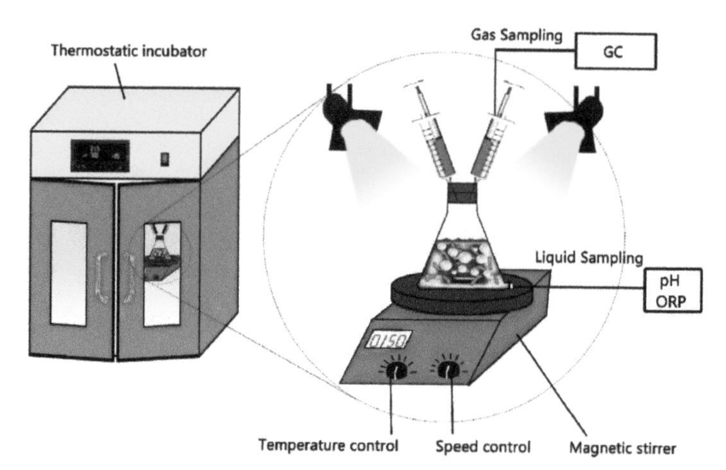

Figure 7. Diagram of photofermentation hydrogen production [122].

A study on the effect of different pretreatments, such as hydrothermal, acid, alkali, acid-heat, and alkali-heat on the structural characteristics, enzymatic saccharification and photo-fermentation of corn straw, has shown that all the treatments effectively destroyed the corn straw structure and improved its enzymatic saccharification potential [135]. The highest cumulative hydrogen yield (137.76 $cm^3 \cdot g_{TS}{}^{-1}$) was obtained with the 2% NaOH pretreated corn straw, while the minimum (44.20 $cm^3 \cdot g_{TS}{}^{-1}$) was obtained with a pretreatment of 4% NaOH-heat.

The addition of iron, molybdenum, and EDTA to the photofermentation of a blend of pre-treated brewery (30%) and restaurant (70%) effluents, was found beneficial for the hydrogen production and bacterial growth by *Rhodobacter sphaeroides* 158 DSM [136]. The additions of Fe at 70 μM, Mo at 14 μM, and co-addition of Fe:Mo at 70 μM:8 μM to the mixture increased the cumulative biohydrogen production of 69%, 27% and 160 93% respectively. The addition of EDTA was able to further increase the hydrogen production, but it is crucial to choose an optimal concentration in order to avoid chelating effects.

To make the process more competitive, the production of interesting chemicals, in addition to hydrogen, can be considered. For example, hydrogen and poly-β-hydroxybutyrate can be contemporary obtained by single-stage photo-fermentation of winery wastewater, by using a purple non-sulfur bacteria mixed consortium [137]. With an initial chemical oxygen demand of 1500 $mg \cdot dm^{-3}$, up to 468 $cm^3 \cdot dm^{-3}$ of hydrogen and 203 $cm^3 \cdot dm^{-3}$ of poly-β-hydroxybutyrate can be produced.

4.3. CO Gas-Fermentation

In CO gas-fermentation hydrogen is produced by reacting carbon monoxide and water in presence of photosynthetic bacteria under anaerobic conditions, through the overall reaction (7) [105]:

$$CO + H_2O = CO_2 + H_2 \ \Delta G^\circ = -201 \ kJ \cdot mol^{-1} \quad (7)$$

Although this technique is promising for hydrogen production, hydrogen consumption by homoacetogenesis still remains the main challenge. In a recent study, the effect of pH and CO loading, on the CO and carbohydrate-rich wastewater co-fermentation was investigated, focusing on the homoacetogenesis behavior [138]. The results showed that the highest hydrogen production was obtained with at pH = 5 with a CO loading of 2000 $cm^3 \cdot d^{-1}$; the hydrogen consumption increased with the pH from 5 to 8, moreover the injection of CO further increased the hydrogen consumption at neutrophil pH = 7–8. In an interesting study, the possibility to produce both thermophilic enzymes and hydrogen at the same time has been demonstrated, by using hyperthermophilic strain Thermococcus onnurineus NA1 156T for CO fermentation, providing the basis for cell factories to upcycle industrial waste gas [139].

4.4. Bio-Photolysis

Bio-photolysis can be defined as a "photonic-driven hydrogen production" process, in which water splitting is obtained by using cyanobacteria and blue-green algae [119]. It is categorized in direct bio-photolysis and indirect bio-photolysis (Figure 8). The direct route consists of a photosynthetic reaction and uses microalgae in presence of solar energy (8) [140]:

$$2H_2O + \text{Light energy} \rightarrow O_2 + 2H_2 \quad (8)$$

The indirect bio-photolysis is a two-step process, the first step is the photosynthesis (9) while the second step hydrogen and CO_2 are generated (10) [140]:

$$12H_2O + 6CO_2 + \text{Solar energy} \rightarrow 6O_2 + C_6H_{12}O_6 \quad (9)$$

$$C_6H_{12}O_6 + 12H_2O + \text{Solar energy} \rightarrow 6CO_2 + 12H_2 \quad (10)$$

Hydrogen is produced by hydrogenase and nitrogenase enzymes.

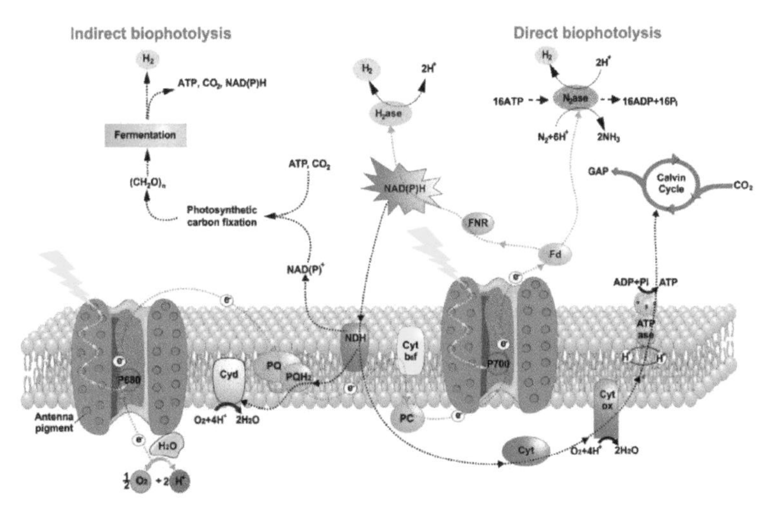

Figure 8. Direct and indirect biophotolysis processes of photosynthetic microorganisms Abbreviation: Cyd, cytochrome bd quinol oxidase; PQH_2/PQ, plastoquinol/plastoquinone; ATPase, ATP synthase; Cyt b_6f, cytochrome b_6f complex; Fd, ferredoxin; FNR, ferredoxin NAD(P) reductase; H_2ase, hydrogenase; NDH, NAD(P)H dehydrogenase; PC, plastocyanin; PQ, plastoquinones; P680, Photosystem II; P700, Photosystem I; N_2ase, nitrogenase; H_2, hydrogen [141].

The hydrogen yield was evaluated in a comparative study with four cyanobacteria strains (*Synechocystis* sp. PCC 6803, *Desertifilum* sp. IPPAS B-1220, *Synechococcus* sp. I12, and *Phormidium corium* B-26) [141]. The maximum hydrogen accumulation was obtained with the wild-type, filamentous, non-heterocystous cyanobacterium *Desertifilum* sp. IPPAS B-1220. The yield was 0.229 $mmol \cdot mg_{chlorophyll}{}^{-1} \cdot h^{-1}$ in the gas phase within 166 h in the light, which, based on the data reported by the authors in the article, is among the highest one reported in the literature, for these cyanobacteria.

4.5. Integrated Systems

The critical aspect of biological processes is certainly the low yield to hydrogen, especially due to the presence of byproducts. One strategy for increasing efficiency lies in combining multiple processes. For example, a process has been reported in which duckweed are used as feedstock for hydrogen production, through dark fermentation, simultaneously using the fermentative waste to produce microalgal lipids [142]. The simultaneous production of hydrogen and the use of waste reduce the costs of microalgae cultivation and wastewater treatment. More attention has received the sequential two-stage dark and photo-fermentation process. The volatile fatty acids produced during the dark-fermentation are used for the hydrogen production in photo-fermentation. Combining the two processes is possible to greatly improve the hydrogen production; as obvious from the previous overall reactions (3) and (6), in the acetate-type fermentation four moles of hydrogen and two moles of acetic acid are obtained from one mole of glucose, subsequently the two moles of acetic acid give eight moles of hydrogen [143]. Some articles, focusing on this double-stage process, have been published in the last year [144]. The dark-fermentation effluents, with low-butyrate and high lactate concentration, were used to maximize the hydrogen yield in photo-fermentation process, using fruit and vegetable waste and cheese whey powder [145]. The strategy was optimizing the C/N ratio to take advantage from higher hydrogen yield obtainable with lactate in photo-fermentation. L-cysteine and Fe_3O_4 nanoparticles can improve the hydrogen production and the electronic distribution from dark-fermentation effluents in photo-fermentation, by acting as reducing agent and by enhancing the bacteria aggregation [146]. Improved hydrogen production was also obtained by adding Fe(II) sulfate concentration during photo-fermentation stage [147]. The addition of enzymatic hydrolysate is able to increase the H_2 yield from 312.54 to 1287.06 $cm^3 \cdot H_2 \cdot g_{TOC}{}^{-1}$ (TOC = total organic carbon), and maximum hydrogen production rate from 2.14 to 10.23 $cm^3 \cdot h^{-1}$, in the photo fermentation from dark fermentation effluents [148]. The efficiency of organic waste conversion to hydrogen was enhanced through the addition of Ca- and Mg-saturated resin and phosphate-laden biochar in a single-stage hybrid dark-photo hydrogen fermentation from food waste [149]. The saturated resins provided the nutrients for the bacteria, by gradual releasing calcium and magnesium, while biochar promoted bacterial adhesive growth and acted as a buffer, due to the presence of phosphate, amino, and carbonate compounds.

To investigate a sequential dark and photofermentation process for the treatment of wastewater and the simultaneous hydrogen production, a circular baffled reactor (Figure 9), operating at ambient temperature was introduced [150]. The highest hydrogen yield (0.4 $dm^3 \cdot g_{COD}{}^{-1}$), a chemical oxygen demand (COD) removal of 82%, and an organic-N removal of 95% were obtained at the hydraulic retention time of 24 h and at the initial pH of 6.5. The techno-economic evaluation highlighted the feasibility when dealing with gelatin-rich wastewater.

Simultaneous hydrogen and poly-β-hydroxybutyrate-PHB productions, under dark, photo, and subsequent dark and photofermentation, by using wastes, has been reported [151]. In the sequential dark and photofermentation the highest hydrogen and PHB were obtained by using rice straw hydrolysate (1.82 ± 0.01 $mol_{H2} \cdot mol_{glucose}{}^{-1}$ and 19.15 ± 0.25 $g \cdot dm_{PHB}{}^{-3}$) at a pH of 7.0, with *Bacillus cereus* (KR809374) and *Rhodopseudomonas rutile*.

In a comparative study, batch, semi-continuous and continuous mode for hydrogen production from dark fermentation effluents were investigated. The results highlighted the best performance of the semi-continuous mode [152].

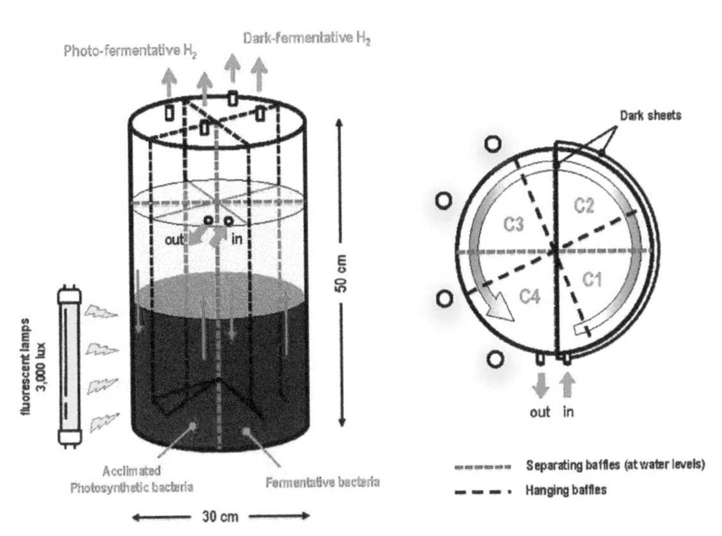

Figure 9. Schematic diagram for the lab-scale dark-photo circular baffled reactor used for anaerobic treatment of gelatinaceous wastewater [150].

4.6. *Comparative Studies*

Comparative studies among photofermentation, dark fermentation and dark-photo co-fermentation methods, by using photosynthetic HAU-M1 bacteria and dark fermentative *Enterobacter aerogenes* bacteria, have demonstrated that photofermentation is the most promising method for hydrogen production from corn stover, due to its maximum cumulative hydrogen yield of 141.42 $cm^3 \cdot g_{TS}^{-1}$, maximum hydrogen production rate of 6.21 $cm^3 \cdot g_{TS}^{-1} \cdot h^{-1}$, maximum hydrogen content of 58.90% and highest energy conversion efficiency of 10.12% [153]. Pretreatment by alkaline-enzymolysis has successfully been used also with cornstalk hydrolysate (CS), to improve the hydrogen yield [154]. The maximum hydrogen yields of 168.9 $cm^3 \cdot g_{CS}^{-1}$, 357.6 $cm^3 \cdot g_{CS}^{-1}$, and 424.3 $cm^3 \cdot g_{CS}^{-1}$ were obtained in dark fermentation, photofermentation, and two-stage fermentation, respectively, with $Ca(OH)_2$ 0.5%, hydrolysis temperature 115 °C, hydrolysis time 1.5 h, cellulase dosage 4000 $U \cdot g_{CS}^{-1}$ and xylanase 4000 $U \cdot g_{CS}^{-1}$.

4.7. *Remarks on Biological and Photonic Methods*

Among the biological methods, the most cost-effective process is dark fermentation [143], due to low production and capital cost and good hydrogen yield. However, if compared with other production methods, the hydrogen yield and production rate are still uncompetitive and the production processes are still at a primordial stage, moreover another critical issue is the availability and pre-treatment of inocula [155]. The design of integrated systems can reduce costs and increase the hydrogen production; the sequential two-stage dark and photo-fermentation process, using wastes as feedstock, seems to significantly improve the hydrogen production, however at the moment this type of technology seems more suitable for small plants, for the local production of hydrogen.

5. Electrical Methods

In the framework of the electrical methods, we basically refer to the water electrolysis, in which hydrogen and oxygen are produced from a molecule of water (11):

$$\text{Overall reaction:} \quad 2\,H_2O \rightarrow 2\,H_2 + O_2 \tag{11}$$

Unfortunately, the water dissociation requires a significant amount of energy (ΔH^0 = 285.8 $kJ \cdot mol^{-1}$ and ΔG^0 = 237.2 $kJ \cdot mol^{-1}$ under standard conditions); therefore, the cell voltage required for water dissociation is 1.23 V under standard conditions [156]. The main limitation of the process is the slow kinetics, to reach a significant hydrogen production rate, cell voltages of 1.8–2.0 V are required, thus increasing the costs, and decreasing the efficiency of the process [156]. There are several water electrolyzes technologies, however the major types of cells commercially available are: alkaline electrolysis cell (AEC), anion exchange membrane electrolysis cell (AEMEC) proton exchange membrane electrolysis cell (PEMEC), and solid oxide electrolysis cells (SOEC) [157].

5.1. Alkaline Electrolysis Cell

In alkaline electrolysis cell, the electrodes are immersed in the liquid electrolyte, typically a 25–30 wt% solution of KOH or NaOH separated by a diaphragm. Anodic (12) and cathodic reactions (13) are the following:

$$\text{Anodic reaction:} \qquad 4\,OH^- \rightarrow O_2 + 2H_2O + 4\,e^- \tag{12}$$

$$\text{Cathodic reaction:} \qquad 2\,H_2O + 2\,e^- \rightarrow H_2 + 2\,OH^- \tag{13}$$

The choice of the electrocatalysts is a critical issue, those based on noble metals are highly efficient, on the other hand, the most convenient material used for the electrodes are Ni-based [158,159], however to improve the performance and reduce the degradation, a series of metals can be used as additives, including cobalt, vanadium iron and selenide [160]. The use of noble metal-based electrodes could provide better performance however, cheaper alternative would be preferred. Recently MoO_2-Ni arrays have been reported to exhibit a Pt-like activity at 25.0 °C [161], its heterogeneous components may avoid agglomeration under high-temperature catalytic conditions.

High efficiency is obtained when the cells operated at low current densities (<0.4 A·cm^{-2}). Historically the most common diaphragm is of porous white asbestos ($Mg_3Si_2O_5(OH)_4$). Due to the toxicity problems related to the use of asbestos, in the last decade considerable efforts have been put into developing hydroxide conducting polymers suitable for alkaline water electrolysis [159].

The chance of using higher current density collides with the gas bubble generation, trapped in porous electrodes, resulting in the reduced accessibility to the active sites. Recently 3D printed electrodes, with a controlled periodic structure, have been reported, which are able to suppress the gas bubble coalescence, jamming and trapping [162]. The 3D-electrodes decorated with carbon-doped NiO can reach a current density of 1000 mA·cm^{-2} in a 1.0 M KOH electrolyte at hydrogen evolution reaction and oxygen evolution reaction overpotentials of 245 and 425 mV, respectively. In a comparative study, the tradeoff between surface area and pore structure, in nickel electrodes—foam, microfiber felt, and nanowire felt—has been studied [163]. The results showed that the microfiber felt is able to maintain a current density of 25 000 mA·cm^{-2} over 100 h without degradation, balancing high surface area with the ability to remove bubbles.

5.2. Anion Exchange Membrane Electrolysis Cell

Unlike the AEC, in the case of anion exchange electrolysis cell, the hydroxyl anions cross a membrane. Typically, metal transition oxides electrocatalysts are used at the anode, and rare-earth metal oxides are used at the cathode [156]. Recently nanostructured nickel-based electrode films have developed, prepared by magnetron sputtering, in an oblique angle configuration, showing good performance [164]. $NiMn_2O_4$ anode catalyst was developed and tested in combination with a commercial FAA3-50 membrane, a durability test was carried out for 1000 h by varying the cell potential between 1 and 1.8 V for the FAA3-50 and $NiMn_2O_4$ based-MEA, showing high stability [165].

Much effort has been focused on the development of highly efficient membranes. Twisted ether-free polyarylene piperidinium, synthesized via acid-catalyzed polycondensation reaction, characterized by efficient ion-conducting channels, which provide an hydroxide conductivity of 37 mS·cm^{-1} at 30 °C, has been proposed [166]. The cell based on this membrane has a high current density of 1064 mA·cm^{-2} at 2.5 V under 1 M KOH and 50 °C and a high frequency resistance of 0.165 Ω·cm^{-2} at 1.8 V. The durability test, performed at current density of 200 mA·cm^{-2} showed a voltage of 2.1 V for more than 500 h.

The effects of ion-exchange capacity and thickness of all-hydrocarbon anion exchange membranes was investigated [167]. The best performance was obtained by using a membrane with a ion-exchange capacity of 2.1–2.5 $meq_{OH^-}g^{-1}$ and a thickness of 50 μm, and the FAA-3 ionomer in a 1 M KOH liquid electrolyte at 60 °C. Under these conditions, a potential of 1.82 V was obtained at a current density of 2 A·cm^{-2} and a cell resistance of 95 mΩ·cm^2.

In a comparative study, the performance of Sustanion®, Aemion™ and A-201 membranes were investigated [168]. Good performance at temperatures up to 60 °C, at KOH concentrations of 0.5–1 M were obtained in all the cases, the use of distilled water led to an increase in the membrane resistance, while the best performance was obtained with the Sustanion®-based membrane electrode assembly (MEA, Figure 10) at all KOH concentrations and temperatures studied.

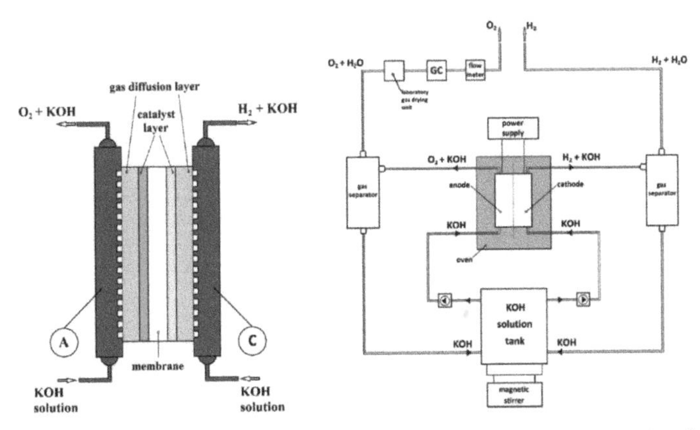

Figure 10. Scheme of the zero-gap cell (left) and the electrolysis cell testing bench [168].

5.3. Proton Exchange Membrane Electrolysis Cell

In the proton exchange electrolysis cell, the anodic and the cathodic compartments are separated by a polymer membrane. In the anodic compartment, water is oxidized according to the reaction (14); the hydrated protons migrate across the membrane to reach the cathode where they are reduced (15):

$$\text{Anodic reaction:} \qquad 2\,H_2O \rightarrow O_2 + 4\,H^+ + 4\,e^- \tag{14}$$

$$\text{Cathodic reaction:} \qquad 2\,H^+ + 2\,e^- \rightarrow H_2 \tag{15}$$

In this case Ir-based oxide are usually used as catalytic anode, while unsupported or of carbon-supported Pt particles are used at the cathode [156]. Unfortunately, iridium is extremely expensive and rare, around 0.5 kg of iridium is required per megawatt installed electrolyzer power [169]. Therefore, it is crucial to obtain a significant reduction of the catalyst loading. IrOx nanofibers have been combined with a conventional nanoparticle based IrOx anodic catalyst layer, resulting in an iridium loading reduction of over 80% while maintaining the same performance. This result has been attributed to a combination among the good electrical contact and high porosity of IrOx nanofibers with the high surface area of IrOx nanoparticles [169].

The most used polymer membranes for proton-exchange membrane water electrolysis are perfluorosulfonated acids, which are characterized by high proton conductivity, mechanical and chemical robustness. On the other hand, they are high-cost materials and are characterized by significant gas permeability, low mechanical stability for temperature higher to 80 °C and present environmental issues due to the presence of fluorine. As alternative, sulfonated poly(phenylene sulfone) (sPPS) combines high proton conductivity with low gas crossover can be used. It has been reported that the performance of the sPPS-MEAs is substantially better than that of Nafion N115-MEAs (3.5 $A \cdot cm^{-2}$ vs 1.5 $A \cdot cm^{-2}$ at 1.8 V) with the same catalyst loading and comparable membrane thickness [170].

The effect of degradation on the performance of a PEM electrolysis cell has been investigated, at low catalyst loading, moreover, the degradation mechanism was investigated by means of ac-impedance spectra and post-operation analyses [171]. The results showed that the mass transfer issues are relevant under steady-state mode, while the catalyst degradation occurs under cycled operations. The membrane thinning depends on the uptime hours at high current density, while the overall cell voltage increase is higher for cycled operations than for steady-state mode, due to a higher decrease of series resistance in steady-state mode.

Porous transport layers (PTL) play a kay role in PEM water electrolysis cells, as it is involved in gas and liquid transport, in thermal and electrical conduction and in the contact between adjacent components [172], particularly those of the PTL surface in contact with the catalyst layer [173]. The actually used PTL are single layer sintered porous Ti-based materials, recently it has been reported that better performance can be achieved with multiple layers [174]. The catalyst loading affects the influence of PTL on the electrolysis performance, for low catalyst loading; moreover, at high current density, the apparent bubble coverage increased with PTL grain size at high catalyst loading [173]. The electrolysis performance are also related to the anode PTL properties, which can significantly impact ohmic, activation and diffusion losses [172]. By modifying the wettability of the PTL by adding hydrophobic additives, diffusion loss, ohmic loss and activation loss significantly increase.

5.4. Bipolar Membrane Electrolysis

One of the major limitations in the production of hydrogen by conventional water electrolysis, derives from the two semi-reactions of evolution of hydrogen and oxygen which, being simultaneous, are interdependent. Generally, the first reaction is fast in acidic conditions where there is an abundance of protons, while the second is slow, unless extremely expensive and rare catalysts such as iridium and ruthenium are used. A possible alternative is the amphoteric water electrolysis that, with a bipolar membrane, can provide optimal pH conditions simultaneously for both cathode and anode, under steady-state operations, without changing the overall thermodynamics of water splitting [175]. It has been shown that the decoupled amphoteric water electrolysis assisted with the redox mediator MnO_2/Mn^{2+}, by separating the production of hydrogen and oxygen into two independent processes, allows to produce hydrogen with a high-power input (up to 1 $A \cdot cm^{-2}$), and low power absorbed oxygen production. Similar results were obtained with liquid water; in particular it has been reported that a bipolar membrane with and without a water splitting catalyst resulted in cell current densities of 450 and 5 $mA \cdot cm^{-2}$ at cell voltages of 2.2 V. Upon moving the bipolar interface directly between the acidic membrane and the high-pH anode, a current density of 9000 $mA \cdot cm^{-2}$ at cell voltages of 2.2 V was achieved [176].

5.5. Solid Oxide Electrolysis Cells

In solid oxide electrolysis cells (SOEC), a solid oxide or ceramic is used as electrolyte; at the cathode side hydrogen is produced from water (17), while the oxygen ions generated, across the electrolyte, reach the anode (16) where they are oxidized to produce oxygen:

$$\text{Anodic reaction:} \quad 2O^{2-} \rightarrow O_2 + 4\,e^- \tag{16}$$

$$\text{Cathodic reaction:} \quad H_2O + 2\,e^- \rightarrow H_2 + O^{2-} \tag{17}$$

The solid oxide electrolysis cells typically operate in the temperature range 500–900 °C [177], which provides a crucial benefit over proton exchange membrane (PEM) and alkaline exchange membrane (AEM) electrolyzers, which operate at a maximum of 100 °C. Unfortunately, the degradation of SOEC is the major limitation to the commercial viability, the aggressive humid condition in the air electrode side, is still a concern to the stability of electrolysis cells [178]. Typically, Ni/yttria-stabilized zirconia (Ni/YSZ) electrodes are used [179], however agglomeration of Ni nanoparticles, low water vapor transmission efficiency and poor durability are serious issues [180].

A series of studies have been focused on the optimization of the electrodes. An A-site cation-deficient $La_{0.4}Sr_{0.55}Co_{0.2}Fe_{0.6}Nb_{0.2}O_{3-d}$ perovskite has been recently reported, exhibiting a high electrolysis current density of 0.956 $A{\cdot}cm^{-2}$ with an applied voltage of 1.3 V at 850 °C and good stability in a high humidity and hydrogen partial pressure environment [180]. The high operating stability of the electrode has been attributed to the strong interaction between the alloy nanoparticles and the perovskite substrate, that suppresses the sintering of the nanoparticles, moreover the SrO phase is able to protect the alloy nanoparticles from oxidation.

A similar effect has been reported with barium doping, $Ba_{0.2}Sr_{1.8}Fe_{1.5}Mo_{0.5}O_{6-\delta}$ double perovskite as fuel electrode, which showed good performance with the appropriate steam amount of 20% [181].

The B-site of $Ba_{0\text{-}5}Sr_{0\text{-}5}Co_{0\text{-}8}Fe_{0\text{-}2}O_{3-\delta}$ of perovskite anode has been partially substituted with a higher valence Ta^{5+} (5, 10, 15 and 20 mol%) to improve the structural stability [182]. The current density of 10 mol% doped catalyst was 8.3 times higher than undoped one at 1.8 V, providing the higher of H_2 production rate, moreover, the degradation rate was 0.0027 $V{\cdot}h^{-1}$, (−0.45 $A{\cdot}cm^{-2}$, 800 °C, steam/H_2 = 70:30). It has been assumed that the Ta^{5+} doping provides a balance between ionic and electronic conductivity in the anode and a better electrochemical performance.

$BaCe_{0.5}Zr_{0.2}Y_{0.1}Yb_{0.1}Gd_{0.1}O_{3-\delta}$ (BCZYYbGd) electrolyte, characterized by high chemical stability and proton conductivity has been coupled with a $PrNi_{0.5}Co_{0.5}O_{3-\delta}$ steam electrode and a Ni-BCYYbGd hydrogen electrode for intermediate temperature operation (Figure 11) [183]. The BCYYbGd electrolyte showed high stability over 200 h at 50 vol % steam in argon and at 600 °C, moreover, high current density of 2.405 $A{\cdot}cm^{-2}$ at a cell voltage of 1.6 V was obtained at 600 °C at 20 vol % of steam in argon.

An effective way to improve the performance of the oxygen electrode is the infiltration, in which the electrocatalysts are introduced into a porous backbone at relatively low temperatures. $La_{0.6}Sr_{0.4}CoO_{3-\delta}$ (LSC) infiltrated gadolinia-doped ceria (CGO) oxygen electrode has been reported; in electrolysis mode, the current density reached 1.07 $A{\cdot}cm^{-2}$ at cell voltage of 1.3 V at 750 °C with a steam 60 vol% [184].

Symmetrical solid oxide cells configuration has also been evaluated, in which $Sr_2Fe_{1.5}Mo_{0.5}O_{6-\delta}$ electrodes are deposited on both sides of YbScSZ tapes previously coated with a $Ce_{1-x}Gd_xO_{1.9}$ [185]. This configuration has shown some advantages such as a reduction of sintering steps or a better thermomechanical compatibility between the electrodes and the electrolyte.

The electrochemical performance of cathode-supported cells having gadolinium doped ceria/yttria-stabilized zirconia (GDC/YSZ), yttria-stabilized zirconia (YSZ) and Sc^{3+}, Ce^{4+}, and Gd^{3+}-doped zirconia (SCGZ) electrolyte was compared, highlighting the highest electrochemical performance of the cathode-supported cell having SCGZ electrolyte (Ni-SCGZ/SCGZ/BSCF) [186].

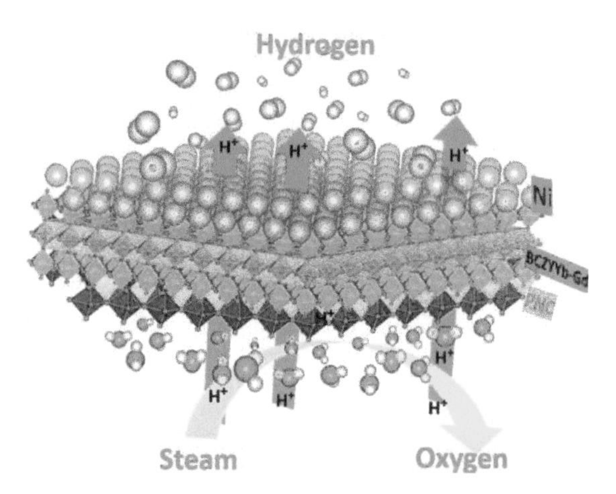

Figure 11. Representation of the solid oxide electrolysis cell [183].

A heterogeneous design for proton-conducting solid oxide electrolysis cells has also been proposed, in which a better stability and higher efficiency of electrolysis cells has been obtained [178]. Yttrium and zirconium co-doped barium cerate-nickel was used as fuel electrode material and yttrium-doped barium zirconate as the electrolyte material. The results showed that the proposed novel design can efficiently improve the proton conductivity of the yttrium-doped barium zirconate electrolyte (from 0.88×10^{-3} S·cm^{-1} to 2.13×10^{-3} S·cm^{-1} at 600 °C) and improve the ionic transport number of the electrolyte (from 0.941 to 0.964 at 600 °C).

5.6. Reamarks on Electrical Methods

The hydrogen production from electrolyzers is a mature technology, and among the electrochemical technology, alkaline water electrolysis is already available for large-scale applications, due to the use of non-expensive materials for the electrodes [187]. The use of renewable energy is a necessary boundary condition, however there is a problem with a partial loading when renewable energy supply is intermittent and unstable. An energy storage device can partially solve the incompatibility between water electrolyzers and renewable energy sources. The process can be efficient if the electrolyser is powered, for example, by solar photovoltaic through a full cell of lithium-ion battery as an energy reservoir [188]. On the other hand, solid oxide electrolysis cells are extremely attractive, due to the possibility to work at high temperatures where higher efficiency can be reached. Although the critical issues related to the stability of these cells have not yet been solved, it has been demonstrated that a plant in which the SOEC is integrated with a parabolic dish solar field, to provide both electricity and thermal energy, necessary for the electrolysis reaction to take place, a nominal solar-to-hydrogen efficiency above 30%, with a SOEC efficiency around 80%, can be reached [189]. In another study it has been calculated that is possible to produce hydrogen in electrolyzers integrated with nuclear plants with an energy cost of 38.83 and 37.55 kWh·$kg_{H2}{}^{-1}$ for protonic and ionic solid oxide electrolyzers, respectively [190].

6. Economic Assessment

Establishing the costs of hydrogen per type of production process is outside the scope of this article, however it is possible to make an approximate assessment based on what is reported in the literature. Evaluating the costs of production processes based on immature techniques is extremely difficult and risky, as the costs are related to diffusion. In any case, however, various costs, capital and operating costs, design, labor, electricity, as well as the costs of raw materials, waste disposal etc. must be considered. In the case of biological and photonic methods, dark fermentation is most promising technology, with estimated hydrogen production costs in the range of 1.02–2.70 USD m^{-3} and a high return on investment (calculated as the ratio between annual profit and fixed capital investment) [155]. The hydrogen production cost via natural gas steam reforming is equal to 0.67 USD m^{-3}, however with a lower return on investment. The hydrogen production cost in the case of electrolysis is strongly dependent on the electricity cost, for example in Germany is around 3.64 €q Kg^{-1}, for alkaline water electrolysers. At the moment, therefore, the lowest costs are found in the case of natural gas reforming, however these assessments lack environmental costs, which could substantially modify the results reported so far.

7. Conclusions and Future Prospective

Hydrogen is the most promising energy carrier; however, it is scarcely present in Nature in molecular form, therefore it must be obtained from primary sources. In this review article, the main characteristics of a number of hydrogen production methods have been listed, focusing primarily on renewable feedstock. The production methods have been grouped into three main sections, based on the type of energy used to sustain the process. For each section, the main limitations to the diffusion of the analysed production processes have been highlighted. As mentioned, the most widespread hydrogen production methods are, still today, natural gas reforming processes, which use feedstock abundantly available at low cost, and use proven processes. However, the need to achieve carbon neutrality requires the use of renewable feedstock and energy to support production processes.

Water is the main source of hydrogen, via water splitting processes, however the enormous amount of energy required to support these constitutes a serious limitation, and therefore the use of energy from renewable sources is a necessary boundary condition. Among the various techniques, alkaline electrolysis of water is now available for large-scale applications. On the other hand, biomasses are extremely attractive; as a variety of materials such as algae, food waste, municipal solid waste, lignocellulosic biomass etc. can be used. Several processes can be used to obtain hydrogen from biomass, such as gasification and fermentation, furthermore bio-oils, obtained by pyrolysis of biomass and the same of biomethane, can be used in reforming processes. The main limitations to their use seem to be the variability of the composition and therefore the unavailability of large quantities at low cost.

The origin of the energy needed to sustain the processes itself plays a crucial role. In this context, electrification can make sustainable the processes that use thermal methods. However, the main problem related to the use of renewable energy is the instability in the supply which, in many cases, is intermittent.

Based on these considerations, a future is outlined increasingly based on the local production and distribution of energy and therefore of hydrogen, in which the decentralization of hydrogen production will play a key role. The energy to support the processes must be generated on site, exploiting the potential of the territory, such as wind power, solar power, etc. Likewise, the feedstock for hydrogen production will have to be available where hydrogen will be produced. A massive use of biomass will make it necessary to implement processes that are tolerant to the variability of the biomass itself, so as to minimize production costs.

Author Contributions: All authors equally contributed to the conceptualization, methodology, software, validation, formal analysis, investigation, resources, data curation, writing-original draft preparation, writing-review and editing, visualization, supervision, project administration and funding acquisition of this manuscript. All authors have read and agreed to the published version of the manuscript.

Funding: This work has received funding from the European Union's Horizon 2020 research and innovation programme under the Marie Skłodowska-Curie grant agreement No 734561.

Acknowledgments: The authors wish to acknowledge Ing. Antonio Corrado for the valuable contribution on the electrification of reforming processes.

Conflicts of Interest: The authors declare no conflict of interest.

References

1. Available online: https://www.un.org/press/en/2012/sgsm14242.doc.htm (accessed on 20 April 2012).
2. Jeuland, M.; Fetter, T.R.; Li, Y.; Pattanayak, S.K.; Usmani, F.; Bluffstone, R.A.; Chávez, C.; Girardeau, H.; Hassen, S.; Jagger, P.; et al. Is energy the golden thread? A systematic review of the impacts of modern and traditional energy use in low- and middle-income countries. *Renew. Sust. Energ. Rev.* **2021**, *135*, 110406. [CrossRef]
3. *Special Report: Global Warming of 1.5 °C*; Intergovernmental Panel on Climate Change: Geneva, Switzerland, 2018. Available online: https://www.ipcc.ch/sr15/ (accessed on 12 March 2020).
4. Marnellos, G.E.; Klassen, T. Welcome to Hydrogen—A New International and Interdisciplinary Open Access Journal of Growing Interest in Our Society. *Hydrogen* **2020**, *1*, 6. [CrossRef]
5. Available online: https://www.europarl.europa.eu/news/en/headlines/society/20190926STO62270/what-is-carbon-neutrality-and-how-can-it-be-achieved-by-2050 (accessed on 8 October 2020).
6. Available online: https://www.e-education.psu.edu/earth104/node/1345 (accessed on 1 February 2021).
7. Abdin, Z.; Zafaranloo, A.; Rafee, A.; Mérida, W.; Lipiński, W.; Khalilpour, K.R. Hydrogen as an energy vector. *Ren. Sust. Energ. Rev.* **2020**, *120*, 109620. [CrossRef]
8. Boyle, R. *Tracts, Containing New Experiments, Touching the Relation betwixt Flame and Air*; Richard Davis: London, UK, 1672.
9. Available online: https://www.thoughtco.com/most-abundant-element-in-known-space-4006866 (accessed on 7 April 2021).
10. Safari, F.; Dincer, I. A review and comparative evaluation of thermochemical water splitting cycles for hydrogen production. *Energy Convers. Manag.* **2020**, *205*, 112182. [CrossRef]
11. Rambhujun, N.; Salman, M.S.; Wang, T.; Pratthana, C.; Sapkota, P.; Costalin, M.; Lai, Q.; Aguey-Zinsou, K.-F. Renewable hydrogen for the chemical industry. *Mrs Energy Sustain.* **2020**, *7*, E33. [CrossRef]
12. Dawood, F.; Anda, M.; Shafiullah, G.M. Hydrogen production for energy: An overview. *Int. J. Hydrogen Energy* **2020**, *45*, 3847–3869. [CrossRef]
13. Shen, Q.; Jiang, Y.; Xia, F.; Wang, B.; Lv, X.; Ye, W.; Yang, G. Hydrogen production by Co-based bimetallic nano-catalysts and their performance in methane steam reforming. *Pet. Sci. Technol.* **2020**, *38*, 618–625. [CrossRef]
14. Chen, L.; Qi, Z.; Zhang, S.; Su, J.; Somorjai, G.A. Catalytic Hydrogen Production from Methane: A Review on Recent Progress and Prospect. *Catalysts* **2020**, *10*, 858. [CrossRef]
15. Palma, V.; Gallucci, F.; Pullumbi, P.; Ruocco, C.; Meloni, E.; Martino, M. Pt/Re/CeO_2 Based Catalysts for CO-Water–Gas Shift Reaction: From Powders to Structured Catalyst. *Catalysts* **2020**, *10*, 564. [CrossRef]
16. Palma, V.; Ruocco, C.; Cortese, M.; Renda, S.; Meloni, E.; Festa, G.; Martino, M. Platinum Based Catalysts in the Water Gas Shift Reaction: Recent Advances. *Metals* **2020**, *10*, 866. [CrossRef]
17. Palma, V.; Ruocco, C.; Cortese, M.; Martino, M. Recent Advances in Structured Catalysts Preparation and Use in Water-Gas Shift Reaction. *Catalysts* **2019**, *9*, 991. [CrossRef]
18. Garbis, P.; Jess, A. Selective CO Methanation in H_2-Rich Gas for Household Fuel Cell Applications. *Energies* **2020**, *13*, 2844. [CrossRef]
19. Jiang, Z.; Liao, M.; Qi, J.; Wang, C.; Chen, Y.; Luo, X.; Liang, B.; Shu, R.; Song, Q. Enhancing hydrogen production from propane partial oxidation via CO preferential oxidation and CO_2 sorption towards solid oxide fuel cell (SOFC) applications. *Renew. Energy* **2020**, *156*, 303–313. [CrossRef]
20. Tosto, E.; Alique, D.; Martinez-Diaz, D.; Sanz, R.; Calles, J.A.; Caravella, A.; Medrano, J.A.; Gallucci, F. Stability of pore-plated membranes for hydrogen production in fluidized-bed membrane reactors. *Int. J. Hydrogen Energy* **2020**, *45*, 7374–7385. [CrossRef]
21. Meloni, E.; Martino, M.; Palma, V. A Short Review on Ni Based Catalysts and Related Engineering Issues for Methane Steam Reforming. *Catalysts* **2020**, *10*, 352. [CrossRef]
22. Gray, J.T.; Che, F.; McEwen, J.-S.; Ha, S. Field-assisted suppression of coke in the methane steam reforming reaction. *Appl. Catal. B-Environ.* **2020**, *260*, 118132. [CrossRef]
23. Palma, V.; Meloni, E.; Renda, S.; Martino, M. Catalysts for Methane Steam Reforming Reaction: Evaluation of CeO_2 Addition to Alumina-Based Washcoat Slurry Formulation. *C J. Carbon Res.* **2020**, *6*, 52. [CrossRef]
24. Han, B.; Wang, F.; Zhang, L.; Wang, Y.; Fan, W.; Xu, L.; Yu, H.; Li, Z. Syngas production from methane steam reforming and dry reforming reactions over sintering-resistant Ni@SiO_2 catalyst. *Res. Chem. Intermed.* **2020**, *46*, 1735–1748. [CrossRef]
25. Balzarotti, R.; Ambrosetti, M.; Beretta, A.; Groppi, G.; Tronconi, E. Investigation of packed conductive foams as a novel reactor configuration for methane steam reforming. *Chem. Eng. J.* **2020**, *391*, 123494. [CrossRef]
26. Renda, S.; Cortese, M.; Iervolino, G.; Martino, M.; Meloni, E.; Palma, V. Electrically driven SiC-based structured catalysts for intensified reforming processes. *Catal. Today* **2020**. [CrossRef]
27. Meloni, E.; Martino, M.; Ricca, A.; Palma, V. Ultracompact methane steam reforming reactor based on microwaves susceptible structured catalysts for distributed hydrogen production. *Int. J. Hydrogen Energy* **2021**, *46*, 13729–13747. [CrossRef]
28. Grasham, O.; Dupont, V.; Cockerill, T.; Alonso Camargo-Valero, M.; Twigg, M.V. Hydrogen via reforming aqueous ammonia and biomethane co-products of wastewater treatment: Environmental and economic sustainability. *Sustain. Energy Fuels* **2020**, *4*, 5835–5850. [CrossRef]

29. Zhang, L.; Yu, Z.; Li, J.; Zhang, S.; Hu, S.; Xiang, J.; Wang, Y.; Liu, Q.; Hu, G.; Hu, X. Steam reforming of typical small organics derived from bio-oil: Correlation of their reaction behaviors with molecular structures. *Fuel* **2020**, *259*, 116214. [CrossRef]
30. Megía, P.J.; Calles, J.A.; Carrero, A.; Vizcaíno, A.J. Effect of the incorporation of reducibility promoters (Cu, Ce, Ag) in Co/CaSBA-15 catalysts for acetic acid steam reforming. *Int. J. Energy Res.* **2021**, *45*, 1685–1702. [CrossRef]
31. Li, J.; Mei, X.; Zhang, L.; Yu, Z.; Liu, Q.; Wei, T.; Wu, W.; Dong, D.; Xu, L.; Hu, X. A comparative study of catalytic behaviors of Mn, Fe, Co, Ni, Cu and ZneBased catalysts in steam reforming of methanol, acetic acid and acetone. *Int. J. Hydrogen Energy* **2020**, *45*, 3815–3832. [CrossRef]
32. Palma, V.; Ruocco, C.; Cortese, M.; Martino, M. Bioalcohol Reforming: An Overview of the Recent Advances for the Enhancement of Catalyst Stability. *Catalysts* **2020**, *10*, 665. [CrossRef]
33. Hou, X.; Qing, S.; Liu, Y.; Li, L.; Gao, Z.; Qin, Y. Enhancing effect of MgO modification of Cu-Al spinel oxide catalyst for methanol steam reforming. *Int. J. Hydrogen Energy* **2020**, *45*, 477–489. [CrossRef]
34. Azenha, C.; Lagarteira, T.; Mateos-Pedrero, C.; Mendes, A. Production of hydrogen from methanol steam reforming using CuPd/ZrO_2 catalysts—Influence of the catalytic surface on methanol conversion and CO selectivity. *Int. J. Hydrogen Energy* **2020**. [CrossRef]
35. López-Martín, A.; Platero, F.; Caballero, A.; Colón, G. Thermo-Photocatalytic Methanol Reforming for Hydrogen Production over a CuPd-TiO_2 Catalyst. *ChemPhotoChem* **2020**, *4*, 630–637. [CrossRef]
36. Yu, C.-L.; Sakthinathan, S.; Hwang, B.-Y.; Lin, S.-Y.; Chiu, T.-W.; Yu, B.-S.; Fan, Y.-J.; Chuang, C. $CuFeO_2$-CeO_2 nanopowder catalyst prepared by self-combustion glycine nitrate process and applied for hydrogen production from methanol steam reforming. *Int. J. Hydrogen Energy* **2020**, *45*, 15752–15762. [CrossRef]
37. Yu, H.; Xu, C.; Li, Y.; Jin, F.; Ye, F.; Li, X. Performance Enhancement of CuO/ZnO by Deposition on the Metal-Organic Framework of Cu-BTC for Methanol Steam Reforming Reaction. *ES Energy Environ.* **2020**, *8*, 65–77. [CrossRef]
38. Kovalskiia, A.M.; Matveev, A.T.; Popov, Z.I.; Volkov, I.N.; Sukhanova, E.V.; Lytkina, A.A.; Yaroslavtsev, A.B.; Konopatsky, A.S.; Leybo, D.V.; Bondarev, A.V.; et al. (Ni,Cu)/hexagonal BN nanohybrids—New efcient catalysts for methanol steam reforming and carbon monoxide oxidation. *Chem. Eng. J.* **2020**, *395*, 125109. [CrossRef]
39. Mateos-Pedreroa, C.; Azenha, C.; Pacheco Tanaka, D.A.; Sousa, J.M.; Mendes, A. The influence of the support composition on the physicochemical and catalytic properties of Cu catalysts supported on Zirconia-Alumina for methanol steam reforming. *Appl. Catal. B-Environ.* **2020**, *277*, 119243. [CrossRef]
40. Khani, Y.; Kamyar, N.; Bahadoran, F.; Safari, N.; Amini, M.M. A520 MOF-derived alumina as unique support for hydrogen production from methanol steam reforming: The critical role of support on performance. *Renew. Energy* **2020**, *156*, 1055–1064. [CrossRef]
41. Fasanya, O.O.; Atta, A.Y.; Myint, M.T.Z.; Dutta, J.; Jibril, B.Y. Effects of synthesis methods on performance of CuZn/MCM-41 catalysts in methanol steam reforming. *Int. J. Hydrogen Energy* **2021**, *46*, 3539–3553. [CrossRef]
42. Song, Q.; Men, Y.; Wang, J.; Liu, S.; Chai, S.; An, W.; Wang, K.; Li, Y.; Tang, Y. Methanol steam reforming for hydrogen production over ternary composite $ZnyCe_1Zr_9O_X$ catalysts. *Int. J. Hydrogen Energy* **2020**, *45*, 9592–9602. [CrossRef]
43. Cai, F.; Juweriah Ibrahim, J.; Fu, Y.; Kong, W.; Zhang, J.; Sun, Y. Low-temperature hydrogen production from methanol steam reforming on Zn-modified Pt/MoC catalysts. *Appl. Catal. B-Environ.* **2020**, *264*, 118500. [CrossRef]
44. Köwitsch, N.; Thoni, L.; Klemmed, B.; Benad, A.; Paciok, P.; Heggen, M.; Köwitsch, I.; Mehring, M.; Eychmüller, A.; Armbrüster, M.; et al. Proving a Paradigm in Methanol Steam Reforming: Catalytically Highly Selective InxPdy/In_2O_3 Interfaces. *ACS Catal.* **2021**, *11*, 304–312. [CrossRef]
45. Ogo, S.; Sekine, Y. Recent progress in ethanol steam reforming using non-noble transition metal catalysts: A review. *Fuel Proces. Technol.* **2020**, *199*, 106238. [CrossRef]
46. Yoo, S.; Park, S.; Song, J.H.; Kim, D.H. Hydrogen production by the steam reforming of ethanol over K-promoted Co/Al_2O_3–CaO xerogel catalysts. *Mol. Catal.* **2020**, *491*, 110980. [CrossRef]
47. Martinelli, M.; Watson, C.D.; Jacobs, G. Sodium doping of Pt/m-ZrO_2 promotes C-C scission and decarboxylation during ethanol steam reforming. *Int. J. Hydrogen Energy* **2020**, *45*, 18490–18501. [CrossRef]
48. Vecchietti, J.; Lustemberg, P. Controlled selectivity for ethanol steam reforming reaction over doped CeO_2 surfaces: The role of gallium. *Appl. Catal. B-Environ.* **2020**, *277*, 119103. [CrossRef]
49. Greluk, M.; Rotko, M.; Turczyniak-Surdacka, S. Enhanced catalytic performance of La_2O_3 promoted Co/CeO_2 and Ni/CeO2 catalysts for effective hydrogen production by ethanol steam reforming. *Renew. Energy* **2020**, *155*, 378–395. [CrossRef]
50. Li, Y.; Zhang, Z.; Jia, P.; Dong, D.; Wang, Y.; Hu, S.; Xiang, J.; Liu, Q.; Hu, X. Ethanol steam reforming over cobalt catalysts: Effect of a range of additives on the catalytic behaviors. *J. Energy Inst.* **2020**, *93*, 165–184. [CrossRef]
51. Dai, R.; Zheng, Z.; Yan, W.; Lian, C.; Wu, X.; An, X.; Xie, X. Dragon fruit-like Pt-Cu@mSiO_2 nanocomposite as an efficient catalyst for low-temperature ethanol steam reforming. *Chem Eng. J.* **2020**, *379*, 122299. [CrossRef]
52. Chen, D.; Liu, C.; Mao, Y.; Wang, W.; Li, T. Efficient hydrogen production from ethanol steam reforming over layer-controlled graphene-encapsulated Ni catalysts. *J. Clean. Prod.* **2020**, *252*, 119907. [CrossRef]
53. Smal, E.A.; Simonov, M.N.; Mezentseva, N.V.; Krieger, T.A.; Larina, T.V.; Saraev, A.A.; Glazneva, T.S.; Ishchenko, A.V.; Rogov, V.A.; Eremeev, N.F.; et al. Spinel-type $MnxCr_{3\text{-}x}O_4$-based catalysts for ethanol steam reforming. *Appl. Catal. B-Environ.* **2021**, *283*, 119656. [CrossRef]

54. Afolabi, A.T.F.; Kechagiopoulos, P.N.; Liu, Y.; Li, C.-Z. Kinetic features of ethanol steam reforming and decomposition using a biochar-supported Ni catalyst. *Fuel Proces. Technol.* **2021**, *212*, 106622. [CrossRef]
55. Charisiou, N.D.; Siakavelas, G.I.; Papageridis, K.N.; Motta, D.; Dimitratos, N.; Sebastian, V.; Polychronopoulou, K.; Goula, M.A. The Effect of Noble Metal (M: Ir, Pt, Pd) on M/Ce_2O_3-γ-Al_2O_3 Catalysts for Hydrogen Production via the Steam Reforming of Glycerol. *Catalysts* **2020**, *10*, 790. [CrossRef]
56. Asmawati Roslan, N.; Zainal Abidin, S.; Ideris, A.; Vo, D.-V.N. A review on glycerol reforming processes over Ni-based catalyst for hydrogen and syngas productions. *Int. J. Hydrogen Energy* **2020**, *45*, 18466–18489. [CrossRef]
57. Charisiou, N.D.; Italiano, C.; Pino, L.; Sebastian, V.; Vita, A.; Goula, M.A. Hydrogen production via steam reforming of glycerol over Rh/ɣ-Al_2O_3 catalysts modified with CeO_2, MgO or La_2O_3. *Renew. Energy* **2020**, *162*, 908–925. [CrossRef]
58. Ghaffari Saeidabad, N.; Noh, Y.S.; Alizadeh Eslami, A.; Song, H.T.; Kim, H.D.; Fazeli, A.; Moon, D.J. A Review on Catalysts Development for Steam Reforming of Biodiesel Derived Glycerol; Promoters and Supports. *Catalysts* **2020**, *10*, 910. [CrossRef]
59. Feng, P.; Huang, K.; Xu, Q.; Qi, W.; Xin, S.; Wei, T.; Liao, L.; Yan, Y. Ni supported on the CaO modified attapulgite as catalysts for hydrogen production from glycerol steam reforming. *Int. J. Hydrogen Energy* **2020**, *45*, 8223–8233. [CrossRef]
60. Yancheshmeh, M.S.; Sahraei, O.A.; Aissaoui, M.; Iliuta, M.C. A novel synthesis of $NiAl_2O_4$ spinel from a Ni-Al mixed-metal alkoxide as a highly efficient catalyst for hydrogen production by glycerol steam reforming. *Appl. Catal. B-Environ.* **2020**, *265*, 118535. [CrossRef]
61. Moogi, S.; Nakka, L.; Prasad Potharaju, S.S.; Ahmed, A.; Farooq, A.; Jung, S.-C.; Hoon Rhee, G.; Park, Y.-K. Copper promoted Co/MgO: A stable and efficient catalyst for glycerol steam reforming. *Int. J. Hydrogen Energy* **2020**. [CrossRef]
62. Wang, R.; Liu, S.; Liu, S.; Li, X.; Zhang, Y.; Xie, C.; Zhou, S.; Qiu, Y.; Luo, S.; Jing, F.; et al. Glycerol steam reforming for hydrogen production over bimetallic MNi/CNTs (M = Co, Cu and Fe) catalysts. *Catal. Today* **2020**, *355*, 128–138. [CrossRef]
63. Conte, F.; Esposito, S.; Dal Santo, V.; Di Michele, A.; Ramis, G.; Rossetti, I. Flame Pyrolysis Synthesis of Mixed Oxides for Glycerol Steam Reforming. *Materials* **2021**, *14*, 652. [CrossRef] [PubMed]
64. Matus, E.V.; Ismagilov, I.Z.; Yashnik, S.A.; Ushakov, V.A.; Prosvirin, I.P.; Kerzhentsev, M.A.; Ismagilov, Z.R. Hydrogen production through autothermal reforming of CH_4: Efficiency and action mode of noble (M = Pt, Pd) and non-noble (M = Re, Mo, Sn) metal additives in the composition of Ni-M/$Ce_{0.5}Zr_{0.5}O_2$/Al_2O_3 catalysts. *Int. J. Hydrogen Energy* **2020**, *45*, 33352–33369. [CrossRef]
65. Osazuwa, O.U.; Abidin, S.Z.; Fan, X.; Nosakhare Amenaghawon, A.; Azizan, T.M. An insight into the effects of synthesis methods on catalysts properties for methane reforming. *J. Environ. Chem. Eng.* **2021**, *9*, 105052. [CrossRef]
66. Mosinska, M.; Stępińska, N.; Maniukiewicz, W.; Rogowski, J.; Mierczynska-Vasilev, A.; Vasilev, K.; Szynkowska, M.I.; Mierczynski, P. Hydrogen Production on Cu-Ni Catalysts via the Oxy-Steam Reforming of Methanol. *Catalysts* **2020**, *10*, 273. [CrossRef]
67. Sun, Z.; Zhang, X.; Li, H.; Sang, S.; Chen, S.; Duan, L.; Zeng, L.; Xiang, W.; Gong, J. Chemical looping oxidative steam reforming of methanol: A new pathway for auto-thermal conversion. *Appl. Catal. B-Environ.* **2020**, *269*, 118758. [CrossRef]
68. Rodríguez Lugo, V.; Mondragon-Galicia, G.; Gutiérrez-Martínez, A.; Gutiérrez-Wing, C.; González, O.R.; López, P.; Salinas-Hernández, P.; Tzompantzi, F.; Valderrama, M.I.R.; Pérez-Hernández, R. Pt–Ni/ZnO-rod catalysts for hydrogen production by steam reforming of methanol with oxygen. *RSC Adv.* **2020**, *10*, 41315. [CrossRef]
69. Ruocco, C.; Palma, V.; Cortese, M.; Martino, M. Noble Metals-Based Catalysts for Hydrogen Production via Bioethanol Reforming in a Fluidized Bed Reactor. *Chem. Proc.* **2020**, *2*, 7543. [CrossRef]
70. Pio, G.; Ruocco, C.; Palma, V.; Salzano, E. Detailed kinetic mechanism for the hydrogen production via the oxidative reforming of ethanol. *Chem. Eng. Sci.* **2021**, *237*, 116591. [CrossRef]
71. Cao, L.; Yu, I.K.M.; Xiong, X.; Tsang, D.C.W.; Zhang, S.; Clark, J.H.; Hu, C.; Ng, Y.H.; Shang, J.; Ok, Y.S. Biorenewable hydrogen production through biomass gasification: A review and future prospects. *Environ. Res.* **2020**, *186*, 109547. [CrossRef]
72. Singh Siwal, S.; Zhang, Q.; Sun, C.; Thakur, S.; Kumar Gupta, V.; Kumar Thakur, V. Energy production from steam gasification processes and parameters that contemplate in biomass gasifier—A review. *Bioresour. Technol.* **2020**, *297*, 122481. [CrossRef] [PubMed]
73. AlNouss, A.; McKay, G.; Al-Ansari, T. Production of syngas via gasification using optimum blends of biomass. *J. Clean. Prod.* **2020**, *242*, 118499. [CrossRef]
74. Wang, C.; Jin, H.; Feng, H.; Wei, W.; Cao, C.; Cao, W. Study on gasification mechanism of biomass waste in supercritical water based on product distribution. *Int. J. Hydrogen Energy* **2020**, *45*, 28051–28061. [CrossRef]
75. Chen, J.; Fan, Y.; Zhao, X.; E, J.; Xu, W.; Zhang, F.; Liao, G.; Leng, E.; Liu, S. Experimental investigation on gasification characteristic of food waste using supercritical water for combustible gas production: Exploring the way to complete gasification. *Fuel* **2020**, *263*, 116735. [CrossRef]
76. Mehrpooya, M.; Habibi, R. A review on hydrogen production thermochemical water-splitting cycles. *J. Clean. Prod.* **2020**, *275*, 123836. [CrossRef]
77. Gao, Y.; Mao, Y.; Song, Z.; Zhao, X.; Sun, J.; Wang, W.; Chen, G.; Chen, S. Efficient generation of hydrogen by two-step thermochemical cycles: Successive thermal reduction and water splitting reactions using equal-power microwave irradiation and a high entropy material. *Appl. Energy* **2020**, *279*, 115777. [CrossRef]
78. Safari, F.; Dincer, I. A study on the Fe-Cl thermochemical water splitting cycle for hydrogen production. *Int. J. Hydrogen Energy* **2020**, *45*, 18867–18875. [CrossRef]

79. Ngoensawata, A.; Tongnana, V.; Laosiripojanab, N.; Kim-Lohsoontornc, P.; Hartley, U.W. Effect of La and Gd substitution in $BaFeO_{3-\delta}$ perovskite structure on its catalytic performance for thermochemical water splitting. *Catal. Commun.* **2020**, *135*, 105901. [CrossRef]
80. Qian, X.; He, J.; Mastronardo, E.; Baldassarri, B.; Wolverton, C.; Haile, S.M. Favorable Redox Thermodynamics of $SrTi_{0.5}Mn_{0.5}O_{3-\delta}$ in Solar Thermochemical Water Splitting. *Chem. Mater.* **2020**, *32*, 9335–9346. [CrossRef]
81. Orfila, M.; Sanz, D.; Linares, M.; Molina, R.; Sanz, R.; Marugan, J.; Botas, J.A. H_2 production by thermochemical water splitting with reticulated porous structures of ceria-based mixed oxide materials. *Int. J. Hydrogen Energy* **2020**. [CrossRef]
82. Wang, Y.; Hong, Z.; Mei, D. A thermally autonomous methanol steam reforming microreactor with porous copper foam as catalyst support for hydrogen production. *Int. J. Hydrogen Energy* **2021**. [CrossRef]
83. Du, W.; Lee, M.-T.; Wang, Y.; Zhao, C.; Li, G.; Li, M. Design of a solar-driven methanol steam reforming receiver/reactor with a thermal storage medium and its performance analysis. *Int. J. Hydrogen Energy* **2020**, *45*, 33076–33087. [CrossRef]
84. da Fonseca Dias, V.; Dias da Silva, J. Mathematical modelling of the solar-driven steam reforming of methanol for a solar thermochemical micro-fuidized bed reformer: Thermal performance and thermochemical conversion. *J. Braz. Soc. Mec. Sci.* **2020**, *42*, 447. [CrossRef]
85. Chen, X.; Zhou, H.; Yu, Z.; Li, W.; Tang, J.; Xu, C.; Ding, Y.; Wan, Z. Thermodynamic and economic assessment of a PEMFC-based micro-CCHP system integrated with geothermal-assisted methanol reforming. *Int. J. Hydrogen Energy* **2020**, *45*, 958–971. [CrossRef]
86. Xu, Y.; Lu, B.; Luo, C.; Chen, J.; Zhang, Z.; Zhang, L. Sorption enhanced steam reforming of ethanol over Ni-based catalyst coupling with high-performance CaO pellets. *Chem. Eng. J.* **2021**, *406*, 126903. [CrossRef]
87. Jia, H.; Xu, H.; Sheng, X.; Yang, X.; Shen, W.; Goldbach, A. High-temperature ethanol steam reforming in PdCu membrane reactor. *J. Memb. Sci.* **2020**, *605*, 118083. [CrossRef]
88. Antonini, C.; Treyer, K.; Streb, A.; van der Spek, M.; Bauer, C.; Mazzotti, M. Hydrogen production from natural gas and biomethane with carbon capture and storage—A techno-environmental analysis. *Sustain. Energy Fuels* **2020**, *4*, 2967–2986. [CrossRef]
89. Lim, Y.; Lee, D.-K.; Kim, S.M.; Park, W.; Cho, S.Y.; Sim, U. Low Dimensional Carbon-Based Catalysts for Efficient Photocatalytic and Photo/Electrochemical Water Splitting Reactions. *Materials* **2020**, *13*, 114. [CrossRef]
90. Li, J.; Jiménez-Calvo, P.; Paineau, E.; Nawfal Ghazzal, M. Metal Chalcogenides Based Heterojunctions and Novel Nanostructures for Photocatalytic Hydrogen Evolution. *Catalysts* **2020**, *10*, 89. [CrossRef]
91. Enesca, A.; Andronic, L. Photocatalytic Activity of S-Scheme Heterostructure for Hydrogen Production and Organic Pollutant Removal: A Mini-Review. *Nanomaterials* **2021**, *11*, 871. [CrossRef]
92. Lettieri, S.; Pavone, M.; Fioravanti, A.; Santamaria Amato, L.; Maddalena, P. Charge Carrier Processes and Optical Properties in TiO_2 and TiO_2-Based Heterojunction Photocatalysts: A Review. *Materials* **2021**, *14*, 1645. [CrossRef] [PubMed]
93. Rusinque, B.; Escobedo, S.; de Lasa, H. Photoreduction of a Pd-Doped Mesoporous TiO_2 Photocatalyst for Hydrogen Production under Visible Light. *Catalysts* **2020**, *10*, 74. [CrossRef]
94. Rusinque, B.; Escobedo, S.; de Lasa, H. Hydrogen Production via Pd-TiO_2 Photocatalytic Water Splitting under Near-UV and Visible Light: Analysis of the Reaction Mechanism. *Catalysts* **2021**, *11*, 405. [CrossRef]
95. Wang, Z.; Fan, J.; Cheng, B.; Yu, J.; Xu, J. Nickel-based cocatalysts for photocatalysis: Hydrogen evolution, overall water splitting and CO_2 reduction. *Mat. Today Phys.* **2020**, *15*, 100279. [CrossRef]
96. Ayala, P.; Giesriegl, A.; Nandan, S.P.; Nagaraju Myakala, S.; Wobrauschek, P.; Cherevan, A. Isolation Strategy towards Earth-Abundant Single-Site Co-Catalysts for Photocatalytic Hydrogen Evolution Reaction. *Catalysts* **2021**, *11*, 417. [CrossRef]
97. Hampel, B.; Pap, Z.; Sapi, A.; Szamosvolgyi, A.; Baia, L.; Hernadi, K. Application of TiO_2-Cu Composites in Photocatalytic Degradation Different Pollutants and Hydrogen Production. *Catalysts* **2020**, *10*, 85. [CrossRef]
98. Potapenko, K.O.; Kurenkova, A.Y.; Bukhtiyarov, A.V.; Gerasimov, E.Y.; Cherepanova, S.V.; Kozlova, E.A. Comparative Study of the Photocatalytic Hydrogen Evolution over Cd1-xMnxS and CdS-β-Mn_3O_4-MnOOH Photocatalysts under Visible Light. *Nanomaterials* **2021**, *11*, 355. [CrossRef]
99. Xing, X.; Tang, S.; Hong, H.; Jin, H. Concentrated solar photocatalysis for hydrogen generation from water by titania-containing gold nanoparticles. *Int. J. Hydrogen Energy* **2020**, *45*, 9612–9623. [CrossRef]
100. Khabirul Islam, A.K.M.; Dunlop, P.S.M.; Hewitt, N.J.; Lenihan, R.; Brandoni, C. Bio-Hydrogen Production from Wastewater: A Comparative Study of Low Energy Intensive Production Processes. *Clean Technol.* **2021**, *3*, 10. [CrossRef]
101. Wei, Z.; Liu, J.; Shangguan, W. A review on photocatalysis in antibiotic wastewater: Pollutant degradation and hydrogen production. *Chin. J. Catal.* **2020**, *41*, 1440–1450. [CrossRef]
102. Armenise, V.; Colella, S.; Fracassi, F.; Listorti, A. Lead-Free Metal Halide Perovskites for Hydrogen Evolution from Aqueous Solutions. *Nanomaterials* **2021**, *11*, 433. [CrossRef] [PubMed]
103. Bahadori, E.; Ramis, G.; Zanardo, D.; Menegazzo, F.; Signoretto, M.; Gazzoli, D.; Pietrogiacomi, D.; Di Michele, A.; Rossetti, I. Photoreforming of Glucose over CuO/TiO_2. *Catalysts* **2020**, *10*, 477. [CrossRef]
104. Al-Madanat, O.; AlSalka, Y.; Ramadan, W.; Bahnemann, D.W. TiO_2 Photocatalysis for the Transformation of Aromatic Water Pollutants into Fuels. *Catalysts* **2021**, *11*, 317. [CrossRef]
105. Akhlaghi, N.; Najafpour-Darzi, G. A comprehensive review on biological hydrogen production. *Int. J. Hydrogen Energy* **2020**, *45*, 22492–22512. [CrossRef]

106. Christopher, F.C.; Kumar, P.S.; Vo, D.-V.N.; Joshiba, G.J. A review on critical assessment of advanced bioreactor options for sustainable hydrogen production. *Int. J. Hydrogen Energy* **2021**. [CrossRef]
107. Dahiya, S.; Chatterjee, S.; Sarkar, O.; Venkata Mohan, S. Renewable hydrogen production by dark-fermentation: Current status, challenges and perspectives. *Biores. Technol.* **2021**, *321*, 124354. [CrossRef]
108. Castello, E.; Nunes Ferraz-Junior, A.D. Stability problems in the hydrogen production by dark fermentation: Possible causes and solutions. *Renew. Sust. Energ. Rev.* **2020**, *119*, 109602. [CrossRef]
109. García-Depraect, O.; Munoz, R.; Rodríguez, E.; Rene, E.R.; Leon-Becerril, E. Microbial ecology of a lactate-driven dark fermentation process producing hydrogen under carbohydrate-limiting conditions. *Int. J. Hydrogen Energy* **2021**. [CrossRef]
110. Dauptain, K.; Trably, E.; Santa-Catalina, G.; Bernet, N.; Carrere, H. Role of indigenous bacteria in dark fermentation of organic substrates. *Biores. Technol.* **2020**, *313*, 123665. [CrossRef]
111. Song, W.; Ding, L.; Liu, M.; Cheng, J.; Zhou, J.; Li, Y.Y. Improving biohydrogen production through dark fermentation of steam-heated acid pretreated Alternanthera philoxeroides by mutant Enterobacter aerogenes ZJU1. *Sci. Total Environ.* **2020**, *716*, 134695. [CrossRef]
112. Li, X.; Guo, L.; Liu, Y.; Wang, Y.; She, Z.; Gao, M.; Zhao, Y. Effect of salinity and pH on dark fermentation with thermophilic bacteria pretreated swine wastewater. *J. Environ. Manag.* **2020**, *271*, 111023. [CrossRef] [PubMed]
113. Atilano-Camino, M.M.; Luévano-Montaño, C.D.; García-González, A.; Olivo-Alani, D.S.; Álvarez-Valencia, L.H.; García-Reyes, R.B. Evaluation of dissolved and immobilized redox mediators on dark fermentation: Driving to hydrogen or solventogenic pathway. *Biores. Technol.* **2020**, *317*, 123981. [CrossRef]
114. Liu, X.; He, D. Freezing in the presence of nitrite pretreatment enhances hydrogen production from dark fermentation of waste activated sludge. *J. Clean. Prod.* **2020**, *248*, 119305. [CrossRef]
115. Zhang, Y.; Xiao, L.; Hao, Q.; Li, X.; Liu, F. Ferrihydrite Reduction Exclusively Stimulated Hydrogen Production by Clostridium with Community Metabolic Pathway Bifurcation. *Acs Sustain. Chem. Eng.* **2020**, *8*, 7574–7580. [CrossRef]
116. Moura, A.G.L.; Rabelo, C.A.B.S.; Okino, C.H.; Maintinguer, S.I.; Silva, E.L.; Varesche, M.B.A. Enhancement of Clostridium butyricum hydrogen production by iron and nickel nanoparticles: Effects on hydA expression. *Int. J. Hydrogen Energy* **2020**, *45*, 28447–28461. [CrossRef]
117. Rambabu, K.; Bharath, G.; Thanigaivelan, A.; Das, D.B.; Loke Show, P.; Banat, F. Augmented biohydrogen production from rice mill wastewater through nano-metal oxides assisted dark fermentation. *Biores. Technol.* **2021**, *319*, 124243. [CrossRef]
118. Sim, Y.-B.; Jung, J.-H.; Park, J.-H.; Bakonyi, P.; Kim, S.-H. Effect of shear velocity on dark fermentation for biohydrogen production using dynamic membrane. *Biores. Technol.* **2020**, *308*, 123265. [CrossRef] [PubMed]
119. Hitam, C.N.C.; Jalil, A.A. A review on biohydrogen production through photo-fermentation of lignocellulosic biomass. *Biomass Convers. Biorefin.* **2020**. [CrossRef]
120. Baeyens, J.; Zhang, H.; Nie, J.; Appels, L.; Dewil, R.; Ansart, R.; Deng, Y. Reviewing the potential of bio-hydrogen production by fermentation. *Renew. Sust. Energ. Rev.* **2020**, *131*, 110023. [CrossRef]
121. Zhang, Z.; Zhang, H.; Li, Y.; Lu, C.; Zhu, S.; He, C.; Ai, F.; Zhang, Q. Investigation of the interaction between lighting and mixing applied during the photo-fermentation biohydrogen production process from agricultural waste. *Biores. Technol.* **2020**, *312*, 123570. [CrossRef] [PubMed]
122. Zhu, S.; Zhang, Z.; Zhang, H.; Jing, Y.; Li, Y.; Zhang, Q. Rheological properties of corn stover hydrolysate and photo-fermentation bio-hydrogen producing capacity under intermittent stirring. *Int. J. Hydrogen Energy* **2020**, *45*, 3721–3728. [CrossRef]
123. Znad, H.; Al-Mohammedawi, H.; Awual, M.R. Integrated pre-treatment stage of biosorbent—sonication for mixed brewery and restaurant effluents to enhance the photo-fermentative hydrogen production. *Biomass Bioenergy* **2021**, *144*, 105899. [CrossRef]
124. Jiang, D.; Zhang, X. Insights into correlation between hydrogen yield improvement and glycerol addition in photo-fermentation of Arundo donax L. *Biores. Technol.* **2021**, *321*, 124467. [CrossRef]
125. Zhang, H.; Li, J.; Zhang, Q.; Zhu, S.; Yang, S.; Zhang, Z. Effect of Substrate Concentration on Photo-Fermentation Bio-Hydrogen Production Process from Starch-Rich Agricultural Leftovers under Oscillation. *Sustainability* **2020**, *12*, 2700. [CrossRef]
126. Jiang, D.; Ge, X.; Zhang, T.; Chen, Z.; Zhang, Z.; He, C.; Zhang, Q.; Li, Y. Effect of alkaline pretreatment on photo-fermentative hydrogen production from giant reed: Comparison of NaOH and $Ca(OH)_2$. *Biores. Technol.* **2020**, *304*, 123001. [CrossRef] [PubMed]
127. Liu, H.; Zhang, Z.; Zhang, H.; Lee, D.-J.; Zhang, Q.; Lu, C.; He, C. Evaluation of hydrogen yield potential from Chlorella by photofermentation under diverse substrate concentration and enzyme loading. *Biores. Technol.* **2020**, *303*, 122956. [CrossRef]
128. Zhanga, Y.; Zhanga, H.; Leeb, D.-J.; Zhanga, T.; Jianga, D.; Zhanga, Z.; Zhang, Q. Effect of enzymolysis time on biohydrogen production from photofermentation by using various energy grasses as substrates. *Biores. Technol.* **2020**, *305*, 123062. [CrossRef] [PubMed]
129. Zhang, X.; Jiang, D.; Zhang, H.; Wang, Y.; Zhang, Z.; Lu, C.; Zhang, Q. Enhancement of the biohydrogen production performance from mixed substrate by photo-fermentation: Effects of initial pH and inoculation volume ratio. *Biores. Technol.* **2021**, *319*, 124153. [CrossRef]
130. Hu, J.; Yang, H.; Wang, X.; Cao, W.; Guo, L. Strong pH dependence of hydrogen production from glucose by Rhodobacter sphaeroides. *Int. J. Hydrogen Energy* **2020**, *45*, 9451–9458. [CrossRef]
131. Lu, C.; Tahir, N.; Li, W.; Zhang, Z.; Jiang, D.; Guo, S.; Wang, J.; Wang, K.; Zhang, Q. Enhanced buffer capacity of fermentation broth and biohydrogen production from corn stalk with Na_2HPO_4/NaH_2PO_4. *Biores. Technol.* **2020**, *313*, 123783. [CrossRef]

132. Guo, S.; Lu, C.; Wang, K.; Wang, J.; Zhang, Z.; Jing, Y.; Zhang, Q. Enhancement of pH values stability and photo-fermentation biohydrogen production by phosphate buffer. *Bioengineered* **2020**, *11*, 291–300. [CrossRef]
133. Hu, B.; Li, Y.; Zhu, S.; Zhang, H.; Jing, Y.; Jiang, D.; He, C.; Zhang, Z. Evaluation of biohydrogen yield potential and electron balance in the photofermentation process with different initial pH from starch agricultural leftover. *Biores. Technol.* **2020**, *305*, 122900. [CrossRef] [PubMed]
134. Lu, C.; Jing, Y.; Zhang, H.; Lee, D.-J.; Tahir, N.; Zhang, Q.; Li, W.; Wang, Y.; Liang, X.; Wang, J.; et al. Biohydrogen production through active saccharification and photofermentation from alfalfa. *Biores. Technol.* **2020**, *304*, 123007. [CrossRef] [PubMed]
135. Zhang, T.; Jiang, D.; Zhang, H.; Lee, D.-J.; Zhang, Z.; Zhang, Q.; Jing, Y.; Zhang, Y.; Xia, C. Effects of different pretreatment methods on the structural characteristics, enzymatic saccharifcation and photo-fermentative bio-hydrogen production performance of corn straw. *Biores. Technol.* **2020**, *304*, 122999. [CrossRef]
136. Al-Mohammedawi, H.H.; Znad, H. Impact of metal ions and EDTA on photofermentative hydrogen production by Rhodobacter sphaeroides using a mixture of pre-treated brewery and restaurant effluents. *Biomass Bioenergy* **2020**, *134*, 105482. [CrossRef]
137. Policastro, G.; Luongo, V.; Fabbricino, M. Biohydrogen and poly-β-hydroxybutyrate production by winery wastewater photofermentation: Effect of substrate concentration and nitrogen source. *J. Environ. Manag.* **2020**, *271*, 111006. [CrossRef]
138. Liu, C.; Shi, Y.; Liu, H.; Ma, M.; Liu, G.; Zhang, R.; Wang, W. Insight of co-fermentation of carbon monoxide with carbohydraterich wastewater for enhanced hydrogen production: Homoacetogenic inhibition and the role of pH. *J. Clean. Prod.* **2020**, *267*, 122027. [CrossRef]
139. Lee, S.H.; Lee, S.-M.; Lee, J.-H.; Lee, H.S.; Kang, S.G. Biological process for coproduction of hydrogen and thermophilic enzymes during CO fermentation. *Biores. Technol.* **2020**, *305*, 123067. [CrossRef] [PubMed]
140. Kumar, R.; Kumar, A.; Pal, A. An overview of conventional and non-conventional hydrogen production methods. *Mater. Today Proc.* **2020**. [CrossRef]
141. Kossalbayev, B.D.; Tomo, T.; Zayadan, B.K.; Sadvakasova, A.K.; Bolatkhan, K.; Alwasel, S.; Allakhverdiev, S.I. Determination of the potential of cyanobacterial strains for hydrogen production. *Int. J. Hydrogen Energy* **2020**, *45*, 2627–2639. [CrossRef]
142. Mu, D.; Liu, H.; Lin, W.; Shukla, P.; Luo, J. Simultaneous biohydrogen production from dark fermentation of duckweed and waste utilization for microalgal lipid production. *Biores. Technol.* **2020**, *302*, 122879. [CrossRef]
143. Das, S.R.; Basak, N. Molecular biohydrogen production by dark and photo fermentation from wastes containing starch: Recent advancement and future perspective. *Bioprocess Biosyst. Eng.* **2021**, *44*, 1–25. [CrossRef]
144. Mıynat, M.E.; Ören, İ.; Özkan, E.; Argun, H. Sequential dark and photo-fermentative hydrogen gas production from agar embedded molasses. *Int. J. Hydrogen Energy* **2020**, *45*, 34730–34738. [CrossRef]
145. Niño-Navarro, C.; Chairez, I.; Christen, P.; Canul-Chan, M.; García-Peña, E.I. Enhanced hydrogen production by a sequential dark and photo fermentation process: Effects of initial feedstock composition, dilution and microbial population. *Renew. Energy* **2020**, *147*, 924–936. [CrossRef]
146. Li, Y.; Zhang, Z.; Lee, D.-J.; Zhang, Q.; Jing, Y.; Yue, T.; Liu, Z. Role of L-cysteine and iron oxide nanoparticle in affecting hydrogen yield potential and electronic distribution in biohydrogen production from dark fermentation effluents by photo-fermentation. *J. Clean. Prod.* **2020**, *276*, 123193. [CrossRef]
147. Pandey, A.; Sinha, P.; Pandey, A. Hydrogen production by sequential dark and photofermentation using wet biomass hydrolysate of Spirulina platensis: Response surface methodological approach. *Int. J. Hydrogen Energy* **2021**, *46*, 7137–7146. [CrossRef]
148. Li, Y.; Zhang, Z.; Zhang, Q.; Tahir, N.; Jing, Y.; Xia, C.; Zhu, S.; Zhang, X. Enhancement of bio-hydrogen yield and pH stability in photo fermentation process using dark fermentation effluent as succedaneum. *Biores. Technol.* **2020**, *297*, 122504. [CrossRef] [PubMed]
149. Rezaeitavabe, F.; Saadat, S.; Talebbeydokhti, N.; Sartaj, M.; Tabatabaei, M. Enhancing bio-hydrogen production from food waste in single-stage hybrid dark-photo fermentation by addition of two waste materials (exhausted resin and biochar). *Biomass Bioenergy* **2020**, *143*, 105846. [CrossRef]
150. Meky, N.; Ibrahim, M.G.; Fujii, M.; Elreedy, A.; Tawfk, A. Integrated dark-photo fermentative hydrogen production from synthetic gelatinaceous wastewater via cost-effective hybrid reactor at ambient temperature. *Energy Convers. Manag.* **2020**, *203*, 112250. [CrossRef]
151. Dinesh, G.H.; Nguyen, D.D.; Ravindran, B.; Chang, S.W.; Vo, D.-V.N.; Bach, Q.-V.; Tran, H.N.; Basu, M.J.; Mohanrasu, K.; Murugan, R.S.; et al. Simultaneous biohydrogen (H_2) and bioplastic (poly-b-hydroxybutyrate-PHB) productions under dark, photo, and subsequent dark and photo fermentation utilizing various wastes. *Int. J. Hydrogen Energy* **2020**, *45*, 5840–5853. [CrossRef]
152. Li, Y.; Zhang, Z.; Xia, C.; Jing, Y.; Zhang, Q.; Li, S.; Zhu, S.; Jin, P. Photo-fermentation biohydrogen production and electrons distribution from dark fermentation effluents under batch, semi-continuous and continuous modes. *Biores. Technol.* **2020**, *311*, 123549. [CrossRef]
153. Zhang, T.; Jiang, D.; Zhang, H.; Jing, Y.; Tahir, N.; Zhang, Y.; Zhang, Q. Comparative study on bio-hydrogen production from corn stover: Photo-fermentation, dark-fermentation and dark-photo co-fermentation. *Int. J. Hydrogen Energy* **2020**, *45*, 3807–3814. [CrossRef]
154. Zhang, Y.; Yuan, J.; Guo, L. Enhanced bio-hydrogen production from cornstalk hydrolysate pretreated by alkaline-enzymolysis with orthogonal design method. *Int. J. Hydrogen Energy* **2020**, *45*, 3750–3759. [CrossRef]
155. Kannah, R.Y.; Kavitha, S.; Preethi; Karthikeyan, O.P.; Kumar, G.; Dai-Viet, N.V.; Banu, J.R. Techno-economic assessment of various hydrogen production methods—A review. *Biores. Technol.* **2021**, *319*, 124175. [CrossRef]

156. Lamy, C.; Millet, P. A critical review on the defnitions used to calculate the energy efficiency coeffcients of water electrolysis cells working under near ambient temperature conditions. *J. Power Sources* **2020**, *447*, 227350. [CrossRef]
157. Grigoriev, S.A.; Fateev, V.N.; Bessarabov, D.G.; Millet, P. Current status, research trends, and challenges in water electrolysis science and technology. *Int. J. Hydrogen Energy* **2020**, *45*, 26036–26058. [CrossRef]
158. Zhou, D.; Li, P.; Xu, W.; Jawaid, S.; Mohammed-Ibrahim, J.; Liu, W.; Kuang, Y.; Sun, X. Recent Advances in Non-Precious Metal-Based Electrodes for Alkaline Water Electrolysis. *ChemNanoMat* **2020**, *6*, 336–355. [CrossRef]
159. Krishnan, S.; Fairlie, M.; Andres, P.; de Groot, T.; Kramer, G.J. Power to gas (H_2): Alkaline electrolysis. In *Technological Learning in the Transition to a Low-Carbon Energy System*; Junginger, M., Louwen, A., Eds.; Academic Press: London, UK, 2020; pp. 165–187. [CrossRef]
160. Song, S.; Yu, L.; Xiao, X.; Qin, Z.; Zhang, W.; Wang, D.; Bao, J.; Zhou, H.; Zhang, Q.; Chen, S.; et al. Outstanding oxygen evolution reaction performance of nickel iron selenide/stainless steel mat for water electrolysis. *Mat. Today Phys.* **2020**, *13*, 100216. [CrossRef]
161. Liu, X.; Guo, R.; Kun, N.; Xia, F.; Niu, C.; Wen, B.; Meng, J.; Wu, P.; Wu, J.; Wu, X.; et al. Reconstruction-Determined Alkaline Water Electrolysis at Industrial Temperatures. *Adv. Mater.* **2020**, *32*, 2001136. [CrossRef] [PubMed]
162. Kou, T.; Wang, S.; Shi, R.; Zhang, T.; Chiovoloni, S.; Lu, J.Q.; Chen, W.; Worsley, M.A.; Wood, B.C.; Baker, S.E.; et al. Periodic Porous 3D Electrodes Mitigate Gas Bubble Trafc during Alkaline Water Electrolysis at High Current Densities. *Adv. Energy Mater.* **2020**, *10*, 2002955. [CrossRef]
163. Yang, F.; Kim, M.J.; Brown, M.; Wiley, B.J. Alkaline Water Electrolysis at 25 A cm^{-2} with a Microfbrous Flow-through Electrode. *Adv. Energy Mater.* **2020**, *10*, 2001174. [CrossRef]
164. López-Fernández, E.; Gil-Rostra, J.; Espinós, J.P.; González-Elipe, A.R.; de Lucas Consuegra, A.; Yubero, F. Chemistry and Electrocatalytic Activity of Nanostructured Nickel Electrodes for Water Electrolysis. *Acs Catal.* **2020**, *10*, 6159–6170. [CrossRef]
165. Carbone, A.; Campagna Zignani, S.; Gatto, I.; Trocino, S.; Aricò, A.S. Assessment of the FAA3-50 polymer electrolyte in combination with a $NiMn_2O_4$ anode catalyst for anion exchange membrane water electrolysis. *Int. J. Hydrogen Energy* **2020**, *45*, 9285–9292. [CrossRef]
166. Yan, X.; Yang, X.; Su, X.; Gao, L.; Zhao, J.; Hu, L.; Di, M.; Li, T.; Ruan, X.; He, G. Twisted ether-free polymer based alkaline membrane for high-performance water electrolysis. *J. Power Sources* **2020**, *480*, 228805. [CrossRef]
167. Fortin, P.; Khoza, T.; Cao, X.; Yngve Martinsen, S.; Oyarce Barnett, A.; Holdcroft, S. High-performance alkaline water electrolysis using Aemion™ anion exchange membranes. *J. Power Sources* **2020**, *451*, 227814. [CrossRef]
168. Pushkareva, I.V.; Pushkarev, A.S.; Grigoriev, S.A.; Modisha, P.; Bessarabov, D.G. Comparative study of anion exchange membranes for low-cost water electrolysis. *Int. J. Hydrogen Energy* **2020**, *45*, 26070–26079. [CrossRef]
169. Hegge, F.; Lombeck, F.; Cruz Ortiz, E.; Bohn, L.; von Holst, M.; Kroschel, M.; Hübner, J.; Breitwieser, M.; Strasser, P.; Vierrath, S. Efficient and Stable Low Iridium Loaded Anodes for PEM Water Electrolysis Made Possible by Nanofiber Interlayers. *ACS Appl. Energy Mater.* **2020**, *3*, 8276–8284. [CrossRef]
170. Klose, C.; Saatkamp, T.; Münchinger, A.; Bohn, L.; Titvinidze, G.; Breitwieser, M.; Kreuer, K.-D.; Vierrath, S. All-Hydrocarbon MEA for PEM Water Electrolysis Combining Low Hydrogen Crossover and High Effciency. *Adv. Energy Mater.* **2020**, *10*, 1903995. [CrossRef]
171. Siracusano, S.; Trocino, S.; Briguglio, N.; Panto, F.; Arico, A.S. Analysis of performance degradation during steady-state and load-thermal cycles of proton exchange membrane water electrolysis cells. *J. Power Sources* **2020**, *468*, 228390. [CrossRef]
172. Kang, Z.; Alia, S.M.; Young, J.L.; Bender, G. Effects of various parameters of different porous transport layers in proton exchange membrane water electrolysis. *Electrochim. Acta* **2020**, *354*, 136641. [CrossRef]
173. Lopata, J.; Kang, Z.; Young, J.; Bender, G.; Weidner, J.W.; Shimpalee, S. Effects of the Transport/Catalyst Layer Interface and Catalyst Loading on Mass and Charge Transport Phenomena in Polymer Electrolyte Membrane Water Electrolysis Devices. *J. Electrochem. Soc.* **2020**, *167*, 064507. [CrossRef]
174. Schuler, T.; Ciccone, J.M.; Krentscher, B.; Marone, F.; Peter, C.; Schmidt, T.J.; Büchi, F.N. Hierarchically Structured Porous Transport Layers for Polymer Electrolyte Water Electrolysis. *Adv. Energy Mater.* **2020**, *10*, 1903216. [CrossRef]
175. Huang, J.; Xie, Y.; Yan, L.; Wang, B.; Kong, T.; Dong, X.; Wang, Y.; Xia, Y. Decoupled amphoteric water electrolysis and its integration with Mn–Zn battery for flexible utilization of renewables. *Energy Environ. Sci.* **2021**, *14*, 883–889. [CrossRef]
176. Mayerhöfer, B.; McLaughlin, D.; Böhm, T.; Hegelheimer, M.; Seeberger, D.; Thiele, S. Bipolar Membrane Electrode Assemblies for Water Electrolysis. *ACS Appl. Energy Mater.* **2020**, *3*, 9635–9644. [CrossRef]
177. Tucker, M.C. Progress in metal-supported solid oxide electrolysis cells: A review. *Int. J. Hydrogen Energy* **2020**, *45*, 24203–24218. [CrossRef]
178. Leia, L.; Zhang, J.; Guan, R.; Liu, J.; Chen, F.; Tao, Z. Energy storage and hydrogen production by proton conducting solid oxide electrolysis cells with a novel heterogeneous design. *Energy Conv. Manag.* **2020**, *218*, 113044. [CrossRef]
179. Trini, M.; Hauch, A.; De Angelis, S.; Tong, X.; Vang Hendriksen, P.; Chen, M. Comparison of microstructural evolution of fuel electrodes in solid oxide fuel cells and electrolysis cells. *J. Power Sources* **2020**, *450*, 227599. [CrossRef]
180. Teng, Z.; Xiao, Z.; Yang, G.; Guo, L.; Yang, X.; Ran, R.; Wang, W.; Zhou, W.; Shao, Z. Efficient water splitting through solid oxide electrolysis cells with a new hydrogen electrode derived from A-site cation-deficient $La_{0.4}Sr_{0.55}Co_{0.2}Fe_{0.6}Nb_{0.2}O_{3-d}$ perovskite. *Mater. Today Energy* **2020**, *17*, 100458. [CrossRef]

181. Kamlungsua, K.; Su, P.-C. Moisture-dependent electrochemical characterization of $Ba_{0.2}Sr_{1.8}Fe_{1.5}Mo_{0.5}O_{6-d}$ as the fuel electrode for solid oxide electrolysis cells (SOECs). *Electrochim. Acta* **2020**, *355*, 136670. [CrossRef]
182. Prasopchokkul, P.; Seeharaj, P.; Kim-Lohsoontorn, P. $Ba_{0.5}Sr_{0.5}(Co_{0.8}Fe_{0.2})_{1-x}Ta_xO_{3-d}$ perovskite anode in solid oxide electrolysis cell for hydrogen production from high-temperature steam electrolysis. *Int. J. Hydrogen Energy* **2021**, *46*, 7023–7036. [CrossRef]
183. Rajendran, S.; Thangavel, N.K.; Ding, H.; Ding, Y.; Ding, D.; Reddy Arava, L.M. Tri-Doped $BaCeO_3-BaZrO_3$ as a Chemically Stable Electrolyte with High Proton-Conductivity for Intermediate Temperature Solid Oxide Electrolysis Cells (SOECs). *ACS Appl. Mater. Interfaces* **2020**, *12*, 38275–38284. [CrossRef]
184. Tong, X.; Ovtar, S.; Brodersen, K.; Vang Hendriksen, P.; Chen, M. Large-area solid oxide cells with $La_{0.6}Sr_{0.4}CoO_{3-\delta}$ infiltrated oxygen electrodes for electricity generation and hydrogen production. *J. Power Sources* **2020**, *451*, 227742. [CrossRef]
185. Bernadet, L.; Moncasi, C.; Torrell, M.; Tarancón, A. High-performing electrolyte-supported symmetrical solid oxide electrolysis cells operating under steam electrolysis and co-electrolysis modes. *Int. J. Hydrogen Energy* **2020**, *45*, 14208–14217. [CrossRef]
186. Temluxame, P.; Puengjinda, P.; Peng-ont, S.; Ngampuengpis, W.; Sirimungkalakul, N.; Jiwanuruk, T.; Sornchamni, T.; Kim-Lohsoontorn, P. Comparison of ceria and zirconia based electrolytes for solid oxide electrolysis cells. *Int. J. Hydrogen Energy* **2020**. [CrossRef]
187. Varela, C.; Mostafa, M.; Zondervan, E. Modeling alkaline water electrolysis for power-to-x applications: A scheduling approach. *Int. J. Hydrogen Energy* **2021**. [CrossRef]
188. Sun, Z.; Wang, G.; Koh, S.W.; Ge, J.; Zhao, H.; Hong, W.; Fei, J.; Zhao, Y.; Gao, P.; Miao, H.; et al. Solar-Driven Alkaline Water Electrolysis with Multifunctional Catalysts. *Adv. Funct. Mater.* **2020**, *30*, 2002138. [CrossRef]
189. Mastropasqua, L.; Pecenati, I.; Giostri, A.; Campanari, S. Solar hydrogen production: Techno-economic analysis of a parabolic dishsupported high-temperature electrolysis system. *Appl. Energy* **2020**, *261*, 114392. [CrossRef]
190. Milewski, J.; Kupecki, J.; Szczęśniak, A.; Uzunow, N. Hydrogen production in solid oxide electrolyzers coupled with nuclear reactors. *Int. J. Hydrogen Energy* **2020**. [CrossRef]

Article

$Pt/Re/CeO_2$ Based Catalysts for CO-Water–Gas Shift Reaction: from Powders to Structured Catalyst

Vincenzo Palma [1], Fausto Gallucci [2], Pluton Pullumbi [3], Concetta Ruocco [1], Eugenio Meloni [1] and Marco Martino [1,*]

1 Department of Industrial Engineering, University of Salerno, Via Giovanni Paolo II 132, 84084 Fisciano (SA), Italy; vpalma@unisa.it (V.P.); cruocco@unisa.it (C.R.); emeloni@unisa.it (E.M.)
2 Inorganic Membranes and Membrane Reactors, Department of Chemical Engineering and Chemistry, Eindhoven University of Technology, 5600 MB Eindhoven, The Netherlands; f.gallucci@tue.nl
3 Air Liquide, Paris Innovation Campus. 1, chemin de la Porte des Loges, 78350 Les Loges en Josas, France; pluton.pullumbi@airliquide.com
* Correspondence: mamartino@unisa.it; Tel.: +39-089-969275

Received: 6 May 2020; Accepted: 18 May 2020; Published: 19 May 2020

Abstract: This work focuses on the development of a $Pt/Re/CeO_2$-based structured catalyst for a single stage water–gas shift process. In the first part of the work, the activity in water–gas shift reactions was evaluated for three $Pt/Re/CeO_2$-based powder catalysts, with Pt/Re ratio equal to 1/1, 1/2 ad 2/1 and total loading ≈ 1 wt%. The catalysts were prepared by sequential dry impregnation of commercial ceria, with the salts precursors of rhenium and platinum; the activity tests were carried out by feeding a reacting mixture with a variable CO/H_2O ratio, equal to 7/14, 7/20 and 7/24, and the kinetic parameters were determined. The model which better described the experimental results involves the water–gas shift (WGS) reaction and CO as well as CO_2 methanation. The preliminary tests showed that the catalyst with the Pt/Re ratio equal to 2/1 had the best performance, and this was selected for further investigations. In the second part of the work, a structured catalyst, obtained by coating a commercial aluminum alloy foam with the chosen catalytic formulation, was prepared and tested in different reaction conditions. The results demonstrated that a single stage water–gas shift process is achievable, obtaining a hydrogen production rate of 18.7 mmol/min at 685 K, at $\tau = 53$ ms, by feeding a simulated reformate gas mixture (37.61 vol% H_2, 9.31 vol% CO_2, 9.31 vol% CO, 42.19 vol% H_2O, 1.37 vol% CH_4).

Keywords: hydrogen; water–gas shift; process intensification; structured catalysts; kinetics; aluminum alloy foam; ceria; platinum; rhenium

---

1. Introduction

The water–gas shift (WGS) reaction [1] is an exothermic reaction, whose process is generally performed through two adiabatic stages, at high (HTS) [2] and low temperatures (LTS) [3]. WGS can be considered as the first syngas purification step to obtain high purity hydrogen, and in fact, it allows for the reduction of the CO percentage and increase of the hydrogen yield in the reformate gas stream [4]. The two-stages process configuration allows us to achieve higher conversions at relatively smaller volumes. The reactors suffer limitations, however, due to the heat of the reaction developed during the process, which induces a temperature gradient on the catalytic bed. The low temperature at the inlet of the bed disfavors the kinetics, while the high temperature at the outlet of the bed thermodynamically limits the conversion. In previous work, the intensification of the WGS reaction through the use of highly thermal conductive structured catalysts has been proposed [5], laying the foundations for the realization of a single stage process [6]. In fact, the use of such a structure [7] assures a redistribution of the heat of the reaction along the catalytic bed, with benefits in terms of CO conversion [8], while minimizing hot-spot occurrence [9]. Moreover, comparative studies have showed that the shape of the selected structure plays a crucial role in determining the performance of a catalyst. In this context, the use of foam structures guarantees an enhanced axial and radial heat dispersion and mass transfer [10]. Concerning the catalytic formulation, the conventionally used catalysts, both in HTS and LTS processes, are not suitable for the single stage process design, since the latter are readily deactivated at high temperatures and are pyrophoric [11], while the former are not active at low temperatures and present waste toxicity problems [12]. A valuable alternative consists in noble metal-based catalysts [13], supported on reducible oxides such as titania or ceria [14], which are highly active in a wide range of temperatures [15,16]. Among the investigated noble metals, platinum has showed the best compromise between catalytic activity, ease of preparation and stability [17]; moreover, it was reported that rhenium doping enhances the activity of platinum-based catalysts for WGS reactions [18]. Some specific studies have been published on the role of rhenium in these catalytic systems [19], however, the mechanisms have not yet been fully clarified [20]; CO chemisorption measurements have highlighted the better platinum dispersion [21] upon rhenium addition, thus achieving the enhancement in the WGS rate and catalyst stability [22]. Transmission electron microscopy (TEM) and X-ray photoelectron spectroscopy (XPS) studies showed the formation of bimetallic surface clusters between rhenium and platinum; in addition, the stabilization of formate species and the acceleration of the H_2 formation rate were attributed to the presence of rhenium [23]. Azzam et al. studied the Pt–Re/TiO_2 catalytic system and speculated that rhenium remains as oxide, providing an additional redox WGS reaction pathway, in which ReOx is reduced by carbon monoxide generating carbon dioxide and oxidized by H_2O to give hydrogen [24]. Del Villar et al. studied a more complex catalytic system (RePt/CeO_2–TiO_2 catalyst) for the WGS reaction under conditions compatible with a membrane reactor [25]. The authors ascribed the enhanced WGS activity and stability to the improvement in the reduction behavior of highly dispersed CeO_2 and ReOx species, as well as to the close contact between Pt and Re species, which could introduce further redox activity sites and prevent the Pt sintering.

In a process intensification view, the deposition of highly active catalytic formulations on proper structured catalysts was shown to be an effective option for minimizing the typical mass and heat transfer issues of packed bed reactors. In this regard, in our previous papers, we demonstrated that the employment of aluminum open cells foams as structured carrier paddles in the direction of HTWGS-LTWGS integration in a single unit. In fact, due to the high thermal conductivity of the selected material, the heat was redistributed along the catalyst from the outlet to the inlet section with a flatter thermal profile and a global increase of CO conversion [5,7,10].

In the first part of this work, a preliminary study on the use of PtRe/CeO_2-based powder catalysts was reported; the activity of three catalysts with a different Pt/Re ratio (Pt/Re ratio equal to 1/1, 1/2 or 2/1) and total loading of 1 wt% was compared in WGS reaction and a kinetic investigation was performed for the three samples. The kinetic parameters were determined based on the results of three dedicated reaction tests carried out at different H_2O/CO ratio (H_2O/CO ratio equal to 24/7, 20/7 or 14/7). In the second part of the work a structured catalyst was prepared by coating an open cell aluminum alloy foam with cerium oxide and loading the resulting structure with the active species in the ratio which gave the best performance in the preliminary study (Pt/Re = 2/1). The structured catalyst performance was evaluated by feeding the reacting mixture, previously used for the powder catalysts, with a H_2O/CO ratio equal to 24/7 in order to evaluate the effect of the scale up from powder to structured catalyst, in terms of activity. Moreover, the as-obtained structured catalyst was also tested under a simulated reformate gas stream as feed, aiming to evaluate the possibility to realize a single stage water–gas shift reactor, which is highly desirable for the realization of a compact fuel processing system.

2. Results and Discussion

2.1. *Powder Catalysts*

2.1.1. Characterization Results

Powder catalysts were characterized by means of a series of analytical techniques: energy dispersive X-ray fluorescence (ED-XRF) was used to check the effective metals loading and the effective Pt/Re ratio; X-ray diffraction (XRD) was used to evaluate the crystallite size; Raman spectroscopy (Raman) was used to provide the structural features; transmission electron microscopy (TEM) was used to evaluate the particle size and shape; the temperature programmed reduction with hydrogen (H_2-TPR) to evaluate the reducibility of the catalysts.

The ED-XRF results stated an effective metal loading of 0.88 wt%–0.90 wt% for the three powder catalysts and confirmed the expected Pt/Re ratio (Table 1). The specific surface area measurements calculated with the Brunauer–Emmett–Teller (B.E.T.) method showed a small decrease in the surface area compared to that of the alone ceria (Table 1).

Table 1. Powder catalysts characterization.

	Chemical Composition (wt%)			$SSA_{B.E.T.}$ (m^2/g)	H_2 Uptake (mmol/g)		Crystallite Size (nm)
	CeO_2	Pt	Re		Exp.	Theor.	
CeO_2	-	-	-	175	-	-	6.10
W1 (2Pt/1Re/CeO_2)	99.12	0.59	0.29	171	0.63	0.13	6.09
W2 (1Pt/1Re/CeO_2)	99.10	0.45	0.45	169	0.84	0.15	6.05
W3 (1Pt/2Re/CeO_2)	99.11	0.30	0.59	168	0.84	0.16	6.15

All the XRD diffractograms of the powder catalysts displayed the cubic fluorite-type for CeO_2 crystal phase [26], and no diffraction peaks corresponding to Pt, Re or the corresponding oxides phase could be identified, due to the low loading and high dispersion of the noble metal phases (Figure 1). Moreover, the calculation of the crystallite sizes by means of the Scherrer equation demonstrates that active species deposition had no effect on the dimension of ceria crystallites (Table 1). For ceria and its solid solutions, Raman spectroscopy contributes to the improved understanding of local structural changes in the nearest and next nearest neighbor shells of Pt and Re dopant cations and oxygen vacancies. Raman is an excellent tool to investigate the defect chemistry of doped and un-doped ceria, due to its sensitivity to non-periodic features. The Raman spectrum of ceria support (Figure 2) showed the typical strong peak at 464 cm^{-1} attributed to the first order F_{2g} mode [27] (Ce^{4+}-O-Ce^{4+} wagging) and the barely visible broad peaks at 258, 595 and 1,179 cm^{-1}, respectively attributed to second-order transverse acoustic (2TA), defect-induced (D) (recently demonstrated to be Ce^{3+}-O-Ce^{4+} stretching [28]) and second-order longitudinal optical modes (2LO) [29]. The Raman spectrum of the W1 catalyst (Pt/Re ratio 2/1) still showed the peak of the first order F_{2g} mode, but instead of D transition, two broad peaks at 575 cm^{-1} and 657 cm^{-1} were visible, attributed to the interaction of Pt with the cerium oxide surface [30]. Moreover, at 829 cm^{-1}, a small peak attributed to the antisymmetric stretching mode of Re-O-Re appeared [31].

The Raman spectrum of the W2 catalyst (Pt/Re ratio 1/1) still showed the peak ascribable to the first order F_{2g} mode; regarding the two broad peaks at 575 cm^{-1} and 657 cm^{-1}, attributed in W1 to the interaction of Pt with cerium oxide surface appeared, the latter seemed to disappear. On the contrary, a further peak was observed at 982 cm^{-1}, which is due to symmetric Re-O stretching [31]. For the W3 catalyst (Pt/Re ratio 1/2), two peaks attributed to Re-O stretching can be observed in the Raman spectrum, while the two broad peaks at 575 cm^{-1} and 657 cm^{-1}, ascribed to the interaction of Pt with cerium oxide surface for the sample W1, disappeared and a new peak at 595 cm^{-1} was observed, linked to defect-induced (D) in ceria. These results suggested that the interaction of platinum with ceria mainly occurs at the D sites. TEM images showed the morphology of the catalysts, suggesting a single phase of spherical type (Figure 3), moreover, the particle size agreed with the crystallite size, calculated by the Scherrer equation. However, the black area may represent groups of accumulated particles. The temperature programmed reduction (TPR) profiles of the three catalysts can be divided in two main areas. The broad peak between 700 K and 750 K was attributed to ceria's surface reduction [32] (Figure 4). The peaks between 500 K and 600 K are ascribable to an overlapping of three reduction peaks of PtOx and ReOx species and the ceria surface, as the result of the spillover effect from metal particles to the support [33].

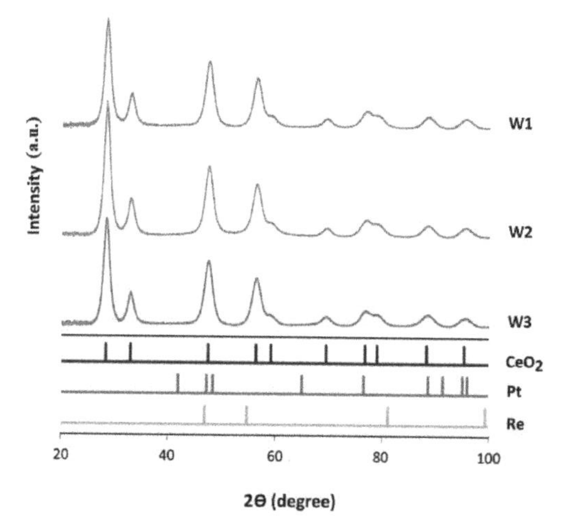

Figure 1. Diffractograms of powder catalysts: W1, W2 and W3.

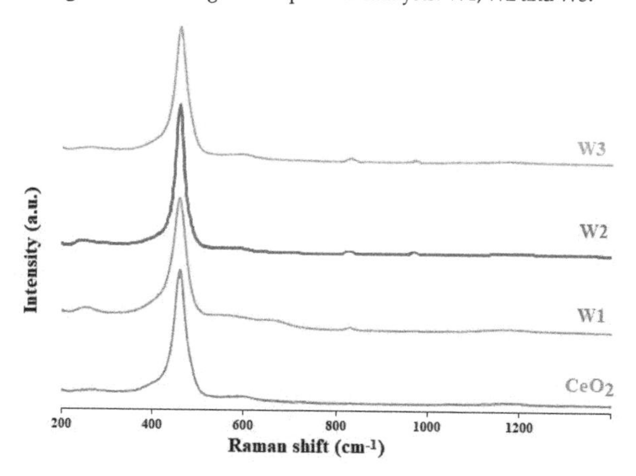

Figure 2. Raman spectra of ceria and powder catalysts: W1, W2 and W3.

Specifically, the TPR profile of W1 catalyst showed a perfect overlapping, with the peak centered at 550 K, while in the case of W2 and W3 TPR profiles, the main peak shifted respectively to 593 K and 608 K, and a shoulder on the left appeared in both the curves, centered between 505 k and 510 K.

These results suggest a correlation between the temperature of the main reduction peak and the Pt/Re ratio, suggesting a strong interaction between platinum and rhenium. The evaluation of hydrogen uptake, calculated by integrating the H_2-TPR profiles, as expected, confirmed the advent of the spillover phenomenon; in fact, the hydrogen consumed in each experiment was more than one order of magnitude higher compared to the expected value (Table 1).

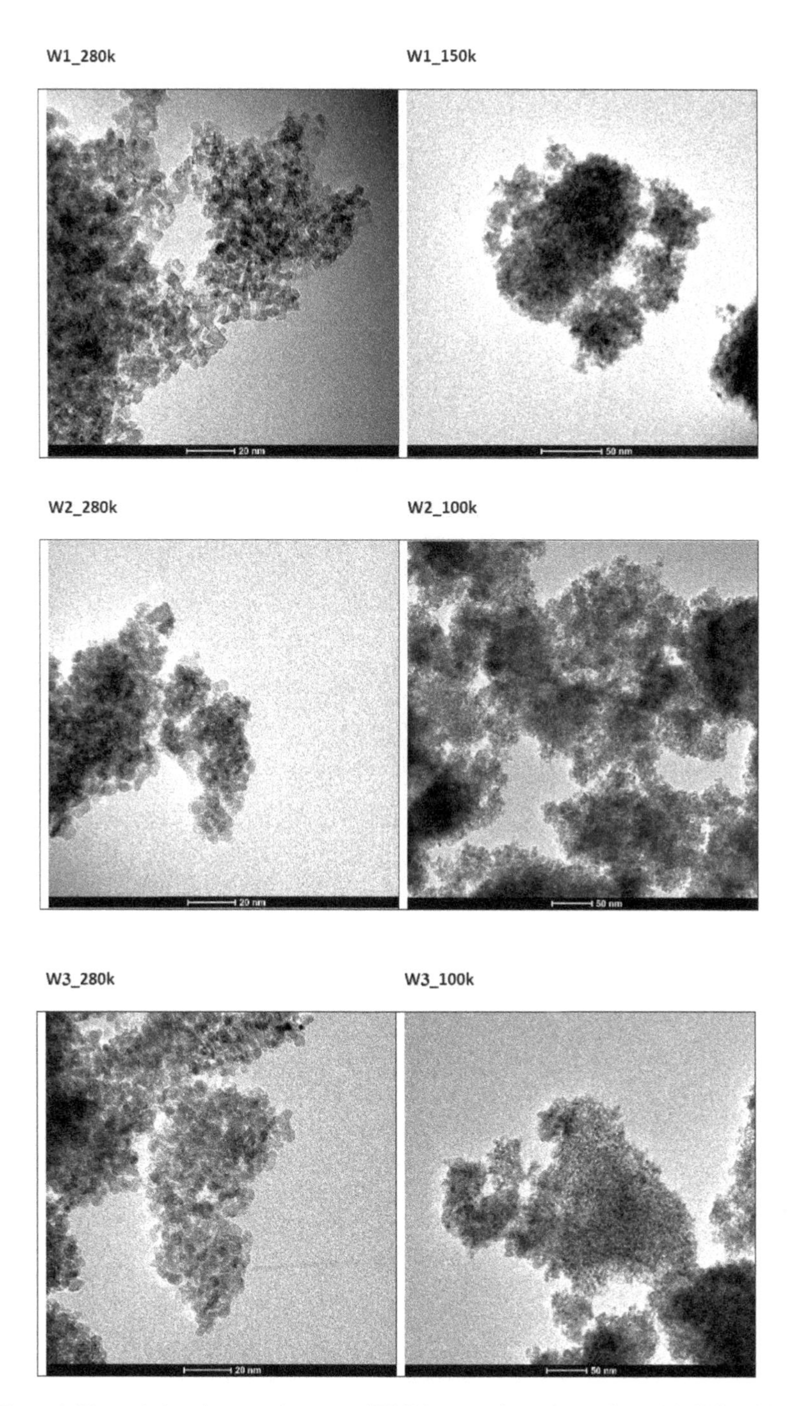

Figure 3. Transmission electron microscopy (TEM) images of powder catalysts: W1, W2 and W3.

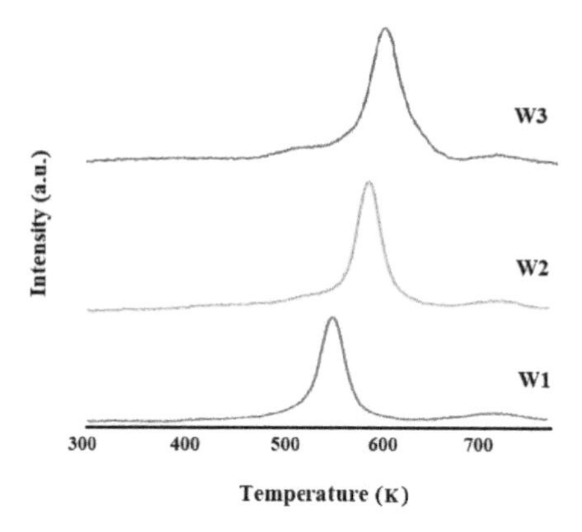

Figure 4. Temperature programmed reduction with hydrogen (H_2-TPR) of powder the catalysts: W1, W2 and W3.

2.1.2. Activity Tests

Figure 5a–c shows the catalytic activity of the three investigated catalysts towards the WGS reaction between 500 and 600 K as a function of the steam to carbon monoxide ratio. In Figure 5d the results obtained at the same H_2O/CO ratio (24/7) are compared.

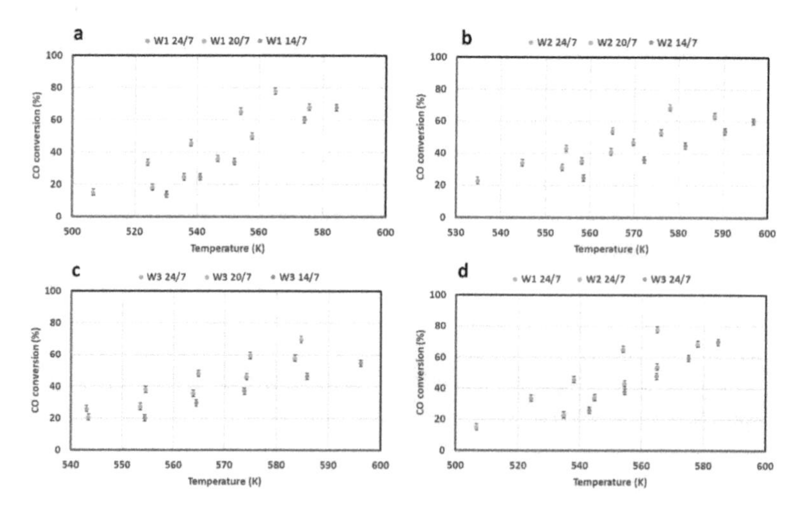

Figure 5. CO conversion as function of temperature for different H_2O/CO ratio (24/7, 20/7, 14/7) for each powder catalyst: W1 (**a**), W2 (**b**) and W3 (**c**), and CO conversion comparison as function of the temperature at the same H_2O/CO ratio (24/7) for the three catalysts (**d**). $\tau = 131$ ms.

As expected, the increase in the H_2O/CO ratio favored the WGS reaction for all the three catalysts. In particular, for the W1 sample, the variation of the H_2O/CO ratio from 20/7 to 24/7 resulted in a more pronounced improvement of the catalytic activity, with the profiles achieved for H_2O/CO=20/7 and 14/7 closer. For the other two catalysts, by changing the feeding conditions, such a strong enhancement in CO conversion was not observed. Moreover, at a fixed H_2O/CO ratio of 24/7, the W1 catalyst showed higher CO conversion than the W2 and W3 catalysts, which essentially displayed similar activity. For example, at 565 K, the W1 reached a conversion of almost 80%, while for the other two samples a value around 50% was recorded. These results demonstrated that the catalyst with a Pt/Re ratio equal to 2/1 is more active than the catalysts with the Pt/Re ratio equal to 1/1 and 1/2.

This trend seems to suggest a direct correlation with the reduction temperature of the PtOx/ReOy system in the H_2-TPR experiments and thus with Pt/Re loading and the catalytic activity. The 2Pt/1Re catalyst, in fact, displayed lower reduction temperatures compared to the other two samples and the strong correlation between the easy reducibility and the activity for WGS was already reported for CeO_2-based catalysts [34]. The most active formulation was deposited on a foam structured carrier and further tested.

2.1.3. Kinetic Measurements

The experimental results obtained over the three Pt/Re catalysts and the analysis of product distribution as a function of reaction temperature, particularly regarding the methane formation, suggested that the reaction pathway over the three Pt/Re catalysts involved water–gas shift reactions and both CO as well as CO_2 methanation. The pre-exponential factors and the apparent activation energies estimated for the three above reactions from the Arrhenius plots over the W1, W2 and W3 sample are shown in Table 2. The kinetic model results were validated by comparing them with the experimental results. In Figure 6, as an example, the comparison related to three catalytic systems in the condition of $H_2O/CO = 14/7$ is reported. The data showed a good agreement between the kinetic model and the experimental results; in particular, it is possible to highlight that the kinetic model was able to predict the behavior of the catalytic systems in the investigated temperature range and, more importantly, it was able—fixing the active species ratio—to take into account the variation of the different feeding conditions investigated in this work (H_2O/CO ratio in the range 2–3.84). The activation energy calculated for the W2 and W3 catalysts, with Pt/Re ratio of 1/1 and 1/2 respectively, were very close, both for WGS reaction and CO and CO_2 methanation.

Table 2. Kinetic parameters for the powder catalysts.

	W1	**W2**	**W3**
$k_{0,WGS}$ (mol/(g·min·atm^2))	1608326	1225116014	1225116014
$Ea_{,WGS}$ (kJ/mol)	78	114	118
$k_{0,CO}$ (mol/(g·min·atm^2))	2989	40000	40000
$Ea_{,CO}$ (kJ/mol)	94	85	85
k_{0,CO_2} (mol/(g·min·atm^2))	0.00712	6.9	6.9
$Ea_{,CO_2}$ (kJ/mol)	16	50	48

On the other hand, the activation energy for the W1 catalyst, with a Pt/Re ratio of 2/1, of WGS reaction was much lower and the CO methanation was higher, suggesting a preference for the WGS reaction. Thus, the results shown in Table 2 confirmed the improved activity of the 2Pt-1Re/CeO_2 catalyst for the water–gas shift reaction (discussed in Paragraph 2.1.2), with an Ea of about 78 kJ/mol: the easier reducibility assured an enhancement in catalyst activity compared to the W2 and W3 samples, which displayed WGS activation energies of 114 and 118 kJ/mol, respectively. The found values agreed well with kinetic data reported in the literature for similar catalytic formulations [35].

2.2. *Structured Catalyst*

2.2.1. Characterization Results

The structured catalyst (denoted as S1) was prepared by coating the commercial aluminum alloy foams, with a relative density of 25%, by means of chemical conversion coating, with cerium oxide. The chemical conversion coating technique is traditionally used in protecting aluminum alloys from corrosion [36], and recently it has been successfully used in catalysis [37]. This technique presents many advantages with respect to the washcoating [38] (the most widely used technique to prepare structured catalysts), including the high resistance to the mechanical stress of the coating, and the usability with highly porous structures, for which the washcoating might occlude the pores. The active components were loaded by wet impregnation with rhenium and platinum precursors. The ED-XRF analysis showed that the total loading of the catalytic formulation on the foams was on average 5.1 wt%, and the Pt/Re ratio was equal to 1.9 (Table 3).

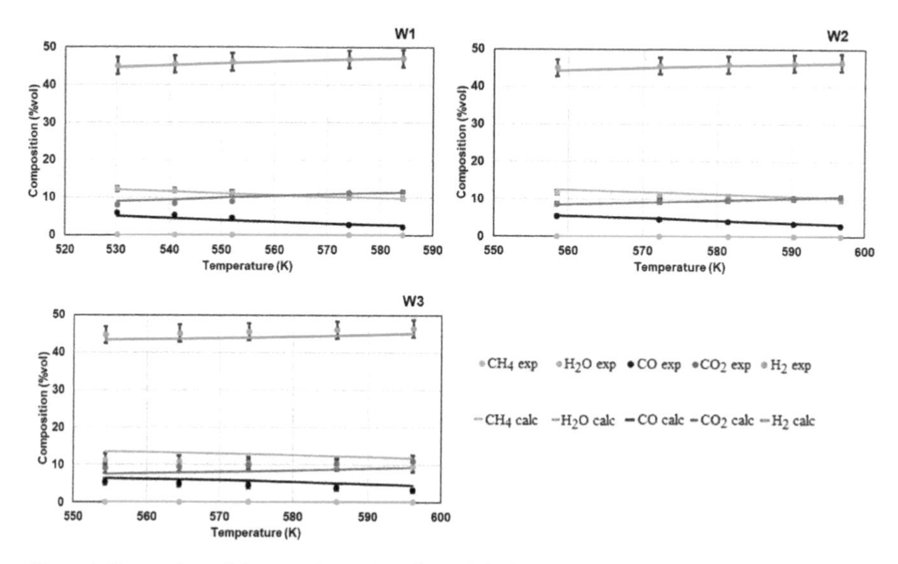

Figure 6. Comparison of the experimental results and the kinetic model results for the three catalytic systems, in the condition $H_2O/CO = 14/7$.

Table 3. Structured catalyst characterization.

	Catalytic Formulation		$SSA_{B.E.T.}$ (m^2/g)	H_2 Uptake (mmol/g)		Relative Density (%)
	Loading (wt%)	Pt/Re Ratio		Exp.	Theor.	
Al-foam	-	-	1	-	-	25
S1	5.1	1.9	3.8	1.2	0.13	-

The specific surface area measurements on the structured catalyst showed an increase of the surface area of the structure due to the ceria coating loading; on the basis of the result, a surface area between 70 m^2/g and 80 m^2/g for the ceria coating was hypothesized.

The hydrogen uptake measured during the TPR experiments was one order of magnitude higher than the theoretical one (Table 3), due to the spillover effect, however, the extension of this phenomenon was higher than in the case of the powder catalyst with the same Pt/Re ratio (Table 1). The H_2-TPR profile of the structured catalyst was very different from the profile of the powder catalyst with the same Pt/Re ratio (Figure 7). The case of the powder catalyst's two main peaks was present, however, the first peak was centered at a lower temperature, while the second peak showed a higher relative intensity, attributed to a much higher reducibility of the ceria surface, in agreement with the higher hydrogen measured uptake. These results suggested a different oxygen storage capacity of the ceria coating [39], attributable mainly to the preparation technique. In Figure 8, the scanning electron microscopy (SEM) images at different magnitude of the structured catalyst S1 are showed (a, c), with the intent to highlight the surface morphology. As is evident, the catalyst surface is extremely irregular, with the presence of numerous cavities and fractures, which justify the increase in surface area obtained due to the coating, and which are beneficial for the catalytic activity.

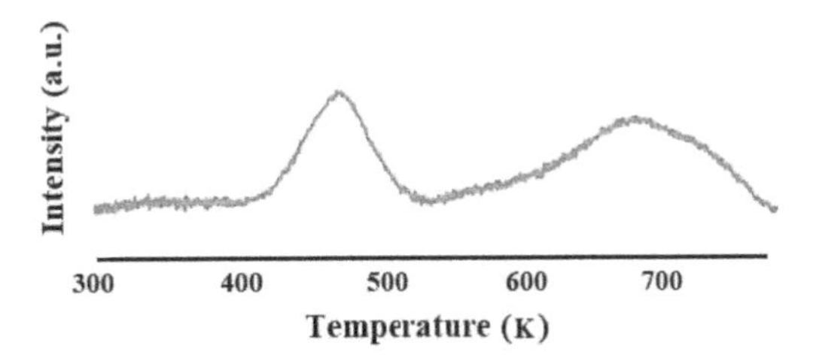

Figure 7. H_2-TPR of the structured catalyst.

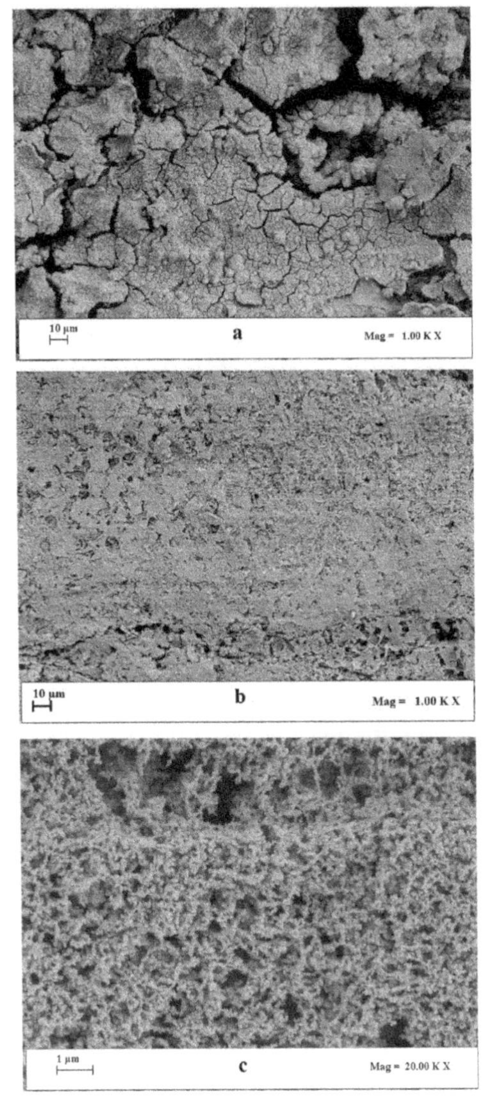

Figure 8. SEM (scanning electron microscopy) image of S1 structured catalyst at different magnitude (**a**,**c**), and a SEM image of a structured catalyst obtained by washcoating and loaded with Pt/Re/$CeZrO_4$ catalytic formulation (**b**).

Moreover, the nanometric nature of the coating is highlighted if compared with the SEM image of a structured catalyst obtained by washcoating (Figure 8b), prepared in previous work, with a similar catalytic formulation (1Pt/1Re/$CeZrO_4$) [8].

2.2.2. Kinetic Measurements

The results of the kinetic study performed on the structured catalysts are summarized in Table 4 in terms of pre-exponential factor and activation energy.

Table 4. Kinetic parameters for the structured catalyst.

	k_0 (mol/(g·min·atm^2))	Ea (kJ/mol)
r_{WGS}	1608326	80
r_{CO}	2500	85
r_{CO2}	1000	75

The kinetic model results were validated by comparing them with the experimental results. In Figure 9, the comparison related to the structured catalyst in the condition of H_2O/CO = 24/7 is reported.

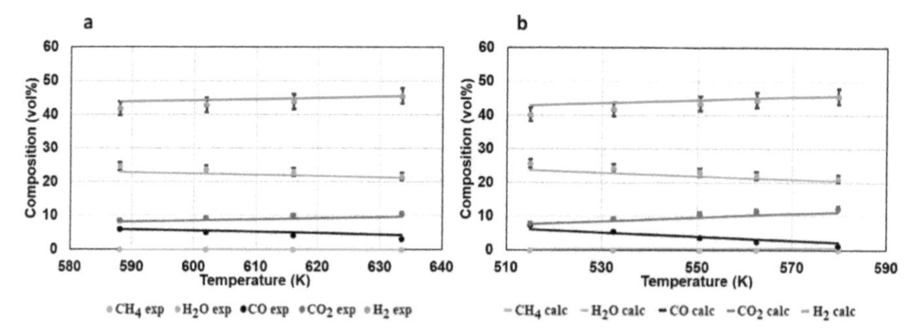

Figure 9. Comparison of the experimental results and the kinetic model results for the structured catalyst, in the condition H_2O/CO = 24/7, at two different contact times, 23 ms (**a**) and 131 ms (**b**).

The data shown in Figure 9 evidence a good agreement between the predicted values and the experimental results, thus demonstrating the feasibility of the kinetic model. The comparison of the kinetic parameters related to the structured (Table 4) and powder catalyst W1 (Table 2) evidenced that the two catalytic systems were characterized by very similar values of the pre-exponential factor and activation energy for WGS, but they had different values regarding the methanation parameters. This result suggested that the two catalytic systems were different, due to the preparation technique, as was also confirmed by the TPR results. However, it is worthwhile noting that, as described above for the powder sample, the activation energy of the WGS is lower than the methanation ones, therefore evidencing a preference for the WGS reaction.

2.2.3. Activity Tests

The activity tests on the structured catalyst were carried out with the aim of testing the feasibility of a single stage WGS process, evaluating the performance for intermediate temperatures with a simulated syngas mixture feeding at a target contact time. In Figure 10a, the CO conversion as a function of the temperature, at three different contact times (τ = 23, 79 and 131 ms) and for a H_2O/CO ratio = 24/7, is shown. The test carried out at 131 ms showed comparable activity with respect to the corresponding powder catalyst, the test were performed at 23 ms, although under a space velocity of more than five times higher compared to the value selected for the powder sample, showed satisfactory results, with a CO conversion of almost 60% at 635 K (Figure 10a).

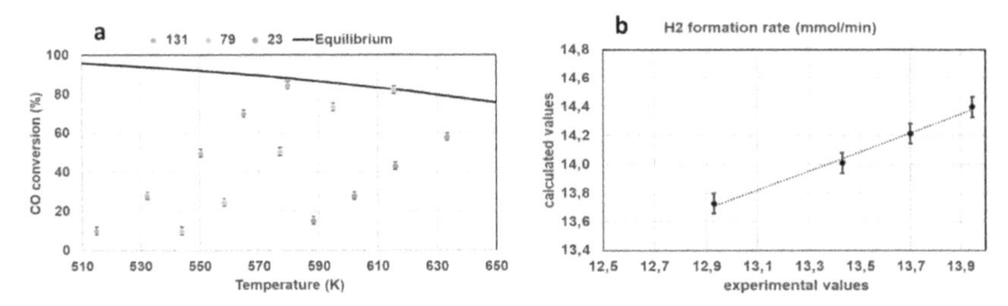

Figure 10. CO conversion as function of temperature (**a**) and comparison between hydrogen formation rate experimental vs. calculated values (**b**) for S1 catalyst with a $H_2O/CO = 24/7$, at a $\tau = 23$ ms.

The experimental hydrogen formation rate was also evaluated, highlighting once again a good agreement with the values predicted by the model (Figure 10b). The structured catalyst was also tested under a simulated reformate gas mixture coming from an ethanol pre-reformer followed by a methane reforming unit [40]. The catalyst activity was investigated in the temperature range 540–690 K and the results, showed in Figure 11, are presented in terms of CO conversion (a) and product gas distribution as a function of outlet reaction temperature (b). CO conversion linearly increased up to 610 K, showing a decreasing trend above 630 K; due to the negative effect of high temperatures on WGS thermodynamics. Only slight variations in methane concentration with reaction temperature were observed, suggesting that carbon oxide methanation was not the dominant pathway under the investigated conditions. A hydrogen production rate of 18.7 mmol/min was recorded at 685 K and this value (properly scaled on the basis of the different feeding flow-rates) was compared with the results of the simulations performed. The hydrogen productivity recorded over the structured catalyst is only 14% lower than the value reported, and this difference is mainly ascribable to the employment of a membrane in the WGS reactor in the reference process scheme. These very promising results allow us to conclude that the S1 structured catalyst is a suitable sample for the realization of a single-stage WGS process.

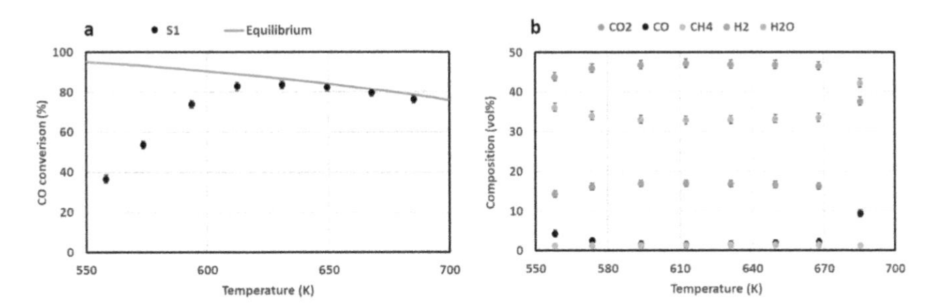

Figure 11. CO conversion (**a**) and product gas distribution as a function of reaction temperature (**b**) for S1 catalyst under a simulated reformate gas (37.61 vol% H_2, 9.31 vol% CO_2, 9.31 vol% CO, 42.19 vol% H_2O, 1.37 vol% CH_4), $\tau = 53$ ms.

3. Materials and Methods

3.1. *Powder Catalysts Preparation*

The powder catalysts were prepared by sequential dry impregnation of Actalys HSA commercial ceria, with the rhenium and platinum precursors respectively, with a total loading of platinum and rhenium equal to 1 wt% with respect to the total weight of the catalyst. Three catalysts with a variable platinum/rhenium ratio of 2/1, 1/1 and 1/2 were prepared. The porous volume of the ceria was measured by mixing 5 g of support with the minimum amount of distilled water needed, then the support was dried at 393 K for 2 h and impregnated with an ammonium perrhenate solution, obtained by dissolving the desired amount of the rhenium salt in a volume of water corresponding to the porous volume of the ceria. The obtained mixture was dried at a temperature of 343 K for 16 h and calcined at 773 K for 1 h. The resulting solid was treated as previously described, in order to evaluate its porous volume, then dried at 393 K for 2 h and impregnated with a solution of tetraamineplatinum (II) nitrate, obtained by dissolving the desired amount of the platinum salt in a volume of water corresponding to the porous volume previously measured. The obtained mixture was dried at a temperature of 343 K for 16 h and calcined at 773 K for 1 h.

3.2. *Structured Catalyst Preparation*

The structure was obtained by assembly 7 circular open cell aluminum alloy foams with a diameter of 1.4 cm, obtained by cutting and shaping 1 cm thick commercial sheet. The as obtained foams were cleaned and corroded by treating with 5 vol% HF solution for 4 min, then washed with distilled water and dried at 393 K for 2 h. The ceria coating was realized by means of chemical conversion coating described in a previous article [41]. The corroded foams were degreased with a 5 wt% solution of NaOH for one minute, washed with distilled water, then etched in a 35 vol% solution of HNO_3, washed with distilled water, then treated with a chemical bath to coating the surface with ceria support. The coating bath consisted of an acidic solution (pH = 2, by HCl) obtained by dissolving $CeCl_3$*$7H_2O$ (1.3 wt%), 35% wt H_2O_2 (3.8 wt%) in distilled water. The foams were treated with this solution for 2 h at 328 K; the treatment was repeated three times. After each treatment, the reacted foams were washed with distilled water and dried for 2 h at 393 K. After the third treatment the foams were calcined at 773 K for 1 h. Rhenium and platinum were loaded by wet impregnation of the coated structures, by immersing in the metal salt precursor solution at 328 K for 20 min, followed by drying at 393 K for 2 h and calcining at 773 K for 1 h.

3.3. *Catalysts Characterization*

The catalysts were characterized with a series of physical–chemical analytical techniques. The B.E.T. specific surface areas measurements were carried out with a Costech Sorptometer 1040 (Costech International, Milano, Italy), by dynamic N_2 adsorption at 77 K. The crystal phases were obtained by X-ray powder diffraction (Rigaku MiniFlex 600). The crystallite sizes were calculated from the diffractograms by applying the Scherrer equation. The chemical composition was checked by means of an ARL^{TM} QUANT'X ED-XRF spectrometer (Thermo Scientific, Rodano, Italy), while the structural features were evaluated with a Raman spectroscopy using an inVia Raman Microscope (Renishaw, Pianezza, Italy), equipped with a 514 nm Ar ion laser operating at 25 mW. TEM images were obtained with a FEI Tecnai 20 (Sphera) microscopy operating at 200 kV LaB6 filament. Samples morphology was observed by a Field Emission Scanning Electron Microscope (FE-SEM, mod. LEO 1525, Carl Zeiss SMT AG, Oberkochen, Germany). The H_2-TPR experiments were carried out in the temperature range of 293–723 K, with a reducing stream of 500 Ncc/min containing 5 vol% of H_2 in Ar, applying a heating rate of 10 K/min, in the same reactor used for the activity tests, as described below.

3.4. *Catalytic Activity Tests*

The powder catalysts were previously compacted and sieved in the 180–355 μm range and diluted with quartz glass at the same total volume. The activity tests were carried out on the reduced catalysts, at atmospheric pressure, in the temperature range 500-620 K, at a contact time τc = 131 ms, with a reacting mixture 43 vol% H_2, 7 vol% CO_2, 7 vol% CO, X vol% H_2O (X = 14, 20 or 24), balance Argon. The structured catalyst was surrounded by a thermo expanding pad with a thickness of 3 mm, and previously reduced in the H_2-TPR experiments. The activity tests were carried out at atmospheric pressure, in the temperature range 500–620 K, at a contact time range τc = 23–131 ms, with two different reacting mixture, the first was 43 vol% H_2, 7 vol% CO_2, 7 vol% CO, 24 vol% H_2O, 19 vol% Argon, the second was 37.61 vol% H_2, 9.31 vol% CO_2, 9.31 vol% CO, 42.19 vol% H_2O, 1.37 vol% CH_4. The catalytic activity tests were performed in a stainless-steel tubular reactor with an internal diameter of 22 mm and a length of 40 cm. The reaction products were dried through a refrigerator Julabo F12 (JULABO Labortechnik GmbH77960 Seelbach, Germany)and sent to a Hiden Analytical mass spectrometer (Hiden Analytical, 420 Europa Blvd, Westbrook, Warrington WA5 7 UN, UK).

3.5. Kinetic Measurements

The kinetic evaluation was performed by numerically analyzing the results of dedicated experimental tests, carried out at atmospheric pressure using mixtures of CO, CO_2, H_2, H_2O, and Ar. The tests were performed as described in the previous paragraph, by varying the H_2O/CO ratio in a wide temperature range. In particular, the data regarding the catalytic performance of the system far from the thermodynamic equilibrium conditions were considered, so that differential reaction conditions could be assumed, with negligible heat and mass transfer effects. The approach described below was followed both for the powder and for the structured catalysts, even if for the structured catalyst only the operating condition corresponding to the H_2O/CO ratio =24/7 (the condition in which the powder catalyst showed the best catalytic performance) was analyzed.

The reactions used for the numeric analysis are:

- The WGS reaction: $CO + H_2O = CO_2 + H_2$
- The CO methanation: $CO + 3H_2 = CH_4 + H_2O$
- The CO_2 methanation: $CO_2 + 4H_2 = CH_4 + 2H_2O$

The reaction rates of the above reported reactions were expressed as follows, considering that (i) for the WGS reaction the keq was calculated using the expression proposed by Moe [42], and (ii) for the two methanation reactions a kinetic expression based on a Langmuir–Hinshelwood approach was used:

- $-r_{CO} = k_{WGS}\left(P_{CO}P_{H_2O} - \frac{P_{CO_2}P_{H_2}}{k_{eq,WGS}}\right)$ with $k_{eq,WGS} = e^{\frac{4577.8}{T} - 4.33}$
- $-r_{CO} = k_{CO}\frac{P_{CO}P_{H_2}}{\left(1+K_{CO}P_{CO}+K_{H_2O}P_{H_2O}\right)^2}$
- $-r_{CO_2} = k_{CO_2}\frac{P_{CO_2}P_{H_2}}{\left(1+K_{CO_2}P_{CO_2}+K_{CO}P_{CO}+K_{H_2O}P_{H_2O}\right)^2}$

In the above reported expressions, the terms have the following meaning:

- Pi is the partial pressure of the component "i".
- *ki (T)* is the reaction rate constant according to the Arrhenius law: $k_i = k_{0,i}e^{-\frac{E_{A,i}}{RT}}$, with $k_{0,i}$ = pre-exponential factor and $E_{A,i}$ = the activation energy.
- *Kj (T)* are the adsorption constants for the component "j", expressed according to the Arrhenius law [41]: $K_j = K_{0,j}e^{-\frac{H_{A,j}}{RT}}$, with $K_{0,j}$ = pre-exponential factor and $\Delta H_{A,j}$ = the adsorption heat.

The material balances on the single components allowed obtaining a set of equations, resolved by applying the Eulero method by means of the Excel software. For each operating condition, the experimental results were compared with the one from the kinetic model, and the Solver of the Excel software allowed the minimization of the function $f = min\left(\sum_{c=1}^{n}\left(x_{exp_c} - x_{mod_c}\right)^2\right)$. The optimization procedure was performed several times with various initial values of the parameters, with the aim to confirm the robustness of the optimization scheme.

4. Conclusions

In this work, a comparative study on Pt/Re/CeO_2 powder catalysts, for CO water-gas shift reaction, has been presented. The activity of three catalysts, obtained by sequential dry impregnation of commercial ceria with the salts precursors of rhenium and platinum, has been evaluated in water–gas shift reaction, as function of the temperature, of the Pt/Re ratio and of the feeding conditions (H_2O/CO ratio). The kinetic parameters have been calculated by means of numerical analysis, considering the water–gas reaction and CO and CO_2 methanation reactions. The results of the activity tests showed the best performance of catalyst with Pt/Re ratio equal to 2/1, which was related to the reduction temperature of the PtOx/ReOy system, observed in the H_2-TPR experiments. This catalyst showed lower reduction temperatures compared to the catalysts with Pt/Re ratio equal to 1/1 and 1/2. In the second part of the work a structured catalyst, loaded with the 2Pt/1Re/CeO_2 catalytic formulation, has been prepared, by coating with ceria a commercial alumina alloy foam by means of chemical conversion coating technique. The SEM images highlighted the nanometric structure and the high rough of the coating. The catalytic activity tests and the kinetic evaluation have been showed a good scale up of the performance from the powder to the structured catalyst. Moreover, the catalytic activity of the structured catalyst has been evaluated in simulated reformate gas feeding conditions, showing that is possible to design a single-stage WGS process, with the prepared structured catalyst.

Author Contributions: Conceptualization, V.P. and M.M.; methodology, V.P., P.P., C.R., E.M. and M.M.; investigation, V.P., F.G., P.P., C.R., E.M. and M.M.; data curation, C.R., E.M. and M.M.; writing—original draft preparation, C.R., E.M. and M.M.; writing—review and editing, V.P., F.G., P.P., C.R., E.M. and M.M.; supervision, V.P.; funding acquisition, V.P., F.G., P.P. All authors have read and agreed to the published version of the manuscript.

Funding: This work has received funding from the European Union's Horizon 2020 research and innovation programme under the Marie Skłodowska-Curie grant agreement No 734561.

Acknowledgments: The authors wish to acknowledge Mariarosa Scognamiglio (Department of Industrial Engineering, University of Salerno) for SEM analysis, and Ing. Giovanni Festa for the valuable contribution.

Conflicts of Interest: The authors declare no conflict of interest.

References

1. Palma, V.; Ruocco, C.; Cortese, M.; Martino, M. Recent Advances in Structured Catalysts Preparation and Use in Water-Gas Shift Reaction. *Catalysts* **2019**, *9*, 991. [CrossRef]
2. Lee, D.-W.; Lee, M.S.; Lee, J.Y.; Kim, S.; Eom, H.-J.; Moon, D.J.; Lee, K.-Y. The review of Cr-free Fe-based catalysts for high-temperature water-gas shift reactions. *Catal. Today* **2013**, *210*, 2–9. [CrossRef]
3. Uchida, H.; Isogai, N.; Oba, M.; Hasegawa, T. The zinc oxide copper catalyst for carbon monoxide-shift conversion. I. The dependency of the catalytic activity on the chemical composition of the catalyst. *Bull. Chem. Soc. Jpn.* **1967**, *40*, 1981–1986. [CrossRef]
4. Cifuentes, B.; Bustamante, F.; Cobo, M. Single and Dual Metal Oxides as Promising Supports for Carbon Monoxide Removal from an Actual Syngas: The Crucial Role of Support on the Selectivity of the Au–Cu System. *Catalysts* **2019**, *9*, 852. [CrossRef]
5. Palma, V.; Pisano, D.; Martino, M. CFD modeling of the influence of carrier thermal conductivity for structured catalysts in the WGS reaction. *Chem. Eng. Sci.* **2018**, *178*, 1–11. [CrossRef]
6. Van Dijk, H.A.J.; Boon, J.; Nyqvist, R.N.; Van Den Brink, R.W. Development of a single stage heat integrated water-gas shift reactor for fuel processing. *Chem. Eng. J.* **2010**, *159*, 182–189. [CrossRef]
7. Palma, V.; Pisano, D.; Martino, M.; Ricca, A.; Ciambelli, P. High Thermal Conductivity Structured Carriers for Catalytic Processes Intensification. *Chem. Eng. Trans.* **2015**, *43*, 2047–2052. [CrossRef]
8. Palma, V.; Pisano, D.; Martino, M. Structured noble metal-based catalysts for the WGS process intensification. *Int. J. Hydrog. Energy* **2018**, *43*, 11745–11754. [CrossRef]
9. Stiegler, T.; Meltzer, K.; Tremel, A.; Baldauf, M.; Wasserscheid, P.; Albert, J. Development of a Structured Reactor System for CO_2 Methanation under Dynamic Operating Conditions. *Energy Technol.* **2019**, *7*, 1900047. [CrossRef]
10. Palma, V.; Pisano, D.; Martino, M. Comparative Study Between Aluminum Monolith and Foam as Carriers for The Intensification of The CO Water Gas Shift Process. *Catalysts* **2018**, *8*, 489. [CrossRef]

11. Kam, R.; Scott, J.; Amal, R.; Selomulya, C. Pyrophoricity and stability of copper and platinum based water-gas shift catalysts during oxidative shut-down/start-up operation. *Chem. Eng. Sci.* **2010**, *65*, 6461–6467. [CrossRef]
12. Zhu, M.; Wachs, I.E. Iron-Based Catalysts for the High-Temperature Water-Gas Shift (HTWGS) Reaction: A Review. *ACS Catal.* **2016**, *6*, 722–732. [CrossRef]
13. Ruettinger, W.; Ilinich, O.; Farrauto, R.J. A new generation of water gas shift catalysts for fuel cell applications. *J. Power Sources* **2003**, *118*, 61–65. [CrossRef]
14. Panagiotopoulou, P.; Kondarides, D.I. Effect of the nature of the support on the catalytic performance of noble metal catalysts for the water–gas shift reaction. *Catal. Today* **2008**, *112*, 49–52. [CrossRef]
15. de León, J.N.D.; Loera-Serna, S.; Zepeda, T.A.; Domínguez, D.; Pawelec, B.; Venezia, A.M.; Fuentes-Moyado, S. Noble metals supported on binary γ-Al_2O_3-α-Ga_2O_3 oxide as potential low temperature water-gas shift catalysts. *Fuel* **2020**, *266*, 117031. [CrossRef]
16. Chein, R.Y.; Lin, Y.H.; Chen, Y.C.; Chyou, Y.P.; Chung, J.N. Study on water-gas shift reaction performance using Pt-based catalysts at high temperatures. *Int. J. Hydrog. Energy* **2014**, *39*, 18854–18862. [CrossRef]
17. Castaño, M.G.; Reina, T.R.; Ivanova, S.; Centeno, M.A.; Odriozola, J.A. Pt vs. Au in water–gas shift reaction. *J. Catal.* **2014**, *314*, 1–9. [CrossRef]
18. Palma, V.; Martino, M. Pt-Re Based Catalysts for the Realization of a Single Stage Water Gas Shift Process. *Chem. Eng. Trans.* **2017**, *57*, 1657–1662. [CrossRef]
19. Sato, Y.; Terada, K.; Hasegawa, S.; Miyao, T.; Naito, S. Mechanistic study of water–gas-shift reaction over TiO_2 supported Pt–Re and Pd–Re catalysts. *Appl. Cat. A-Gen.* **2005**, *296*, 80–89. [CrossRef]
20. Carrasquillo-Flores, R.; Gallo, J.M.R.; Hahn, K.; Dumesic, J.A.; Mavrikakis, M. Density Functional Theory and Reaction Kinetics Studies of the Water–Gas Shift Reaction on Pt–Re Catalysts. *ChemCatChem* **2013**, *5*, 3690–3699. [CrossRef]
21. Iida, H.; Igarashi, A. Structure characterization of Pt-Re/TiO_2 (rutile) and Pt-Re/ZrO_2 catalysts for water gas shift reaction at low-temperature. *Appl. Cat. A-Gen.* **2006**, *303*, 192–198. [CrossRef]
22. Choung, S.Y.; Ferrandon, M.; Krause, T. Pt-Re bimetallic supported on CeO_2-ZrO_2 mixed oxides as water–gas shift catalysts. *Catal. Today* **2005**, *99*, 257–262. [CrossRef]
23. Sato, Y.; Terada, K.; Soma, Y.; Miyao, T.; Naito, S. Marked addition effect of Re upon the water gas shift reaction over TiO_2 supported Pt, Pd and Ir catalysts. *Catal. Commun.* **2006**, *7*, 91–95. [CrossRef]
24. Azzam, K.G.; Babich, I.V.; Seshan, K.; Lefferts, L. Role of Re in Pt–Re/TiO2 catalyst for water gas shift reaction: A mechanistic and kinetic study. *Appl. Cat. B-Environ.* **2008**, *80*, 129–140. [CrossRef]
25. del Villar, V.; Barrio, L.; Helmi, A.; Van Sint Annaland, M.; Gallucci, F.; Fierro, J.L.G.; Navarro, R.M. Effect of Re addition on the WGS activity and stability of Pt/CeO_2–TiO_2 catalyst for membrane reactor applications. *Catal. Today* **2016**, *268*, 95–102. [CrossRef]
26. Palma, V.; Ruocco, C.; Meloni, E.; Ricca, A. Renewable Hydrogen from Ethanol Reforming over CeO_2-SiO_2 Based Catalysts. *Catalysts* **2017**, *7*, 226. [CrossRef]
27. Cooper, A.; Davies, T.E.; Morgan, D.J.; Golunski, S.; Taylor, S.H. Influence of the Preparation Method of Ag-K/CeO_2-ZrO_2-Al_2O_3 Catalysts on Their Structure and Activity for the Simultaneous Removal of Soot and NOx. *Catalysts* **2020**, *10*, 294. [CrossRef]
28. Xu, Y.; Wang, F.; Liu, X.; Liu, Y.; Luo, M.; Teng, B.; Fan, M.; Liu, X. Resolving a Decade-Long Question of Oxygen Defects in Raman Spectra of Ceria-Based Catalysts at Atomic Level. *J. Phys. Chem. C* **2019**, *123*, 18889–18894. [CrossRef]
29. Wu, Z.; Li, M.; Howe, J.; Mayer III, H.M.; Overbury, S.H. Probing Defect Sites on CeO_2 Nanocrystals with Well-Defined Surface Planes by Raman Spectroscopy and O_2 Adsorption. *Langmuir* **2010**, *26*, 16595–16606. [CrossRef]
30. Brogan, M.S.; Dines, T.J.; Cairns, J.A. Raman spectroscopic study of the Pt-CeO_2 interaction in the Pt/Al_2O_3-CeO_2 catalyst. *J. Chem. Soc. Faraday Trans.* **1994**, *90*, 1461–1466. [CrossRef]
31. Hardcastle, F.D.; Wachs, I.E.; Horsley, J.A.; Via, G.H. The structure of surface rhenium oxide on alumina from laser raman spectroscopy and x-ray absorption near-edge spectroscopy. *J. Mol. Catal.* **1988**, *46*, 15–36. [CrossRef]
32. Rocchini, E.; Trovarelli, A.; Llorca, J.; Graham, G.W.; Weber, W.H.; Maciejewski, M.; Baiker, A. Relationships between Structural/Morphological Modifications and Oxygen Storage–Redox Behavior of Silica-Doped Ceria. *J. Catal.* **2000**, *194*, 461–478. [CrossRef]

33. Prins, R. Hydrogen spillover. Facts and fiction. *Chem. Rev.* **2012**, *112*, 2714–2738. [CrossRef]
34. Jeong, D.-W.; Potdar, H.S.; Shim, J.-O.; Jang, W.-J.; Roh, H.-S. H_2 production from a single stage water-gas shift reaction over Pt/CeO_2, Pt/ZrO_2, and Pt/$Ce_{(1-x)}Zr_{(x)}O_2$ catalysts. *Int. J. Hydrog. Energy* **2013**, *38*, 4502–4507. [CrossRef]
35. Radhakrishnan, R.; Willigan, R.R.; Dardas, Z.; Vanderspurt, T.H. Water gas shift activity and kinetics of Pt/Re catalysts supported on ceria-zirconia oxides. *Appl. Cat. B-Environ.* **2006**, *66*, 23–28. [CrossRef]
36. Dabalà, M.; Armelao, L.; Buchberger, A.; Calliari, I. Cerium-based conversion layers on aluminum alloys. *Appl. Surf. Sci.* **2001**, *172*, 312–322. [CrossRef]
37. Palma, V.; Goodall, R.; Thompson, A.; Ruocco, C.; Renda, S.; Leach, R.; Martino, M. Ceria-coated Replicated Aluminium Sponges as Catalysts for the CO-Water Gas Shift Process. *Int J. Hydrog. Energy* **2020**, in press. [CrossRef]
38. Cristiani, C.; Visconti, C.G.; Finocchio, E.; Gallo Stampino, P.; Forzatti, P. Towards the rationalization of the washcoating process conditions. *Catal. Today* **2009**, *147*, S24–S29. [CrossRef]
39. Lia, P.; Chen, X.; Li, Y.; Schwank, J.W. A review on oxygen storage capacity of CeO_2-based materials: Influence factors, measurement techniques, and applications in reactions related to catalytic automotive emissions control. *Catal. Today* **2019**, *327*, 90–115. [CrossRef]
40. Mosca, L.; Medrano Jimenez, J.A.; Assefa Wassie, S.; Gallucci, F.; Palo, E.; Colozzi, M.; Taraschi, S.; Galdieri, G. Process design for green hydrogen production. *Int. J. Hydrog. Energy* **2020**, *45*, 7266–7277. [CrossRef]
41. Palma, V.; Martino, M.; Truda, L. Nano-CeO_2 Coating on Aluminum Foam Carriers for Structured Catalysts Preparation. *Chem. Eng. Trans.* **2019**, *73*, 127–132. [CrossRef]
42. Ratnasamy, C.; Wagner, J. Water Gas Shift Catalysis. *Catal. Rev.-Sci. Eng.* **2009**, *51*, 325–440. [CrossRef]

Article

Enhanced Electrocatalytic Activity of Stainless Steel Substrate by Nickel Sulfides for Efficient Hydrogen Evolution

Jong-Sang Youn [1], Sangmin Jeong [2], Inhwan Oh [2], Sunyoung Park [2], Hien Duy Mai [2] and Ki-Joon Jeon [2,*]

1 Department of Environmental Engineering, The Catholic University of Korea, Bucheon 14662, Korea; jsyoun@catholic.ac.kr
2 Department of Environmental Engineering, Inha University, Incheon 22212, Korea; smjeong3268@gmail.com (S.J.); oh910611@naver.com (I.O.); 12171148@inha.edu (S.P.); hien.fasdas@gmail.com (H.D.M.)
* Correspondence: kjjeon@inha.ac.kr; Tel.: +82-32-860-7509

Received: 15 October 2020; Accepted: 2 November 2020; Published: 3 November 2020

Abstract: Water splitting is one of the efficient ways to produce hydrogen with zero carbon dioxide emission. Thus far, Pt has been regarded as a highly reactive catalyst for the hydrogen evolution reaction (HER); however, the high cost and rarity of Pt significantly hinder its commercial use. Herein, we successfully developed an HER catalyst composed of NiS_x (x = 1 or 2) on stainless steel (NiS_x/SUS) using electrodeposition and sulfurization techniques. Notably, the electrochemical active surface area(ECSA) of NiS_x/SUS was improved more than two orders of magnitude, resulting in a considerable improvement in the electrochemical charge transfer and HER activity in comparison with stainless steel (SUS). The long-term HER examination by linear scan voltammetry (LSV) confirmed that NiS_x/SUS was stable up to 2000 cycles.

Keywords: sulfurization; NiS-NiS_2; stainless steel 304; hydrogen evolution

---

1. Introduction

Hydrogen production via electrochemical water splitting is regarded as an ideal energy source that can potentially substitute conventional fossil fuel because of its large energy density and zero carbon dioxide emission [1–3]. Developing efficient and durable electrocatalysts with cost- effectiveness is highly desirable for the practical usage of water splitting devices. Pt-based catalysts are highly reactive toward the hydrogen evolution reaction (HER) due to the optimal binding energy with adsorbed hydrogen species formed during the HER process [4]; however, the high cost and rarity of Pt significantly hinder its commercial use. Therefore, tremendous effort has been devoted to develop earth-abundant materials that can substitute Pt in HER [5,6]. Among those, TMS (transition metal sulfides), such as nickel sulfides, appear to be a promising class of HER electrocatalysts owing to their earth abundance, remarkable HER performance, and long-term operational stability in both acidic and alkaline solutions [7–12]. For example, T. F. Hung et al. reported that the nickel sulfide nanostructure showed high electrochemical performance in HER and supercapacitors [13,14]. Nanostructured nickel sulfide was also superior to the bulk counterpart as reported by P. Liu et al. [14]. Although a number of synthetic methods for nickel sulfide (e.g., solvothermal and hydrothermal [7,9], and electrospinning [13] have been reported, it is still very important to establish an effective and simple approach to fabricate the nickel sulfide-based electrodes for HER application.

Electrode substrates are equally important in terms of HER efficiency as for the catalytically active phase deposited on the electrode surface since the electrochemical performance and operational stability of HER electrodes can be substantially affected by the substrates used [15,16]. In this regard, stainless steel (SUS) has potential for efficient hydrogen evolution as a substrate because of its mechanical strength, corrosion resistance, cost-effectiveness, and good electrical conductivity compared to other substrates [17,18]. However, growing electrochemically active materials on the surface of SUS remains a challenging task. This is mainly because SUS is composed of many kinds of metals such as iron, nickel, molybdenum, chromium, and magnesium, and these metals can inevitably diffuse to the surface at high pressure and temperature [19]. This prevents researchers from controlling the material composition and in understanding the composition-HER performance correlation. Further, the slippery surface of SUS and the generation of hydrogen bubbles during HER result in the undesired detachment of catalysts from the SUS and the electrode eventually to lose its reactivity [20].

Herein, we directly fabricated nickel sulfide (NiS_x) nanostructures on SUS (NiS_x/SUS) with a facile method to enhance the stability of electrocatalysts on SUS and improve the electrocatalytic activity. Nickel electrodeposition was carried out to introduce a nickel overlay on the surface of SUS. The following sulfurization at relatively low temperature, while preventing the metal diffusion from SUS to the surface catalytic phase, was sufficient for the direct growth of NiS_x on the SUS. The reported synthetic approach is important to prevent the undesired detachment of NiS_x from the SUS surface during the HER process. The crystallinity, chemical composition and oxidation state, and morphologies of NiS_x/SUS were fully characterized by XRD (X-ray diffraction), XPS (X-ray photoelectron spectroscopy), and SEM (scanning electron microscopy). HER examination shows that the NiS_x/SUS led to a substantial improvement in the electrochemical activity and long-term stability in comparison with SUS.

2. Results

2.1. Characteristics of the NiS_x/SUS Electrode

Figure 1 depicts a schematic illustration of the electrode preparation via two steps. The first step, the nickel electrodeposition of SUS, was optimized at −5 mA cm^{-2} for 30 min in 1 M nickel sulfate solution for a uniform covering of nickel on SUS. The sulfurization (step 2) involved the chemical reaction between vaporized sulfur and nickel on SUS to generate NiS_x/SUS. The sulfurization temperature was set at 300 °C because lower or higher temperatures only resulted in inefficient sulfurization or electrode damage, respectively. This synthetic approach is very facile and can be easily scaled up. For comparison, SUS was also sulfurized under the same condition (see more details in the experimental section). After the sulfurization, the colors of the electrodes changed to black (NiS_x/SUS) and grey (S-SUS).

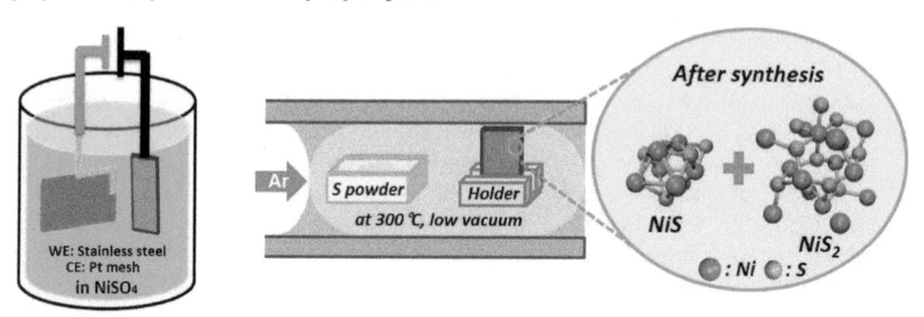

Figure 1. Schematic illustration of two-step sample preparation. Step (1) nickel electrodeposition in 1 M nickel sulfate solution and step (2) sulfurization for NiS_x/SUS at 300 °C for 30 min.

The morphological properties of the prepared samples are revealed in Figure 2. The morphology of S-SUS sparsely shows nanorod shapes lying horizontally on the SUS (Figure 2a,c). On the other hand, NiS_x/SUS was densely covered by the NiS_x nanoparticles with mean size of approx. 100 nm and thickness of approx. 24 μm. There are many more exposed active sites than for that of S-SUS (Figure 2b,d, Figure S1). These morphologies from sulfurization are consistent with previous studies [21].

Figure 2. Typical SEM images of S-SUS (**a**,**c**) and NiS_x/SUS (**b**,**d**) at scale of 1 μm and 500 nm, respectively.

The crystallinity of the prepared electrodes was investigated using X-ray diffraction (XRD) analysis performed in the 2θ range from 20 to 90°. Figure 3 shows the XRD results of SUS, S-SUS, and NiS_x/SUS. SUS shows normal FCC (face centered cubic) crystal structure of (111), (200), and (220) at 43.5, 50.7, and 74.5° respectively, indicating consistency with previous studies [22], and there is no significant difference from XRD results between SUS and S-SUS, suggesting that SUS is relatively passive toward sulfur vapor at low temperatures. After the sulfurization of nickel on SUS, the evolution of a new set of diffraction peaks can be assigned to NiS ((100), (101), (102), and (110) at 30.1, 34.6, 45.7, and 53.5°, respectively) [7] and NiS_2 ((200), (210), (211), (220), and (311) at 31.4, 35.3, 38.8, 45, and 53.4°, respectively) [14]. The peak intensity corresponding to those characteristic of SUS is seen to lessen to some extent, implying that the surface of SUS is covered with a thick layer of NiS_x.

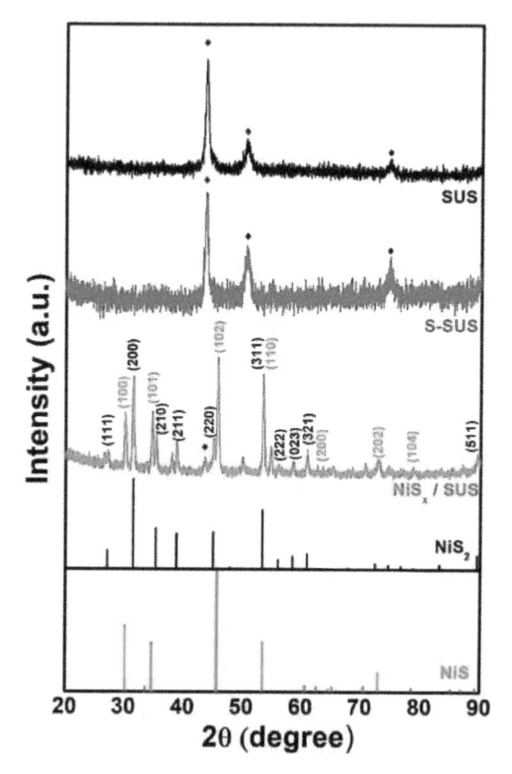

Figure 3. X-ray diffraction analysis of SUS (black), S-SUS (blue), and NiS_x/SUS (red) with JCPDS No. 01-089-7142 (NiS_2) and 03-065-3419 (NiS).

In order to investigate the chemical composition and binding states of the prepared electrodes, XPS analysis was carried out as shown in Figure 4. The XPS survey spectra of SUS, which mainly consists of Fe, Ni, Cr, and Mn, shows high intensity of O1s and C1s and very low intensity of Fe 2p indicating metal oxide film on SUS (Figure 4a) [23]. In case of S-SUS, because the intensity of Fe 2p and Cr 2p was too small to quantify the metal sulfides, S 2p was analyzed by the four peaks at 163.98 eV, 163.08 eV, 162.3 eV, and 161.2 eV (S_0, S_n^{2-}, S_2^{2-}, and S^{2-}) indicating the metal-S bonds (Figure S2) [24–26]. Therefore, it was elucidated that small amounts of metals were combined with sulfur on the surface of S-SUS. The survey spectra of NiS_x/SUS shows higher intensity of Ni 2p and S 2p compared to others. The low intensity of Fe 2p and Mn 2p is ascribed to minor metal diffusion during the synthesis. In order to confirm the binding states of NiS_x, the XPS result of NiS_x/SUS was presented by Ni 2p and S 2p deconvolution. The Ni 2p spectrum of the spin-orbit doublet was deconvoluted into six well-resolved peaks. As shown in Figure 4b, two major peaks are observed at 854.2 eV and 872.6 eV designated to Ni^{2+}, while the two peaks at 856.1 eV and 875.9 eV are attributed to Ni^{3+} in the Ni $2p_{1/2}$ and Ni $2p_{3/2}$ [27]. The elemental contents of SUS, S-SUS, and NiS_x/SUS are shown in Table S1. From this result, it was confirmed that the surface of the electrode was covered by NiS_x and no metallic nickel remained. The high resolution scan of S 2p is exhibited in Figure 4c. The binding states of S^{2-} and S_2^{2-} are observed at 161.2 eV and 162.3 eV respectively, which is indicative of NiS and NiS_2 [24]. These results are highly consistent with the XRD results of NiS_x/SUS, implying that the sulfurization was successful in synthesizing the hybrids of NiS and NiS_2 without metal diffusion from SUS during the synthesis.

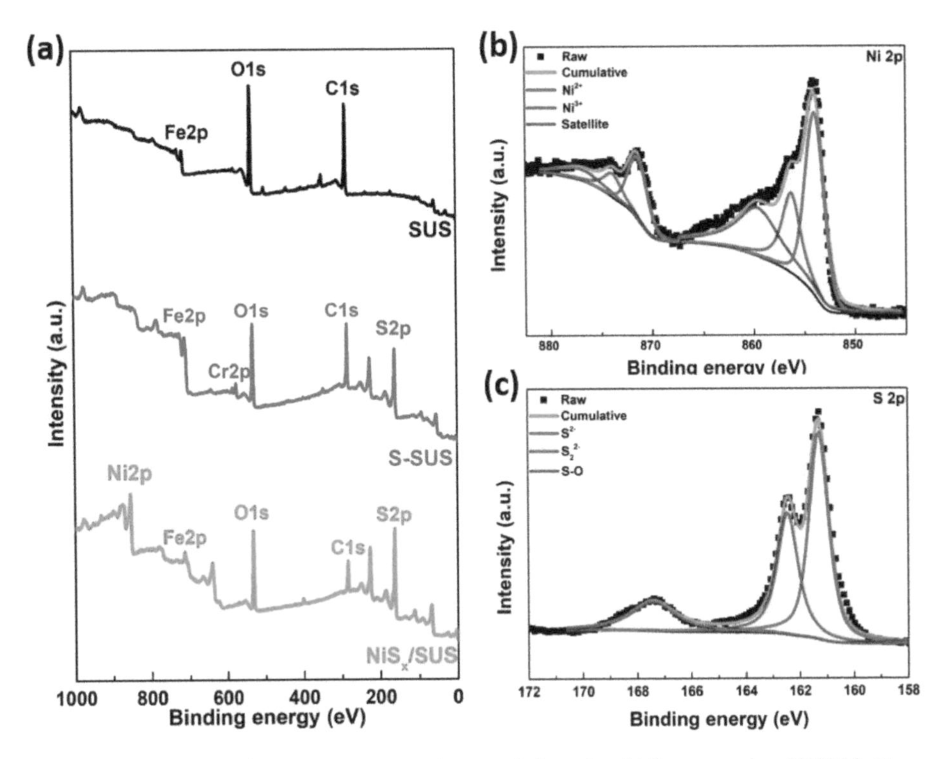

Figure 4. X-ray photoelectron spectroscopy of prepared electrodes. (**a**) Survey spectra of SUS (black), S-SUS (blue), and NiS_x/SUS (red); (**b**) high resolution XPS scans of Ni 2p; and (**c**) S 2p from NiS_x/SUS.

2.2. Electrochemical Results and Analysis

To investigate the HER performance of the prepared electrodes, a conventional three electrode experiment was performed in 1 M KOH. NiS_x/SUS was used as a working electrode, Pt and the saturated calomel electrode (SCE) were used as a counter electrode and reference electrode respectively. In order to see how much the HER performance was improved, SUS and S-SUS were used for the working electrodes as well. The polarization curve of NiS_x/SUS was considerably improved compared to SUS and S-SUS. For specific comparison we observed the η_{10} (overpotential at −10 mA cm^{-2}, mV vs reversible hydrogen electrode (RHE)) of each electrode (Figure 5a). The η_{10} of NiS_x/SUS was 258 mV, which is much smaller than those of S-SUS (494 mV) and SUS (457 mV). Tafel slope was used to understand the HER kinetics (Figure 5b,c). In general, the mechanism of HER in alkaline medium involves a two-step process [28]. The first step is the Volmer reaction ($H_2O + e^- \rightarrow H_{ads} + OH^-$) associated with the hydrogen adsorption on the electrocatalysts. The second step is the Heyrovsky ($H_2O + e^- + H_{ads} \rightarrow H_2 + OH^-$) or Tafel ($H_{ads} + H_{ads} \rightarrow H_2$) reaction which explain the H_2 dissociation from the electrocatalysts. Figure 5b shows that the Tafel slope of NiS_x/SUS was determined as 100 mV dec^{-1} which shows the improved electrocatalytic activity of NiS_x from the SUS. The Tafel slope reveals that the HER mechanism of NiS_x/SUS can be elucidated by the Volmer–Heyrovsky reaction as shown in Figure 5c. SUS and S-SUS show low Tafel slopes of 170 and 154 mV dec^{-1}, respectively, indicating poor electrocatalytic activity. As shown in the linear scan voltammetry (LSV) and Tafel plot, the HER performance of S-SUS was barely improved from the SUS even after sulfurization, which means the enhanced HER performance of NiS_x/SUS mainly stems from the NiS_x on the SUS. To better understand the electrocatalytic activities of the prepared electrode, double layer capacitances (C_{dl}) and electrochemical impedances were measured by cyclic voltammetry (CV) and electrochemical impedance spectroscopy (EIS) experiments (Figure 5d,e). C_{dl} has a linear relationship with ECSA, which is correlated with the electrochemical performance of a given electrode (Figure 5d). Figure S3 shows the CV result obtained in the non-faradaic region at different scan rates (20 to 260 mV s^{-1}) for NiS_x/SUS, S-SUS, and SUS. S-SUS has 0.2 mF cm^{-2} for C_{dl} which is ten times higher than that of SUS (0.018 mF cm^{-2}) which is indicative of the similar trend with the LSV graph (Figure 5a). NiS_x/SUS show the highest C_{dl} value (2.63 mF cm^{-2}) which is 140 times higher than SUS and 13 times higher than S-SUS. For analysis of EIS, the Nyquist plot was used as shown in Figure 5e. All electrodes show the semicircles of the Nyquist plot and have similar R_s (solution resistance) around ~3 Ω in 1 M KOH. On the other hand, R_{ct} of NiS_x/SUS has the smallest value of 11.1 Ω, however, SUS and S-SUS show very high impedances (231.5 and 186.1 Ω). It was confirmed that the interfacial charge transfer reaction during the HER process occurs much more rapidly on the NiS_x/SUS than on the other electrodes.

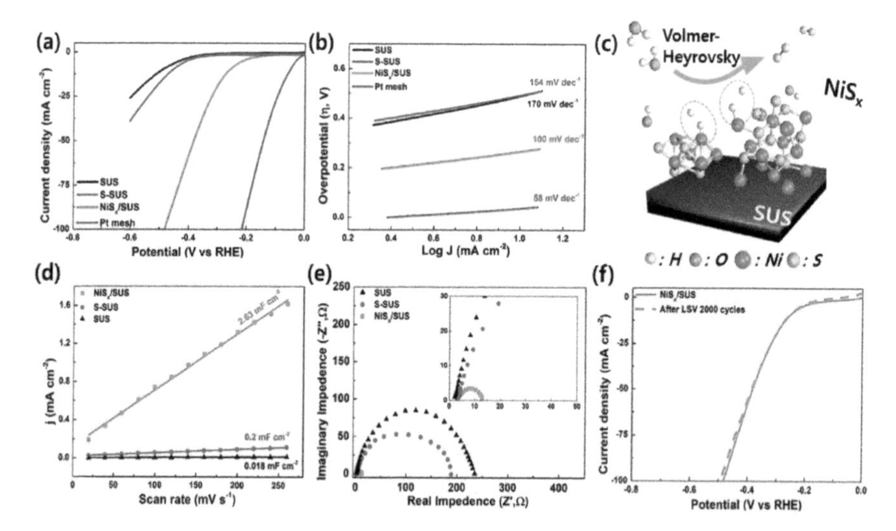

Figure 5. Hydrogen evolution reaction (HER) performances of prepared electrodes. (**a**) Linear scan voltammetry (LSV) in the range of 0 to −0.6 V vs reversible hydrogen electrode (RHE) in 1 M KOH; (**b**) Tafel plots; (**c**) illustration for hydrogen evolution reaction of NiS_x/SUS; (**d**) increased current density in non-faradaic region depending on the increasing scan rates for double layer capacitances of electrodes; (**e**) Nyquist plot at −0.4 V vs RHE; and (**f**) LSV 2000 cycles of NiS_x/SUS for 33 h.

In order to evaluate the electrochemical stability of NiS_x/SUS, the electrode was subjected to 2000 cycles of LSV (0 to −0.6 V vs RHE at 10 mV s^{-1}) for 33 h. As seen in Figure 5f, there is no significant difference in the LSV before and even after 2000 cycles, indicating the excellent long-term stability of the NiS_x/SUS. In addition to this, XPS results revealed that the NiS_x on NiS_x/SUS is stable after 2000 cycles of LSV in alkaline condition (Figure S4). These results imply high ECSA and low R_{ct} of NiS_x high corrosion resistance of SUS not only enhances the stability but also improves the HER performance in alkaline solution. Because SUS has slippery surface, electrodeposition was used for attachment of nickel on SUS for the stability of the electrode. To overcome the electrochemically poor activity of SUS, the ECSA of the electrode was considerably increased and R_{ct} was decreased by sulfurization of nickel on the SUS.

3. Materials and Methods

3.1. Preparation of Electrodes

Commercial stainless steel 304 (SUS, 10 mm × 20 mm × 250 μm) was used as a substrate for HER. DI water, ethanol, and acetone were used to clean the SUS subsequently. Nickel electrodeposition was carried out at −5 mA cm^{-2} for 30 min in 1 M nickel sulfate solution (Kanto, Tokyo, Japan). The deposited Ni on SUS was placed in the electric furnace for sulfurization. Ar gas (100 sccm) and sulfur powder (300 mg, Sigma Aldrich, St. Louis, MO, USA) were used as a carrier gas and an S precursor respectively. Sulfurization was carried out at 300 °C for 60 min in a low vacuum environment and the cooling temperature was controlled at a rate of 20 °C·min^{-1}. After synthesis, the electrode (denoted as NiS_x/SUS) was washed with acetone and dried at room temperature for 10 min. For comparison, SUS was sulfurized and dried, without electrochemically deposited nickel (denoted as S-SUS).

3.2. Characterizations

Field emission-scanning electron microscopy (FE-SEM; Hitachi, S-4300, Tokyo, Japan) was used to analyze the morphologies of NiS_x/SUS and S-SUS at the scale of 1 μm and 500 nm, respectively. The chemical composition and binding state of the electrode surface were confirmed through an XPS (Thermo Fisher Scientific Co, Waltham MA, USA) which had a micro-focused Al-Kα source. XRD (X'Pert PRO MRD, Phillips, Eindhoven, The Netherlands) analysis was carried out in the 2θ range of 20 to 90°.

3.3. Electrochemical Measurements

HER performance was evaluated by using potentiostat/galvanostat (Vertex, IVIUM Technology, Eindhoven, Netherlands) for a conventional three electrode system. To see the hydrogen evolution from alkaline solution, 1 M KOH was used as the electrolyte. The prepared electrodes, Pt and a saturated calomel electrode (SCE, Hg/Hg_2Cl_2) were used as working electrodes, a counter electrode, and a reference electrode respectively. LSV data were acquired in the range of 0 to −0.8 V vs RHE (versus reversible hydrogen electrode) at a scan rate of 10 mV s^{-1}, and CV experiments were carried at scan rates for every 20 mV s^{-1} increments up to a scan rate of 300 mV s^{-1} in the non-faradaic regions. The Nyquist plot was used for the analysis of EIS at the −0.4 V vs RHE in the frequency from 100k to 0.5 Hz. For a stability test, LSV 2000 cycles were carried out in the range of 0 to −0.6 V vs RHE at a scan rate of 10 mV s^{-1} for 33 h.

4. Conclusions

In this research, we overcame the limitations of SUS as a substrate and directly synthesized nickel sulfides through electrodeposition and sulfurization. The reported synthetic paradigm is very facile and straightforward, rendering the fabrication of the water splitting electrode at a large scale. Importantly, the NiS_x/SUS shows substantially improved HER kinetic performance in alkaline solution compared with SUS and S-SUS. The enhanced HER activity and long-term operational durability mainly stem from the increased ECSA of NiS_x/SUS and the high corrosion resistance of SUS. These results indicate that SUS substrate can be used for efficient HER with various electrocatalysts and nickel sulfides also can be synthesized on other substrates with the same methods.

Supplementary Materials: The following are available online at http://www.mdpi.com/2073-4344/10/11/1274/s1, contains supporting XPS and CV data. Figure S1: The thickness of the SEM image of NiS_x/SUS. Figure S2: Deconvolution of the XPS peaks of S-SUS in the S 2p region. Table S1: Surface chemical analysis of samples by X-ray photoelectron spectroscopy (XPS). Figure S3: CV of the prepared electrodes in the non-faradaic regions at different scan rates (20 mV s^{-1} to 260 mV s^{-1}). Figure S4: XPS of NiS_x/SUS after 2000 cycles in 1M KOH.

Author Contributions: K.-J.J. proposed the research direction and guided the project. J.-S.Y. carried out the conceptualization, methodology, data curation, Writing—original draft. S.J. carried out the conceptualization, methodology, data curation, writing—review and editing. I.O. and S.P. carried out the methodology, writing—review and editing. H.D.M. carried out the data curation and writing—review and editing. All authors have read and agreed to the published version of the manuscript.

Funding: This work was supported by INHA UNIVERSITY Research Grant.

Conflicts of Interest: The authors declare no conflict of interest.

References

1. Shabanian, S.R.; Edrisi, S.; Khoram, F.V. Prediction and optimization of hydrogen yield and energy conversion efficiency in a non-catalytic filtration combustion reactor for jet A and butanol fuels. *Korean J. Chem. Eng.* **2017**, *34*, 2188–2197. [CrossRef]
2. Loipersböck, J.; Lenzi, M.; Rauch, R.; Hofbauer, H. Hydrogen production from biomass: The behavior of impurities over a CO shift unit and a biodiesel scrubber used as a gas treatment stage. *Korean J. Chem. Eng.* **2017**, *34*, 2198–2203. [CrossRef]
3. Kim, S.; Song, J.; Lim, H. Conceptual feasibility studies of a COX-free hydrogen production from ammonia decomposition in a membrane reactor for PEM fuel cells. *Korean J. Chem. Eng.* **2018**, *35*, 1509–1516. [CrossRef]
4. Roger, I.; Shipman, M.A.; Symes, M.D. Earth-abundant catalysts for electrochemical and photoelectrochemical water splitting. *Nat. Rev. Chem.* **2017**, *1*, 3. [CrossRef]
5. Wang, J.; Liu, J.; Zhang, B.; Ji, X.; Xu, K.; Chen, C.; Miao, L.; Jiang, J. The mechanism of hydrogen adsorption on transition metal dichalcogenides as hydrogen evolution reaction catalyst. *Phys. Chem. Chem. Phys.* **2017**, *19*, 10125–10132. [CrossRef]
6. Zhang, G.; Liu, H.; Qu, J.; Li, J. Two-dimensional layered MoS2: Rational design, properties and electrochemical applications. *Energy Environ. Sci.* **2016**, *9*, 1190–1209. [CrossRef]
7. Wei, W.; Mi, L.; Gao, Y.; Zheng, Z.; Chen, W.; Guan, X. Partial ion-exchange of nickel-sulfide-derived electrodes for high performance supercapacitors. *Chem. Mater.* **2014**, *26*, 3418–3426. [CrossRef]
8. Wang, Y.; Pang, H. Nickel-Based Sulfide Materials for Batteries. *ChemistrySelect* **2018**, *3*, 12967–12986. [CrossRef]
9. Wang, P.; Zhang, X.; Zhang, J.; Wan, S.; Guo, S.; Lu, G.; Yao, J.; Huang, X. Precise tuning in platinum-nickel/nickel sulfide interface nanowires for synergistic hydrogen evolution catalysis. *Nat. Commun.* **2017**, *8*, 1–9. [CrossRef]
10. Jayaramulu, K.; Masa, J.; Tomanec, O.; Peeters, D.; Ranc, V.; Schneemann, A.; Zboril, R.; Schuhmann, W.; Fischer, R.A. Nanoporous Nitrogen-Doped Graphene Oxide/Nickel Sulfide Composite Sheets Derived from a Metal-Organic Framework as an Efficient Electrocatalyst for Hydrogen and Oxygen Evolution. *Adv. Funct. Mater.* **2017**, *27*, 1–10. [CrossRef]
11. Wu, Y.; Gao, Y.; He, H.; Zhang, P. Novel electrocatalyst of nickel sulfide boron coating for hydrogen evolution reaction in alkaline solution. *Appl. Surf. Sci.* **2019**, *480*, 689–696. [CrossRef]
12. Palapati, N.K.R.; Demir, M.; Harris, C.T.; Subramanian, A.; Gupta, R.B. Enhancing the electronic conductivity of Lignin-sourced, sub-micron carbon particles. In Proceedings of the 2015 IEEE Nanotechnology Materials and Devices Conference (NMDC), Anchorage, AK, USA, 13–16 September 2015; pp. 1–2. [CrossRef]
13. Hung, T.F.; Yin, Z.W.; Betzler, S.B.; Zheng, W.; Yang, J.; Zheng, H. Nickel sulfide nanostructures prepared by laser irradiation for efficient electrocatalytic hydrogen evolution reaction and supercapacitors. *Chem. Eng. J.* **2019**, *367*, 115–122. [CrossRef]
14. Liu, P.; Li, J.; Lu, Y.; Xiang, B. Facile synthesis of NiS2 nanowires and its efficient electrocatalytic performance for hydrogen evolution reaction. *Int. J. Hydrogen Energy* **2018**, *43*, 72–77. [CrossRef]
15. TONG, S.S.; WANG, X.J.; LI, Q.C.; HAN, X.J. Progress on Electrocatalysts of Hydrogen Evolution Reaction Based on Carbon Fiber Materials. *Chinese J. Anal. Chem.* **2016**, *44*, 1447–1457. [CrossRef]
16. Benck, J.D.; Pinaud, B.A.; Gorlin, Y.; Jaramillo, T.F. Substrate selection for fundamental studies of electrocatalysts and photoelectrodes: Inert potential windows in acidic, neutral, and basic electrolyte. *PLoS ONE* **2014**, *9*. [CrossRef] [PubMed]
17. Pu, N.W.; Shi, G.N.; Liu, Y.M.; Sun, X.; Chang, J.K.; Sun, C.L.; Der Ger, M.; Chen, C.Y.; Wang, P.C.; Peng, Y.Y.; et al. Graphene grown on stainless steel as a high-performance and ecofriendly anti-corrosion coating for polymer electrolyte membrane fuel cell bipolar plates. *J. Power Source* **2015**, *282*, 248–256. [CrossRef]

18. Kim, M.; Ha, J.; Shin, N.; Kim, Y.T.; Choi, J. Self-activated anodic nanoporous stainless steel electrocatalysts with high durability for the hydrogen evolution reaction. *Electrochim. Acta* **2020**, *364*, 137315. [CrossRef]
19. Kovendhan, M.; Kang, H.; Youn, J.S.; Cho, H.; Jeon, K.-J. Alternative cost-effective electrodes for hydrogen production in saline water condition. *Int. J. Hydrogen Energy* **2018**, *44*, 5090–5098. [CrossRef]
20. Kovendhan, M.; Kang, H.; Jeong, S.; Youn, J.S.; Oh, I.; Park, Y.K.; Jeon, K.J. Study of stainless steel electrodes after electrochemical analysis in sea water condition. *Environ. Res.* **2019**, *173*, 549–555. [CrossRef]
21. Zhu, W.; Yue, X.; Zhang, W.; Yu, S.; Zhang, Y.; Wang, J.; Wang, J. Nickel sulfide microsphere film on Ni foam as an efficient bifunctional electrocatalyst for overall water splitting. *Chem. Commun.* **2016**, *52*, 1486–1489. [CrossRef] [PubMed]
22. Quan, C.; He, Y. Properties of nanocrystalline Cr coatings prepared by cathode plasma electrolytic deposition from trivalent chromium electrolyte. *Surf. Coatings Technol.* **2015**, *269*, 319–323. [CrossRef]
23. Jung, R.H.; Tsuchiya, H.; Fujimoto, S. XPS characterization of passive films formed on Type 304 stainless steel in humid atmosphere. *Corros. Sci.* **2012**, *58*, 62–68. [CrossRef]
24. Limaye, M.V.; Chen, S.C.; Lee, C.Y.; Chen, L.Y.; Singh, S.B.; Shao, Y.C.; Wang, Y.F.; Hsieh, S.H.; Hsueh, H.C.; Chiou, J.W.; et al. Understanding of sub-band gap absorption of femtosecond-laser sulfur hyperdoped silicon using synchrotron-based techniques. *Sci. Rep.* **2015**, *5*, 1–12. [CrossRef]
25. Demir, M.; Farghaly, A.A.; Decuir, M.J.; Collinson, M.M.; Gupta, R.B. Supercapacitance and oxygen reduction characteristics of sulfur self-doped micro/mesoporous bio-carbon derived from lignin. *Mater. Chem. Phys.* **2018**, *216*, 508–516. [CrossRef]
26. Altinci, O.C.; Demir, M. Beyond Conventional Activating Methods, a Green Approach for the Synthesis of Biocarbon and Its Supercapacitor Electrode Performance. *Energy Fuels* **2020**, *34*, 7658–7665. [CrossRef]
27. Nis, S.D.; Catalysis, E. Solvothermally Doping NiS2 Nanoparticles on Carbon with Ferric Ions for E ffi cient Oxygen Evolution Catalysis. *Catalysts* **2019**, *9*, 458. [CrossRef]
28. Tian, X.; Zhao, P.; Sheng, W. Hydrogen Evolution and Oxidation: Mechanistic Studies and Material Advances. *Adv. Mater.* **2019**, *31*, 1–7. [CrossRef] [PubMed]

Review

Recent Advances in Noble Metal Catalysts for Hydrogen Production from Ammonia Borane

Mengmeng Liu [1,2], Liu Zhou [1,2], Xianjin Luo [3], Chao Wan [1,2,3,4,*] and Lixin Xu [1,2,3,*]

1 Hexian Chemical Industrial Development Institute, Engineering Research Institute, School of Chemistry and Chemical Engineering, Anhui University of Technology, Ma'anshan 243002, China; lmm0530@outlook.com (M.L.); iLiu97@hotmail.com (L.Z.)
2 Ahut Chemical Science & Technology Co., Ltd., Ma'anshan 243002, China
3 Anhui Haide Chemical Technology Co., Ltd., Ma'anshan 243002, China; AHUT.XJLuo@outlook.com
4 College of Chemical and Biological Engineering, Zhejiang University, Hangzhou 310027, China
* Correspondence: wanchao@ahut.edu.cn (C.W.); lxxu@ahut.edu.cn (L.X.); Tel.: +86-555-231-1807 (C.W.); +86-555-231-1521 (L.X.)

Received: 29 June 2020; Accepted: 13 July 2020; Published: 15 July 2020

Abstract: Interest in chemical hydrogen storage has increased, because the supply of fossil fuels are limited and the harmful effects of burning fossil fuels on the environment have become a focus of public concern. Hydrogen, as one of the energy carriers, is useful for the sustainable development. However, it is widely known that controlled storage and release of hydrogen are the biggest barriers in large-scale application of hydrogen energy. Ammonia borane (NH_3BH_3, AB) is deemed as one of the most promising hydrogen storage candidates on account of its high hydrogen to mass ratio and environmental benignity. Development of efficient catalysts to further improve the properties of chemical kinetics in the dehydrogenation of AB under appropriate conditions is of importance for the practical application of this system. In previous studies, a variety of noble metal catalysts and their supported metal catalysts (Pt, Pd, Au, Rh, etc.) have presented great properties in decomposing the chemical hydride to generate hydrogen, thus, promoting their application in dehydrogenation of AB is urgent. We analyzed the hydrolysis of AB from the mechanism of hydrogen release reaction to understand more deeply. Based on these characteristics, we aimed to summarize recent advances in the development of noble metal catalysts, which had excellent activity and stability for AB dehydrogenation, with prospect towards realization of efficient noble metal catalysts.

Keywords: ammonia borane; noble metal catalysts; chemical hydrogen; hydrogen production

1. Introduction

Energy is considered as a substantial material basis for survival and development of human society [1–3]. The rapid growth of world population and continuous improvement of living standard provokes excessive consumption of traditional fossil fuels, especially oil, coal and natural gas, as a matter of fact that in the future, these traditional resources will be inadequate to support the constantly progress of human civilization. Apart from the fact that fossil fuel resources are in short supply, many problems were exposed because of fossil fuel utilization [4,5], such as air pollution, water pollution and excessive carbon dioxide emissions leading to global warming. From this point of view, these problems have reached a serious point nowadays, thus, the exploitation of clean and renewable energy is receiving more attention.

Hydrogen, which is clean, effective and environmental benign nature [6–9], is regarded as an ideal alternative to fossil energy and the prospective secondary green energy in the 21st century. However, storage and transportation of large quantities of hydrogen are the difficult challenges [10–12], which limit its advance and application. Hydrogen storage materials (mainly including chemical hydrides [13], adsorption hydrogen storage materials [14], metal hydrides [15], etc.) have been widely concerned and investigated [16–18]. In recent years, chemical hydride (NH_3BH_3, N_2H_4, etc.), due to high hydrogen storage density, volume, have been proved to be a practical hydrogen source. Among them, ammonia borane is one of the most employed chemical hydrides, with providing safe and efficient alternative [19–21]. Ammonia borane (NH_3BH_3, AB), has a high capacity of hydrogen (19.6 wt%) [22,23], low molecular mass (30.7 g/mol) [24,25] and high solubility, which is a stable solid at normal atmospheric temperature, making it attractive candidate for hydrogen storage application. Moreover, AB with low condition of hydrogen production and various forms of decomposition has attracted intensive attention to explore.

The hydrolysis rate of AB is closely related to the selected catalysts [26]. The highly dispersed monometallic particles can improve the sites to achieve the effect of rapid hydrogen release. So far, more and more metal catalysts have been produced to accelerate the hydrolysis of AB. In this review, the progress of noble metal catalysts for the hydrolysis of AB was reviewed. At the same time, we introduced three kinds of decomposition methods of AB, including pyrolysis, alcoholysis and hydrolysis. Hydrolysis reaction is more widely used in the practical application for hydrogen production from AB. The development of new advance in single noble metal catalysts and composite catalysts of noble metals and non-precious metals were discussed, and the influence of using carbon, graphene, carbon nanotubes, silica, ceria and titanium supported catalysts on the hydrolysis of AB were explored. Finally, the inspiration for the future development of catalysts was given in this field.

2. The Processes of Hydrogen Production on NH_3BH_3 (AB)

2.1. The Methods of Producing Hydrogen from NH_3BH_3(AB)

Through research and exploration, there are three main ways to release hydrogen from AB, pyrolysis, alcoholysis and hydrolysis, respectively [27–34].

The pyrolysis process of solid phase thermal decomposition of AB [35–37] is shown in the following table (Table 1). Each reaction releases about one-third of the content, producing polyaminoborane, polyiminoborane, boron amide and other substances. When the temperature reaches 110 °C, AB begins to release its first equivalent of hydrogen, about 6.5 wt%. Intramolecular polymerization takes place at around 125 °C. When the temperature reaches 150 °C, the reaction releases a second amount of hydrogen. Finally, BN compounds are formed when the reaction temperature exceeds 500 °C [38]. In all, only the first and second processes are considered throughout the reaction, producing about 13% by weight of hydrogen. It is easy to produce borane, ammonia, bororazine and other harmful substances at high temperature, which is not conducive to environmental protection and practice application [39,40]. Contemporary, pyrolysis reaction as high temperature, high energy consumption and slow dehydrogenation power, is not suitable for mass manufacture.

Table 1. Ammonia borane decomposition of hydrogen.

Thermal Decomposition Step	Chemical Equation	Processes	Ref.
The first step (110 °C)	$NH_3BH_3 \rightarrow NH_2BH_2 + H_2$	The first yield of hydrogen	[35]
The second step (125 °C)	$nNH_2BH_2 \rightarrow (NH_2BH_2)_n$	Intramolecular polymerization	[36]
The third step (150 °C)	$(NH_2BH_2)_n \rightarrow (NHBH)_n + nH$	The second yield of hydrogen	[37]
The remaining step (500 °C)	$(NHBH)_n \rightarrow nBN + nH_2$	generation of excess hydrogen	[38]

Compared with pyrolysis, adding an appropriate catalyst, the alcoholysis of AB can release three equivalents of hydrogen at room temperature (Equation (1)) [41]. Chen et al. [42] prepared amorphous Co nanoparticles as catalyst for hydrogen liberation of AB alcoholysis, which displayed the intended hydrogen production performance. After ten tests of catalytic cycles, the turnover frequency (TOF) value of cobalt nanoparticles was still up to 515 mol_{H_2} mol_{metal}^{-1} h^{-1}. Yu et al. [43] synthesized CuNi nanoparticles with a diameter of 16 nm by liquid phase method, which were successfully loaded onto graphene, and then obtained G-CuNi catalyst after being treated with tert-butylamine. It was surprising that the TOF value and activation energy of the catalyst reached up to 49.1 mol_{H_2} mol_{CuNi}^{-1} min^{-1} and 24.4 KJ/mol, respectively. Özhava et al. [44] reported that a stable Ni nanoparticles catalyst with polyvinylpyrrolidone (PVP) could be separated from the reaction solution by centrifugation, which had the advantages of simple preparation, high activity and high cost effectiveness. The TOF value of the catalyst utilized for the methanolysis of AB was 12.1 min^{-1}. The high production cost of alcoholysis even though adding catalysts can increase the rate of hydrogen releasing has drawn attention to solve the practice difficulties in promoting production.

$$NH_3BH_3 + 4CH_3OH \rightarrow NH_4B(OCH_3)_4 + 3H_2 \quad (1)$$

AB hydrolysis is a process in which AB reacts with water in the presence of the suitable catalyst to release the hydrogen contained in the molecule [45]. The hydrolysis of AB to produce hydrogen can be carried out at room temperature, clean and pollution-free. Compared with the other two methods of hydrogen production, hydrolysis has obvious advantages, such as lower production cost, faster hydrogen liberation rate than alcoholysis and lower reaction temperature than pyrolysis [46]. This is an environmentally friendly and efficient hydrogen release method. AB reaction equation is as follows (Equation (2)). In the reaction process, the two hydrogen atoms in the hydrogen come from the AB molecule and the water molecule, and the hydrolysate is pollution-free. As a result, hydrogen production by hydrolysis of AB is considered to be an attractive approach to meet the low-cost and environmentally-friendly market needs.

$$NH_3BH_3 + 2H_2O \rightarrow NH_4BO_2 + 3H_2 \quad (2)$$

2.2. *The Hydrolysis Mechanism of NH_3BH_3(AB)*

The research on the hydrolysis mechanism of AB is mainly concluded by theoretical calculation and reasoning. Figure 1 showed the mechanism of hydrogen production by hydrolysis of AB. During the hydrolysis reaction, AB interacts with the catalyst surface to form a complex containing the H bond, and then in the attack of water, AB and H_2O each lose a hydrogen atom to form hydrogen [47]. Therefore, it can be found that the key step of catalytic water interpretation of hydrogen is that water molecules attack the M-H of the transition state. The properties of the metal catalyst M directly affect the formation of the transition state M-H and the speed of water interpretation of hydrogen. Thus, the catalysts are the key to whole hydrolysis reaction [48]. According to numerous literatures, the hydrolysis rate of AB is related to the amount of catalyst added, the amount of AB and the reaction temperature. Among them, temperature has the greatest influence on the hydrolysis rate, and the rate change can be obviously seen in the test of hydrogen release at different temperatures [49,50]. Generally speaking, the hydrolysis rate of AB has a zero-order relation or quasi-zero-order relation with the concentration of AB [51]. Highly dispersed metal particles can quite improve the hydrolysis rate, however the metal particles are easily agglomerated resulting in decreased catalytic activity. Mechanism studies show that AB interacts with metal particles and then dissociates the B-N bond in water to form BO_2^- and H_2 [28].

The development of high performance, low cost and easy recovery catalysts are the key to realize the application of AB hydrolysate to obtain hydrogen. Xu [27] studied the performance of many metal catalyst systems on the generation of hydrogen by the hydrolysis of AB. The results presented that both noble metals and non-noble metals such as Pt, Ru, Pd, Rh, Cu, Co, Ni, etc. could be catalysts for the hydrolysis of AB [6]. Among them, noble metals were supposed to be potential candidate materials on account of their high chemical activity and stability, which were prominent in the hydrolysis reaction.

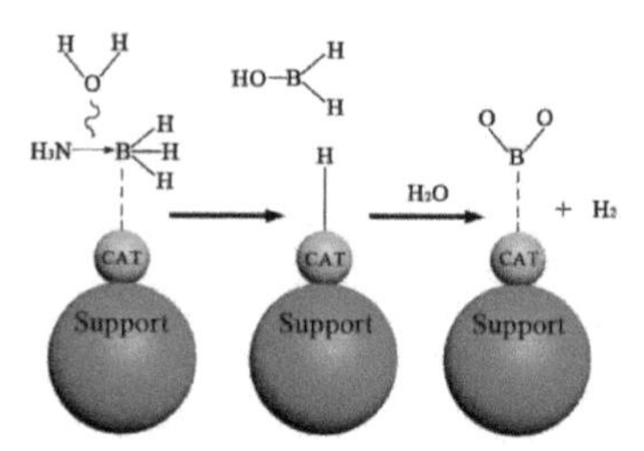

Figure 1. Mechanism of hydrogen production by ammonia borane (AB) hydrolysis.

Although non-noble metals have low costs, they exhibit low hydrogen generation rates in contrast to noble metals. What is more, their catalyst activities decline dramatically upon usage [52,53]. Hence, the utilization of noble metals is a consequence for high kinetics and stable hydrogen generation rate. Since there have been reports focused on catalytic hydrolysis of AB with noble metals [27], a number of studies have been conducted on monometallic and polymetallic catalysts [10]. Based on the above points, we mainly summarized the progress of high efficiency noble metal catalysts, composite catalysts and supported catalysts for AB dehydrogenation in the following papers.

3. The Development of Catalysts for NH_3BH_3 (AB) Dehydrogenation

In general, catalytic activity is evaluated based on the TOF value. TOF value is considered as the molecule that reacts per unit of active area in per unit time in the light of International Union of Pure and Apple Chemistry (IUPAC) [54]. In fact, the life of the catalyst is determined by the chemical, thermal and mechanical stability of the catalyst. However, as time goes on, the continuous accumulation of impurities or the loss of active particles on the surface of the catalyst will eventually lead to a decrease in the activity of the catalyst [55]. Therefore, it is desirable to find catalysts with excellent performance, which will remain stable and minimize deactivation after multiple reactions.

3.1. *Noble Metal Catalysts*

Noble metal catalyst systems mainly refer to rare transition metal elements or multi-metal catalysts, as well as noble metal composite materials supported on other support materials. Pd [56], Pt [10,57–60], Ru [61–69] and Rh [70,71] are the main noble metal elements, which exhibit excellent performance in catalytic hydrolysis to produce hydrogen.

At the initial research, Xu's research team [27] found that noble metal showed high activity and stability in the AB hydrolysis catalyze process, which was due to the fact that the empty d-electron-orbitals contained in these elements were easy to absorb negatively charged protons and form intermediates to increase the reaction rate [72,73]. Among the many monometallic catalysts, the noble metals include Pd, Pt, Ru, Rh, showing high catalytic activity. On the other hands, Ru and Rh metals are often an alternative to the choice of precious metals, much slightly cheaper than that of noble Pt and Pd metals, at the same time, their catalytic effect can be comparable to noble metals.

Researchers are trying to improve the utilization rate and catalytic activity of noble metals to solve the problems of resource shortage and high price. Recently, investigators focused on preparing better and stable supported noble metal catalysts by selecting suitable carrier and adding additives [74–76]. On the basis of single metal, the dispersion of metal particles is strong, and the rate of hydrogen discharge increases significantly, by loading metal particles onto the carrier [77,78]. Poly(N-vinyl-2-pyrrolidone) (PVP) protected palladium rhodium nanoparticles with a size of 2.5 nm have been reported to be used as efficient catalysts [79], which provided the process of catalytic activity in the AB hydrolysis process and analyzed the generation of hydrogen. In the presence of PVP, the admixture of potassium tetrachloropropionate and rhodium chloride trihydrate was reduced by ethanol to Pd-Rh@PVP nanoparticles in ethanol water mixtures at reflux temperature. Among them, PVP was used as stabilizer and reducing agent. By comparing the catalytic activity of various types (single metal Pd and Rh nanoparticles, their 1:1 physical mixture and 1:1 Pd-Rh bimetallic nanoparticles) in AB hydrolysis reaction, the formation of alloy type Pd-Ph@PVP nanoparticles has not been determined to be a physical mixture of single metal nanoparticles. Pd-Rh@PVP nanoparticles could be considered as a promising catalyst with the highest activity in realistic applications and could be used for proton exchange membrane fuel cell AB hydrolysis to produce hydrogen. Pd@PVP nanoparticles, Rh@PVP nanoparticle, 1:1 physical mixture Pd@PVP and Rh@PVP nanoparticles as well as Pd-Rh@PVP bimetallic nanoparticles respectively provided TOF values with growth trend: 182, 228, 430 and 1333. Figure 2 indicated that the hydrogen production presented linear relationship, starting immediately without induction period and continuing until complete hydrolysis of AB. It was worth noting that Pd-Rh@PVP nanoparticles with a concentration of 0.3 mM lead to the complete release of hydrogen in AB hydrolysis within 45 s, which was equivalent to an average TOF value of 1333 mol H_2 (mol cat)$^{-1}$ min^{-1} at 25.0 °C. Compared with the physical mixture of Pd and Rh nanoparticles, Pd-Rh@PVP nanoparticles had higher catalytic performance, which was due to the effect of the synergy of Pd and Rh and the reduction of catalyst particle size. In addition, the Pd-Rh@PVP nanoparticle's catalyst maintained 78% of initial catalytic activity in AB hydrolysis, even after the fifth reaction.

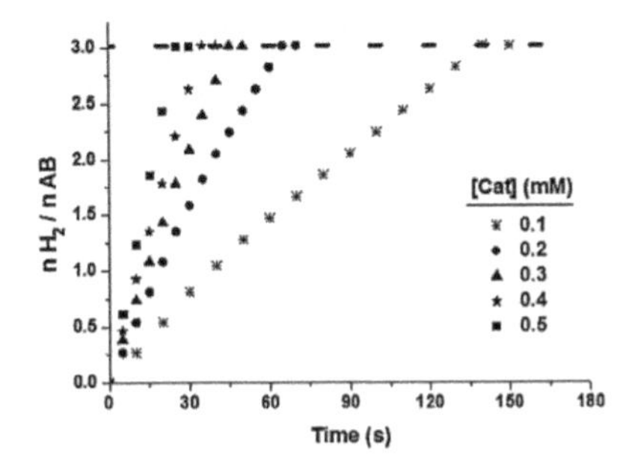

Figure 2. Plot of mol H_2/mol AB versus time for the hydrolysis of 100 mM AB solutions in the presence of Pd-Rh@PVP nanoparticles in different catalyst concentrations (0.1, 0.2, 0.3, 0.4 and 0.5 mM) at 25.0 ±0.1 °C; copyright (2014), Catalysis.

Because of its high hydrogen production activity in transition metal catalysts, platinum has attracted widespread attention [80,81], and been deeply studied by scholars. Kinetic studies and model calculations [57] show that Pt (111) facets are the main active surface. The particle size of about 1.8 nm is the optimal size of Pt. At the same time, the durability of the catalyst is closely related to the particle size of Pt [82–87]. The smaller the Pt particle size is, the lower its durability is, which may be related to the more obvious adsorption of B-containing species on the Pt surface and the easier the change of Pt particle size and shape. The results of this study paved the way for the rational design of highly active and durable platinum based catalysts for hydrogen production. Wang et al. [88] proposed a simple and gentle one-pot method to prepare porous PtPd bimetallic nanoparticles (NPs) with reverse structure under the adjustment of 1-hexadecyl-3-methylimidazolium chloride ([C_{16}mim] Cl) in an aqueous solution. The composition and morphological concentration of PtPd NPs could be easily adjusted by changing the initial molar ratio of precursor and IL. In addition, by simply changing the content of glycine, they also found that it was possible to change the structure of porous PtPd NPs from Pt-on-Pt to Pt-on-Pd. Figure 3 revealed the possible growth mechanism of porous PtPd NPs in the presence of [C_{16}mim] Cl. As shown in Figure 4, using various samples as catalysts, the hydrogen equivalent generated per mole of AB varies with the reaction time. It could be seen that AB maintained stability without hydrolysis in aqueous solution without any catalyst. Among all the catalysts used, porous $Pt_{25}Pd_{75}$ NPs displayed the highest catalytic activity for AB dehydrogenation and hydrogenation. The reaction took only twelve minutes to complete and the hydrogen yield was as high as 97.3%. Porous $Pt_{25}Pd_{75}$ NPs with a Pd-on-Pt structure could better combine with the B atom in AB to activate the electronic B-H bond, making it easier to break the attack with H_2O [89]. Due to its high specific surface area, porous layered structure (including mesopores and micropores) and possible electronic effects between Pt and Pd, porous $Pt_{25}Pd_{75}$ NPs (Pd-on-Pt structure) had outstanding catalytic activity and the higher the stability of AB hydrolysis to produce hydrogen.

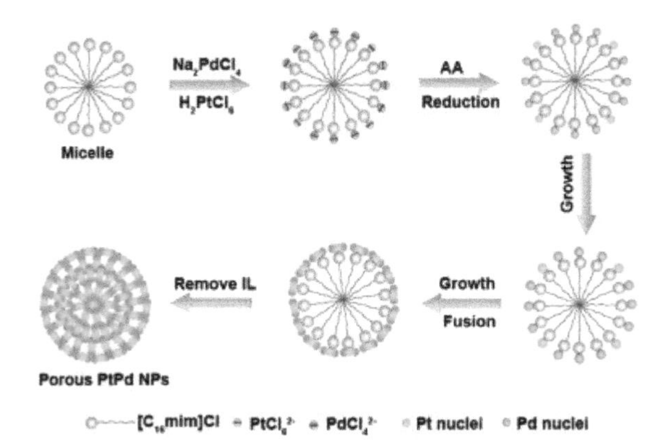

Figure 3. Illustration of the possible growth and assembly of the porous PtPd nanoparticles (NPs) in the presence of $[C_{16}mim]$ Cl; copyright (2020), the Royal Society of Chemistry.

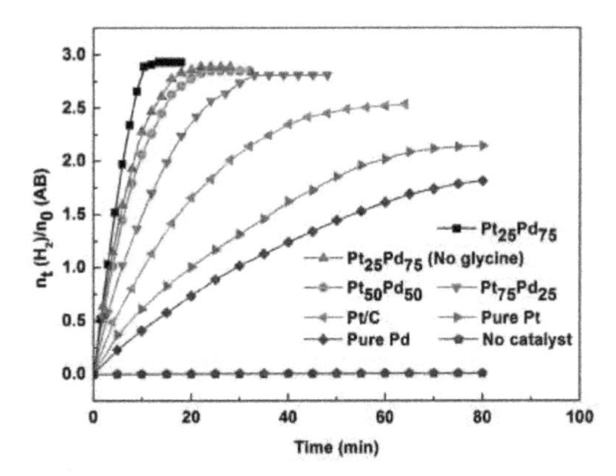

Figure 4. The curves of H_2 equivalents produced per mole of ammonia borane as a function of reaction time at 25 °C with various samples as catalysts; copyright (2020), the Royal Society of Chemistry.

Compared with platinum and palladium, ruthenium and rhodium are slightly cheaper, however their catalytic performance can be similar to those of them, which is ruthenium and rhodium being often used to replace platinum and palladium. In recent years, scholars have made extensive exploration on ruthenium and rhodium. Martina K, et al. [90] presented new Ru compounds having PNP amido chelate ligands, which could undergo reversible hydrogenation/dehydrogenation reaction both at the N functionality and the ethylene backbone. The reactivity of the ruthenium complexes was utilized for the homogeneous catalytic dehydrogenation of AB with excellent activities. Moreover, the catalysis of gold nanoparticles has attracted increasing attention because supported gold catalysts have been found having surprisingly high activity in oxidation. L.Wen et al. [91] successfully synthesized ultrafine Ru NPs deposited on MCM-41via using a simple liquid impregnation reduction method. Furthermore, they determined the effect of different Ru content attached to MCM-41 on hydrolysis dehydrogenation of AB (0.52, 0.70, 0.90 and 1.12 wt%, respectively) (Figure 5). Among all results, 1.12 wt% Ru/MCM-41 presented the highest catalytic activity, with TOF value of 288 min^{-1}. In summary, Ru/MCM-41 appeared excellent catalytic activity toward hydrolysis of AB owing to the unique structure of MCM-41 with ordered hexagonal pores and the strong synergistic effect.

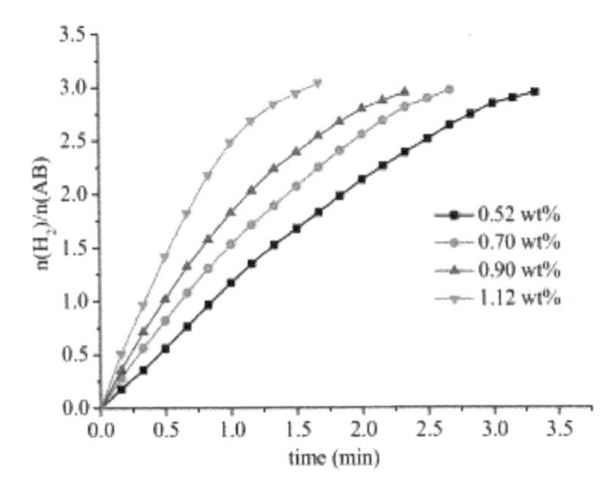

Figure 5. Hydrogen generation from aqueous AB in the presence of Ru/MCM-41 catalysts at room temperature. Ru/AB (molar ratio) = 0.0026, 0.0044, 0.0045 and 0.0055 at Ru loadings of 0.52, 0.70, 0.90 and 1.12 wt%; copyright (2015), Chinese Chemical Letters.

In the literature on catalytic hydrolysis of AB with support Ru catalysts, their catalytic activities are related to many factors, including particle diameter and location distribution, as well as preparation method and selection of load carrier [92,93]. Late transition metal nanoparticles (NPs) are more prone to aggregate in solution [94,95]. Although it can be alleviated by adding or supporting substances, the effective area of NPs will be reduced by retention matching, which will affect the catalytic chemical reaction [96]. Abo-Hamed E K. et al. [97] reported a monodisperse metastable ruthenium nanoparticle (Ru NPs), which could easily solve these problems. Their report represented the case in which Ru NPs remained stable without protective ligands or carriers while simultaneously exhibited high catalytic activity. Metastable Ru NPs have been proved to be a promising catalytic active material for production of hydrogen via hydrolysis of AB at room temperature. Shen et al. [76] have produced a one-step in situ route for synthesis Rh nanoparticles supported on graphene by using methylamine borane (MeAB) (Figure 6a,b). Compared with conventional carriers, prepared Rh/NPs supported on graphene exhibited better catalytic activity on hydrolysis of AB, with high TOF values of 325 min^{-1} and low activation energy (Ea) values of 16.4 KJ mol^{-1}. As clearly shown from transmission electron microscope (TEM) image of Rh/graphene NPs, there were no significant changes in morphology of Rh/NPs on graphene, and no noticeable aggregation of Rh/NPs after the fifth run durability (Figure 6c,d).

3.2. *Noble Metal and Non-Precious Metal Composite Catalysts*

Noble metal catalysts exhibit high catalytic capability, but the high cost and low utilization of precious metals have greatly limited their application in commercial production [98]. Whereas non-precious metals are favored by researchers on account of the abundant resources on earth and low price, even if their catalytic property are not as good as noble metals. To address effectiveness of catalysts [99], it is better to add non-precious metals to the catalysts without significantly reducing the catalytic activity to manufacture noble metal and non-precious metal composite catalysts. Although the hydrolyzing rate of AB can be improved obviously by using non-noble metal materials or composites as catalysts, the catalytic performance of non-precious metal catalysts has yet to be advanced compared with noble metals. Now, how to improve the catalytic activity of cheap non-precious metals has become a popular topic. Pure noble metals and non-precious metals have not been able to meet the requirements of market, thus, the combination of noble metals and non-precious metals or mutual fusion for solution provides ideas to solve this matter. Then binary or multivariate catalysts gradually emerged [100,101], which not only reduce the material cost, but also enhance catalyst activity.

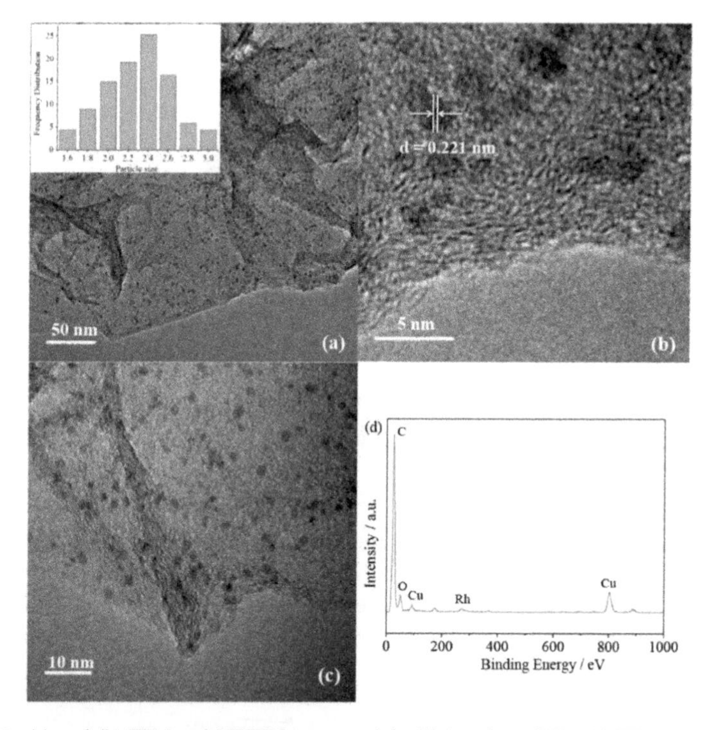

Figure 6. (**a**) and (**b**) TEM and HRTEM images of the Rh/graphene NPs; (**c**) TEM mages of the Rh/graphene NPs after five cycles; (**d**) EDS spectrum of Rh/graphene NPs; and (**a**) inset: particle size distribution of Rh/graphene NPs; copyright (2014), Hydrogen Energy.

Manna et al. [56] used cobalt-ferrite coated by polydopamine (PDA) as the carrier, which was the coating layer formed by the long-term mixing of dopamine hydrochloride and cobalt ferrite, and then loaded palladium particles onto the carrier to synthesize Pd/PDA-$CoFe_2O_4$. The results displayed that Pd/PDA-$CoFe_2O_4$ with Pd loading of 1.08 wt% was a highly active and reusable catalyst. Moreover, it was worth mentioning that Pd still maintained its initial catalytic activity after ten times of catalysis. Zhang J et al. [102] fabricated Pd-Cu nanocrystals by a facile and flexible protocol, with diverse morphological features to enhance catalytic and durability. As a result that elements Pd (2.20) and Cu (1.91) were significantly different in the electronegativity, it was easier to generate electrons from one element to another after forming the alloy, resulting in the electron coupling effect. Through the characterization technology, X-ray photoelectron spectroscopy (XPS) analysis results are confirmed in Figure 6. As could be seen from Figure 7, there was indeed an electric coupling or synergistic effect in Pd-Cu alloy, which was conducive to the adsorption of H to form metal-H species, thereby promoting the catalytic effect of AB hydrolysis. In theory, it was the accumulation of electron clouds around the nucleus of Pd that promoted the adsorption of H and facilitated the formation of metal-H species to accelerate the dehydrogenation of AB [103]. Furthermore, the catalytic activity of AB was investigated in detail using Pd-Cu alloy nanocrystals as catalysts. The hydrogen equivalent produced from AB hydrolysis by six different Pd-Cu alloy nanocrystals (Pd-Cu NAs-1, Pd-Cu NAs-2, Pd-Cu NAs-3, Pd-Cu NAs -4, Pd-Cu NAs -5, Pd-Cu NAs -6) prepared them and metal nanocrystals of two elements (Pd-NPs, Cu-NPs) were shown in Figure 8. For all the catalysts investigated, the catalytic capacity of Pd-NPs was the strongest, while that of Cu-NPs was the weakest. In bimetallic alloy nanocrystals,

Pd-Cu NAs-1 held the highest catalytic ability. For bimetal Pd-Cu alloy nanocrystal catalytic capacity follows the following order: Pd-Cu NAs-1 > Pd-Cu NAs-5 > Pd-Cu NAs-2 > Pd-Cu NAs-3 > Pd-Cu NAs-4 > Pd-Cu NAs-5. By this treating, experimental data confirmed, actually, the fact that catalytic performance of bimetallic catalysts was superior to that of their single metal elements, which has been recorded in other literature on other alloy catalysts [104,105].

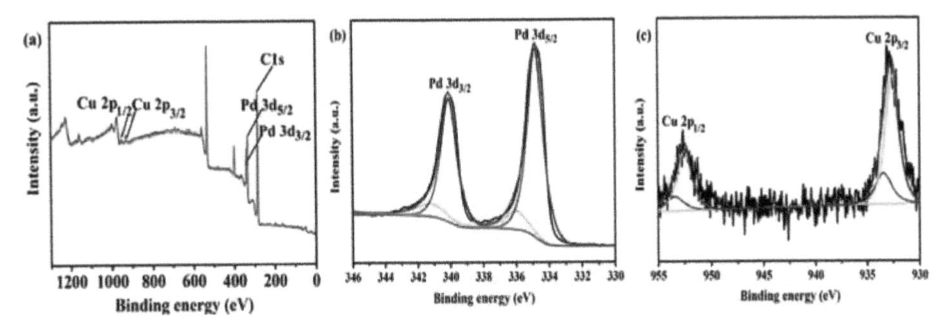

Figure 7. XPS spectra for the Pd-Cu NAs-1 sample. (**a**) Survey scan; (**b**) Pd 3d; (**c**) Cu 2p; copyright (2019), Hydrogen Energy.

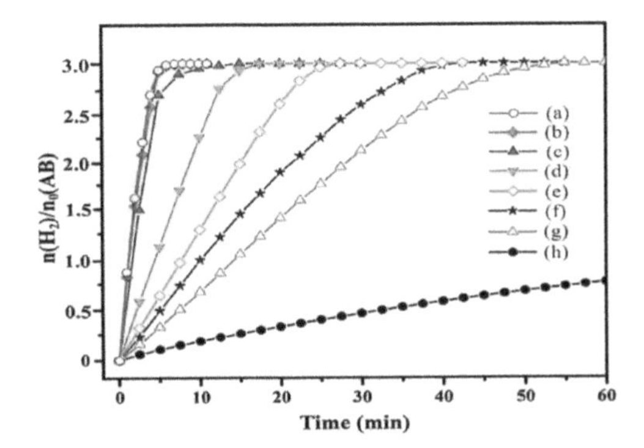

Figure 8. H_2 equivalents generated from AB hydrolysis catalyzed by the different Pd-Cu alloy nanoparticles at 298 K. (**a**) Pd-NPs; (**b**) Pd-Cu NAs-1; (**c**) Pd- Cu NAs-5; (**d**) Pd-Cu NAs-2; (**e**) Pd-Cu NAs-3; (**f**) Pd-Cu NAs-4; (**g**) Pd-Cu NAs-6; (**h**) Cu-NPs; copyright (2019), Hydrogen Energy.

Compared to monometallic catalysts, much better selectivity and catalytic activity have been demonstrated for bimetallic catalysts [106–110]. The bimetallic catalysts formed by Pt and non-precious metal have been proved to be superior to other catalysts. Yang's group [111] reported that Pt_xNi_{1-x} (x = 0, 0.35, 0.44, 0.65, 0.75 and 0.93) nanoparticles were used as catalyst for hydrogen generation from hydrolysis of AB. They studied the catalytic activity of Pt_xNi_{1-x} (x = 0, 0.35, 0.44, 0.65, 0.75 and 0.93) the hydrolysis and dehydrogenation of AB solution. The results exhibited that the particle size of Pt_xNi_{1-x} nanoparticles were about 2–4 nm, and the contents of Pt atom in the catalyst were 35%, 44%, 65%, 75% and 93%, respectively. It was found that catalytic activity of AB hydrolysis was related to the composition of Pt_xNi_{1-x} catalyst and $Pt_{0.65}Ni_{0.35}$ nanoparticles have the highest catalytic activity. The activity of the synthesized PtxNi1-x catalyst was better than that of pure Pt or Ni catalyst. The TOF value and the Ea of the reaction were 4784.7 mL min^{-1} g^{-1} and 39.0 KJ/mol. Gao M et al. [112] prepared monodisperse PtCu alloy NPs and explored the action of their catalysts for hydrolysis AB under mild condition. Among different composition PtCu NPs, the $Cu_{50}Pt_{50}$ NPs as optimum catalyst demonstrated the highest catalytic performance with an initial TOF of 102.5. These experimental results exhibited that the validity of partly replacing Pt by a first-row transition metal on designing superior property heterogeneous nanocatalysts for AB hydrolytic. Chen et al. [113] designed Pt-WO_3 double active site catalyst to boost the catalytic hydrolysis of AB. Figure 9 showed the hydrolysis mechanism of AB on Pt double activity. Pt-WO_3/CNT could significantly improve hydrogen production activity and durability, which was attributed to the double active center of Pt-WO_3 and the sacrifice site of WO_3.

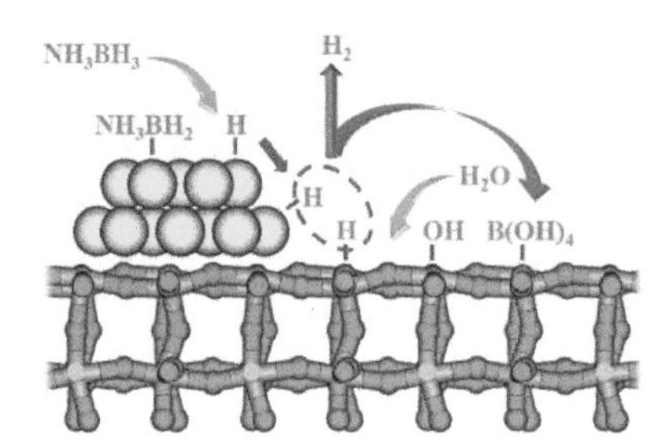

Figure 9. A Proposed Mechanism for Ammonia Borane Hydrolysis over Pt-WO_3 Dual Metal Sites; copyright (2020), Elsevier.

Transition metal nanoparticles are widely applied in the hydrolysis AB [114,115]. However, in the catalytic process, metal particles can aggregate to form clumps, which will lead to the inactivation of catalyst, the existence of instability and low efficiency [116]. In order to avoid the influence of low repetition rate, the nanoparticle's catalyst was prepared by using magnetic powder as the catalyst active metal carrier so as to become the magnetic separable catalyst in the liquid phase reaction, which can enhance the rate of catalysts utilization [117–119]. Hence, Akbayrak S et al. [120] reported three metal(0) nanoparticle catalysts formed by loading ruthenium(III), rhodium(III) and palladium(II) onto magnetic carriers of iron coated with carbon(C-Fe) at room temperature as transition metal nanoparticles for hydrolysis of AB, M^0/C-Fe NP(M=Ru, Rh and Pd). Using the XPS, energy dispersive X-ray detector (EDX), TEM techniques, the results of tests showed that Rh^0/C-Fe (0.45% wt. Rh), Ru^0/C-Fe (1.59% wt. Ru) and Pd^0/C-Fe (2.0% wt. Pd) nanoparticles gave TOF of 83, 93 and 29 min^{-1}, respectively. In the repeatability tests, M^0/C-Fe remained premier activity even after hydrolysis reactions, as shown in Figure 10. In this review, M^0/C-Fe nanoparticles revealed outstanding reusability and activity.

Zhou Q et al. [121] prepared nanoporous ruthenium (NP-Ru), which was consisted of an interconnected nanoscaled ligament by one-step mild etching of RuAl alloy. NP-Ru showed high catalytic activity at room temperature and had a long life to hydrolyze AB. In addition, it was found that even after five runs, NP-Ru still had excellent reusability and recyclability, and its original catalytic activity was 67%. Wei Z et al. [122] proposed a simple method for preparing CoRu nanoalloy catalysts (CoRu@N-C) by encapsulating the alloy material into carbon layer. With this strategy, CoRu nanoalloy catalysts could effectively prevent the alloy from accumulating in the corrosive medium and facilitate the catalytic reaction on the surface. Moreover, CoRu@N-C exhibited excellent sustainability and high catalytic performance for hydrolysis of AB.

It has been dedicated to the exploration of bimetallic catalysts for AB catalytic dehydrogenation [123–127]. A variety of bimetallic catalysts have better catalytic performance than single metal catalysts, however, the catalytic efficiency of AB as a hydrogen storage candidate is still far from meeting the needs of practical applications. Au-containing hybrid materials have received great attention from scholars due to their unique synergy. For example, the contact between Au NPs and metal oxides improved the high catalytic activity of inert gold [128–131]. Introducing Au clusters into vulnerable parts or subsurface of Pt and Pt-TM nanocatalysts greatly improved the electrochemical stability of the catalyst [132–135]. By introducing Au into the Pt-TM nanocatalysts, it was reasonable to speculate that it has superior catalytic performance for AB catalytic dehydrogenation. Zhai et al. [136] succeeded in obtaining a PtAuCo trimetallic nano-alloy with a single-phase structure through a sequential digestion and reduction strategy. Figure 11 illustrated the formation mechanism of single-phase PtAuCo trimetallic alloy catalyst. In order to evaluate the performance of the catalyst for AB hydrolysis, they tested the activity of the prepared PtAuCo nanocatalyst with single metals (Pt, Au, Co) and bimetals ($Pt_{85}Au_{15}$, $Pt_{86}Co_{14}$, Au@Co) based on 298 K. As shown in Figure 12a,b, the activity of bimetallic nanocatalysts was better than that of single metal nanoparticles. Moreover, the reinforcement of the nanoalloy structure was superior to that of the core-shell structure. $Pt_{85}Au_{15}$ and $Pt_{84}Co_{16}$ have almost the same metal ratio, nevertheless, the TOF value of the former (~137 mol H_2 min^{-1} (mol metal) $^{-1}$) was much higher than the latter (~66 mol H_2 min^{-1} (mol metal)$^{-1}$), indicating that the activity enhancement efficiency of Au was very high. Figure 12c,d showed that when about 10% of Pt in $Pt_{84}Co_{16}$ was replaced by Au to form a $Pt_{76}Au_{12}Co_{12}$ trimetallic nanoalloy, the hydrogen evolution reaction of AB was completed within 36s and the TOF value was increased to 450 mol H_2 min^{-1} (mol metal)$^{-1}$. In all synthetic samples, $Pt_{76}Au_{12}Co_{12}$ showed the best catalytic performance for the catalytic hydrolysis of AB. In addition, they used XPS technology to systematically measure the trimetallic catalysts with different compositions and corresponding monometallic and bimetallic counterparts, in order to explore the underlying mechanism. The results exhibited that the main reason for the excellent catalytic performance of $Pt_{76}Au_{12}Co_{12}$ catalyst was the modified electronic interaction and enhanced charge transfer ability.

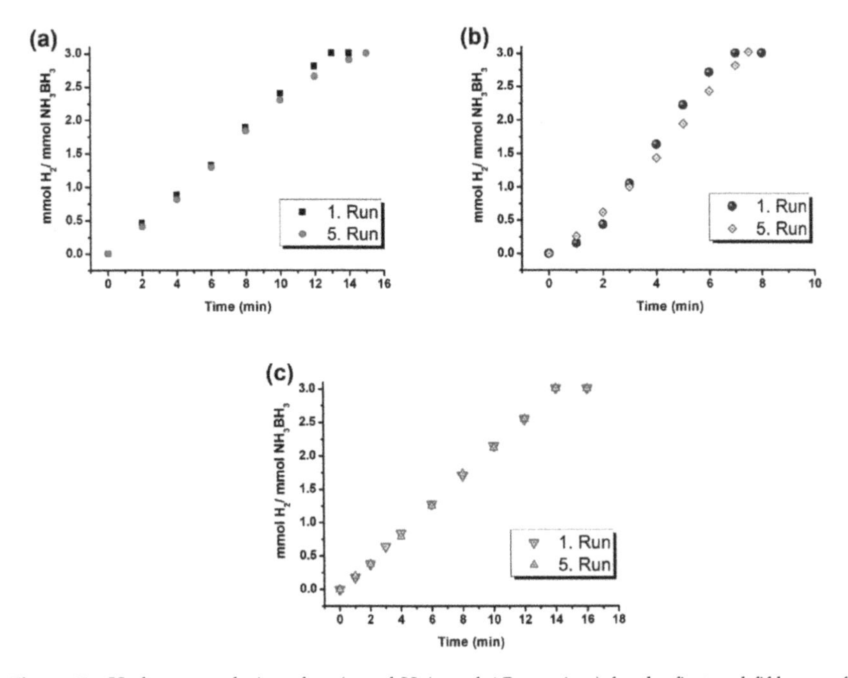

Figure 10. Hydrogen evolution plots (mmol H_2/mmol AB vs. time) for the first and fifth run of hydrolysis starting with (**a**)Rh^0/C-Fe NPs; (**b**) Ru^0/C-Fe NPs; (**c**) Pd^0/C-Fe NPs at 25.0 °C; copyright (2020), Hydrogen Energy.

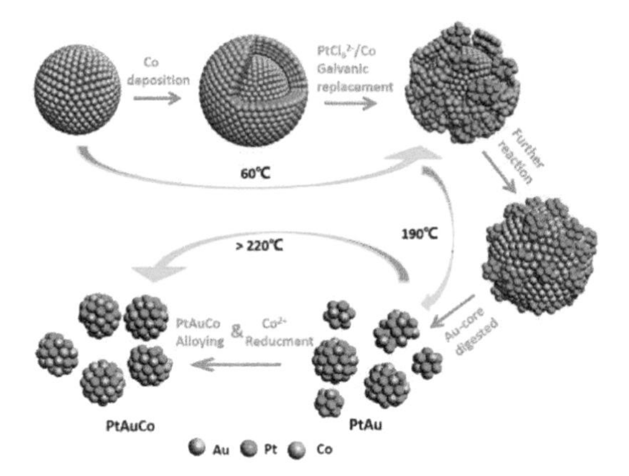

Figure 11. Fabrication strategy of the PtAuCo trimetallic alloys; copyright (2020), American Chemical Society.

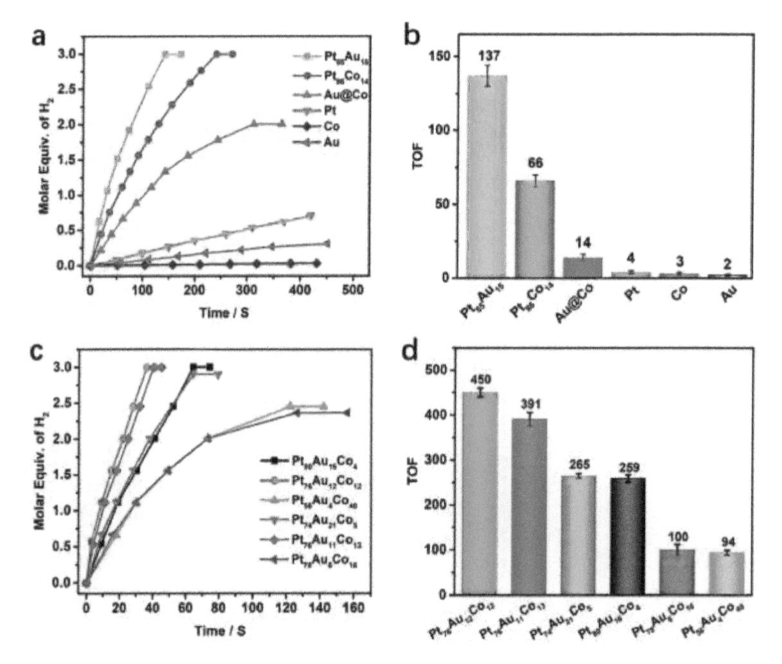

Figure 12. (**a**,**c**) Plots of time vs. H_2 generation and (**b**,**d**) the corresponding TOF value of the AB hydrolysis reaction at 298 K catalyzed by (**a**,**b**) the monometallic (Pt, Au, Co) and bimetallic ($Pt_{85}Au_{15}$, Pt_86Co_{14}, Au@Co) catalysts and (**c**,**d**) the trimetallic catalysts of different composition; copyright (2020), American Chemical Society.

In summary, according to numerous literatures, the hydrolysis of noble metals for the dehydrogenation of AB revealed excellent activity. However, the defects of noble metals have restricted their widespread adoption. It is worth noting that the combination of noble metals and non-precious metals to form a supported catalyst can not only settle the resource shortage of precious metals, but also reduce the material cost [137–139]. Furthermore, the supported catalysts greatly improve the catalytic capacity of AB due to adjust the electronic structure and surface geometry to adjust the catalytic performance [140–142]. Therefore, the supported catalysts of AB catalytic hydrolysis are the focus of research and development in the future.

3.3. *Catalytic Activities of Supported Metal Catalysts in NH_3BH_3(AB) Hydrolysis*

On the basis of the metal catalyst, by loading the metal particles on the support, the dispersion of the metal particles is improved, and the hydrogen release rate is significantly increased [143]. Due to its porous structure, the carrier materials increase the specific surface area between the metal nanoparticles and delay the formation of impurities on the nanoparticles, ultimately preventing the agglomeration of the nanoparticles [144]. The supports commonly used are graphene, carbon, carbon nanotubes (CNT), silicon, cerium, titanium [145–150].

3.3.1. Graphene Material Supported Metal Catalysts

Graphene is a single-atom honeycomb lattice-like carbon material with a surface area of 2600 $m^2\ g^{-1}$ [151]. In addition, it is suitable for supporting materials based on the characteristics of high mechanical strength, excellent electrical conductivity [152], outstanding thermal stability and chemical stability [153]. Through burdening metal nanoparticles on graphene, it can prevent its polymerization and improve its catalytic activity. In this way, in addition to increasing the superficial area of the catalysts, the accelerated charge transfer at the graphene metal interface is also conducive to promote catalytic activity [154]. Nowadays, there were a lot of literatures about the application of graphene loaded metal catalysts in AB hydrolysis. Chemical derived graphene (CDG) was synthesized by reduction of graphene oxide with hydrazine hydrate and used as the carrier of palladium nanoparticles (Pd NPs) [145]. By advanced analytical technology, Pd NPs keeping particle size dispersion and stability loaded onto CDG was used as catalyst for AB dehydrogenation and hydrolysis. Using CDG-Pd as a catalyst, the AB dehydrogenation and hydrolysis hydrogen production processes were tested, and it was found that CDG-Pd had high activity in both dehydrogenation and hydrolysis reactions. Figure 13 showed the graph of the molar H_2/mol AB ratio with time during the catalytic dehydrogenation and hydrolysis of 2.0 mmol AB solution in taking advantage of CDG-Pd catalyst (2.1% wt Pd) at 25 °C. Under existence of CDG-Pd catalyst, the dehydrogenation of AB produced one equivalent of hydrogen, while the hydrolysis of AB produced three equivalent of hydrogen. In the presence of CDG-Pd catalyst, the calculated values of AB for the initial TOF of dehydrogenation and hydrolysis were 170 h^{-1} and 933 h^{-1}, respectively (Figure 13). These values were comparable to the AB dehydrogenation and hydrolysis catalyst system. In addition to its high activity and stability, CDG-Pd was also found to be a reusable catalyst in dehydrogenation and hydrolysis. After the 5th and 10th runs, the hydrolysis of AB remained its initial activity of 85% and 95%, respectively, which made CDG-Pd have broad application prospects in noble metals to be used as a catalyst to develop an available portable hydrogen production system employing AB as a solid hydrogen storage and release material.

Ked et al. [155] reported a new type of high-efficiency catalyst for the hydrolysis of AB to produce hydrogen by embedding Pt-Co nanoparticles in nanoporous graphene sheets. In order to expound the formation mechanism of Pt-Co@PG catalyst, Figure 14 showed the steps of preparing Pt-Co NP aggregated on nanoporous graphene (PG) sheets, which included two important steps: preparing nanoporous graphene sheet by carbothermal metal oxide etching method, and then uniformly embedding Pt-Co nanoparticles into plane and holes defects. They studied the catalytic properties of Pt-Co bimetallic NPs supported in nanoporous graphene (Pt-Co@PG) and the catalytic dehydrogenation performance in AB aqueous solution. It was critical important to study the preparation of NP embedded in porous graphene and the synergistic effect between them. They used prepared nanoporous graphene loaded pure Pt NP, pure Co NP and Pt-Co NP to catalyze the hydrolysis of AB (1.5 mmol, 6 mL), respectively. Figure 15a displayed the amount of hydrogen produced during the process of hydrolytic dehydrogenation using the prepared Pt_xCo_{1-x}@PG NP. Obviously, pure PG had no catalytic activity for AB hydrolysis, and the hydrogen release rate of Pt-Co@PG NPs was much higher than that of pure Co@PG NPs, which indicated that Pt was a more effective element for AB hydrolysis. The catalytic mechanism of Pt-Co bimetallic NPs could be attributed to the synergistic effect of Co and Pt, which was triggered by the charge interaction between Pt Co NPs and PG carrier and the reduced particle size (providing rich active sites) [156]. Therefore, in the catalysis of Pt Co bimetallic NPs heterogeneous reaction, the catalyst with the best ratio of Pt and Co showed the highest catalytic activity. $Pt_{0.1}Co_{0.9}$@PG achieved the best performance, which had obvious high catalytic activity to release hydrogen in three minutes hydrolysis AB, with TOF value as high as 461.17 mol_{H_2} min^{-1} $mol_{Pt}{}^{-1}$. As shown in Figure 15b, compared with the bare $Pt_{0.1}Co_{0.9}$ catalyst and the reduced graphene oxide supported $Pt_{0.1}Co_{0.9}$ catalyst, supporting function of nanoporous graphene had been clearly demonstrated. Recent studies showed that the enhanced catalytic activity of graphene supported metal nanoparticles was attributed to the interface interaction between metal nanoparticles and graphene materials [33]. The nanoporous graphene sheet provided more edges related to the presence of pores and more anchoring agents to stabilize Pt-Co nanoparticles with uniform dimensions. The simple synthesis, excellent catalytic performance revealed that the Pt-Co@PG nanohybrid material was a promising candidate material for the development of highly efficient and portable AB hydrogen production system.

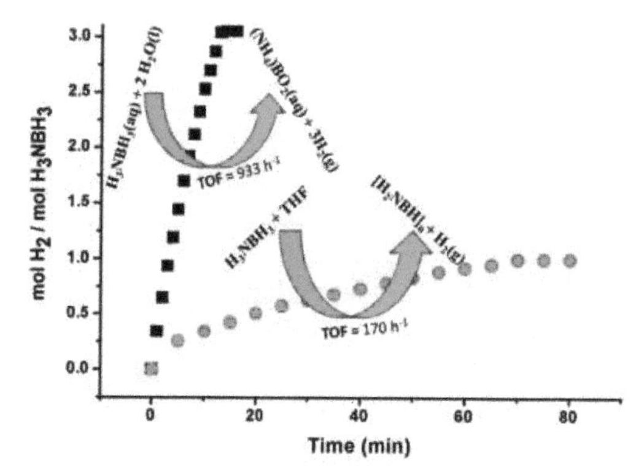

Figure 13. The mol H_2/mol H_3NBH_3 versus time plots for the dehydrogenation and hydrolysis of AB catalyzed by CDG Pd (2.1% wt Pd); copyright (2012), Elsevier.

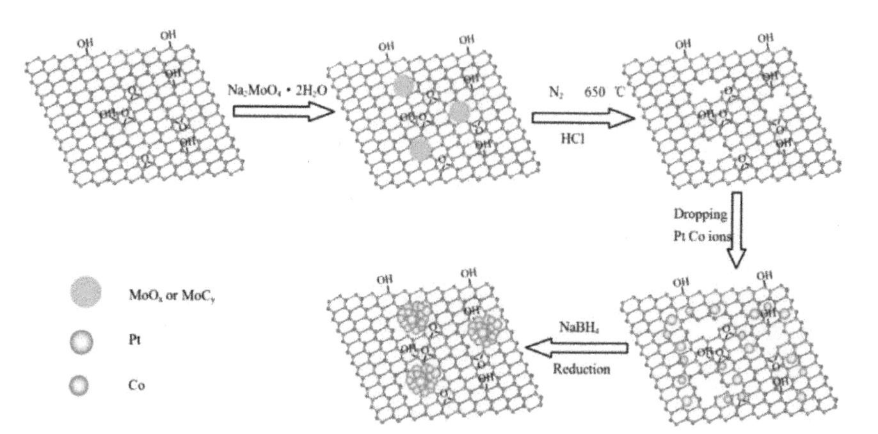

Figure 14. Mechanism of Pt-Co@PG catalyst preparation; copyright (2017), Elsevier.

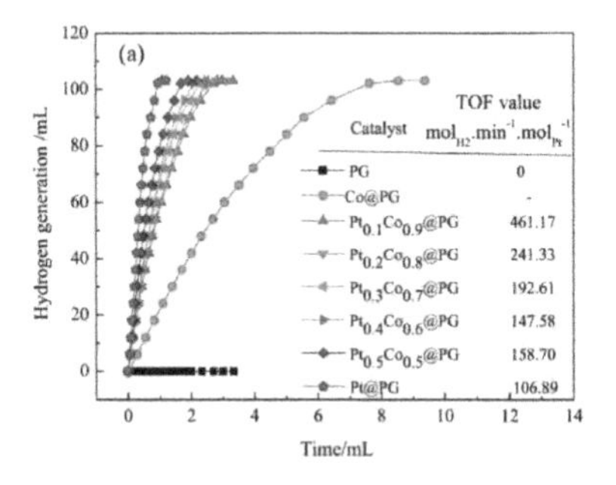

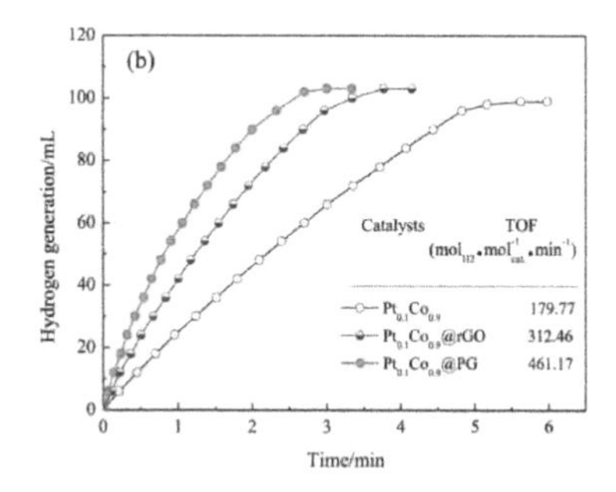

Figure 15. (**a**) Hydrolysis of aqueous NH_3BH_3 solution under ambient atmosphere catalyzed by pure Pt@PG NPs, pure Co@PG NPs and Pt-Co@PGNPs with different ratios; (**b**) hydrolysis of aqueous NH_3BH_3 solution catalyzed by $Pt_{0.1}Co_{0.9}$ NPs, Pt0.1$Co_{0.9}$@GO NPs and $Pt_{0.1}Co_{0.9}$@PG; copyright (2017), Elsevier.

3.3.2. Carbon Material Supported Metal Catalysts

Due to these performance that the modified surface chemistry [60] (such as defects and oxygen groups), texture characteristics (such as aperture), outstanding thermal conductivity and resistance to acid and alkaline environment [157], scholars are keen on exploiting carbon-based materials as catalyst carriers. Carbon is a good maintaining material because it has excellent interaction with metal, chemical inert structure and easy to produce various forms and porosity [158]. The metal catalysts distributed on the carbon support are mostly employed in the hydrolysis of AB. Lu et al. [146] prepared ultrafine homogeneous Ru nanoparticles on phosphorus-doped carbon(PPC) carriers to synthesize Ru/PPC catalyst through in-situ reduction method. Figure 16 illustrated procedure for preparing of PPC and Ru^{3+}/PPC. The Ru/PPC material could be utilized as catalyst to promote the hydrolysis of AB to produce hydrogen. In order to investigate the effect of Ru loading on the property of Ru/PPC catalyst, the hydrolysis of AB was carried out with different Ru content (1.5, 2.5, 3.5, 4.5 wt%) at 25 °C by maintaining the total concentration of Ru at 0.7 mM. It could be seen from the experimental results that Ru/PPC with a load of 3.5 wt% displayed the highest catalytic capacity with TOF value of 413 mol H_2 $(mol_{Ru}\ min)^{-1}$. Ru concentration also affected the performance of the catalyst. With the increased of Ru concentration (between 0.3 and 0.9 mM), the hydrogen production rate was increasing gradually. The high activity of AB hydrolysis on Ru/PPC was attributed to the super refinement and high dispersion of Ru NCS, which provided more surface active centers for the reaction. This review showed that the PPC had promising catalyst support for hydrogen generation from AB hydrolysis.

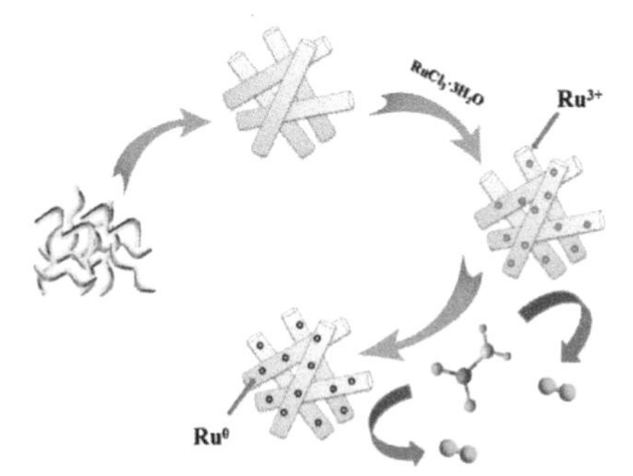

Figure 16. The procedure for preparing Ru^{3+}/PPC and the hydrolysis of AB over Ru/PPC; copyright (2018), Hydrogen Energy.

On the other side, the performance of surface charge distribution and energy storage/release can be regulated and ameliorated by doped miscellaneous elements [159]. The catalyst of porous carbon material containing B, N, P heteroatoms reveal excellent catalytic performance in AB hydrolysis reaction [160,161]. So far, the method of doping nitrogen or phosphorus in carbon materials is that additional N or P sources (NH_3, PH_3) are required during the material preparation process [162]. The existing preparation process is cumbersome and dangerous to a certain extent, and it is rare to obtain N, P co-doped nanoporous carbon directly from existing and frequently used materials. Therefore, it is a challenge to develop a practical, valid and single technique for preparing multi-element doped (such as N, P) nanoporous carbon. Herein, Fan et al. [163] explored a simple and effective method for preparing N, P-doped nanocarbon as metal nanoparticles (MNP), in which N or P doped carbon as MNP carrier enhanced the catalytic activity of AB decomposition. They chose adenosine triphosphate (ATP) as the ideal material for N-rich and P-rich raw materials due to the high content of N and P provided by an adenine structure and three phosphate groups. Consequently, they utilized ATP-derived N, P co-doped carbon materials to fix Rh NPs in porous carbon for the catalytic dehydrogenation of AB. As shown in Figure 17, ATP-C was prepared from ATP via a one-step heat treatment procedure. ATP-C-700, ATP-C-800, ATP-C-900 were synthesized to explore the effect of temperature on surface area and pore size. It was worth mentioning that the specific surface area and average pore diameter of ATP-C-800 were the largest, 154.2 m^2 g^{-1} and 6.83 nm, respectively. According to the above results, 800 °C was taken for the optimal carbonization temperature. The TOF of Rh/ATP-C-800 catalyst hydrolyzing AB at 25 °C was 566 mol H_2 min^{-1} (mol Rh)$^{-1}$, which was higher than that of RH based catalyst reported in most reports [164–166]. This proves the fact that the surface metal atoms and heteroatoms are connected to each other, and the carbon material doped with heteroatoms will change the catalytic performance of the catalyst, thereby increasing the dehydrogenation rate. [167]. Thus, they speculated that the role of ATP-C enrichment of N and P atoms not only disperses Rh NPs and resides in the aggregation of metal NP, but also makes Rh NP have more accessible surface active sites.

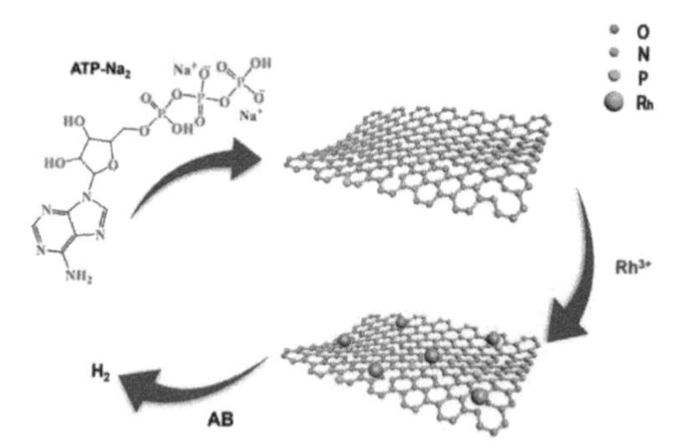

Figure 17. Illustration of the synthesis process of Rh/ATP-C; copyright (2020), Nanoscale Advances.

Carbon dots (CDs) has various structures, low price, easy doping (including N, B, s, P, etc.) and non-toxic. CDs are called excellent catalyst supports due to their special electron transfer properties and high specific surface area [168]. Their surfaces have many catalytically favorable positions and can support a variety of surface functional groups (such as -NH_2, -OH and -COOH). CDs doped with heteroatoms are conducive to adsorb hydrogen intermediates through transforming the electronic structure of the catalytically active center [169]. It is essential to promote catalytic performance, that is, by accelerating the intermolecular electron transfer to enhance the influence of the interaction between multi-component nanostructures. Lu et al. [170] prepared RuP_2 nanoparticles doped with nitrogen CDs as a catalyst for AB hydrolysis reaction, called RuP_2/CDs. The RuP_2/CDs nanocomposites were successfully prepared by simple physical mixing of CDs, phytic acid and ruthenium ions. Figure 18 illustrated the preparation process of RuP_2/CDs nanocomposites. In view of characterization and testing of catalytic performance, a simple synergistic mechanism could roughly explain the catalytic performance of this composite material. Firstly, AB molecules were adsorbed on RuP_2 NPs, which activated the breaking of B-H and O-H bonds; secondly, adjacent C and N atoms could simultaneously activate water to promote the transfer of protons from RuP_2 to the water inside the carbon nanosheets. The mechanism of hydrogen production by AB was that the dissociated H atoms in the B-H bond of AB molecules could combine with protonated water molecules to form hydrogen molecules. RuP_2 NPs and nitrogen-doped CDs could act as bifunctional active sites, activating AB and water molecules, thus, significantly accelerating the release of hydrogen. In addition, the nanosheets morphology of CDs further increased the catalytic activity of hydrogen production by strengthening the utilization rate of active centers.

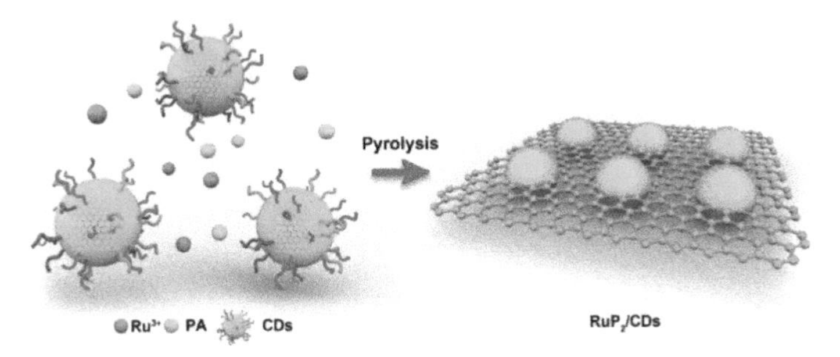

Figure 18. Illustration of the synthesis of the RuP2/carbon dots (CDs) nanocomposites; copyright 2020, American Chemical Society.

3.3.3. Carbon Nanotubes Material Supported Metal Catalysts

CNT is a new kind of honeycomb lattice graphene layer, whose outer diameter ranges from 1 to 100 nm [171]. According to the number of graphene layers, carbon nanotubes can be single-wall (SWCNT) and multi-wall (MWCNT) [172]. Carbon nanotubes are very attractive as catalyst carriers because they have a high surface area and provide a high proportion of nanoparticles. Moreover, the contact surface between the reactant and the active region is greatly increased [173]. Their mesoporous structure is suitable for increasing the mass transfer rate between the reactant and the active center, so it has a significant impact on the catalytic activity [174].

Recently, except for engineering the surface chemistry of CNT support, adding surface ligands could be another potential strategy to engineer metal electronic properties owing to its flexible capacity for demand of stability [60]. Along this line, Fu W et al. [175] proposed a new strategy to engineer the surface of catalysts and electronic properties of Pt/CNT using polyoxometalates (POMs) as the ligands in 2019. They designed three types of POMs including silicotungstic acid (STA), phosphotungstic acid (PTA) and molybdphosphoric acid (PMA), respectively, which were established and analyzed by a combination of kinetic and isotopic analyses with various characterization techniques. It could be obviously showed in Figure 19 that the rate of hydrogen generation was extremely sensitive to the kinds of POMs following the order of STA-Pt/CNT > PTA-Pt/CNT > Pt/CNT > PMA-Pt/CNT, with the STA-Pt/CNT having the highest rate of production hydrogen, which indicated the promotion effects of STA on the catalytic reaction. According to all characterization techniques, the analytic results indicated that the STA compared to the PTA and the PMA acted as a good receptor to increase the binding energy of Pt in order to improve hydrogen production efficiency and catalyst durability. From these experiment consequences, choosing POMs based on their electron-absorbing/donating properties was of vital importance to adjust the electronic nature of catalysts.

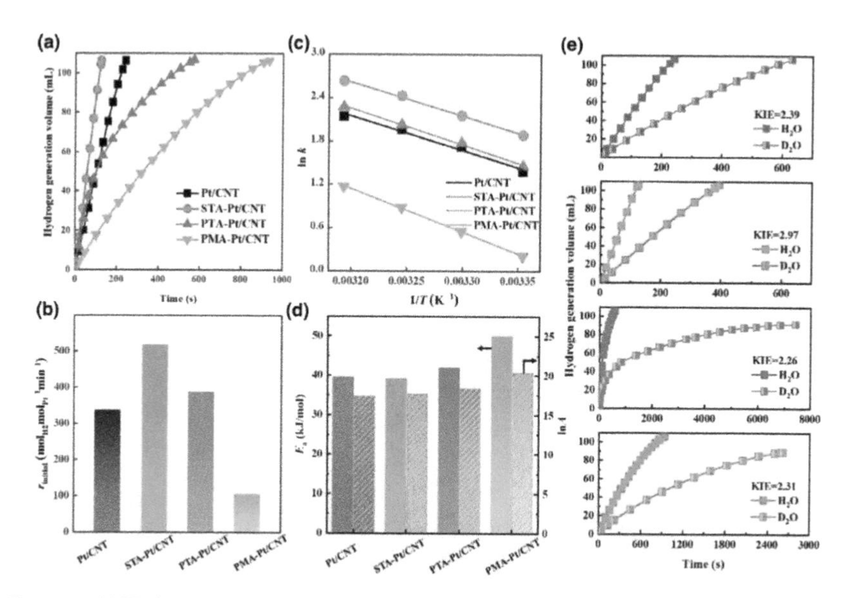

Figure 19. (**a**) Hydrogen generation volume as a function of time; (**b**) the initial hydrogen generation rate ($r_{initial}$) over Pt/CNT, silicotungstic acid (STA)-Pt/CNT, phosphotungstic acid (PTA)-Pt/CNT and molybdphosphoric acid (PMA)-Pt/CNT catalysts. Reaction conditions: 30 °C, n_{Pt}: n_{AB}: $n_{W/Mo}$ = 1:420:40, m_{cat} = 0.025 g, c_{AB} = 0.01gmL^{-1}; (**c**) ink as a function of 1/T; (**d**) the corresponding activation energy (Ea) and the logarithm of pre-exponential factor (ln A); (**e**) hydrogen generation volume as a function of time at 30 °C over the four catalysts using H_2O or D_2O as the reactant; copyright (2020), Journal of Energy Chemistry.

Akbayrak S et al. [173] reported the in situ formation of ruthenium (0) nanoparticles supported on MWCNT catalyst during AB hydrolysis at room temperature. Ru (III) ions were impregnated on MWCNT surface from aqueous solution of Ru(III) chloride, and then reduced by AB to form multi walled carbon nanotubes, referred to as Ru(0)@MWCNT. The results showed that Ru nanoparticles were well dispersed on the walls of carbon nanotubes in the range of 2.0–3.0 nm. They used Ru(III)@MWCNT sample with various Ru loading (0.73, 1.47, 1.91, 2.26, 2.83 wt%) to provide the same ruthenium concentration. The catalytic activity of Ru(0)@MWCNT with Ru loading of 1.91 wt% Ru was the highest at 25 °C. As the further increase of Ru loading, the catalytic activity of Ru(0)@MWCNT decreases, which was probably due to the agglomeration of nanoparticles, resulting in the decrease of specific surface area and accessibility of active sites. It was expected that the catalytic activity of Ru(0)@MWCNTs was still 41% of its initial catalytic activity even after the fourth operation. The easy preparation and high catalytic performance of Ru(0)@MWCNT revealed that the Ru(0)nanoparticle catalyst supported on MWCNTs was a promising catalyst for the development of efficient and portable hydrogen production system.

3.3.4. Silicon Dioxide Material Supported Metal Catalysts

In recent years, metal nanoparticles in porous silica shell have attracted people's attention because of the possibility of obtaining nano scale monodisperse particles [176]. The obtained core-shell structure can fully prevent the aggregation of metal nanoparticles through the protection of porous silica shell, so as to improve the stability of metal nanoparticles and the long-term use performance of metal nanoparticles [143].

The core-shell metal NPs not only have the characteristic of heterogeneous metals, but also present distinct chemical and physical performances [177,178]. Moreover, it is particularly noted that a one-pot reduction technology is indeed required to prepare the core-shell heterometallic catalyst. Hu J et al. [148] reported a one-pot synthesis of core-shell-type nanospheres $Pt@SiO_2$, which displayed excellent stability and performance after recycle test for hydrogen generation from AB at room temperature. The detailed measurement and characterization of the nanoparticles were carried out through SEM and TEM techniques. As shown in Figure 20a, The SEM image shown that the average diameter of the prepared $Pt@SiO_2$ was 25 nm, and the particle size was uniform. The monodisperse spherical morphology of $Pt@SiO_2$ could make further efforts to be affirmed by TEM images. It was found in Figure 20b that a single Pt NP as a core with a diameter of 4 nm was availably embedded in the silica nanospheres. $Pt@SiO_2$ NPs and Pt/SiO_2 were investigated for their catalytic activity in AB hydrolysis at room temperature. The results were shown in Figure 21a that the hydrogen precipitation was completed in 7.72, 55.98 and 111, in existence of as-synthesized $Pt@SiO_2$, Pt/SiO_2 and Pt Nps, respectively. Moreover, the hydrogen production rate of AB was $Pt@SiO_2$ > Pt/SiO_2 >Pt > SiO_2. Among all the catalysts, $Pt@SiO_2$ emerged the highest catalytic activity for generation hydrogen in AB aqueous solution while the TOF value can reach 158.6 mol H_2 (mol Pt min)$^{-1}$. Compared with other core-shell catalysts, Pt-based nanocatalysts had higher catalytic activity in the same reaction. In the whole reaction process, after five times of operation, the catalytic activity of $Pt@SiO_2$ NPs catalyst did not decrease significantly owing to the metal core covered by the silicon shell as Figure 21b shown, which exhibited the excellent stability of the nucleocapsid structure catalyst. In NP-5/cyclohexane reverse micelle system, $Ru@SiO_2$ core-shell nanospheres were successfully prepared [69]. The results of TEM and EDX exhibited that with the increase of Ru loading, the amount of RuNP in SiO_2 spherical particles increased. At room temperature, the synthesized $Ru@SiO_2$ catalyst had excellent catalytic activity and good durability for the aqueous solution of the AB. The activation energy of $Ru@SiO_2$ was estimated to be about 38.2 KJ/mol, which was lower than that of many different Ru-based and other noble metal catalysts for the hydrolysis of AB, indicating that these core-shell nanospheres had excellent catalytic performance.

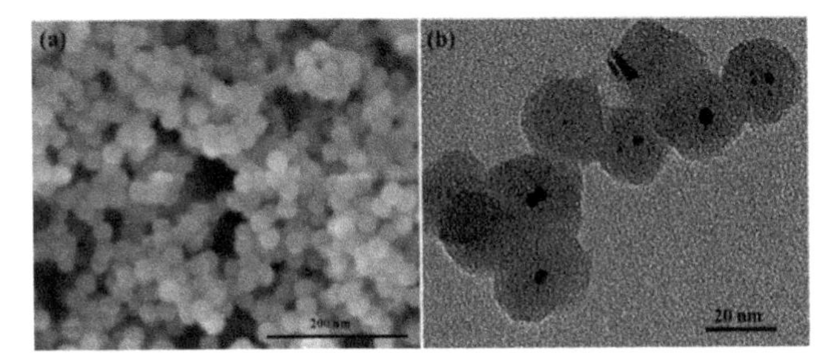

Figure 20. (**a**) SEM, (**b**) TEM images of $Pt@SiO_2$; copyright (2015), Elsevier.

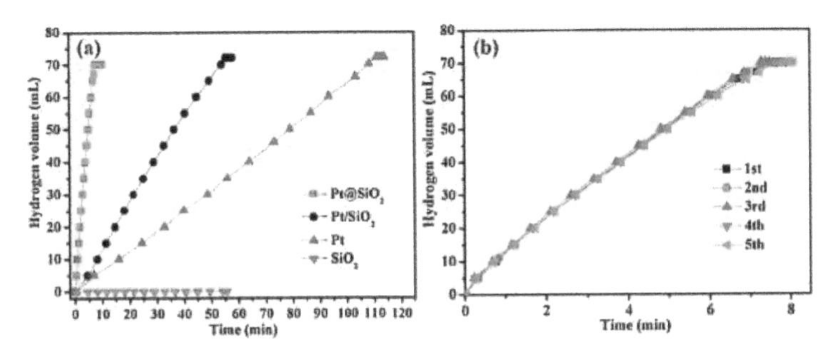

Figure 21. (**a**) Plots of the volume of hydrogen generation from AB (100 mM, 10 mL) hydrolysis as a function of time catalyzed by $Pt@SiO_2$, Pt/SiO_2, Pt and SiO_2 at 25 °C, respectively; (**b**) recyclability of $Pt@SiO_2$; copyright (2015), Elsevier.

Platinum based catalysts have attracted considerable attention according to higher hydrogen generation activities among the transition metal catalysts [179]. Ye et al. [180] synthesized a supported catalyst SiO_2@Pt@NGO, in which nanometer graphene oxidem (NGO) was coated with a layer of 1 nm and the average size of the supported Pt nanoparticles was 1.9 nm (Figure 22). Through the experimental comparative analysis, the activity and stability of AB hydrolyzed hydrogen production would be improved by the increased content of NGO. The enhanced catalytic performance of SiO_2@Pt@NGO could be attributed to the synergistic effects among NGO, Pt nanoparticles and SiO_2, especially the modified electronic structure of Pt nanoparticles by NGO coating.

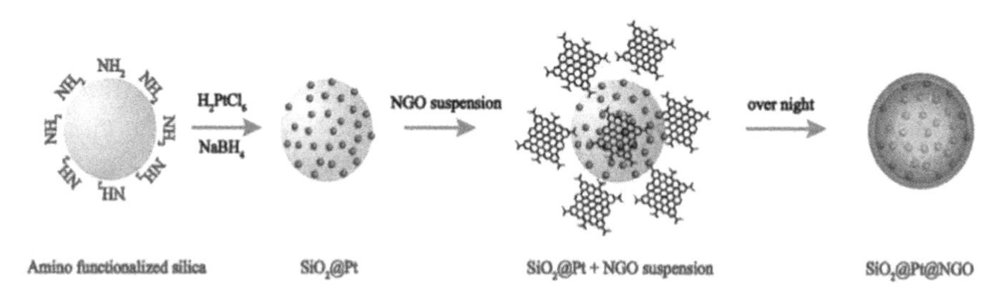

Figure 22. Illustration of the formation process of SiO_2@Pt@NGO; copyright (2017), Sustainable Energy and Fuels.

3.3.5. Cerium Dioxide Material Supported Metal Catalysts

Transition metal nanoparticles tend to aggregate to larger particles, which eventually lead to shorter lifetime. However, reducible oxide supports such as cerium (CeO_2) combined with metal nanoparticles have high activity in many reactions [181–183]. Cerium oxide has cerium (III) defect, which is easy to form due to its favorable large positive reduction potential of $Ce^{4+} \rightarrow Ce^{3+}$ (1.76 v [184] in acid solution). Under the catalytic reaction conditions, the two oxidation states of cerium(IV) and cerium(III) can be mutually converted, that is, cerium oxide can be redox cycled in an aqueous solution [185]. The formation of cerium(III) causes excessive negative charges to accumulate on the surface of the oxide, which enhances the coordination between metal(0) nanoparticles and the oxide surface, thus enhancing the catalytic activity through more favorable substrate metal interaction [186]. Therefore, ceria has been used to improve the catalytic performance of transition metals through strong metal-support interactions, especially electron-rich post-transition metal nanoparticles [187–189]. Therefore, more and more attention has been paid to cerium as a support material to stabilize the anti-aggregation of metal nanoparticles. Tonbul's group [149] reported the preparation, characterization and catalytic application of palladium(0) nanoparticles supported on cerium, Pd^0/CeO_2. Palladium(II) ion impregnated on the surface of nanospheres with an average particle size of 25 nm were reduced to Pd^0/CeO_2 by sodium borohydride, which was used as a catalyst for hydrogen generation from the hydrolysis of AB [190]. The high catalytic activity of Pd^0/CeO_2 was attributed to the reducibility of cerium, that is to say, two kinds of cerium(IV) and cerium(III) transformed each other under catalytic reaction. The formation of cerium(III) led to the accumulation of excessive negative charges on the oxidation surface, which enhanced the ligand interaction between the metal nanoparticles and the oxidation surface. Pd^0/CeO_2 samples with Palladium loading capacity of 1.18 wt% exhibited the highest activity in AB hydrolysis at room temperature.

Özkar et al. [70] prepared rhodium (0) nanocatalyst with cerium (CeO_2), silicon (SiO_2), alumina (Al_2O_3), titanium (TiO_2), zirconia (ZrO_2) and hafnium (HfO_2) as carriers for the hydrolysis of AB under the same conditions, and then investigated the influence of various oxygenate carriers on the catalytic activity of rhodium boron nanoparticles in the hydrogen produced by AB hydrolysis. It was easy to see from the figure description in Figure 23 that in the tested catalyst, rhodium (0) nanoparticles supported on nanoceria revealed the highest catalytic activity in the hydrogen produced by the hydrolysis of AB at room temperature. The resulting Rh^0/CeO_2 with a metal loading of 0.1 wt% Rh had excellent catalytic activity for the hydrogen production from hydrolysis of AB with TOF of 2010 min^{-1}. Rh^0/CeO_2 was a reusable catalyst that retained 67% of its initial catalytic activity, even after the fifth use of hydrogen produced by the hydrolysis of AB at room temperature (TOF = 1350). Rh^0/CeO_2 was a very attractive catalyst for hydrogen generation due to its simple preparation and high catalytic activity as a solid hydrogen storage material.

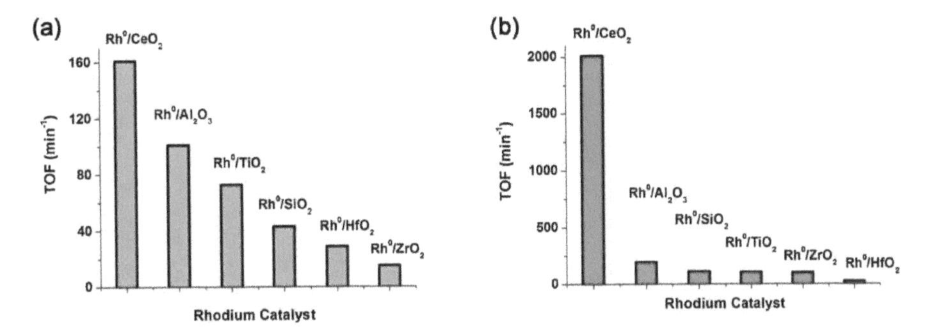

Figure 23. Comparison of TOF (turnover frequency in mol H_2/(mol Rh × min)) values of rhodium nanoparticles supported on different oxides at (**a**) high and (**b**) low rhodium loadings of catalysts used in hydrogen generation from the hydrolysis of ammonia borane (10 mL,100 mM) at 25.0 °C; copyright (2016), Elsevier.

Heterogeneous catalytic liquid phase selective hydrogenation is widely used in chemical synthesis in industry. However, active nanoparticles (such as Pd) have high conversion and selectivity, especially under mild conditions, while preventing aggregation/leaching [191]. Li et al. [192] prepared CeO_2 nanotubes/Pd@MIL-53 (Al) sandwich structure catalyst to solve these problems, in which MIL-53 (Al) porous shell can effectively stabilize Pd nanoparticles. The CeO_2 nanotubes/million Pd-53(Al) were synthesized under mild conditions without any surfactant or carrier surface modification, as shown in Figure 24. Compared with CeO_2 nanotubes/Pd and Pd/MIL-53 (Al), due to the promotion effect of CeO_2 and the enrichment/sieving effect of MIL-53(Al), CeO_2 nanotubes/Pd@MIL-53(Al) exhibited the highest catalytic performance in terms of conversion rate and selectivity.

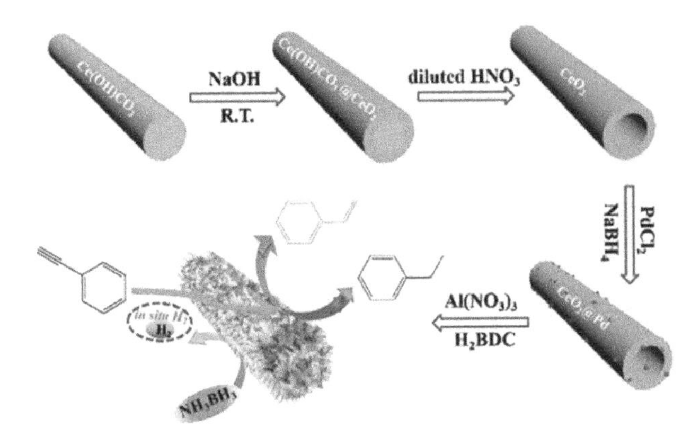

Figure 24. Illustration of the formation of CeO_2 nanotube/Pd@MIL–53(Al); copyright 2020, Wiley.

3.3.6. Titanium Dioxide Material Supported Metal Catalysts

The specific surface area of porous titanium dioxide is between 10–300 m^2/g, which can be used as a carrier to avoid the diffusion problem in porous materials [193]. In recent years, people pay more and more attention to the research of titanium dioxide as precious metal carrier material, because its chemical stability, interesting optical, antibacterial and catalytic properties, titanium dioxide has been widely used in the fields of filler, catalyst carriers and photocatalysts [194,195].

Akbayrak S et al. [150] reported that nanotitanium supported ruthenium (0) nanoparticles as catalysts for AB hydrolysis to produce hydrogen. Ru(0)/TiO_2 exhibited high catalytic activity in hydrogen generation from the hydrolysis of AB its TOF value showing as high as 241 min^{-1} at room temperature. This catalytic activity was due to the dispersion of small nanoparticles on the large outer surface of TiO_2 nanoparticles. Ru(0)/TiO_2 was reusable catalysts, because it provided complete hydrolysis of AB generating three mol H_2 per mole of AB in the third run, but the catalytic activity had no significant change. Furthermore, Ru(0)/TiO_2 was a long-life catalyst, which could provide 71,500 cycles of hydrogen production by hydrolysis of AB at 25.0 °C. M. Rakap et al. [196] prepared Pd-activated TiO_2-supported Co-Ni-P ternary alloy catalyst (Co-Ni-P/Pd-TiO_2) by chemical deposition method. Cobalt based catalysts were more active and expensive than nickel based catalysts in the hydrolysis of AB. In order to obtain a cheap catalyst with activity around the pure cobalt catalyst, they prepared alloy catalysts by changing the cobalt nickel ratio, and tested their catalytic activity in AB hydrolysis. Figure 25 showed that there are alloy type catalysts with different cobalt nickel ratio in the time curve of hydrogen volume generated by hydrolysis of AB solution. As seen from Figure 26, Co-Ni-P/Pd-TiO_2 catalyst revealed good durability in recycling. Even in the fifth cycle, it exhibited the same catalytic activity as the first cycle. Co-Ni-P/Pd-TiO_2 catalyst had the advantages of high efficiency, low cost and reusability, which made them a promising candidate in the hydrolysis of AB to produce hydrogen.

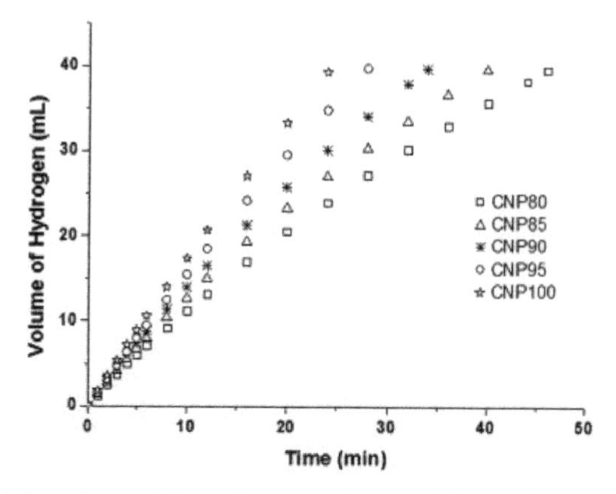

Figure 25. Plot of the volume of H_2 (mL) versus time (min) for the hydrolysis of AB (31.8 mg, 50 mM) catalyzed by Co-Ni-P/Pd-TiO_2 catalysts (25 mg) with different compositions; copyright (2010), Hydrogen Energy.

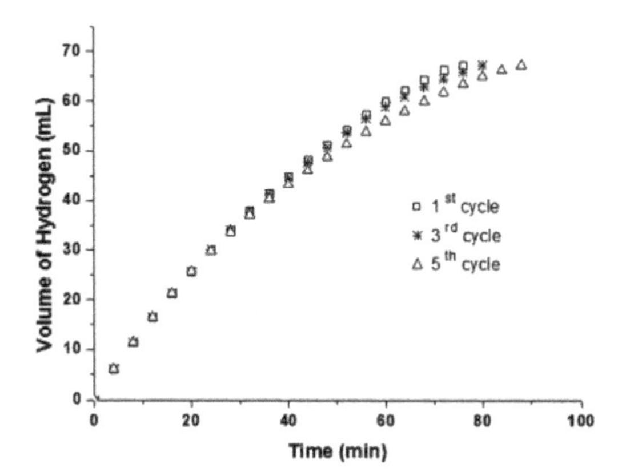

Figure 26. Reusability tests of the Co-Ni-P/Pd-TiO_2 catalyst in the hydrolysis AB (31.8 mg, 50 mM); copyright (2010), Hydrogen Energy.

4. Direction of Development

Ammonia borane is a potential candidate material owing to its unique properties, which has the advantages of high theoretical hydrogen capacity, good solubility, stability, environmental safety, etc. The aqueous solution of AB is very stable at room temperature, but in the presence of a metal catalyst, AB can quickly liberate hydrogen from water. The catalysts for AB hydrolytic dehydrogenation are principally transition metal nanoparticles, including noble metal and non-precious metal catalysts. The noble metals are mainly Pd, Pt, Ru, Rh, while the non-precious are mainly Cu, Fe, Ni and Co.

Noble metals have always displayed outstanding catalytic performance in the catalytic hydrolysis of AB on the basis of existing literature reviews. However, in the practical application, due to the limited precious metal resources and high price, the cost of hydrogen production keeps increasing, which limits the mass production and application of noble metals. In terms of non-precious metals, as abundant resources and low price transition metal, have also been experimentally attested certain level of catalytic capacity for dehydrogenation of AB [197,198]. It has been proved by many related studies that the performance of the catalyst is largely determined by the phase composition, microstructure and surface morphology of the bimetallic or multimetallic catalyst [199–201]. The coordination of electron configuration and geometry among metal components and the effective coordination can make the bimetal or polymetal composites with stronger catalytic capacity [202–205]. In order to highlight the advantages of noble metals and non-precious metals, we reviewed the bimetallic or polymetallic composite catalysts formed by noble metals and non-precious metals. Therefore, the low-cost and effective catalysts composed of precious metals and non-noble metals are worthy of expectation and further intensive study [102]. In addition, we also explore the catalytic effect of noble metal catalysts supported on various support materials for AB hydrolysis. It was found that the supported catalysts can significantly increase the rate of hydrogen evolution under the support of the carriers. Therefore, it is particularly important to find and investigate suitable carriers.

On the other hand, the lattice irregularity is closely related to the catalytic activity and the metal surface properties such as chemisorption and electron transfer. To make ammonia borane hydrogen production widely used, the efficient catalyst can be explored from the aspects of adjusting the degree of chemisorption and electron transfer between catalyst and molecule of AB. It is believed that with the continuous efforts of researchers, metal-catalyzed hydrogen production from AB will be more and more in practical applications.

At present, the biggest challenge in the research of AB hydrogen storage is how to realize efficient regeneration and recycling. Although some evolution has been developed in allusion to recycling of AB from the by-products of pyrolysis, alcoholysis and hydrolytic dehydrogenation, respectively, in this regard, the regeneration yield of AB still needs to be further improved and studied. Compared with the large number of studies on the decomposition of AB to produce hydrogen, there are few studies on the regeneration of AB. Therefore, the regeneration of AB will be the key research direction in the future. In addition, the current market price of ab is relatively expensive, so it is significant to develop a technology suitable for large-scale industrial production of AB in the future, so as to effectively reduce the cost of using AB as hydrogen storage material.

5. Conclusions

Currently, with the continuous progress of science and technology, the consumption of fossil energy causes a serious of environmental problems, thus the development of efficient and clean energy has attracted more and more attention to replace traditional fuels. Hydrogen, as an ideal energy carrier, plays the main representatives role in the field of future new energy. Ammonia Borane, being not only hydrogen storage but hydrogen production, is considered as a potential hydrogen storage material on account of its high theoretical hydrogen content, environmental friendliness, good cycling performance and excellent stability. We analyzed and compared the advantages and disadvantages of three AB decomposition methods, including pyrolysis, alcoholysis and hydrolysis. In these ways of producing hydrogen, compared with alcoholysis, producing many by-products, and pyrolysis under the high temperature condition, hydrolysis has obvious advantages. Nowadays, the research on the hydrolysis of AB primarily focuses on the synthesis of simple catalysts with high stability and good cycling ability, which provides the possibility for the practical application of AB. Furthermore, we explored the effects of using carbon, graphene, carbon nanotubes (CNT), silica, ceria and titanium supported catalysts on the hydrolytic dehydrogenation of AB. Simultaneously, the activity, reusability and turnover frequency (TOF) value of these catalysts for AB dehydrogenation reaction were introduced, and the future development prospects of precious metal catalysts were prospected.

Author Contributions: Conceptualization, C.W. and X.L.; writing—original draft preparation, M.L., L.Z., and X.L.; writing—review and editing, M.L., X.L. and L.X.; supervision, C.W. All authors have read and agreed to the published version of the manuscript.

Funding: This research was supported by Anhui Provincial Natural Science Foundation (1908085QB68), the Natural Science Foundation of the Anhui Higher Education Institutions of China (KJ2019A0072), the Major Science and Technology Project of Anhui Province (201903a05020055), the Foundation of Zhejiang Provincial Key Laboratory of Advanced Chemical Engineering Manufacture Technology (ZJKL-ACEMT-1802), the China Postdoctoral Science Foundation (2019M662060), and the Research Fund for Young Teachers of Anhui University of Technology (QZ201610).

Conflicts of Interest: The authors declare no conflicts of interest.

References

1. Adeniran, B.; Mokaya, R. Compactivation: A mechanochemical approach to carbons with superior porosity and exceptional performance for hydrogen and CO_2 storage. *Nano Energy* **2015**, *16*, 173–185. [CrossRef]
2. Zhou, L.; Sun, L.; Xu, L.X.; Wan, C.; An, Y.; Ye, M.F. Recent Developments of Effective Catalysts for Hydrogen Storage Technology Using N-Ethylcarbazole. *Catalysts* **2020**, *10*, 648. [CrossRef]
3. Zhu, Q.L.; Xu, Q. Liquid organic and inorganic chemical hydrides for high-capacity hydrogen storage. *Energy Environ. Sci.* **2015**, *8*, 478–512.
4. Niaz, S.; Manzoor, T.; Pandith, A.H. Hydrogen storage: Materials, methods and perspectives. *Renew. Sustain. Energy Rev.* **2015**, *50*, 457–469. [CrossRef]
5. Ali, N.; Hussain, A.; Ahmed, R.; Wang, M.K.; Zhao, C.; Haq, B.U.; Fu, Y.Q. Advances in nanostructured thin film materials for solar cell applications. *Renew. Sustain. Energy Rev.* **2016**, *59*, 726–737. [CrossRef]
6. Wan, C.; Zhou, L.; Sun, L.; Xu, L.X.; Cheng, D.G.; Chen, F.Q.; Zhan, X.L.; Yang, Y.R. Boosting visible-light-driven hydrogen evolution from formic acid over AgPd/2D g-C_3N_4 nanosheets Mott-Schottky photocatalyst. *Chem. Eng. J.* **2020**, *396*, 125229.

7. Liu, Q.B.; Zhang, S.J.; Liao, J.Y.; Feng, K.J.; Zheng, Y.Y.; Pollet, B.G.; Li, H. $CuCo_2O_4$ nanoplate film as a low-cost, highly active and durable catalyst towards the hydrolytic dehydrogenation of ammonia borane for hydrogen production. *J. Power Sources* **2017**, *355*, 191–198. [CrossRef]
8. Wan, C.; Cheng, D.G.; Chen, F.Q.; Zhan, X.L. Fabrication of CeO_2 nanotube supported Pt catalyst encapsulated with silica for high and stable performance. *Chem. Commun.* **2015**, *51*, 9785–9788. [CrossRef]
9. Wan, C.; An, Y.; Xu, G.H.; Kong, W.J. Study of catalytic hydrogenation of N-ethylcarbazole over ruthenium catalyst. *Int. J. Hydrog. Energy* **2012**, *37*, 13092–13096. [CrossRef]
10. Wan, C.; Sun, L.; Xu, L.X.; Cheng, D.G.; Chen, F.Q.; Zhan, X.L.; Yang, Y.R. Novel NiPt alloy nanoparticle decorated 2D layered g-C_3N_4 Nanosheets: A highly efficient catalyst for hydrogen generation from hydrous hydrazine. *J. Mater. Chem. A* **2019**, *7*, 8798–8804. [CrossRef]
11. Li, X.F.; Ma, X.F.; Zhang, J.; Akiyama, E.; Wang, Y.F.; Song, X.L. Review of Hydrogen Embrittlement in Metals: Hydrogen Diffusion, Hydrogen Characterization, Hydrogen Embrittlement Mechanism and Prevention. *Acta Metall. Sin.* **2020**, *33*, 1–15. [CrossRef]
12. Zhang, M.Y.; Liu, L.; Lu, S.; Xu, L.X.; An, Y.; Wan, C. Facile Fabrication of NiPt/CNTs as an Efficient Catalyst for Hydrogen Production from Hydrous Hydrazine. *ChemistrySelect* **2019**, *4*, 10494–10500. [CrossRef]
13. Zhu, M.; Xu, L.; Du, L.; An, Y.; Wan, C. Palladium Supported on Carbon Nanotubes as a High-Performance Catalyst for the Dehydrogenation of Dodecahydro-N-ethylcarbazole. *Catalysts* **2018**, *8*, 638. [CrossRef]
14. Attia, N.F.; Lee, S.M.; Kim, H.J.; Geckeler, K.E. Nanoporous polypyrrole: Preparation and hydrogen storage properties. *Int. J. Energy Res.* **2014**, *38*, 466–476. [CrossRef]
15. Oumellal, Y.; Courty, M.; Rougier, A.; Nazri, G.A.; Aymard, L. Electrochemical reactivity of magnesium hydride toward lithium: New synthesis route of nano-particles suitable for hydrogen storage. *Int. J. Hydrog. Energy* **2014**, *39*, 5852–5857. [CrossRef]
16. Kojima, Y. Hydrogen storage materials for hydrogen and energy carriers. *Int. J. Hydrog. Energy* **2019**, *44*, 18179–18192. [CrossRef]
17. Andersson, J.; Grönkvist, S. Large-scale storage of hydrogen. *Int. J. Hydrog. Energy* **2019**, *23*, 11901–11919. [CrossRef]
18. Yao, F.; Li, X.; Wan, C.; Xu, L.X.; An, Y.; Ye, M.F.; Lei, Z. Highly efficient hydrogen release from formic acid using a graphitic carbon nitride-supported AgPd nanoparticle catalyst. *Appl. Surf. Sci.* **2017**, *426*, 605–611. [CrossRef]
19. Semiz, L. Hydrogen generation from ammonia borane by chemically dealloyed platinum nanoparticles. *React. Kinet. Mech. Catal.* **2020**, *129*, 205–218. [CrossRef]
20. Staubitz, A.; Robertson, A.P.; Manners, L. Ammonia-borane and related compounds as dihydrogen sources. *Chem. Rev.* **2010**, *110*, 4079–4124. [CrossRef]
21. Yao, Q.L.; Lu, Z.H.; Huang, W.; Chen, X.S.; Zhu, J. High Pt-like activity of the Ni–Mo/graphene catalyst for hydrogen evolution from hydrolysis of ammonia borane. *J. Mater. Chem. A* **2016**, *4*, 8579–8583. [CrossRef]
22. Sutton, A.D.; Burrell, A.K.; Dixon, D.A.; Garner, E.B.; Gordon, J.C.; Nakagawa, T.; Ott, K.C.; Robinson, P.; Vasiliu, M. Regeneration of ammonia borane spent fuel by direct reaction with hydrazine and liquid ammonia. *Science* **2011**, *331*, 1426–1429. [CrossRef]
23. Smythe, N.C.; Gordon, J.C. Ammonia borane as a hydrogen carrier: Dehydrogenation and regeneration. *Eur. J. Inorg. Chem.* **2010**, 509–521. [CrossRef]
24. Sun, Q.M.; Wang, N.; Bai, R.S.; Hui, Y.; Zhang, T.J.; Do, D.A.; Zhang, P.; Song, L.J.; Miao, S.; Yu, J.H. Synergetic Effect of Ultrasmall Metal Clusters and Zeolites Promoting Hydrogen Generation. *Adv. Sci.* **2019**, *6*, 1802350. [CrossRef]
25. Amali, A.J.; Aranishi, K.; Uchida, T.; Xu, Q. PdPt Nanocubes: A High-Performance Catalyst for Hydrolytic Dehydrogenation of Ammonia Borane. *Part. Part. Syst. Charact.* **2013**, *30*, 888–892. [CrossRef]
26. Roy, B.; Hajari, A.; Kumar, V.; Manna, J.; Sharma, P. Kinetic model analysis and mechanistic correlation of ammonia borane thermolysis under dynamic heating conditions. *Int. J. Hydrog. Energy* **2018**, *43*, 10386–10395. [CrossRef]
27. Yan, J.M.; Zhang, X.B.; Shioyama, H.; Xu, Q. Room temperature hydrolytic dehydrogenation of ammonia borane catalyzed by Co nanoparticles. *J. Power Sources* **2010**, *195*, 1091–1094. [CrossRef]
28. Chandra, M.; Xu, Q. A high-performance hydrogen generation system: Transition metal-catalyzed dissociation and hydrolysis of ammonia-borane. *J. Power Sources* **2006**, *156*, 190–194. [CrossRef]

29. Sun, D.H.; Mazumder, V.; Metin, O.; Sun, S.H. Hethanolysis of ammonia borane by CoPd nanoparticles. *ACS Catal.* **2012**, *2*, 1290–1295. [CrossRef]
30. Marder, T.B. Will we soon be fueling our automobiles with ammonia–borane? *Angew. Chem. Int. Ed.* **2007**, *46*, 8116–8118. [CrossRef]
31. Gutowska, A.; Li, L.Y.; Shin, Y.S.; Wang, C.M.; Li, X.H.; Linehan, J.C.; Smith, R.S.; Kay, B.D.; Schmid, B.; Shaw, W.; et al. Nanoscaffold mediates hydrogen release and the reactivity of ammonia borane. *Angew. Chem. Int. Ed.* **2005**, *44*, 3578–3582. [CrossRef]
32. Li, Z.Y.; Zhu, G.Z.; Lu, G.Q.; Qiu, S.L.; Yao, X.D. Ammonia borane confined by a metal– organic framework for chemical hydrogen storage: Enhancing kinetics and eliminating ammonia. *J. Am. Chem. Soc.* **2010**, *132*, 1490–1491. [CrossRef] [PubMed]
33. Stephens, F.H.; Baker, R.T.; Matus, M.H.; Grant, D.J.; Dixon, D.A. Acid initiation of ammonia–borane dehydrogenation for hydrogen storage. *Angew. Chem. Int. Ed.* **2007**, *46*, 746–749. [CrossRef] [PubMed]
34. Wang, L.B.; Li, H.L.; Zhang, W.B.; Zhao, X.; Qiu, J.X.; Li, A.W.; Zheng, X.S.; Hu, Z.P.; Si, R.; Zeng, J. Supported rhodium catalysts for ammonia–borane hydrolysis: Dependence of the catalytic activity on the highest occupied state of the single rhodium atoms. *Angew. Chem. Int. Ed.* **2017**, *56*, 4712–4718. [CrossRef]
35. Hamilton, C.W.; Baker, R.T.; Staubitz, A.; Manners, I. B–N compounds for chemical hydrogen storage. *Chem. Soc. Rev.* **2009**, *38*, 279–293. [CrossRef]
36. Diwan, M.; Diakov, V.; Shafirovich, E.; Varma, A. Noncatalytic hydrothermolysis of ammonia borane. *Int. J. Hydrog. Energy* **2008**, *33*, 1135–1141. [CrossRef]
37. Diwan, M.; Hanna, D.; Varma, A. Method to release hydrogen from ammonia borane for portable fuel cell applications. *Int. J. Hydrog. Energy* **2010**, *35*, 577–584. [CrossRef]
38. Demirci, U.B.; Akdim, O.; Miele, P. Ten-year efforts and a no-go recommendation for sodium borohydride for on-board automotive hydrogen storage. *Int. J. Hydrog. Energy* **2009**, *34*, 2638–2645. [CrossRef]
39. Hu, M.G.; Geanangel, R.A.; Wendlandt, W.W. The thermal decomposition of ammonia borane. *Thermochim. Acta* **1978**, *23*, 249–255. [CrossRef]
40. Nakagawa, Y.; Ikarashi, Y.; Isobe, S.; Hino, S.; Ohnuki, S. Ammonia borane–metal alanate composites: Hydrogen desorption properties and decomposition processes. *RSC Adv.* **2014**, *4*, 20626–20631. [CrossRef]
41. Dong, H.L.; Berke, H. A mild and efficient rhenium-catalyzed transfer hydrogenation of terminal olefins using alcoholysis of amine–borane adducts as a reducing system. *J. Organomet. Chem.* **2011**, *696*, 1803–1808. [CrossRef]
42. Chen, H.; Yu, Z.J.; Xu, D.D.; Li, Y.; Wang, M.M.; Xia, L.M.; Shu-Ping, L. In-Situ Formed Amorphous Co Nanoparticles for Efficiently Catalytic Hydrogen Production from the Methanolysis of Ammonia Borane. *Chin. J. Inorg. Chem.* **2019**, *35*, 141–148.
43. Yu, C.; Fu, J.J.; Muzzio, M.; Shen, T.L.; Su, D.; Zhu, J.J.; Sun, S.H. CuNi nanoparticles assembled on graphene for catalytic methanolysis of ammonia borane and hydrogenation of nitro/nitrile compounds. *Chem. Mater.* **2017**, *29*, 1413–1418. [CrossRef]
44. Özhava, D.; Kılıçaslan, N.Z.; Özkar, S. PVP-stabilized nickel (0) nanoparticles as catalyst in hydrogen generation from the methanolysis of hydrazine borane or ammonia borane. *Appl. Catal. B Environ.* **2015**, *162*, 573–582. [CrossRef]
45. Heldebrant, D.J.; Karkamkar, A.; Hess, N.J.; Bowden, M.; Rassat, S.; Zheng, F.; Kenneth, R.; Autrey, T. The effects of chemical additives on the induction phase in solid-state thermal decomposition of ammonia borane. *Chem. Mater.* **2008**, *20*, 5332–5336. [CrossRef]
46. Rueda, M.; Sanz-Moral, L.M.; Segovia, J.J.; Martín, Á. Improvement of the kinetics of hydrogen release from ammonia borane confined in silica aerogel. *Microporous Mesoporous Mater.* **2017**, *237*, 189–200. [CrossRef]
47. Zhang, X.Y.; Kam, L.; Trerise, R.; Williams, T.J. Ruthenium-catalyzed ammonia borane dehydrogenation: Mechanism and utility. *Acc. Chem. Res.* **2017**, *50*, 86–95. [CrossRef]
48. Gil-San-Millan, R.; Grau-Atienza, A.; Johnson, D.T.; Rico-Francés, S.; Serrano, E.; Linares, N.; Garcia-Martinez, J. Improving hydrogen production from the hydrolysis of ammonia borane by using multifunctional catalysts. *Int. J. Hydrog. Energy* **2018**, *43*, 17100–17111. [CrossRef]
49. Du, X.Q.; Yang, C.L.; Zeng, X.; Wu, T.; Zhou, Y.H.; Cai, P.; Cheng, G.Z.; Luo, W. Amorphous NiP supported on rGO for superior hydrogen generation from hydrolysis of ammonia borane. *Int. J. Hydrog. Energy* **2017**, *42*, 14181–14187. [CrossRef]

50. Zhang, M.Y.; Xiao, X.; Wu, Y.; An, Y.; Xu, L.X.; Wan, C. Hydrogen Production from Ammonia Borane over PtNi Alloy Nanoparticles Immobilized on Graphite Carbon Nitride. *Catalysts* **2019**, *9*, 1009. [CrossRef]
51. Kim, S.K.; Han, W.S.; Kim, T.J.; Kim, T.Y.; Nam, S.W.; Mitoraj, M.; Piekos, L.; Michalak, A.; Hwang, S.J.; Kang, S.O. Palladium catalysts for dehydrogenation of ammonia borane with preferential B–H activation. *J. Am. Chem. Soc.* **2010**, *132*, 9954–9955. [CrossRef] [PubMed]
52. Yousef, A.; Brooks, R.M.; El-Halwany, M.M.; Abutaleb, A.; El-Newehy, M.H.; Al-Deyab, S.S.; Kim, H.Y. Electrospun $CoCr_7C_3$-supported C nanofibers: Effective, durable, and chemically stable catalyst for H_2 gas generation from ammonia borane. *Mol. Catal.* **2017**, *434*, 32–38. [CrossRef]
53. Figen, A.K.; Filiz, B.C. Polymeric and metal oxide structured nanofibrous composites fabricated by electrospinning as highly efficient hydrogen evolution catalyst. *J. Colloid Interface Sci.* **2019**, *533*, 82–94. [CrossRef] [PubMed]
54. Alpaydın, C.Y.; Gülbay, S.K.; Colpan, C.O. A review on the catalysts used for hydrogen production from ammonia borane. *Int. J. Hydrog. Energy* **2020**, *45*, 3414–3434. [CrossRef]
55. Brockman, A.; Zheng, Y.; Gore, J. A study of catalytic hydrolysis of concentrated ammonia borane solutions. *Int. J. Hydrog. Energy* **2010**, *35*, 7350–7356. [CrossRef]
56. Manna, J.; Akbayrak, S.; Özkar, S. Palladium (0) nanoparticles supported on polydopamine coated $CoFe_2O_4$ as highly active, magnetically isolable and reusable catalyst for hydrogen generation from the hydrolysis of ammonia borane. *Appl. Catal. B Environ.* **2017**, *208*, 104–115. [CrossRef]
57. Xu, L.X.; Yao, F.; Luo, J.L.; Wan, C.; Ye, M.F.; Cui, P.; An, Y. Facile synthesis of amine-functionalized SBA-15-supported bimetallic Au–Pd nanoparticles as an efficient catalyst for hydrogen generation from formic acid. *RSC Adv.* **2017**, *7*, 4746–4752. [CrossRef]
58. Xu, Q.; Chandra, M. A portable hydrogen generation system: Catalytic hydrolysis of ammonia–borane. *J. Alloy. Compd.* **2007**, *446*, 729–732. [CrossRef]
59. Chandra, M.; Xu, Q. Room temperature hydrogen generation from aqueous ammonia-borane using noble metal nano-clusters as highly active catalysts. *J. Power Sources* **2007**, *168*, 135–142. [CrossRef]
60. Chen, W.Y.; Ji, J.; Duan, X.Z.; Qian, G.; Li, P.; Zhou, X.G.; Chen, D.; Yuan, W.K. Unique reactivity in Pt/CNT catalyzed hydrolytic dehydrogenation of ammonia borane. *Chem. Commun.* **2014**, *50*, 2142–2144. [CrossRef] [PubMed]
61. Durap, F.; Zahmakıran, M.; Özkar, S. Water soluble laurate-stabilized ruthenium (0) nanoclusters catalyst for hydrogen generation from the hydrolysis of ammonia-borane: High activity and long lifetime. *Int. J. Hydrog. Energy* **2009**, *34*, 7223–7230. [CrossRef]
62. Cao, N.; Luo, W.; Cheng, G.Z. One-step synthesis of graphene supported Ru nanoparticles as efficient catalysts for hydrolytic dehydrogenation of ammonia borane. *Int. J. Hydrog. Energy* **2013**, *38*, 11964–11972. [CrossRef]
63. Liang, H.Y.; Chen, G.Z.; Desinan, S.; Rosei, R.; Rosei, F.; Ma, D.L. In situ facile synthesis of ruthenium nanocluster catalyst supported on carbon black for hydrogen generation from the hydrolysis of ammonia-borane. *Int. J. Hydrog. Energy* **2012**, *37*, 17921–17927. [CrossRef]
64. Du, C.; Ao, Q.; Cao, N.; Yang, L.; Luo, W.; Cheng, G.Z. Facile synthesis of monodisperse ruthenium nanoparticles supported on graphene for hydrogen generation from hydrolysis of ammonia borane. *Int. J. Hydrog. Energy* **2015**, *40*, 6180–6187. [CrossRef]
65. Wen, L.; Su, J.; Wu, X.J.; Cai, P.; Luo, W.; Cheng, G.Z. Ruthenium supported on MIL-96: An efficient catalyst for hydrolytic dehydrogenation of ammonia borane for chemical hydrogen storage. *Int. J. Hydrog. Energy* **2014**, *39*, 17129–17135. [CrossRef]
66. Yang, K.Z.; Zhou, L.Q.; Yu, G.F.; Xiong, X.; Ye, M.L.; Li, Y.; Lu, D.; Pan, Y.X.; Chen, M.H.; Zhang, L.; et al. Ru nanoparticles supported on MIL-53 (Cr, Al) as efficient catalysts for hydrogen generation from hydrolysis of ammonia borane. *Int. J. Hydrog. Energy* **2016**, *41*, 6300–6309. [CrossRef]
67. Fan, G.Y.; Liu, Q.Q.; Tang, D.M.; Li, X.J.; Bi, J.; Gao, D.J. Nanodiamond supported Ru nanoparticles as an effective catalyst for hydrogen evolution from hydrolysis of ammonia borane. *Int. J. Hydrog. Energy* **2016**, *41*, 1542–1549. [CrossRef]
68. Wu, Z.J.; Duan, Y.L.; Ge, S.H.; Yip, A.C.; Yang, F.; Li, Y.F.; Dou, T. Promoting hydrolysis of ammonia borane over multiwalled carbon nanotube-supported Ru catalysts via hydrogen spillover. *Catal. Commun.* **2017**, *91*, 10–15. [CrossRef]

69. Yao, Q.L.; Shi, W.M.; Feng, G.; Lu, Z.H.; Zhang, X.L.; Tao, D.J.; Kong, D.J.; Chen, X.S. Ultrafine Ru nanoparticles embedded in SiO_2 nanospheres: Highly efficient catalysts for hydrolytic dehydrogenation of ammonia borane. *J. Power Sources* **2014**, *257*, 293–299. [CrossRef]
70. Akbayrak, S.; Tonbul, Y.; Özkar, S. Ceria supported rhodium nanoparticles: Superb catalytic activity in hydrogen generation from the hydrolysis of ammonia borane. *Appl. Catal. B Environ.* **2016**, *198*, 162–170. [CrossRef]
71. Özhava, D.; Özkar, S. Rhodium (0) nanoparticles supported on nanosilica: Highly active and long lived catalyst in hydrogen generation from the methanolysis of ammonia borane. *Appl. Catal. B Environ.* **2016**, *181*, 716–726. [CrossRef]
72. Roy, S.; Pachfule, P.; Xu, Q. High Catalytic Performance of MIL-101-Immobilized NiRu Alloy Nanoparticles towards the Hydrolytic Dehydrogenation of Ammonia Borane. *Eur. J. Inorg. Chem.* **2016**, *2016*, 4353–4357. [CrossRef]
73. Yao, Q.L.; Lu, Z.H.; Wang, Y.Q.; Chen, X.S.; Feng, G. Synergetic catalysis of non-noble bimetallic Cu–Co nanoparticles embedded in SiO_2 nanospheres in hydrolytic dehydrogenation of ammonia borane. *J. Phys. Chem. C* **2015**, *119*, 14167–14174. [CrossRef]
74. Fernandes, R.; Patel, N.; Edla, R.; Bazzanella, N.; Kothari, D.C.; Miotello, A. Ruthenium nanoparticles supported over carbon thin film catalyst synthesized by pulsed laser deposition for hydrogen production from ammonia borane. *Appl. Catal. A Gen.* **2015**, *495*, 23–29. [CrossRef]
75. Park, J.W.; Lai, S.W.; Cho, S.O. Catalytic hydrogen generation from hydrolysis of ammonia borane using octahedral Au@ Pt nanoparticles. *Int. J. Hydrog. Energy* **2015**, *40*, 16316–16322. [CrossRef]
76. Shen, J.F.; Yang, L.; Hu, K.; Luo, W.; Cheng, G.Z. Rh nanoparticles supported on graphene as efficient catalyst for hydrolytic dehydrogenation of amine boranes for chemical hydrogen storage. *Int. J. Hydrog. Energy* **2015**, *40*, 1062–1070. [CrossRef]
77. Xia, B.Q.; Liu, C.; Wu, H.; Luo, W.; Cheng, G.Z. Hydrolytic dehydrogenation of ammonia borane catalyzed by metal-organic framework supported bimetallic RhNi nanoparticles. *Int. J. Hydrog. Energy* **2015**, *40*, 16391–16397. [CrossRef]
78. Shang, N.Z.; Feng, C.; Gao, S.T.; Wang, C. Ag/Pd nanoparticles supported on amine-functionalized metal–organic framework for catalytic hydrolysis of ammonia borane. *Int. J. Hydrog. Energy* **2016**, *41*, 944–950. [CrossRef]
79. Rakap, M. The highest catalytic activity in the hydrolysis of ammonia borane by poly (N-vinyl-2-pyrrolidone)-protected palladium–rhodium nanoparticles for hydrogen generation. *Appl. Catal. B Environ.* **2015**, *163*, 129–134. [CrossRef]
80. Sullivan, J.A.; Herron, R.; Phillips, A.D. Towards an understanding of the beneficial effect of mesoporous materials on the dehydrogenation characteristics of NH_3BH_3. *Appl. Catal. B Environ.* **2017**, *201*, 182–188. [CrossRef]
81. Zhang, T.R.; Yang, X.J.; Yang, S.Q.; Li, D.X.; Cheng, F.Y.; Tao, Z.L.; Chen, J. Silica hollow nanospheres as new nanoscaffold materials to enhance hydrogen releasing from ammonia borane. *Phys. Chem. Chem. Phys.* **2011**, *13*, 18592–18599. [CrossRef] [PubMed]
82. Xu, L.X.; Liu, N.; Hong, B.; Cui, P.; Cheng, D.G.; Chen, F.Q.; An, Y.; Wan, C. Nickel–platinum nanoparticles immobilized on graphitic carbon nitride as highly efficient catalyst for hydrogen release from hydrous hydrazine. *RSC Adv.* **2016**, *6*, 31687–31691. [CrossRef]
83. Kuhn, J.N.; Huang, W.Y.; Tsung, C.K.; Zhang, Y.W.; Somorjai, G.A. Structure sensitivity of carbon−nitrogen ring opening: Impact of platinum particle size from below 1 to 5 nm upon pyrrole hydrogenation product selectivity over monodisperse platinum nanoparticles loaded onto mesoporous silica. *J. Am. Chem. Soc.* **2008**, *130*, 14026–14027. [CrossRef] [PubMed]
84. Tsung, C.K.; Kuhn, J.N.; Huang, W.Y.; Aliaga, C.; Hung, L.I.; Somorjai, G.A.; Yang, P. Sub-10 nm platinum nanocrystals with size and shape control: Catalytic study for ethylene and pyrrole hydrogenation. *J. Am. Chem. Soc.* **2009**, *131*, 5816–5822. [CrossRef]
85. Arenz, M.; Mayrhofer, K.J.; Stamenkovic, V.; Blizanac, B.B.; Tomoyuki, T.; Ross, P.N.; Markovic, N.M. The effect of the particle size on the kinetics of CO electrooxidation on high surface area Pt catalysts. *J. Am. Chem. Soc.* **2005**, *127*, 6819–6829. [CrossRef] [PubMed]

86. Allian, A.D.; Takanabe, K.; Fujdala, K.L.; Hao, X.H.; Truex, T.J.; Cai, J.; Buda, C.; Neurock, M.; Iglesia, E. Chemisorption of CO and mechanism of CO oxidation on supported platinum nanoclusters. *J. Am. Chem. Soc.* **2011**, *133*, 4498–4517. [CrossRef]
87. Chin, Y.H.; Buda, C.; Neurock, M.; Iglesia, E. Reactivity of chemisorbed oxygen atoms and their catalytic consequences during CH_4–O_2 catalysis on supported Pt clusters. *J. Am. Chem. Soc.* **2011**, *133*, 15958–15978. [CrossRef] [PubMed]
88. Yao, K.S.; Zhao, C.C.; Wang, N.; Li, T.J.; Lu, W.W.; Wang, J.J. An aqueous synthesis of porous PtPd nanoparticles with reversed bimetallic structures for highly efficient hydrogen generation from ammonia borane hydrolysis. *Nanoscale* **2020**, *12*, 638–647. [CrossRef]
89. Zhou, M.; Wang, H.L.; Vara, M.; Hood, Z.D.; Luo, M.; Yang, T.H.; Bao, S.X.; Chi, M.F.; Xiao, P.; Zhang, Y.H.; et al. Quantitative analysis of the reduction kinetics responsible for the one-pot synthesis of Pd–Pt bimetallic nanocrystals with different structures. *J. Am. Chem. Soc.* **2016**, *138*, 12263–12270. [CrossRef]
90. Käß, M.; Friedrich, A.; Drees, M.; Schneider, S. Ruthenium complexes with cooperative PNP ligands: Bifunctional catalysts for the dehydrogenation of ammonia–borane. *Angew. Chem. Int. Ed.* **2009**, *48*, 905–907. [CrossRef]
91. Wen, L.; Zheng, Z.; Luo, W.; Cai, P.; Cheng, G.Z. Ruthenium deposited on MCM-41 as efficient catalyst for hydrolytic dehydrogenation of ammonia borane and methylamine borane. *Chin. Chem. Lett.* **2015**, *26*, 1345–1350. [CrossRef]
92. Xu, C.X.; Su, J.X.; Xu, X.H.; Liu, P.P.; Zhao, H.J.; Tian, F.; Ding, Y. Low temperature CO oxidation over unsupported nanoporous gold. *J. Am. Chem. Soc.* **2007**, *129*, 42–43. [CrossRef] [PubMed]
93. Yu, L.B.; Shi, Y.Y.; Zhao, Z.; Yin, H.B.; Wei, Y.C.; Liu, J.; Kang, W.B.; Jiang, T.S.; Wang, A.L. Ultrasmall silver nanoparticles supported on silica and their catalytic performances for carbon monoxide oxidation. *Catal. Commun.* **2011**, *12*, 616–620. [CrossRef]
94. Huang, L.; Zou, J.S.; Ye, J.Y.; Zhou, Z.Y.; Lin, Z.; Kang, X.W.; Jain, P.; Chen, S.W. Synergy between Plasmonic and Electrocatalytic Activation of Methanol Oxidation on Palladium–Silver Alloy Nanotubes. *Angew. Chem.* **2019**, *131*, 8886–8890. [CrossRef]
95. Wu, Y.; Jiao, L.; Xu, W.Q.; Gu, W.L.; Zhu, C.Z.; Du, D.; Lin, Y.H. Polydopamine-Capped Bimetallic AuPt Hydrogels Enable Robust Biosensor for Organophosphorus Pesticide Detection. *Small* **2019**, *15*, 1900632. [CrossRef]
96. Lv, H.; Sun, L.Z.; Zou, L.; Xu, D.D.; Yao, H.Q.; Liu, B. Size-dependent synthesis and catalytic activities of trimetallic PdAgCu mesoporous nanospheres in ethanol electrooxidation. *Chem. Sci.* **2019**, *10*, 1986–1993. [CrossRef]
97. Abo-Hamed, E.; Pennycook, T.; Vaynzof, Y.; Toprakcioglu, C.; Koutsioubas, A.; Scherman, O.A. Highly active metastable ruthenium nanoparticles for hydrogen production through the catalytic hydrolysis of ammonia borane. *Small* **2014**, *10*, 3145–3152. [CrossRef]
98. Wan, C.; An, Y.; Chen, F.Q.; Cheng, D.G.; Wu, F.Y.; Xu, G.H. Kinetics of N-ethylcarbazole hydrogenation over a supported Ru catalyst for hydrogen storage. *Int. J. Hydrogen Energy* **2013**, *38*, 7065–7069. [CrossRef]
99. Du, J.; Cheng, F.Y.; Si, M.; Liang, J.; Tao, Z.L.; Chen, J. Nanoporous Ni-based catalysts for hydrogen generation from hydrolysis of ammonia borane. *Int. J. Hydrog. Energy* **2013**, *38*, 5768–5774. [CrossRef]
100. Li, Y.; Dai, Y.; Tian, X.K. Controlled synthesis of monodisperse Pd_xSn_{100-x} nanoparticles and their catalytic activity for hydrogen generation from the hydrolysis of ammonia-borane. *Int. J. Hydrog. Energy* **2015**, *40*, 9235–9243.
101. Güngörmez, K.; Metin, Ö. Composition-controlled catalysis of reduced graphene oxide supported CuPd alloy nanoparticles in the hydrolytic dehydrogenation of ammonia borane. *Appl. Catal. A Gen.* **2015**, *494*, 22–28. [CrossRef]
102. Zhang, J.; Dong, Y.; Liu, Q.X.; Zhou, M.; Mi, G.; Du, X.G. Hierarchically alloyed Pd–Cu microarchitecture with tunable shapes: Morphological engineering, and catalysis for hydrogen evolution reaction of ammonia borane. *Int. J. Hydrog. Energy* **2019**, *44*, 30226–30236. [CrossRef]
103. Deka, J.R.; Saikia, D.; Chen, P.H.; Chen, K.T.; Kao, H.M.; Yang, Y.C. Palladium nanoparticles encapsulated in carboxylic acid functionalized periodic mesoporous organosilicas as efficient and reusable heterogeneous catalysts for hydrogen generation from ammonia borane. *Materials Research Bulletin* **2020**, *125*, 110786. [CrossRef]

104. Wang, K.; Zhang, J.G.; Man, T.T.; Wu, M.; Chen, C.C. Recent process and development of metal aminoborane. *Chem. Asian J.* **2013**, *8*, 1076–1089. [CrossRef] [PubMed]
105. Zhou, X.; Meng, X.F.; Wang, J.M.; Shang, N.Z.; Feng, T.; Gao, Z.Y.; Zhang, H.X.; Ding, X.L.; Gao, S.T.; Feng, C.; et al. Boron nitride supported NiCoP nanoparticles as noble metal-free catalyst for highly efficient hydrogen generation from ammonia borane. *Int. J. Hydrog. Energy* **2019**, *44*, 4764–4770. [CrossRef]
106. Singh, A.K.; Xu, Q. Synergistic catalysis over bimetallic alloy nanoparticles. *ChemCatChem* **2013**, *5*, 652–676. [CrossRef]
107. Yang, L.; Su, J.; Meng, X.Y.; Luo, W.; Cheng, G.Z. In situ synthesis of graphene supported Ag@CoNi core–shell nanoparticles as highly efficient catalysts for hydrogen generation from hydrolysis of ammonia borane and methylamine borane. *J. Mater. Chem. A* **2013**, *1*, 10016–10023. [CrossRef]
108. El-Sayed, M.A. Some interesting properties of metals confined in time and nanometer space of different shapes. *Acc. Chem. Res.* **2001**, *34*, 257–264. [CrossRef]
109. Chen, H.M.; Liu, R.S. Architecture of metallic nanostructures: Synthesis strategy and specific applications. *J. Phys. Chem. C* **2011**, *115*, 3513–3527. [CrossRef]
110. Chen, S.W.; Yang, Y.Y. Magnetoelectrochemistry of gold nanoparticle quantized capacitance charging. *J. Am. Chem. Soc.* **2002**, *124*, 5280–5281. [CrossRef]
111. Yang, X.J.; Cheng, F.Y.; Liang, J.; Tao, Z.L.; Chen, J. Pt_xNi_{1-x} nanoparticles as catalysts for hydrogen generation from hydrolysis of ammonia borane. *Int. J. Hydrog. Energy* **2009**, *34*, 8785–8791. [CrossRef]
112. Gao, M.Y.; Yang, W.W.; Yu, Y.S. Monodisperse PtCu alloy nanoparticles as highly efficient catalysts for the hydrolytic dehydrogenation of ammonia borane. *Int. J. Hydrog. Energy* **2018**, *43*, 14293–14300. [CrossRef]
113. Chen, W.Y.; Fu, W.Z.; Qian, G.; Zhang, B.S.; Chen, D.; Duan, X.Z.; Zhou, X.G. Synergistic Pt-WO_3 Dual Active Sites to Boost Hydrogen Production from Ammonia Borane. *Iscience* **2020**, *23*, 100922. [CrossRef]
114. Akbayrak, S.; Özkar, S. Ammonia borane as hydrogen storage materials. *Int. J. Hydrog. Energy* **2018**, *43*, 18592–18606. [CrossRef]
115. Zahmakiran, M.; Özkar, S. Transition metal nanoparticles in catalysis for the hydrogen generation from the hydrolysis of ammonia-borane. *Top. Catal.* **2013**, *56*, 1171–1183. [CrossRef]
116. Özkar, S.; Finke, R.G. Nanocluster formation and stabilization fundamental studies: Ranking commonly employed anionic stabilizers via the development, then application, of five comparative criteria. *J. Am. Chem. Soc.* **2002**, *124*, 5796–5810. [CrossRef]
117. Shylesh, S.; Schuenemann, V.; Thiel, W.R. Magnetically separable nanocatalysts: Bridges between homogeneous and heterogeneous catalysis. *Angew. Chem. Int. Ed.* **2010**, *49*, 3428–3459. [CrossRef]
118. Baig, R.N.; Varma, R.S. Magnetically retrievable catalysts for organic synthesis. *Chem. Commun.* **2013**, *49*, 752–770. [CrossRef]
119. Wang, D.; Astruc, D. Fast-growing field of magnetically recyclable nanocatalysts. *Chem. Rev.* **2014**, *114*, 6949–6985. [CrossRef]
120. Akbayrak, S.; Çakmak, G.; Öztürk, T.; Özkar, S. Rhodium (0), Ruthenium (0) and Palladium (0) nanoparticles supported on carbon-coated iron: Magnetically isolable and reusable catalysts for hydrolytic dehydrogenation of ammonia borane. *Int. J. Hydrog. Energy* **2020**. [CrossRef]
121. Zhou, Q.X.; Yang, H.X.; Xu, C.X. Nanoporous Ru as highly efficient catalyst for hydrolysis of ammonia borane. *Int. J. Hydrog. Energy* **2016**, *41*, 12714–12721. [CrossRef]
122. Wei, Z.H.; Liu, Y.; Peng, Z.K.; Song, H.Q.; Liu, Z.Y.; Liu, B.Z.; Li, B.Z.; Yang, B.; Lu, S.Y. Cobalt-ruthenium nanoalloys parceled in porous nitrogen-doped graphene as highly efficient difunctional catalysts for hydrogen evolution reaction and hydrolysis of ammonia borane. *ACS Sustain. Chem. Eng.* **2019**, *7*, 7014–7023. [CrossRef]
123. Guo, L.T.; Cai, Y.Y.; Ge, J.M.; Zhang, Y.N.; Gong, L.H.; Li, X.H.; Wang, K.X.; Ren, Q.Z.; Su, J.; Chen, J.S. Multifunctional Au–Co@CN nanocatalyst for highly efficient hydrolysis of ammonia borane. *ACS Catal.* **2015**, *5*, 388–392. [CrossRef]
124. Liu, Y.; Wang, Q.H.; Wu, L.L.; Long, Y.; Li, J.; Song, S.Y.; Zhang, H.J. Tunable bimetallic Au–Pd@CeO_2 for semihydrogenation of phenylacetylene by ammonia borane. *Nanoscale* **2019**, *11*, 12932–12937. [CrossRef] [PubMed]
125. Wang, Q.; Fu, F.Y.; Yang, S.; Martinez Moro, M.; Ramirez, M.D.L.A.; Moya, S.; Moya, S.; Salmon, L.; Ruiz, J.; Astruc, D. Dramatic synergy in CoPt nanocatalysts stabilized by "Click" dendrimers for evolution of hydrogen from hydrolysis of ammonia borane. *ACS Catal.* **2018**, *9*, 1110–1119. [CrossRef]

126. Zhou, Q.X.; Qi, L.; Yang, H.X.; Xu, C.X. Hierarchical nanoporous platinum–copper alloy nanoflowers as highly active catalysts for the hydrolytic dehydrogenation of ammonia borane. *J. Colloid Interface Sci.* **2018**, *513*, 258–265. [CrossRef]
127. Yang, X.; Li, Q.L.; Li, L.L.; Lin, J.; Yang, X.J.; Yu, C.; Liu, Z.Y.; Fang, Y.; Huang, Y.F.; Tang, C.C. CuCo binary metal nanoparticles supported on boron nitride nanofibers as highly efficient catalysts for hydrogen generation from hydrolysis of ammonia borane. *J. Power Sources* **2019**, *431*, 135–143. [CrossRef]
128. Haruta, M.; Daté, M. Advances in the catalysis of Au nanoparticles. *Appl. Catal. A Gen.* **2001**, *222*, 427–437. [CrossRef]
129. Haruta, M.; Yamada, N.; Kobayashi, T.; Iijima, S. Gold catalysts prepared by coprecipitation for low-temperature oxidation of hydrogen and of carbon monoxide. *J. Catal.* **1989**, *115*, 301–309. [CrossRef]
130. Jiang, H.L.; Umegaki, T.; Akita, T.; Zhang, X.B.; Haruta, M.; Xu, Q. Bimetallic Au–Ni nanoparticles embedded in SiO_2 nanospheres: Synergetic catalysis in hydrolytic dehydrogenation of ammonia borane. *Chem. A Eur. J.* **2010**, *16*, 3132–3137. [CrossRef]
131. Rej, S.; Hsia, C.F.; Chen, T.Y.; Lin, F.C.; Huang, J.S.; Huang, M.H. Facet-Dependent and Light-Assisted Efficient Hydrogen Evolution from Ammonia Borane Using Gold–Palladium Core-Shell Nanocatalysts. *Angew. Chem. Int. Ed.* **2016**, *55*, 7222–7226. [CrossRef] [PubMed]
132. Lv, H.F.; Xi, Z.; Chen, Z.Z.; Guo, S.J.; Yu, Y.S.; Zhu, W.L.; Li, Q.; Zhang, X.; Pan, M.; Lu, G.; et al. A new core/shell NiAu/Au nanoparticle catalyst with Pt-like activity for hydrogen evolution reaction. *J. Am. Chem. Soc.* **2015**, *137*, 5859–5862. [CrossRef] [PubMed]
133. Gong, K.P.; Su, D.; Adzic, R.R. Platinum-monolayer shell on $AuNi_{0.5}Fe$ nanoparticle core electrocatalyst with high activity and stability for the oxygen reduction reaction. *J. Am. Chem. Soc.* **2010**, *132*, 14364–14366. [CrossRef] [PubMed]
134. Kang, J.X.; Chen, T.W.; Zhang, D.F.; Guo, L. PtNiAu trimetallic nanoalloys enabled by a digestive-assisted process as highly efficient catalyst for hydrogen generation. *Nano Energy* **2016**, *23*, 145–152. [CrossRef]
135. Yamauchi, Y.; Tonegawa, A.; Komatsu, M.; Wang, H.J.; Wang, L.; Nemoto, Y.; Suzuki, N.; Kuroda, K. Electrochemical synthesis of mesoporous Pt–Au binary alloys with tunable compositions for enhancement of electrochemical performance. *J. Am. Chem. Soc.* **2012**, *134*, 5100–5109. [CrossRef]
136. Fu, L.L.; Zhang, D.F.; Yang, Z.; Chen, T.W.; Zhai, J. PtAuCo Trimetallic Nanoalloys as Highly Efficient Catalysts toward Dehydrogenation of Ammonia Borane. *ACS Sustain. Chem. Eng.* **2020**, *8*, 3734–3742. [CrossRef]
137. Shui, J.L.; Chen, C.; Li, J.C. Evolution of nanoporous Pt–Fe alloy nanowires by dealloying and their catalytic property for oxygen reduction reaction. *Adv. Funct. Mater.* **2011**, *21*, 3357–3362. [CrossRef]
138. Cui, C.H.; Gan, L.; Li, H.H.; Yu, S.H.; Heggen, M.; Strasser, P. Octahedral PtNi nanoparticle catalysts: Exceptional oxygen reduction activity by tuning the alloy particle surface composition. *Nano Lett.* **2012**, *12*, 5885–5889. [CrossRef]
139. Xia, B.Y.; Wu, H.B.; Li, N.; Yan, Y.; Lou, X.W.; Wang, X. One-pot synthesis of Pt-Co alloy nanowire assemblies with tunable composition and enhanced electrocatalytic properties. *Angew. Chem. Int. Ed.* **2015**, *54*, 3797–3801. [CrossRef]
140. Zhang, Z.C.; Luo, Z.M.; Chen, B.; Wei, C.; Zhao, J.; Chen, J.Z.; Zhang, X.; Lai, Z.C.; Fan, Z.X.; Tan, C.L.; et al. One-pot synthesis of highly anisotropic five-fold-twinned PtCu nanoframes used as a bifunctional electrocatalyst for oxygen reduction and methanol oxidation. *Adv. Mater.* **2016**, *28*, 8712–8717. [CrossRef]
141. Bu, L.Z.; Shao, Q.; Bin, E.; Guo, J.; Yao, J.L.; Huang, X.Q. PtPb/PtNi intermetallic core/atomic layer shell octahedra for efficient oxygen reduction electrocatalysis. *J. Am. Chem. Soc.* **2017**, *139*, 9576–9582. [CrossRef] [PubMed]
142. Meng, C.; Ling, T.; Ma, T.Y.; Wang, H.; Hu, Z.P.; Zhou, Y.; Mao, J.; Du, X.W.; Jaroniec, M.; Qiao, S.Z. Atomically and electronically coupled Pt and CoO hybrid nanocatalysts for enhanced electrocatalytic performance. *Adv. Mater.* **2017**, *29*, 1604607. [CrossRef] [PubMed]
143. Xu, C.L.; Hu, M.; Wang, Q.; Fan, G.Y.; Wang, Y.; Zhang, Y.; Gao, D.J.; Bi, J. Hyper-cross-linked polymer supported rhodium: An effective catalyst for hydrogen evolution from ammonia borane. *Dalton Trans.* **2018**, *47*, 2561–2567. [CrossRef]
144. Zhong, F.Y.; Wang, Q.; Xu, C.L.; Yang, Y.C.; Wang, Y.; Zhang, Y.; Gao, D.J.; Bi, J.; Fan, G. Ultrafine and highly dispersed Ru nanoparticles supported on nitrogen-doped carbon nanosheets: Efficient catalysts for ammonia borane hydrolysis. *Appl. Surf. Sci.* **2018**, *455*, 326–332. [CrossRef]

145. Metin, Ö.; Kayhan, E.; Özkar, S.; Schneider, J.J. Palladium nanoparticles supported on chemically derived graphene: An efficient and reusable catalyst for the dehydrogenation of ammonia borane. *Int. J. Hydrog. Energy* **2012**, *37*, 8161–8169. [CrossRef]
146. Lu, R.; Xu, C.L.; Wang, Q.; Wang, Y.; Zhang, Y.; Gao, D.J.; Bi, J.; Fan, G.Y. Ruthenium nanoclusters distributed on phosphorus-doped carbon derived from hypercrosslinked polymer networks for highly efficient hydrolysis of ammonia-borane. *Int. J. Hydrog. Energy* **2018**, *43*, 18253–18260. [CrossRef]
147. Yao, Q.L.; Lu, Z.H.; Jia, Y.S.; Chen, X.S.; Liu, X. In situ facile synthesis of Rh nanoparticles supported on carbon nanotubes as highly active catalysts for H_2 generation from NH_3BH_3 hydrolysis. *Int. J. Hydrog. Energy* **2015**, *40*, 2207–2215. [CrossRef]
148. Hu, Y.J.; Wang, Y.Q.; Lu, Z.H.; Chen, X.S.; Xiong, L.H. Core–shell nanospheres Pt@SiO_2 for catalytic hydrogen production. *Appl. Surf. Sci.* **2015**, *341*, 185–189. [CrossRef]
149. Tonbul, Y.; Akbayrak, S.; Özkar, S. Palladium (0) nanoparticles supported on ceria: Highly active and reusable catalyst in hydrogen generation from the hydrolysis of ammonia borane. *Int. J. Hydrog. Energy* **2016**, *41*, 11154–11162. [CrossRef]
150. Akbayrak, S.; Tanyıldızı, S.; Morkan, İ.; Özkar, S. Ruthenium (0) nanoparticles supported on nanotitania as highly active and reusable catalyst in hydrogen generation from the hydrolysis of ammonia borane. *Int. J. Hydrog. Energy* **2014**, *39*, 9628–9637. [CrossRef]
151. Garaj, S.; Hubbard, W.; Reina, A.; Kong, J.; Branton, D.; Golovchenko, J.A. Graphene as a subnanometre trans-electrode membrane. *Nature* **2010**, *467*, 190–193. [CrossRef] [PubMed]
152. Choi, B.G.; Hong, J.; Park, Y.C.; Jung, D.H.; Hong, W.H.; Hammond, P.T.; Park, H. Innovative polymer nanocomposite electrolytes: Nanoscale manipulation of ion channels by functionalized graphenes. *ACS Nano* **2011**, *5*, 5167–5174. [CrossRef] [PubMed]
153. Novoselov, K.S.; Geim, A.K.; Morozov, S.V.; Jiang, D.; Zhang, Y.; Dubonos, S.V.; Grigorieva, I.V.; Firsov, A.A. Electric field effect in atomically thin carbon films. *Science* **2004**, *306*, 666–669. [CrossRef] [PubMed]
154. Guo, S.J.; Sun, S.H. FePt nanoparticles assembled on graphene as enhanced catalyst for oxygen reduction reaction. *J. Am. Chem. Soc.* **2012**, *134*, 2492–2495. [CrossRef]
155. Ke, D.D.; Wang, J.; Zhang, H.M.; Li, Y.; Zhang, L.; Zhao, X.; Han, S.M. Fabrication of Pt–Co NPs supported on nanoporous graphene as high-efficient catalyst for hydrolytic dehydrogenation of ammonia borane. *Int. J. Hydrog. Energy* **2017**, *42*, 26617–26625. [CrossRef]
156. Wang, S.; Zhang, D.; Ma, Y.Y.; Zhang, H.; Gao, J.; Nie, Y.T.; Sun, X.H. Aqueous solution synthesis of Pt–M (M = Fe, Co, Ni) bimetallic nanoparticles and their catalysis for the hydrolytic dehydrogenation of ammonia borane. *ACS Appl. Mater. Interfaces* **2014**, *6*, 12429–12435. [CrossRef]
157. Wang, J.M.; Ma, X.; Yang, W.R.; Sun, X.P.; Liu, J.Q. Self-supported Cu $(OH)_2$@$Co_2CO_3(OH)_2$ core–shell nanowire array as a robust catalyst for ammonia-borane hydrolysis. *Nanotechnology* **2016**, *28*, 045606. [CrossRef] [PubMed]
158. Cheng, W.; Zhao, X.; Luo, W.X.; Zhang, Y.; Wang, Y.; Fan, G.Y. Bagasse-derived carbon supported Ru nanoparticles catalyst for efficient dehydrogenation of ammonia borane. *ChemNanoMat* **2020**, *6*, 1–10.
159. Men, Y.N.; Su, J.; Du, X.Q.; Liang, L.J.; Cheng, G.Z.; Luo, W. CoBP nanoparticles supported on three-dimensional nitrogen-doped graphene hydrogel and their superior catalysis for hydrogen generation from hydrolysis of ammonia borane. *J. Alloy. Compd.* **2018**, *735*, 1271–1276. [CrossRef]
160. Zhang, F.W.; Ma, C.; Zhang, Y.; Li, H.; Fu, D.Y.; Du, X.Q.; Zhang, X.M. N-doped mesoporous carbon embedded Co nanoparticles for highly efficient and stable H_2 generation from hydrolysis of ammonia borane. *J. Power Sources* **2018**, *399*, 89–97. [CrossRef]
161. Hou, C.C.; Li, Q.; Wang, C.J.; Peng, C.Y.; Chen, Q.Q.; Ye, H.F.; Fu, W.F.; Che, C.M.; López, N.; Chen, Y. Ternary Ni–Co–P nanoparticles as noble-metal-free catalysts to boost the hydrolytic dehydrogenation of ammonia-borane. *Energy Environ. Sci.* **2017**, *10*, 1770–1776. [CrossRef]
162. Panchakarla, L.S.; Subrahmanyam, K.S.; Saha, S.K.; Govindaraj, A.; Krishnamurthy, H.R.; Waghmare, U.V.; Rao, C.N.R. Synthesis, structure, and properties of boron-and nitrogen-doped graphene. *Adv. Mater.* **2009**, *21*, 4726–4730. [CrossRef]
163. Luo, W.L.; Zhao, X.; Cheng, W.; Zhang, Y.; Wang, Y.; Fan, G.F. A simple and straightforward strategy for synthesis of N, P co-doped porous carbon: An efficient support for Rh nanoparticles for dehydrogenation of ammonia borane and catalytic application. *Nanoscale Adv.* **2020**, *2*, 1685–1693. [CrossRef]

164. Hu, M.; Ming, M.X.; Xu, C.L.; Wang, Y.; Zhang, Y.; Gao, D.J.; Bi, J.; Fan, G.Y. Towards High-Efficiency Hydrogen Production through in situ Formation of Well-Dispersed Rhodium Nanoclusters. *ChemSusChem* **2018**, *11*, 3253–3258. [CrossRef]
165. Tonbul, Y.; Akbayrak, S.; Özkar, S. Group 4 oxides supported Rhodium (0) catalysts in hydrolytic dehydrogenation of ammonia borane. *Int. J. Hydrog. Energy* **2019**, *44*, 14164–14174. [CrossRef]
166. Chen, J.Q.; Hu, M.; Ming, M.; Xu, C.L.; Wang, Y.; Zhang, Y.; Wu, J.T.; Gao, G.J.; Bi, J.; Fan, G.Y. Carbon-supported small Rh nanoparticles prepared with sodium citrate: Toward high catalytic activity for hydrogen evolution from ammonia borane hydrolysis. *Int. J. Hydrog. Energy* **2018**, *43*, 2718–2725. [CrossRef]
167. Ai, W.; Zhou, W.W.; Du, Z.Z.; Chen, Y.; Sun, Z.P.; Wu, C.; Zou, C.J.; Li, C.M.; Huang, W.; Yu, T. Nitrogen and phosphorus codoped hierarchically porous carbon as an efficient sulfur host for Li-S batteries. *Energy Storage Mater.* **2017**, *6*, 112–118. [CrossRef]
168. Qin, Q.; Jang, H.; Chen, L.L.; Nam, G.; Liu, X.; Cho, J. Low loading of RhxP and RuP on N, P codoped carbon as two trifunctional electrocatalysts for the oxygen and hydrogen electrode reactions. *Adv. Energy Mater.* **2018**, *8*, 1801478. [CrossRef]
169. Liu, Y.; Yong, X.; Liu, Z.Y.; Chen, Z.M.; Kang, Z.H.; Lu, S.Y. Unified catalyst for efficient and stable hydrogen production by both the electrolysis of water and the hydrolysis of ammonia borane. *Adv. Sustain. Syst.* **2019**, *3*, 1800161. [CrossRef]
170. Song, H.Q.; Cheng, Y.J.; Li, B.J.; Fan, Y.P.; Liu, B.Z.; Tang, Z.Y.; Lu, S.Y. Carbon dots and RuP_2 nanohybrid as an efficient bifunctional catalyst for electrochemical hydrogen evolution reaction and hydrolysis of ammonia borane. *ACS Sustain. Chem. Eng.* **2020**, *8*, 3995–4002. [CrossRef]
171. Lv, Y.A.; Cui, Y.H.; Xiang, Y.Z.; Wang, J.G.; Li, X.N. Modulation of bonding between noble metal monomers and CNTs by B-, N-doping. *Comput. Mater. Sci.* **2010**, *48*, 621–625. [CrossRef]
172. Singh, P.; Samorì, C.; Toma, F.M.; Bussy, C.; Nunes, A.; Al-Jamal, K.T.; Menard-Moyon, C.; Kostarelos, K.; Bianco, A. Polyamine functionalized carbon nanotubes: Synthesis, characterization, cytotoxicity and siRNA binding. *J. Mater. Chem.* **2011**, *21*, 4850–4860. [CrossRef]
173. Akbayrak, S.; Özkar, S. Ruthenium (0) nanoparticles supported on multiwalled carbon nanotube as highly active catalyst for hydrogen generation from ammonia–borane. *ACS Appl. Mater. Interfaces* **2012**, *4*, 6302–6310. [CrossRef]
174. Li, S.F.; Guo, Y.H.; Sun, W.W.; Sun, D.L.; Yu, X.B. Platinum nanoparticle functionalized CNTs as nanoscaffolds and catalysts to enhance the dehydrogenation of ammonia-borane. *J. Phys. Chem. C* **2010**, *114*, 21885–21890. [CrossRef]
175. Fu, W.; Han, C.; Li, D.; Chen, W.; Ji, J.; Qian, G.; Yuan, W.; Duan, X.; Zhou, X. Polyoxometalates-engineered hydrogen generation rate and durability of Pt/CNT catalysts from ammonia borane. *J. Energy Chem.* **2020**, *41*, 142–148. [CrossRef]
176. Lu, Z.H.; Jiang, H.L.; Yadav, M.; Aranishi, K.; Xu, Q. Synergistic catalysis of Au-Co@SiO_2 nanospheres in hydrolytic dehydrogenation of ammonia borane for chemical hydrogen storage. *J. Mater. Chem.* **2012**, *22*, 5065–5071. [CrossRef]
177. Metin, Ö.; Dinc, M.; Eren, Z.S.; Özkar, S. Silica embedded cobalt (0) nanoclusters: Efficient, stable and cost effective catalyst for hydrogen generation from the hydrolysis of ammonia borane. *Int. J. Hydrog. Energy* **2011**, *36*, 11528–11535. [CrossRef]
178. Umegaki, T.; Yan, J.M.; Zhang, X.B.; Shioyama, H.; Kuriyama, N.; Xu, Q. Co–SiO_2 nanosphere-catalyzed hydrolytic dehydrogenation of ammonia borane for chemical hydrogen storage. *J. Power Sources* **2010**, *195*, 8209–8214. [CrossRef]
179. Roy, B.; Manna, J.; Pal, U.; Hajari, A.; Bishnoi, A.; Sharma, P. An in situ study on the solid state decomposition of ammonia borane: Unmitigated by-product suppression by a naturally abundant layered clay mineral. *Inorg. Chem. Front.* **2018**, *5*, 301–309. [CrossRef]
180. Ye, W.Y.; Ge, Y.Z.; Gao, Z.M.; Lu, R.W.; Zhang, S.F. Enhanced catalytic activity and stability of Pt nanoparticles by surface coating of nanosized graphene oxide for hydrogen production from hydrolysis of ammonia–borane. *Sustain. Energy Fuels* **2017**, *1*, 2128–2133. [CrossRef]
181. Ciftci, A.; Eren, S.; Ligthart, D.M.; Hensen, E.J. Platinum-Rhenium Synergy on Reducible Oxide Supports in Aqueous-Phase Glycerol Reforming. *ChemCatChem* **2014**, *6*, 1260–1269. [CrossRef]

182. Badwal, S.P.S.; Fini, D.; Ciacchi, F.T.; Munnings, C.; Kimpton, J.A.; Drennan, J. Structural and microstructural stability of ceria–gadolinia electrolyte exposed to reducing environments of high temperature fuel cells. *J. Mater. Chem. A* **2013**, *1*, 10768–10782. [CrossRef]
183. Le Gal, A.; Abanades, S.; Bion, N.; Le Mercier, T.; Harleé, V. Reactivity of doped ceria-based mixed oxides for solar thermochemical hydrogen generation via two-step water-splitting cycles. *Energy Fuels* **2013**, *27*, 6068–6078. [CrossRef]
184. Hoare, J.P. *Standard Potentials in Aqueous Solution*; Marcel Dekker: New York, NY, USA, 1985; p. 49.
185. Lee, S.S.; Song, W.S.; Cho, M.J.; Puppala, H.L.; Nguyen, P.; Zhu, H.G.; Segatori, L.; Colvin, V.L. Antioxidant properties of cerium oxide nanocrystals as a function of nanocrystal diameter and surface coating. *ACS Nano* **2013**, *7*, 9693–9703. [CrossRef] [PubMed]
186. He, L.; Liang, B.L.; Li, L.; Yang, X.F.; Huang, Y.Q.; Wang, A.Q.; Wang, X.D.; Zhang, T. Cerium-oxide-modified nickel as a non-noble metal catalyst for selective decomposition of hydrous hydrazine to hydrogen. *ACS Catal.* **2015**, *5*, 1623–1628. [CrossRef]
187. Cargnello, M.; Doan-Nguyen, V.V.; Gordon, T.R.; Diaz, R.E.; Stach, E.A.; Gorte, R.J.; Fornasiero, P.; Murray, C.B. Control of metal nanocrystal size reveals metal-support interface role for ceria catalysts. *Science* **2013**, *341*, 771–773. [CrossRef]
188. Sun, C.W.; Li, H.; Chen, L.Q. Nanostructured ceria-based materials: Synthesis, properties, and applications. *Energy Environ. Sci.* **2012**, *5*, 8475–8505. [CrossRef]
189. Zhang, Z.; Lu, Z.H.; Tan, H.; Chen, X.; Yao, Q. CeO_x-modified RhNi nanoparticles grown on rGO as highly efficient catalysts for complete hydrogen generation from hydrazine borane and hydrazine. *J. Mater. Chem. A* **2015**, *3*, 23520–23529. [CrossRef]
190. Li, Y.T.; Zhang, X.L.; Peng, Z.K.; Liu, P.; Zheng, X.-C. Highly efficient hydrolysis of ammonia borane using ultrafine bimetallic RuPd nanoalloys encapsulated in porous g-C_3N_4. *Fuel* **2020**, *277*, 118243. [CrossRef]
191. Chen, S.J.; Meng, L.; Chen, B.X.; Chen, W.Y.; Duan, X.Z.; Huang, X.; Zhang, B.S.; Fu, H.B.; Wan, Y. Poison tolerance to the selective hydrogenation of cinnamaldehyde in water over an ordered mesoporous carbonaceous composite supported Pd catalyst. *ACS Catal.* **2017**, *7*, 2074–2087. [CrossRef]
192. Li, X.Y.; Song, L.H.; Gao, D.W.; Kang, B.T.; Zhao, H.Q.; Li, C.C.; Hu, X.; Chen, G.Z. Tandem of Ammonia Borane Dehydrogenation and Phenylacetylene Hydrogenation Catalyzed by CeO_2 Nanotube/Pd@MIL-53 (Al). *Chem. A Eur. J.* **2020**, *26*, 4419–4424. [CrossRef]
193. Ruiz, A.M.; Sakai, G.; Cornet, A.; Shimanoe, K.; Morante, J.R.; Yamazoe, N. Microstructure control of thermally stable TiO_2 obtained by hydrothermal process for gas sensors. *Sens. Actuators B Chem.* **2004**, *103*, 312–317. [CrossRef]
194. Yousef, A.; Barakat, N.A.; Kim, H.Y. Electrospun Cu-doped titania nanofibers for photocatalytic hydrolysis of ammonia borane. *Appl. Catal. A Gen.* **2013**, *467*, 98–106. [CrossRef]
195. Carp, O.; Huisman, C.L.; Reller, A. Photoinduced reactivity of titanium dioxide. *Prog. Solid State Chem.* **2004**, *32*, 33–177. [CrossRef]
196. Rakap, M.; Kalu, E.E.; Özkar, S. Hydrogen generation from the hydrolysis of ammonia borane using cobalt-nickel-phosphorus (Co–Ni–P) catalyst supported on Pd-activated TiO_2 by electroless deposition. *Int. J. Hydrog. Energy* **2011**, *36*, 254–261. [CrossRef]
197. Shu, H.F.; Lu, L.l.; Zhu, S.F.; Liu, M.M.; Zhu, Y.; Ni, J.Q.; Ruan, Z.H.; Liu, Y. Ultra small cobalt nanoparticles supported on MCM41: One-pot synthesis and catalytic hydrogen production from alkaline borohydride. *Catal. Commun.* **2019**, *118*, 30–34. [CrossRef]
198. Lu, L.L.; Zhang, H.J.; Zhang, S.W.; Li, F.L. A family of high-efficiency hydrogen-generation catalysts based on ammonium species. *Angew. Chem. Int. Ed.* **2015**, *54*, 9328–9332. [CrossRef] [PubMed]
199. Zhang, Y.; Wang, M.S.; Zhu, E.B.; Zheng, Y.B.; Huang, Y.; Huang, X.Q. Seedless growth of palladium nanocrystals with tunable structures: From tetrahedra to nanosheets. *Nano Lett.* **2015**, *15*, 7519–7525. [CrossRef] [PubMed]
200. Chen, G.X.; Xu, C.F.; Huang, X.Q.; Ye, J.Y.; Gu, L.; Li, G.; Tang, Z.C.; Wu, B.H.; Yang, H.Y.; Zhao, Z.P.; et al. Interfacial electronic effects control the reaction selectivity of platinum catalysts. *Nat. Mater.* **2016**, *15*, 564–569. [CrossRef] [PubMed]
201. Huang, X.Q.; Tang, S.H.; Mu, X.L.; Dai, Y.; Chen, G.X.; Zhou, Z.Y.; Ruan, F.X.; Yang, Z.Y.; Zheng, N.F. Freestanding palladium nanosheets with plasmonic and catalytic properties. *Nat. Nanotechnol.* **2011**, *6*, 28. [CrossRef]

202. Zhang, H.; Jin, M.S.; Xia, Y.N. Noble-metal nanocrystals with concave surfaces: Synthesis and applications. *Angew. Chem. Int. Ed.* **2012**, *51*, 7656–7673. [CrossRef]
203. Song, J.; Gu, X.J.; Cheng, J.; Fan, N.; Zhang, H.; Su, H.Q. Remarkably boosting catalytic H_2 evolution from ammonia borane through the visible-light-driven synergistic electron effect of non-plasmonic noble-metal-free nanoparticles and photoactive metal-organic frameworks. *Appl. Catal. B Environ.* **2018**, *225*, 424–432. [CrossRef]
204. Huang, X.Q.; Tang, S.H.; Zhang, H.H.; Zhou, Z.Y.; Zheng, N.F. Controlled formation of concave tetrahedral/trigonal bipyramidal palladium nanocrystals. *J. Am. Chem. Soc.* **2009**, *131*, 13916–13917. [CrossRef] [PubMed]
205. Xie, S.F.; Zhang, H.; Lu, N.; Jin, M.S.; Wang, J.G.; Kim, M.J.; Xie, Z.X.; Xia, Y.N. Synthesis of rhodium concave tetrahedrons by collectively manipulating the reduction kinetics, facet-selective capping, and surface diffusion. *Nano Lett.* **2013**, *13*, 6262–6268. [CrossRef] [PubMed]

MDPI

Review

A Review of Hydrogen Purification Technologies for Fuel Cell Vehicles

Zhemin Du [1,2,†], Congmin Liu [2,†], Junxiang Zhai [2], Xiuying Guo [2], Yalin Xiong [3], Wei Su [1,*] and Guangli He [2,*]

1 Tianjin Key Laboratory of Membrane and Desalination Technology, School of Chemical Engineering and Technology, Tianjin University, Tianjin 300350, China; duzhemin1997@tju.edu.cn
2 National Institute of Clean-and-Low-Carbon Energy, Future Science City, Changping District, Beijing 102211, China; congmin.liu@chnenergy.com.cn (C.L.); junxiang.zhai.a@chnenergy.com.cn (J.Z.); xiuying.guo.e@chnenergy.com.cn (X.G.)
3 China Energy Hydrogen Technology, Guohua Investment Building, 3 South Street, Dongcheng District, Beijing 100007, China; yalin.xiong@chnenergy.com.cn
* Correspondence: suweihb@tju.edu.cn (W.S.); guangli.he@chnenergy.com.cn (G.H.); Tel.: +86-022-2740-3389 (W.S.); +86-010-5733-9646 (G.H.)
† These authors contributed equally to this work.

Abstract: Nowadays, we face a series of global challenges, including the growing depletion of fossil energy, environmental pollution, and global warming. The replacement of coal, petroleum, and natural gas by secondary energy resources is vital for sustainable development. Hydrogen (H_2) energy is considered the ultimate energy in the 21st century because of its diverse sources, cleanliness, low carbon emission, flexibility, and high efficiency. H_2 fuel cell vehicles are commonly the end-point application of H_2 energy. Owing to their zero carbon emission, they are gradually replacing traditional vehicles powered by fossil fuel. As the H_2 fuel cell vehicle industry rapidly develops, H_2 fuel supply, especially H_2 quality, attracts increasing attention. Compared with H_2 for industrial use, the H_2 purity requirements for fuel cells are not high. Still, the impurity content is strictly controlled since even a low amount of some impurities may irreversibly damage fuel cells' performance and running life. This paper reviews different versions of current standards concerning H_2 for fuel cell vehicles in China and abroad. Furthermore, we analyze the causes and developing trends for the changes in these standards in detail. On the other hand, according to characteristics of H_2 for fuel cell vehicles, standard H_2 purification technologies, such as pressure swing adsorption (PSA), membrane separation and metal hydride separation, were analyzed, and the latest research progress was reviewed.

Keywords: hydrogen energy and fuel cells; impurity; hydrogen purification

Citation: Du, Z.; Liu, C.; Zhai, J.; Guo, X.; Xiong, Y.; Su, W.; He, G. A Review of Hydrogen Purification Technologies for Fuel Cell Vehicles. *Catalysts* **2021**, *11*, 393. https://doi.org/10.3390/catal11030393

Academic Editors: Marco Martino and Concetta Ruocco

Received: 26 February 2021
Accepted: 18 March 2021
Published: 19 March 2021

1. Introduction

Energy resource depletion and global warming are severe challenges of our modern society. The transportation industry plays an essential role in energy consumption and greenhouse gas emissions. According to the International Energy Agency (IEA), it was responsible for 29% of global energy consumption in 2017 and 25% of global carbon dioxide emission in 2016 [1]. Hydrogen (H_2) fuel cells provide zero pollutant discharge. The authorities in many countries have strongly supported the production of fuel cell vehicles, and this initiative will inevitably become the future developmental direction in the automotive industry. The USA was the first country that set H_2 energy and fuel cells as a long-term energy strategy. There were 5899 fuel cell vehicles in the USA by the end of 2018 [2]. Concerning the promotion of H_2 fuel cell vehicles, Japanese and South Korean companies were pioneers in large-scale mass production, successfully launching various mass-produced vehicles, such as Toyota Mirai, Honda Clarity, and Hyundai Nexo [3]. Since then, four automobile group alliances have gradually been formed, including Daimler, Ford, and Renault–Nissan, General Motors and Honda, Bayerische Motoren Werke (BMW) and Toyota, and Audi and Hyundai. The alliances invested joint effort in developing H_2 fuel cell vehicle technologies, and accelerated their commercialization. The Shanghai Automotive Industry Corporation of China launched the fourth fuel cell vehicle using a Roewe 950 vehicle with a 400 km driving range without refueling, demonstrating its capacity for small-scale production [4].

H_2 fuel cells mainly include phosphoric acid fuel cells (PAFCs), molten carbonate fuel cells (MCFCs), solid oxide fuel cells (SOFCs), alkaline fuel cells (AFCs), and proton-exchange membrane fuel cells (PEMFCs) [5]. PEMFCs are dominant since they possess a high power density, low-temperature start, and compact structure, representing an ideal power source for H_2 fuel cell vehicles. On the other side, PEMFCs require high-purity H_2. Otherwise, the fuel cell performance and running life may be severely affected [6]. Currently, H_2 production technologies, such as coal gasification, natural gas steam reforming, methanol reforming, and water electrolysis, are very well established in China [7]. According to statistical data from the China Hydrogen Alliance and China National Petroleum and Chemical Planning Institute, the current H_2 production capacity in China is approximately 41 million tons/year, with a yield of 33.42 million tons. Specifically, the yield of H_2 as an independent component (synthetic gas not containing H_2), which meets the quality standards of H_2 for industrial use and can be directly sold as industrial gas, is about 12.7 million tons/year. Among these, the H_2 yield produced from coal is the highest (21.24 million tons), accounting for 63.54%, followed by H_2 produced from by-product gas (7.08 million tons), natural gas (4.6 million tons), and electrolyzed water (0.5 million tons). However, the H_2 contributions from supercritical steam coal [8], photocatalytic water decomposition with solar energy [9], and biological H_2 production [10] are still in the research and developmental stage (Table 1). Different raw materials yield large differences in the composition and impurity contents of H_2 produced using various technologies. Thus, efficient H_2 purification technologies that enable the removal of impurities from H_2 and provide high-qualify H_2 for fuel cell vehicles are of the utmost importance for developing the H_2 fuel cell vehicle industry.

Table 1. Emerging H_2 production methods.

H_2 Production Method	Technical Feature
H_2 production from supercritical steam coal	In this technology, supercritical water [namely, temperature and pressure are at or above the critical values (374.3 °C and 2.1 MPa)] is used as a medium that provides a homogeneous and high-speed reaction because of its special physical and chemical properties, so that the chemical energy of coal is directly and efficiently converted into hydrogen energy [8].
H_2 production from water photocatalytically decomposed by solar energy	Photocatalyst powders or electrodes can produce photo-generated carriers by absorbing solar energy, so they decompose water into H_2 and O_2. The photocatalytic H_2 production can be subdivided mainly into heterogeneous photocatalytic (HPC) H_2 production and photo-electrochemical (PEC) H_2 production [9].
Biological H_2 production	H_2 is a product of microorganisms' metabolism using biomass and organic wastewater as raw materials. Based on the type of microorganisms and their metabolic mechanisms, the biological H_2 production technology includes water splitting H_2 production, photo-fermentative H_2 production, dark fermentative H_2 production, and H_2 production combined with photo-fermentation and dark fermentation [10].

To support large-scale applications of H_2 energy in the transportation field, novel and high-efficient purification technologies for the production of low-cost and high-quality H_2 should be urgently developed. In this study, the characteristics of standard H_2 purification technologies, such as pressure swing adsorption (PSA), membrane separation and metal hydride separation, were analyzed. Research progress was reviewed according to the characteristics of H_2 for fuel cell vehicles. To further improve the separation efficiency, it is necessary to continuously conduct studies on the novel and highly selective adsorption materials, long-lasting and low-cost membrane materials, anti-poisoning metal hydride materials with a low regeneration energy consumption, as well as new separation and coupling processes based on the materials mentioned above.

2. H_2 Standards for Fuel Cell Vehicles

The International Organization for Standardization (ISO) issued the ISO 14687-2:2012 standard in 2012, and the Society of Automotive Engineers (SAE) issued the SAE J2719-201511 standard in 2015, presenting the same requirements for H_2 quality for PEMFCs. Until 2019, China was following the GB/T 3634.2-2011 Hydrogen Part 2: Pure Hydrogen, High-Pure Hydrogen, and Ultrapure Hydrogen standard. However, this standard was aimed at industrial H_2 use, limiting the impurity content only partly, without specific regulations on other impurities that may affect H_2 fuel cells' performance. Therefore, by the end of 2018, China set the GB/T 37244-2018 Fuel Specification for Proton Exchange Membrane Fuel Cell Vehicles—Hydrogen standard, which was in line with ISO 14687-2:2012 and SAE J2719-201511 standards, regulating the concentration of fourteen impurities: water (H_2O), total hydrocarbon (HC) (by methane), oxygen (O_2), helium (He), nitrogen (N_2), argon (Ar), carbon dioxide (CO_2), carbon monoxide (CO), total sulfide (by H_2S), formaldehyde (HCHO), formic acid (HCOOH), ammonia (NH_3), total halide (by halide ions), and maximum particulate matter. The PEMFC technology was remarkably improved, e.g., a lower Pt usage, thinner electrolyte membrane, and higher operating electric current density and lower humidity, so it is necessary to reconsider the previously set impurity limit in H_2. The ISO technical committee for H_2 energy, IOS/TC 197, issued the ISO 14687:2019 standard in November 2019, combining and revising three H_2 fuel cell-related standards, namely, ISO 14687-1, ISO 14687-2, and ISO 14687-3. Meanwhile, according to the ISO 14687:2019 standard, the SAE issued the SAE J2719-202003 standard in March 2020, extending the limit of CH_4, N_2, Ar, and HCHO impurities. Table 2 shows the requirements for the impurity content in H_2 for fuel cells, including previous and new standards in China and abroad.

Table 2. Requirements for the impurity content in H_2 for fuel cells in previous and new standards in China and abroad.

Component	**GB/T 3634.2-2011**			**ISO 14687-2:2012 SAE J2719-201511 GB/T 37244-2018**	**ISO 14687:2019 SAE J2719-202003**
	Pure H_2	**High Pure H_2**	**Ultrapure H_2**		
H_2 purity (mole fraction)	99.99%	99.999%	99.9999%	99.97%	99.97%
Total non-hydrogen gases	-	10 ppm	1 ppm	300 ppm	300 ppm
H_2O	10 ppm	3 ppm	0.5 ppm	5 ppm	5 ppm
Total HC (by methane)	-	-	-	2 ppm	-
Non-methane HC (by C_1)	-	-	-	-	2 ppm
Methane	10 ppm	1 ppm	0.2 ppm	-	100 ppm
O_2	5 ppm	1 ppm	0.2 ppm	5 ppm	5 ppm
He	-	-	-	300 ppm	300 ppm
N_2 and Ar	-	-	-	100 ppm	-
N_2	60 ppm	5 ppm	0.4 ppm	-	300 ppm
Ar	Agreed by supply and demand	Agreed by supply and demand	0.2 ppm	-	300 ppm
CO_2	5 ppm	1 ppm	0.1 ppm	2 ppm	2 ppm
CO	5 ppm	1 ppm	0.1 ppm	0.2 ppm	0.2 ppm

Table 2. *Cont.*

Component	GB/T 3634.2-2011			ISO 14687-2:2012 SAE J2719-201511 GB/T 37244-2018	ISO 14687:2019 SAE J2719-202003
	Pure H_2	High Pure H_2	Ultrapure H_2		
Total sulfide (by H_2S)	-	-	-	0.004 ppm	0.004 ppm
HCHO	-	-	-	0.01 ppm	0.2 ppm
HCOOH	-	-	-	0.2 ppm	0.2 ppm
NH_3	-	-	-	0.1 ppm	0.1 ppm
Total halide (by halide ion)	-	-	-	0.05 ppm	0.05 ppm
The concentration of maximum particulate matter	-	-	-	1 mg/kg	1 mg/kg

3. The Impact of Impurities on Fuel Cells

As shown in Table 1, compared with H_2 for industrial applications, the requirements for H_2 purity for fuel cells are not high. Still, the impurity content is strictly controlled, determined by fuel cells' structure and operating characteristics. For instance, even a low CO content may cause irreversible damage to the performance and running life of fuel cells. Table 3 shows the impact of excessive impurities on fuel cells.

Table 3. Impact of impurities on the performance of fuel cells.

Impurity	Damage Induced by Excessive Impurities
H_2O	H_2O can transport water-soluble impurities, such as Na^+ and K^+, and reduce the membrane proton conductivity. Excessive H_2O induced corrosion of metal parts [11].
HC	Most HCs adsorbed onto the catalyst layer will decrease catalytic performance. Methane does not pollute fuel cells, but it dilutes H_2 and hampers performance [12].
O_2	O_2 in specific concentrations negatively affects the performance of metal hydride, a type of H_2 storage material [11].
Inert gas	Dilution and diffusion of He, Ar, and N_2 in H_2 decrease the electric potential of fuel cells [13].
CO_2	CO_2 has a dilution effect on H_2. CO_2 in high concentrations can be converted into CO through a reverse water gas shift reaction, thereby leading to catalyst poisoning [14].
CO	CO closely binds to the active site of Pt catalysts, decreasing the effective electrochemical surface area used for H_2 adsorption and oxidation [15].
Sulfide	The adsorption of sulfides on the active catalyst sites prevents H_2 adsorption on the catalyst surface. The adsorbed sulfides react with Pt catalysts to form stable Pt sulfides, irreversibly degrading the fuel cell performance [16].
HCHO and HCOOH	HCHO and HCOOH are adsorbed on catalysts to form CO, thereby leading to catalyst poisoning [17].
NH_3	NH_4^+ can reduce the proton conductivity of the ionic polymer. NH_3 adsorbed on the surface of the catalyst blocks the active sites [18].
Halide	Halide adsorbed on the catalyst layer decreases the superficial area of catalysts. Chloride ions are deposited in the fuel cell membrane by forming soluble chlorides, leading to the Pt catalyst's dissolution [19].
Particulate matter	Particulate matters adsorbed on the active site of catalysts of fuel cells prevent the H_2 adsorption on the catalyst surface, blocking the filter and destroying the full cell components [20].

4. H_2 Purification Technology

H_2 purification technology is a crucial link from H_2 production to H_2 utilization. Stable, reliable, and low-cost H_2 sources represent a base for large-scale applications of fuel cell vehicles. Thus, high-efficient and low-power H_2 purification technologies for fuel cell vehicles play an underlying role in the development of the H_2 energy industry.

A fuel cell power system can operate efficiently only if high-quality H_2 is provided. H_2 produced in coal gasification, natural gas reforming, by-product H_2, or from water electrolysis, is collectively referred to as crude hydrogen. It cannot be directly used for fuel cell vehicles without purification according to the existing standards. The composition of different types of crude H_2 is listed in Table 4. The H_2 purification methods can be mainly classified as physical and chemical methods [21]. The former include adsorption methods [PSA, temperature swing adsorption (TSA), and vacuum adsorption], low-temperature separation methods (cryogenic distillation and low-temperature adsorption), and membrane separation methods (inorganic membrane and organic membrane), while the latter involve a metal hydride separation and catalysis method (Figure 1). The selection of an appropriate H_2 purification method is closely related to the hydrogen supply mode and gas source. For H_2 production by centralized large-scale coal gasification and natural gas reforming with an H_2 supply amount $\geq$10,000 Nm^3/h, PSA purification is primarily adopted after transformation, desulfurization, and decarbonization. The PSA technology has been around for a while, and is characterized by low operation costs and a long service life. However, the H_2 for fuel cell vehicles produced via traditional PSA with a standard impurity content results in a decreased recovery rate and yield. It is also not cost-efficient due to low requirements for specific impurity removal (e.g., CO $\leq$ 0.2 ppm). Cryogenic distillation is also applicable to large-scale production, but standard H_2 purity is 85–99%, which does not satisfy the application requirements. For H_2 production by centralized by-product mode with an H_2 supply of 1000–10,000 Nm^3/h, versatile processes should be applied based on different impurities to improve the H_2 recovery efficiency. For example, an organic membrane combined with a PSA process is used for obtaining methanol purge gas, while a two-stage or multi-stage PSA process is adopted for obtaining coke oven gas and by-product gas from the refinery. Concerning such small-scale on-site distributed H_2 production scenarios, with the H_2 supply $\leq$ 1000 Nm^3/h and vehicle H_2 supply, traditional PSA separation shows the disadvantages of large floor area, inflexibility, and low adaptability. Hence, low-temperature adsorption, metal hydride, and metal membrane separations are available processes according to the types and amounts of impurities. Low-temperature adsorption can effectively eliminate multiple impurities, such as sulfide, HCHO, and HCOOH. However, it requires high energy consumption, and it is a complex process suitable for special small-scale and cold source applications [22]. Metal hydride separation and palladium (Pd) membrane separation methods are reasonably effective in separating gas sources with a high content of inert components. At the same time, their inherent disadvantage is that purified materials react with impure gas during the H_2 recovery, reducing the purification efficiency [23]. New membrane technologies, such as carbon molecular sieve membranes (CMSMs) [24], ionic liquid membranes [25], and electrochemical H_2 pump membranes [26], are currently hot spots in scientific research. However, their industrial-scale implementation is still hard to foresee.

Table 4. Composition of different types of crude hydrogen.

Component (%)	H_2	CO	CO_2	CH_4	N_2	Ar	Total Sulfur	H_2O	O_2	Others
Coal gasification [27]	25–35	35–45	15–25	0.1–0.3	0.5–1	-	0.2–1	15–20	-	-
Natural gas reforming [28]	70–75	10–15	10–15	1–3	0.1–0.5	-	-	-	-	-
Methanol reforming [29]	75–80	0.5–2	20–25	-	-	-	-	-	-	-
Coke oven gas [30]	45–60	5–10	2–5	25–30	2–5	-	0.01–0.5	-	0.2–0.5	2–5
Methanol purge gas [31]	70–80	4–8	5–10	2–8	5–15	0.1–2	-	-	-	-
Synthetic ammonia tail gas [32]	60–75	-	-	-	15–20	-	-	1–3	10–15	-
Biomass gasification [33]	25–35	30–40	10–15	10–20	1	-	0.2–1	-	0.3–1	-

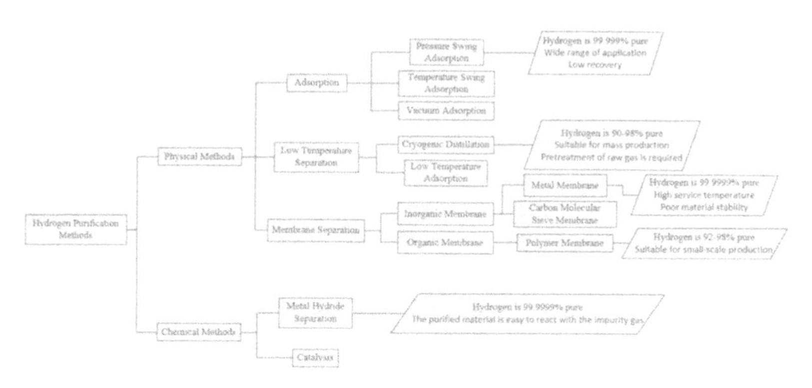

Figure 1. Classification of hydrogen purification technologies.

4.1. PSA Methods

Gas separation and purification by PSA are implemented by periodical pressure changing based on the difference in the adsorbent capacity for different gases. The PSA separation effect primarily depends on the type of adsorbent and the technical process used. H_2 significantly differs from the majority of gas molecules, such as CO_2, CO, and CH_4, in terms of static capacity, so it is very suitable for PSA separation and purification [34]. Air Product, Air Liquid, and other renowned global gas companies, are already established and very successful examples for the industrial application of H_2.

Traditional adsorbents include zeolite molecular sieves, activated carbon, activated alumina, and silica gel. Modifications and innovations of these adsorbents were reported in regard to different impurities, and most of the studies focused on CO_2 removal. Lively et al. **[35]** used hollow fibers as an adsorbent to investigate the CO_2 removal in PSA experimental devices. The purity of the obtained H_2 was 99.2% pure, with a recovery rate of 88.1%, and this needs to be optimized. Shamsudin et al. **[36]** increased the H_2 purity to about 100% and the recovery rate to 88.43% via the strong CO_2 absorption of palm shell charcoal. He et al. **[37]** reported a structured activated carbon system applied to rapid PSA (RPSA) using a dip-coated Ni foam framework. Under the working conditions of 0.4 MPa and 200 mL/min, the adsorption rate constant K was 0.0029 s^{-1}, which was about two times higher than that of traditional adsorbents. The material exhibited a better CO_2 adsorption effect in H_2. Moreover, Kuroda et al. **[38]** applied hydroxyl aluminum silicate clay (HAS-Clay) to purify the H_2 produced by biomass, and found a relatively high adsorption selectivity to CO_2. This adsorbent is also applicable for the adsorption and separation of H_2S. Metal-organic frameworks (MOFs) are recently developed materials with easily adjustable structures and properties, and they are ideal novel adsorption materials. Agueda et al. **[39]** utilized UTSA-16 as an adsorbent to remove the CO_2 impurities, and simulated the PSA process of the steam methane reforming of tail gas. The results revealed a H_2 purity up to 99.99–99.999%, a recovery rate of 93–96%, and a yield of 2–2.8 mol/kg/h.

Researchers are trying to develop novel adsorbents for the simultaneous removal of multiple impurities in H_2. Brea et al. [40] synthesized a raw material NaX molecular sieve within CaX and MgX molecular sieves using an ion-exchange method. They conducted an adsorption simulation for the $H_2/CH_4/CO/CO_2$ gas mixture and showed that these three adsorbents could yield H_2 with a purity higher than 99.99%. The CaX molecular sieve application exhibited the highest recovery rate and yield of H_2. Besides this, Banu et al. [41] compared the performances of four kinds of MOF adsorbents (UiO-66(Zr), UiO-66(Zr)-Br, UiO-67(Zr), and Zr-Cl_2AzoBDC) and discovered that UiO-66(Zr)-Br had the best purification effect on H_2 produced via methane steam reforming. Relvas et al. [42] prepared a novel Cu-AC-2 adsorbent to process the $H_2/CH_4/CO/CO_2$ gas mixture. The H_2 purity exceeded 99.97%, while the CO content declined to 0.17 ppm, reaching the H_2 standards required for fuel cell vehicles.

The improvement and optimization of the PSA process are crucial ways to increase the H_2 purification efficiency. The flow scheme of the classical PSA system is shown in Figure 2. Ahn et al. [43] used a two-bed PSA and a four-bed PSA to recover H_2 from coal gas with N_2 as a major impurity. The four-bed PSA process' performance was superior to that of the two-bed PSA process, yielding a H_2 purity of 96–99.5% and a recovery rate of 71–85%. Abdeljaoued et al. [44] established a four-bed PSA theoretical model and performed a twelve-step four-bed PSA experiment at room temperature. They investigated the removal of impurities from H_2 produced by ethanol steam reforming for fuel cell vehicles. Further optimization was expected to increase the H_2 recovery rate above 75%, providing a CO concentration lower than 20 ppm. Moon et al. [45] investigated the eight-bed PSA process. Considering that a H_2 purity of 99.99%, the highest H_2 recovery rate was 89.7%, which was approximately 11% higher than that of a four-bed PSA process. Moreover, Zhang et al. [46] established the five-step one-bed and six-step two-bed PSA cycle models, and compared them in terms of purity, recovery rate, and yield of H_2. The recovery rate and the yield of the two-bed PSA process were 11% and 1 mol/kg/h, respectively, higher than those of the one-bed PSA process. Li et al. [47] explored the effects of adsorption pressure, adsorption time, and P/F ratio on PSA. They designed a six-step two-bed PSA process for the purification of H_2 produced by methane steam reforming, and the produced H_2 exhibited a purity and a rate of more than 99.95% and 80%, respectively. When the CH_4 concentration in impurities was high, it was necessary to increase the adsorption pressure to ensure the purity of the H_2. Yáñez et al. [32] developed a four-bed PSA device with a 5 Å molecular sieve as an adsorbent for purification of H_2 from synthetic ammonia tail gas (H_2:N_2:CH_4:Ar = 58:25:15:2). It was indicated that the H_2 purity was up to 99.25–99.97%, while the recovery rate was 55.5–75.3%.

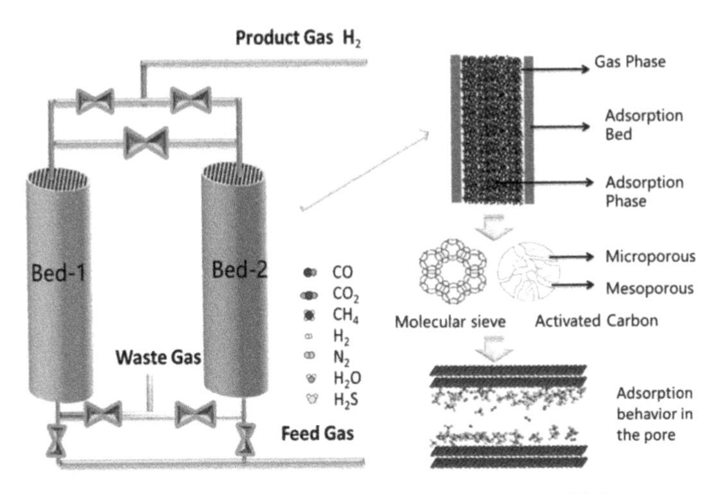

Figure 2. Flow scheme of the classical pressure swing adsorption (PSA) system.

The improvement of process flow based on conventional PSA is an important research direction for increasing the H_2 purity and recovery rate. Vacuum PSA (VPSA) can forcibly desorb impurities with a strong adsorption capacity from adsorbents via vacuum pumping in order to regenerate adsorbents. You et al. [48] showed that the VPSA and PSA could produce H_2 with a similar purity level under the same conditions, but the recovery rate was increased by about 10% during VPSA. Besides this, Lopes et al. [49] performed an experiment on rapid VPSA (RVPSA), and the results illustrated that RVPSA was able to improve the H_2 yield by nearly 410% compared with PSA. Golmakani et al. [50] conducted a comparative study on PSA, VPSA, and TSA processes, and discovered that the VPSA process could yield the H_2 for fuel cell vehicles with a reasonable cost-efficiency and recovery rate, proposing it as the best option among the three investigated processes. Furthermore, Golmakani et al. [51] established a sixteen-step four-bed VPSA model. They studied the influence of N_2 on the process' performance so as to enhance the recovery rate of H_2 produced by VPSA and decrease the energy consumption.

Many novel PSA devices are currently in the development or implementation phase for different impurities and H_2 utilization demands. Thus, Air Products [52] has investigated and developed a novel PSA device, named Sour PSA, to capture the acidic impurities, such as CO_2 and sulfides, in H_2. Majlan et al. [53] designed a compact PSA (CPSA) system with a rapid circulation rate to provide H_2 continuously without the need for adsorbent regeneration. Moreover, it could reduce CO concentration in the $H_2/CO/CO_2$ gas mixture from 4000 to 1.4 ppm, and the CO_2 concentration from 5% to 7 ppm, yielding H_2 with a purity of 99.999%. Zhu et al. [54] proposed a seven-step two-bed elevated-temperature PSA (ET-PSA) system to separate the feed gas $CO/CO_2/H_2O/H_2$ under proper conditions. In this way, H_2 with a purity of 99.9991% and a recovery rate of 99.6% was obtained. As such, the system was applicable to the removal of CO and CO_2 from H_2. A summary of the PSA technology is shown in Table 5.

Table 5. Summary of PSA technology.

Adsorbent	Process Flow	Feed Gas	H_2 Purity	H_2 Recovery Rate	Reference
Hollow fiber sorbent	RCPSA	CO_2:H_2 = 25:75	99.2%	88.1%	[34]
Palm kernel shell activated carbon	Two-column PSA	CO_2:H_2 = 15:85	About 100%	88.43%	[35]
UTSA-16	Four-column PSA	Steam methane reforming off-gas	99.99–99.999%	93–96%	[38]
CaX zeolite	Four-column PSA	H_2:CH_4:CO:CO_2 = 75.89:4.01:3.03:17.07	+99.99%	69.6%	[39]
Cu-AC-2	Four-column PSA	H_2:CO_2:CH_4:CO = 70:25:4:1	+99.97%	+75%	[41]
Activated carbon/zeolite 5A	Four-bed PSA	H_2:CO_2:CH_4:CO:N_2 = 38:50:1:1:10	96–99.5%	71–85%	[42]
Activated carbon	Twelve-step four-column PSA	CO:CO_2:CH_4:H_2 = 1:25:5:69	99.999%	+75%	[43]
Activated carbon/zeolite LiX	Eight-layered bed PSA	H_2:CO_2:CO:N_2:Ar = 88.75:2.12:2.66:5.44:1.03	99.99%	89.7%	[44]
Activated carbon/zeolite 5A	Six-step layered two-bed PSA	H_2:CH_4:CO:CO_2 = 72.9:3.6:4.5:19	+99.95%	+80%	[46]
5A zeolite	Four-column PSA	H_2:N_2:CH_4:Ar = 58:25:15:2	99.25–99.97%	55.5–75.3%	[32]
Activated carbon	CPSA	H_2:CO:CO_2 = 94.6:0.4:5	99.999%	-	[52]
Potassium-promoted layered doubleoxide	Two-column seven-step ET-PSA	CO:CO_2:H_2O:H_2 = 1:1:10:88	99.9991%	99.6%	[53]

4.2. Membrane Separation Methods

As an emerging gas separation technology, membrane separation has the advantages of flexible and simple operation, compact structure, low energy consumption, and environmental friendliness. In the membrane separation technology with a perm-selective membrane as a separation medium, the raw material components can selectively permeate the membrane under the action of driving forces (pressure difference, concentration difference, and potential difference), thereby achieving separation and purification [55]. The performance of membrane materials is the most critical factor determining the H_2 separation and purification effects of the membrane. Commonly used membrane materials primarily include metal and polymer membranes, and novel membrane materials, such as nanomaterial membrane, CMSM, and MOF membranes, may exhibit preferable separation performance. Therefore, the performance of these membrane materials in the H_2 purification is analyzed and evaluated below.

4.2.1. Metal Membranes

H_2 is catalyzed to protons and electrons on the compact structure of metal membranes. The protons pass through the metal membrane and bind electrons on the other side to form H_2 again. However, the metal membrane blocks CO_2, N_2, CH_4, and O_2 gas molecules, thereby achieving the selective permeation of H_2. Pd membranes are currently the most commonly used metal membranes, due to their excellent H_2 permeability, and their high resistance to H_2 fluidity and auto-catalytic hydrogenolysis reactions [56]. However, the Pd membrane is related to high manufacturing costs, and it is prone to H_2 embrittlement at a low temperature.

A Pd alloy membrane can be formed by adding other metal elements (Ag, Au, Cu, Ni, Y, etc.) into the Pd membrane to solve the H_2 embrittlement problem, enlarge the Pd lattice, and increase the H_2 permeation rate at the same time. Nayebossadri et al. [57] studied the performance of H_2 in natural gas separated by Pd, $PdCu_{53}$, and $PdAg_{24}$ membrane materials at different concentrations. They found that the H_2 permeability of the $PdAg_{24}$ membrane is better than that of the other two membranes. Zhao et al. [58] prepared a bilayer bcc–PdCu alloy membrane by the alternative electrodeposition of Pd and Cu on a ceramic support membrane. The membrane exhibited excellent low-temperature tolerance and H_2 permeability, and it is a candidate membrane material for H_2 separation at ambient temperature.

Both pure Pd membranes and Pd alloy membranes are self-supporting membranes. Their thickness is limited from several tens to several hundreds of micrometers to assure sufficient mechanical strength. When the membrane thickness is too high, it increases the total cost and lowers the H_2 permeation rate. As such, it is possible to deposit a Pd membrane or a Pd alloy membrane on the surface of a porous material to prepare a supported Pd composite membrane. The support increases the mechanical strength of the Pd membrane and decreases the Pd amount and membrane thickness, which is beneficial to the total cost and the H_2 permeation rate, as described above. Kong et al. [59] deposited a nanoscale Pd membrane on polybenzimidazole-4,4′-(hexafluoroisopropylidene)-bis(benzoic acid) (PBI-HFA) using the vacuum electroless plating (VELP) technique. The novel Pd/PBI-HFA composite membrane completely prevented CO permeation and exhibited good H_2/N_2 and H_2/CO_2 selectivity. Kiadehi et al. [60] deposited a NaY molecular sieve and a Pd membrane on porous stainless-steel substrates. The permeation of the H_2 and N_2 mixture into the prepared Pd/NaY/PSS composite membrane was tested, showing that the membrane's H_2/N_2 selectivity was 736 at 450 °C. Moreover, Iulianelli et al. [61] prepared a supported Pd_{70}-Cu_{30}/γ-Al_2O_3 thin membrane using the metal vapor synthesis method. The membrane showed H_2/N_2 and H_2/CO_2 selectivity of 1800 and 6500, respectively, at 400 °C and 50 kPa. Huang et al. [62] used natural mineral Nontronite-15A as a surface coating material of porous Al_2O_3 to prepare a Pd/Nontronite-15A/Al_2O_3 membrane, lowering the production cost compared to other composite membranes that provide high H_2 permeability.

The permeation ability of Pd H_2 is not the strongest among metals. It has been recently indicated that vanadium group metals, V, Nb, and Ta, have different bcc lattice structures, higher H_2 permeability and mechanical strength, and weaker H_2 dissociation and adsorption ability than Pd [63]. However, a compact oxide layer forms on the surface, preventing H_2 permeation. As a result, the H_2 permeation rate of the thin membrane is not very high, although vanadium group metals have a strong lattice H_2 permeation ability. Besides this, these metals are more susceptive to H_2 embrittlement than Pd. A useful approach toward this problem was depositing an extremely thin Pd layer plated on both sides of the vanadium group metals to form symmetric composite membranes. In that way, the H_2 adsorption and dissociation ability of the Pd membrane was combined with the H_2 permeation ability of the vanadium group metals, lowering the total cost. Dolan et al. [64] prepared a Pd-coated vanadium membrane with a tubular structure, and this revealed a high H_2 permeability and stability, suitable for H_2 separation for fuel cell vehicles. Fasolin et al. [65] applied high-power pulse magnetron sputtering technology to prepare a $Pd/V_{93}Pd_7/Pd$ multilayer membrane with a total thickness less than 7 μm on the surface of porous alumina. Besides this, research studies have demonstrated that such V-based thin membranes have similar permeability and higher resistance to H_2 embrittlement than Pd-based membranes. Alimov et al. [66] prepared a thin-walled seamless tubular membrane using V–Pd and V–Fe alloys. Furthermore, they manufactured a membrane module by welding 18 membranes, which was applied to extract ultra-pure H_2. Jo et al. [67] adopted a Pd/Ta composite membrane for ammonia dehydrogenation, overcoming H_2 embrittlement and producing H_2 with a purity over 99.9999%, while the NH_3 concentration was reduced below 1 ppm. Additionally, they applied the Pd/Ta composite membrane to purify H_2 from the CO, CO_2, H_2O, and H_2 gas mixture, yielding a H_2 purity of more than 99.999% and a CO content of 10 ppm [68]. Budhi et al. [69] investigated the separation of H_2 from the H_2 and N_2 mixture using a Pd/α-Al_2O_3 membrane. They achieved a higher H_2 recovery rate by adjusting the feed gas flow rate to make the membrane operate under non-steady-state conditions.

4.2.2. Polymer Membranes

The working principle of polymer membrane separation is based on the different permeation rates of gases through the polymer membrane. Nowadays, polysulfone (PSF), polyimide (PI), and polyamide are commonly used as polymer membrane materials [70]. An ideal polymer membrane material should possess high selectivity, permeability, thermal stability, and good mechanical performance. However, as a rule of thumb, a highly permeable polymer membrane has low selectivity, and vice versa [71]. Since the trade-off between selectivity and permeability limits the use of polymer membranes, researchers attempted to prepare mixed matrix membranes (MMMs) by adding zeolite, silicon dioxide, CMS, and other inorganic materials into the polymer to improve the overall performance [72]. Rezakazemi et al. [73] added 4A zeolite nanoparticles into a polydimethylsiloxane (PDMS) substrate to prepare the PDMS/4A MMMs. The prepared MMMs experimentally exhibited higher H_2/CH_4 selectivity and H_2 permeability than the pure PDMS membrane. Peydayesh et al. [74] introduced Deca-dodecasil 3R (DDR) zeolite into a Matrimid® 5218 PI substrate to prepare the Matrimid® 5218-DDR MMM, yielding H_2 permeability and H_2/CH_4 that were increased by 100 and 189%, respectively.

In addition, polymer blending could also improve the performance of polymer membranes. Hamid et al. [75] synthesized a PSF/PI membrane that possessed higher H_2 permeability and H_2/CO_2 selectivity (4.4) than a single PSF or PI membrane, with a H_2 purification efficiency of 80%. Meanwhile, the PSF/PI membrane exhibited more stable physical and chemical properties, yielding a novel polymer membrane with excellent performance. Structurally, the mechanical performance and specific surface area of a hollow fiber membrane are superior to those of a traditional plate membrane. These findings are also used as a developmental direction of the gas separation membrane. Naderi et al. [76] developed a bilayer hollow fiber membrane with a polybenzimidazole (PBI) and sulfonated polyphenylenesulfone (sPPSU) mixture as an outer selection layer, and PSF as an inner support layer. The experimental results indicated a H_2 permeability in the membrane of 16.7 GPU, and a H_2/CO_2 selectivity of 9.7 at 90 °C and 14 atm. Therefore, the membrane was suitable for H_2 and CO_2 separation at high temperatures.

All the polymer membranes mentioned above have very high H_2 selectivity. Furthermore, researchers have developed separation membranes with CO_2 selectivity to remove CO_2 from H_2 efficiently. Figure 3 schematically illustrates two selective membranes. Compared with the H_2-selective membrane, the CO_2-selective membrane requires a smaller area during separation and generates H_2 as a product in a high-pressure state, significantly reducing the mechanical energy loss [77]. As the CO_2 molecular diameter is larger than that of H_2, the polymer membranes should have a particular CO_2 affinity to achieve negative selectivity [55]. Abedini et al. [78] prepared a poly(4-methyl-1-pentene) (PMP)/MIL 53 (Al) MMMs membrane by adding MIL 53(Al) MOF into a poly(4-methyl-1-pentene) (PMP) substrate. It was experimentally indicated that the MMMs possessed higher CO_2/H_2 negative selectivity and thermal stability than pure PMP membranes. In the meantime, the negative selectivity of MMMs was enhanced with the increase in feed pressure, and it was capable of overcoming the Robeson upper limit. Cao et al. [79] introduced a covalent organic framework (COF) into polyvinyl amine to prepare the PVAM/COF MMMs, with a CO_2/H_2 selectivity of 15 and a CO_2 permeation rate of 396 GPU. Moreover, Salim et al. [80] prepared novel oxidatively stable membranes containing quaternary ammonium hydroxide, fluoride, and tetrafluoroborate using a crosslinked polyvinyl alcohol–polysiloxane substrate, with a CO_2 permeation rate of 100 GPU and CO_2/H_2 selectivity greater than 100, and such membranes were expected to be applied to purify H_2 for fuel cell vehicles. Nigiz et al. [81] added graphene oxide (GO) into PDMS to prepare nanocomposite membranes, increasing the CO_2 permeation rate and CO_2/H_2 selectivity. At a GO content of 0.5% and transmembrane pressure of 0.2 MPa, the CO_2/H_2 selectivity rose from 7.1 to 11.7, and the CO_2 permeability reached 3670 Barrer. Besides this, Chen et al. [77] prepared ZIF-8-TA nanoparticles using a hydrophilic modification of ZIF-8 with tannic acid (TA). The nanoparticles were introduced into a hydrophilic polyvinyl amine substrate to obtain an MMM. Under the feed pressure of 0.12 MPa, the CO_2 permeability and CO_2/H_2 selectivity were 987 GPU and 31, respectively, providing a preferable CO_2/H_2 separation performance.

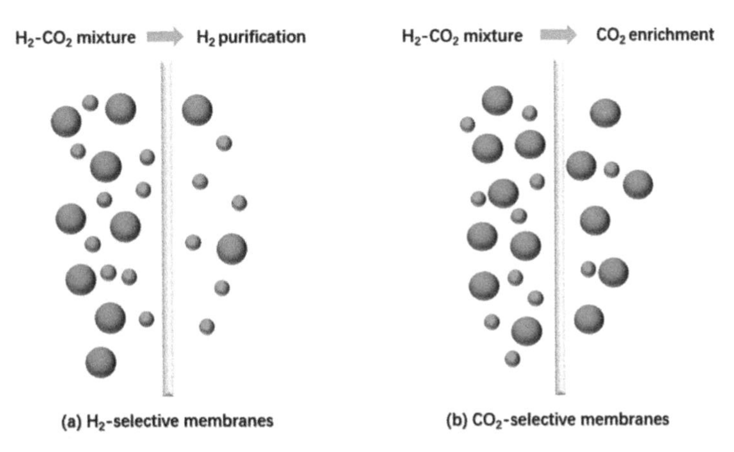

Figure 3. Schematic representation of (**a**) H_2-selective and (**b**) CO_2-selective membranes.

4.2.3. Carbon-Based Membranes

Carbon molecular sieve (CMS) membranes (CMSMs) with an amorphous microporous structure are the most common carbon-based membranes, usually obtained by the carbonization or pyrolysis of polymer precursors in the inert gas or vacuum environment. Common polymer precursors include polyimide and its derivatives, polyfurfuryl alcohols, and phenolic resins [55]. An in-depth exploration was conducted to improve the permeability and selectivity of CMS membranes. Tanco et al. [82] prepared composite alumina–CMS membranes (Al-CMSMs) with tubular porous alumina as a carrier, achieving a H_2 and CH_4 separation performance considerably better than the Robeson upper limit for polymer membranes at 30 °C. The H_2 extracted from an H_2/CH_4 gas mixture possessed a purity of 99.4%. Xu et al. [83] prepared CMSMs with ultra-high selectivity by decomposing polyetherketone–cardo polymers at a high temperature. The reported permeability of CMS membranes prepared by carbonization at 700 °C was 5260 Barrer, while the H_2/CH_4, H_2/N_2, and H_2/CO selectivity was 311, 142, and 75, respectively. When the carbonization was performed at 900 °C, the H_2/CH_4 selectivity reached 1859.

Graphene-based membranes, as a new type of carbon-based membranes, have attracted extensive attention in the gas separation field. Graphene and GO exhibit a single-atom thickness, high mechanical strength, and good chemical stability [84]. Keeping in mind that the membrane thickness is inversely proportional to its permeability, graphene-based membranes have become ideal membranes with minimum transmission resistance and maximum permeation flux because of their ultra-low thickness. However, most graphene-based materials do not have suitable natural pores, so they cannot be directly used for gas separation. Therefore, to improve the gas separation performance of graphene-based membranes, the emphasis of the research is on imparting uniformly distributed nanopores with an appropriate size and shape and high porosity in graphene sheets [85]. By designing two dumbbell-shaped porous γ-graphene monolayers containing γ-graphyne N_2 (γ-GYN) and γ-graphyne H_2 (γ-GYH), respectively, Sang et al. [86] simulated the membrane performance to separate H_2 from an H_2, H_2O, CO_2, N_2, CO, and CH_4 gas mixture. The γ-GYN membranes exhibited better selectivity and H_2 permeability, and they could be an ideal choice for H_2 purification from the gas mixture. Silva et al. [87] proved that g-C_3N_4 graphene-like two-dimensional nanomaterials could effectively purify H_2 from CO_2 and CH_4. Theoretical analyses suggested that H_2 permeability might be improved, by enlarging the pore area by applying 2.5 and 5% biaxial strains to the membranes, without affecting the H_2/CO_2 and H_2/CH_4 selectivity. Moreover, Wei et al. [88] used density functional theory (DFT) to study the performance of 3N-PG and 6N-PG monolayers composed of porous graphene (PG) membranes and nitrogen in separating H_2 from the H_2, CO, N_2, and CH_4 gas mixture. It was also revealed that 3N-PG monolayers and 6N-PG monolayers possessed better H_2 permeability than PG membranes, providing a novel membrane material for H_2 purification. Sun et al. [84] studied the nano-PG (NPG) membranes and found that the H_2 permeability reached 106 GPU, which was much higher than that of polymer membranes. At the same time, the H_2/CH_4 selectivity was 225, similar to that of polymer membranes. Meanwhile, NPG membranes are more cost-efficient than polymer membranes under the same separation conditions and purification requirements. Zeynali et al. [89] prepared GO nanocomposite membranes on modified alumina tubes, indicating their good H_2 permeability, favorable H_2/CO_2 and H_2/N_2 selectivity, and stability, accompanied with lower costs than Pd membranes. In addition, Liu et al. [90] simulated the reaction path of gas molecules through nano-graphene C216 and proved that H_2 could penetrate C216 membranes with a diffusion barrier of 0.65 eV. The H_2 selectivity to O_2, N_2, NO, H_2O, CO, and CO_2 was up to 1033, higher than that of PG and polymer membranes.

4.2.4. MOF Membranes

MOFs generally represent a novel class of organic–inorganic hybrid porous solid materials with regular geometric and crystal structures. They are composed of metal ions or metal ion clusters connected by organic connectors. Compared with other porous materials, MOFs have the advantages of structural variability, ultra-high porosity, uniform and adjustable apertures, adjustable inner surface properties, etc. [91]. Wang et al. [92] prepared dense and defect-free Mg-MOF-74 membranes with MgO crystal seeds and modified them with ethylenediamine. The results indicated significantly improved H_2/CO_2 separation performance, while the H_2/CO_2 selectivity increased from 10.5 to 28 at room temperature. Jin et al. [93] prepared novel CAU-10-H MOF membranes and reported their good H_2 permeability. The maximum separation coefficients of H_2/CO_2 and H_2/H_2O were 11.1 and 5.67, respectively. They also found that such membranes could retain their structure and H_2 selectivity under long-term hydrothermal conditions, suggesting that they are suitable for H_2 separation in ethanol steam reforming. Liu et al. [94] synthesized a novel heterogeneous MIL-121/118 MOF membrane. The mixed H_2/CO_2, H_2/CH_4, and H_2/N_2 separation coefficients were 10.7, 8.9, and 7.5, respectively, at 293K and 1 bar. The average H_2 permeability was 7.83×10^{-8} mol·m^{-2}·s^{-1}·Pa^{-1}. Meanwhile, MIL-121/118 exhibited high thermal stability and durability, showing a good application prospect.

4.3. Metal Hydride Separation Method

The metal hydride separation method refers to purifying H_2 using H_2 storage alloys to absorb and desorb H_2 reversibly. H_2 molecules decompose into H atoms catalyzed by H_2 storage alloys by lowering the temperature and increasing the pressure. Then, metal hydrides are generated via diffusion, phase transition, combination reaction, and other processes, while impurity gases are trapped among metal particles. After the temperature is elevated and pressure is lowered, the impurity gases discharge from the metal particles, and then H_2 comes out from the crystal lattice. H_2 storage alloys can be divided into rare earth alloys, titanium alloys, zirconium alloys, and magnesium alloys based on the type of the main element. Furthermore, they can also be classified into AB_5-type alloys, AB_2-type alloys, AB-type alloys, and A_2B-type alloys according to the main elements' atomic ratio [95]. The performance of the H_2 storage alloys determines the efficiency of H_2 purification, so the chemical stability and tolerance of H_2 storage alloys can be improved and the influences of impurity gases can be reduced by modifying H_2 storage alloys. Dunikov et al. [96] used two kinds of AB_5-type alloys to separate the H_2/CO_2 mixture. They found that for the low-pressure $LaNi_{4.8}Mn_{0.3}Fe_{0.1}$ alloy, H_2 can be purified from the mixture containing 59% H_2 with a recovery rate of 94%, supporting the operation of PEMFCs. Yang et al. [97] carried out cyclic experiments on the $LaNi_{4.3}Al_{0.7}$ H_2 storage alloy in the high CO concentration environment. The H_2 storage capacity of this alloy slowly decreased at 363 K or higher temperatures, maintaining a relatively high kinetic rate so that it can be used for H_2 separation and purification in different applications. Besides this, Hanada et al. [98] studied the effects of CO_2 on the H_2 absorption performance of AB_2-type alloys to develop metal hydrides for H_2 purification and storage. The results showed that Fe and Co addition could improve the alloys' tolerance to CO_2, while Ni addition had the opposite effect. Zhou et al. [99] found that MgH_2 catalyzed by nano VTiCr easily reacted with low-pressure H_2, and is recycled in mixed gas. Therefore, the material showed H_2 separation and purification potential.

4.4. *Cryogenic Distillation*

The principle of cryogenic distillation is to separate and purify H_2 by utilizing the difference in the relative volatility of different components in feed gases. Compared with CH_4 and other light HCs, H_2 has relatively high volatility, such that HCs, CO, N_2, and other gases condense before H_2 with temperature reductions [100]. This process is usually used for H_2–HC separation. The low-temperature separation method assures a high H_2 recovery rate, but it is challenging to adapt the method for treating different feed gases. As such, it is necessary to remove CO_2, H_2O, and other impurities from the feed gases before the separation so as to avoid equipment blockage at a low temperature. Besides this, high costs and energy consumption accompany the requirements for gas compressors and cooling equipment in the actual operation. Although most impurities are liquefied at a low temperature, some remain in the gas phase as saturated steam, so it is difficult to directly obtain H_2 that meets the purity standards of fuel cell vehicles.

5. Conclusions and Prospects

Compared with industrial H_2, the purity of H_2 for fuel cell vehicles is not sufficient, although the requirements for the impurity content in H_2 are stringent. According to the existing standards, an impurity level above the limit may damage the fuel cell's performance. Thus, removing specific impurities is the focus of future research on H_2 purification for fuel cell vehicles. PSA is a universal method that can be applied to remove most contaminants. H_2-permeable membranes are often used to remove CO, CO_2, N_2, CH_4, H_2O, and other gas impurities, while CO_2-permeable membranes enable only CO_2 removal. Owing to CO, CO_2, and H_2O sensitivity, metal hydrides can be used to remove N_2, Ar, and other inert gases. However, all the existing H_2 purification methods are limited, and it is difficult to achieve the H_2 impurity level standards for fuel cell vehicles by using only one separation and purification method. Since there are many different H_2 sources, two or even more H_2 purification technologies should be adopted.

Author Contributions: conceptualization, Z.D., C.L. and W.S.; writing—original draft preparation, C.L. and G.H.; writing—review and editing, J.Z., C.L., Y.X. and X.G.; All authors have read and agreed to the published version of the manuscript.

Funding: This research was funded by the National Key Research and Development Program of China (Grant No. 2019YFB1505000) and the Technology Innovation Project of China Energy Investment (Project Number: GJNY-19-136).

Conflicts of Interest: The authors declare no conflict of interest.

References

1. Zhao, F.; Mu, Z.; Hao, H.; Liu, Z.; He, X.; Przesmitzki, S.V.; Amer, A.A. Hydrogen Fuel Cell Vehicle Development in China: An Industry Chain Perspective. *Energy Technol.* **2020**, *8*. [CrossRef]
2. Meng, X.; Gu, A.; Wu, X.; Zhou, L.; Zhou, J.; Liu, B.; Mao, Z. Status quo of China hydrogen strategy in the field of transportation and international comparisons. *Int. J. Hydrogen Energy* **2020**. [CrossRef]
3. Olabi, A.; Wilberforce, T.; Abdelkareem, M.A. Fuel cell application in the automotive industry and future perspective. *Energy* **2021**, *214*, 118955. [CrossRef]
4. Zhongfu, T.; Chen, Z.; Pingkuo, L.; Reed, B.; Jiayao, Z. Focus on fuel cell systems in China. *Renew. Sustain. Energy Rev.* **2015**, *47*, 912–923. [CrossRef]
5. Manoharan, Y.; Hosseini, S.E.; Butler, B.; Alzhahrani, H.; Senior, B.T.F.; Ashuri, T.; Krohn, J. Hydrogen Fuel Cell Vehicles; Current Status and Future Prospect. *Appl. Sci.* **2019**, *9*, 2296. [CrossRef]
6. Zhao, Y.; Mao, Y.; Zhang, W.; Tang, Y.; Wang, P. Reviews on the effects of contaminations and research methodologies for PEMFC. *Int. J. Hydrogen Energy* **2020**, *45*, 23174–23200. [CrossRef]
7. Wang, Q.; Xue, M.; Lin, B.-L.; Lei, Z.; Zhang, Z. Well-to-wheel analysis of energy consumption, greenhouse gas and air pollutants emissions of hydrogen fuel cell vehicle in China. *J. Clean. Prod.* **2020**, *275*, 123061. [CrossRef]
8. Sun, J.; Feng, H.; Xu, J.; Jin, H.; Guo, L. Investigation of the conversion mechanism for hydrogen production by coal gasification in supercritical water. *Int. J. Hydrogen Energy* **2021**, *46*, 10205–10215. [CrossRef]
9. Tasleem, S.; Tahir, M. Current trends in strategies to improve photocatalytic performance of perovskites materials for solar to hydrogen production. *Renew. Sustain. Energy Rev.* **2020**, *132*, 110073. [CrossRef]
10. Cao, L.; Yu, I.K.; Xiong, X.; Tsang, D.C.; Zhang, S.; Clark, J.H.; Hu, C.; Ng, Y.H.; Shang, J.; Ok, Y.S. Biorenewable hydrogen production through biomass gasification: A review and future prospects. *Environ. Res.* **2020**, *186*, 109547. [CrossRef]
11. Ligen, Y.; Vrubel, H.; Girault, H. Energy efficient hydrogen drying and purification for fuel cell vehicles. *Int. J. Hydrogen Energy* **2020**, *45*, 10639–10647. [CrossRef]
12. Chugh, S.; Meenakshi, S.; Sonkar, K.; Sharma, A.; Kapur, G.; Ramakumar, S. Performance evaluation of PEM fuel cell stack on hydrogen produced in the oil refinery. *Int. J. Hydrogen Energy* **2020**, *45*, 5491–5500. [CrossRef]
13. Murugan, A.; Brown, A.S. Review of purity analysis methods for performing quality assurance of fuel cell hydrogen. *Int. J. Hydrogen Energy* **2015**, *40*, 4219–4233. [CrossRef]
14. Díaz, M.A.; Iranzo, A.; Rosa, F.; Isorna, F.; López, E.; Bolivar, J.P. Effect of carbon dioxide on the contamination of low temperature and high temperature PEM (polymer electrolyte membrane) fuel cells. Influence of temperature, relative humidity and analysis of regeneration processes. *Energy* **2015**, *90*, 299–309. [CrossRef]
15. Pérez, L.C.; Koski, P.; Ihonen, J.; Sousa, J.M.; Mendes, A. Effect of fuel utilization on the carbon monoxide poisoning dynamics of Polymer Electrolyte Membrane Fuel Cells. *J. Power Sources* **2014**, *258*, 122–128. [CrossRef]
16. Lopes, T.; Paganin, V.A.; Gonzalez, E.R. The effects of hydrogen sulfide on the polymer electrolyte membrane fuel cell anode catalyst: H2S–Pt/C interaction products. *J. Power Sources* **2011**, *196*, 6256–6263. [CrossRef]
17. Viitakangas, J.; Ihonen, J.; Koski, P.; Reinikainen, M.; Aarhaug, T.A. Study of Formaldehyde and Formic Acid Contamination Effect on PEMFC. *J. Electrochem. Soc.* **2018**, *165*, F718–F727. [CrossRef]
18. Gomez, Y.A.; Oyarce, A.; Lindbergh, G.; Lagergren, C. Ammonia Contamination of a Proton Exchange Membrane Fuel Cell. *J. Electrochem. Soc.* **2018**, *165*, F189–F197. [CrossRef]
19. Li, H.; Wang, H.; Qian, W.; Zhang, S.; Wessel, S.; Cheng, T.T.; Shen, J.; Wu, S. Chloride contamination effects on proton exchange membrane fuel cell performance and durability. *J. Power Sources* **2011**, *196*, 6249–6255. [CrossRef]
20. Terlip, D.; Hartmann, K.; Martin, J.; Rivkin, C. Adapted tube cleaning practices to reduce particulate contamination at hydrogen fueling stations. *Int. J. Hydrogen Energy* **2019**, *44*, 8692–8698. [CrossRef]
21. Aasadnia, M.; Mehrpooya, M.; Ghorbani, B. A novel integrated structure for hydrogen purification using the cryogenic method. *J. Clean. Prod.* **2021**, *278*, 123872. [CrossRef]
22. Schorer, L.; Schmitz, S.; Weber, A. Membrane based purification of hydrogen system (MEMPHYS). *Int. J. Hydrogen Energy* **2019**, *44*, 12708–12714. [CrossRef]
23. Dunikov, D.; Borzenko, V.; Malyshenko, S. Influence of impurities on hydrogen absorption in a metal hydride reactor. *Int. J. Hydrogen Energy* **2012**, *37*, 13843–13848. [CrossRef]
24. Hamm, J.B.; Ambrosi, A.; Griebeler, J.G.; Marcilio, N.R.; Tessaro, I.C.; Pollo, L.D. Recent advances in the development of supported carbon membranes for gas separation. *Int. J. Hydrogen Energy* **2017**, *42*, 24830–24845. [CrossRef]
25. Sasikumar, B.; Arthanareeswaran, G.; Ismail, A. Recent progress in ionic liquid membranes for gas separation. *J. Mol. Liq.* **2018**, *266*, 330–341. [CrossRef]
26. Bernardo, G.; Araújo, T.; Lopes, T.D.S.; Sousa, J.; Mendes, A. Recent advances in membrane technologies for hydrogen purification. *Int. J. Hydrogen Energy* **2020**, *45*, 7313–7338. [CrossRef]

27. Liszka, M.; Malik, T.; Manfrida, G. Energy and exergy analysis of hydrogen-oriented coal gasification with CO_2 capture. *Energy* **2012**, *45*, 142–150. [CrossRef]
28. Martinez, I.; Romano, M.C.; Chiesa, P.; Grasa, G.; Murillo, R. Hydrogen production through sorption enhanced steam reforming of natural gas: Thermodynamic plant assessment. *Int. J. Hydrogen Energy* **2013**, *38*, 15180–15199. [CrossRef]
29. Zhou, S.; Yuan, Z.; Wang, S. Selective CO oxidation with real methanol reformate over monolithic Pt group catalysts: PEMFC applications. *Int. J. Hydrogen Energy* **2006**, *31*, 924–933. [CrossRef]
30. Onozaki, M.; Watanabe, K.; Hashimoto, T.; Saegusa, H.; Katayama, Y. Hydrogen production by the partial oxidation and steam reforming of tar from hot coke oven gas. *Fuel* **2006**, *85*, 143–149. [CrossRef]
31. Khalilpourmeymandi, H.; Mirvakili, A.; Rahimpour, M.R.; Shariati, A. Application of response surface methodology for optimization of purge gas recycling to an industrial reactor for conversion of CO 2 to methanol. *Chin. J. Chem. Eng.* **2017**, *25*, 676–687. [CrossRef]
32. Yáñez, M.; Relvas, F.; Ortiz, A.; Gorri, D.; Mendes, A.; Ortiz, I. PSA purification of waste hydrogen from ammonia plants to fuel cell grade. *Sep. Purif. Technol.* **2020**, *240*, 116334. [CrossRef]
33. Kırtay, E. Recent advances in production of hydrogen from biomass. *Energy Convers. Manag.* **2011**, *52*, 1778–1789. [CrossRef]
34. Sircar, S.; Golden, T.C. Purification of Hydrogen by Pressure Swing Adsorption. *Sep. Sci. Technol.* **2000**, *35*, 667–687. [CrossRef]
35. Lively, R.P.; Bessho, N.; Bhandari, D.A.; Kawajiri, Y.; Koros, W.J. Thermally moderated hollow fiber sorbent modules in rapidly cycled pressure swing adsorption mode for hydrogen purification. *Int. J. Hydrogen Energy* **2012**, *37*, 15227–15240. [CrossRef]
36. Shamsudin, I.; Abdullah, A.; Idris, I.; Gobi, S.; Othman, M. Hydrogen purification from binary syngas by PSA with pressure equalization using microporous palm kernel shell activated carbon. *Fuel* **2019**, *253*, 722–730. [CrossRef]
37. He, B.; Liu, J.; Zhang, Y.; Zhang, S.; Wang, P.; Xu, H. Comparison of structured activated carbon and traditional adsorbents for purification of H2. *Sep. Purif. Technol.* **2020**, *239*, 116529. [CrossRef]
38. Kuroda, S.; Nagaishi, T.; Kameyama, M.; Koido, K.; Seo, Y.; Dowaki, K. Hydroxyl aluminium silicate clay for biohydrogen purification by pressure swing adsorption: Physical properties, adsorption isotherm, multicomponent breakthrough curve modelling, and cycle simulation. *Int. J. Hydrogen Energy* **2018**, *43*, 16573–16588. [CrossRef]
39. Agueda, V.I.; Delgado, J.A.; Uguina, M.A.; Brea, P.; Spjelkavik, A.I.; Blom, R.; Grande, C. Adsorption and diffusion of H_2, N_2, CO, CH_4 and CO_2 in UTSA-16 metal-organic framework extrudates. *Chem. Eng. Sci.* **2015**, *124*, 159–169. [CrossRef]
40. Brea, P.; Delgado, J.; Águeda, V.I.; Gutiérrez, P.; Uguina, M.A. Multicomponent adsorption of H_2, CH_4, CO and CO_2 in zeolites NaX, CaX and MgX. Evaluation of performance in PSA cycles for hydrogen purification. *Microporous Mesoporous Mater.* **2019**, *286*, 187–198. [CrossRef]
41. Banu, A.-M.; Friedrich, D.; Brandani, S.; Dueren, T. A Multiscale Study of MOFs as Adsorbents in H2 PSA Purification. *Ind. Eng. Chem. Res.* **2013**, *52*, 9946–9957. [CrossRef]
42. Relvas, F.; Whitley, R.D.; Silva, C.M.; Mendes, A. Single-Stage Pressure Swing Adsorption for Producing Fuel Cell Grade Hydrogen. *Ind. Eng. Chem. Res.* **2018**, *57*, 5106–5118. [CrossRef]
43. Ahn, S.; You, Y.-W.; Lee, D.-G.; Kim, K.-H.; Oh, M.; Lee, C.-H. Layered two- and four-bed PSA processes for H_2 recovery from coal gas. *Chem. Eng. Sci.* **2012**, *68*, 413–423. [CrossRef]
44. Abdeljaoued, A.; Relvas, F.; Mendes, A.; Chahbani, M.H. Simulation and experimental results of a PSA process for production of hydrogen used in fuel cells. *J. Environ. Chem. Eng.* **2018**, *6*, 338–355. [CrossRef]
45. Moon, D.-K.; Park, Y.; Oh, H.-T.; Kim, S.-H.; Oh, M.; Lee, C.-H. Performance analysis of an eight-layered bed PSA process for H_2 recovery from IGCC with pre-combustion carbon capture. *Energy Convers. Manag.* **2018**, *156*, 202–214. [CrossRef]
46. Zhang, N.; Xiao, J.; Bénard, P.; Chahine, R. Single- and double-bed pressure swing adsorption processes for H2/CO syngas separation. *Int. J. Hydrogen Energy* **2019**, *44*, 26405–26418. [CrossRef]
47. Li, H.; Liao, Z.; Sun, J.; Jiang, B.; Wang, J.; Yang, Y. Modelling and simulation of two-bed PSA process for separating H2 from methane steam reforming. *Chin. J. Chem. Eng.* **2019**, *27*, 1870–1878. [CrossRef]
48. You, Y.-W.; Lee, D.-G.; Yoon, K.-Y.; Moon, D.-K.; Kim, S.M.; Lee, C.-H. H2 PSA purifier for CO removal from hydrogen mixtures. *Int. J. Hydrogen Energy* **2012**, *37*, 18175–18186. [CrossRef]
49. Lopes, F.V.; Grande, C.A.; Rodrigues, A.E. Fast-cycling VPSA for hydrogen purification. *Fuel* **2012**, *93*, 510–523. [CrossRef]
50. Golmakani, A.; Fatemi, S.; Tamnanloo, J. Investigating PSA, VSA, and TSA methods in SMR unit of refineries for hydrogen production with fuel cell specification. *Sep. Purif. Technol.* **2017**, *176*, 73–91. [CrossRef]
51. Golmakani, A.; Nabavi, S.A.; Manović, V. Effect of impurities on ultra-pure hydrogen production by pressure vacuum swing adsorption. *J. Ind. Eng. Chem.* **2020**, *82*, 278–289. [CrossRef]
52. Hufton, J.; Golden, T.; Quinn, R.; Kloosterman, J.; Wright, A.; Schaffer, C.; Hendershot, R.; White, V.; Fogash, K. Advanced hydrogen and CO2 capture technology for sour syngas. *Energy Procedia* **2011**, *4*, 1082–1089. [CrossRef]
53. Majlan, E.H.; Daud, W.R.W.; Iyuke, S.E.; Mohamad, A.B.; Kadhum, A.A.H.; Mohammad, A.W.; Takriff, M.S.; Bahaman, N. Hydrogen purification using compact pressure swing adsorption system for fuel cell. *Int. J. Hydrogen Energy* **2009**, *34*, 2771–2777. [CrossRef]
54. Zhu, X.; Shi, Y.; Li, S.; Cai, N. Elevated temperature pressure swing adsorption process for reactive separation of CO/CO_2 in H_2-rich gas. *Int. J. Hydrogen Energy* **2018**, *43*, 13305–13317. [CrossRef]
55. Li, P.; Wang, Z.; Qiao, Z.; Liu, Y.; Cao, X.; Li, W.; Wang, J.; Wang, S. Recent developments in membranes for efficient hydrogen purification. *J. Membr. Sci.* **2015**, *495*, 130–168. [CrossRef]

56. Rahimpour, M.; Samimi, F.; Babapoor, A.; Tohidian, T.; Mohebi, S. Palladium membranes applications in reaction systems for hydrogen separation and purification: A review. *Chem. Eng. Process. Process. Intensif.* **2017**, *121*, 24–49. [CrossRef]
57. Nayebossadri, S.; Speight, J.D.; Book, D. Hydrogen separation from blended natural gas and hydrogen by Pd-based membranes. *Int. J. Hydrogen Energy* **2019**, *44*, 29092–29099. [CrossRef]
58. Zhao, C.; Sun, B.; Jiang, J.; Xu, W. H2 purification process with double layer bcc-PdCu alloy membrane at ambient temperature. *Int. J. Hydrogen Energy* **2020**, *45*, 17540–17547. [CrossRef]
59. Kong, S.Y.; Kim, D.H.; Henkensmeier, D.; Kim, H.-J.; Ham, H.C.; Han, J.; Yoon, S.P.; Yoon, C.W.; Choi, S.H. Ultrathin layered Pd/PBI–HFA composite membranes for hydrogen separation. *Sep. Purif. Technol.* **2017**, *179*, 486–493. [CrossRef]
60. Kiadehi, A.D.; Taghizadeh, M. Fabrication, characterization, and application of palladium composite membrane on porous stainless steel substrate with NaY zeolite as an intermediate layer for hydrogen purification. *Int. J. Hydrogen Energy* **2019**, *44*, 2889–2904. [CrossRef]
61. Iulianelli, A.; Ghasemzadeh, K.; Marelli, M.; Evangelisti, C. A supported Pd-Cu/Al2O3 membrane from solvated metal atoms for hydrogen separation/purification. *Fuel Process. Technol.* **2019**, *195*. [CrossRef]
62. Huang, Y.; Liu, Q.; Jin, X.; Ding, W.; Hu, X.; Li, H. Coating the porous Al2O3 substrate with a natural mineral of Nontronite-15A for fabrication of hydrogen-permeable palladium membranes. *Int. J. Hydrogen Energy* **2020**, *45*, 7412–7422. [CrossRef]
63. Liguori, S.; Kian, K.; Buggy, N.; Anzelmo, B.H.; Wilcox, J. Opportunities and challenges of low-carbon hydrogen via metallic membranes. *Prog. Energy Combust. Sci.* **2020**, *80*, 100851. [CrossRef]
64. Dolan, M.D.; Viano, D.M.; Langley, M.J.; Lamb, K.E. Tubular vanadium membranes for hydrogen purification. *J. Membr. Sci.* **2018**, *549*, 306–311. [CrossRef]
65. Fasolin, S.; Barison, S.; Boldrini, S.; Ferrario, A.; Romano, M.; Montagner, F.; Miorin, E.; Fabrizio, M.; Armelao, L. Hydrogen separation by thin vanadium-based multi-layered membranes. *Int. J. Hydrogen Energy* **2018**, *43*, 3235–3243. [CrossRef]
66. Alimov, V.; Bobylev, I.; Busnyuk, A.; Kolgatin, S.; Kuzenov, S.; Peredistov, E.Y.; Livshits, A. Extraction of ultrapure hydrogen with V-alloy membranes: From laboratory studies to practical applications. *Int. J. Hydrogen Energy* **2018**, *43*, 13318–13327. [CrossRef]
67. Jo, Y.S.; Cha, J.; Lee, C.H.; Jeong, H.; Yoon, C.W.; Nam, S.W.; Han, J. A viable membrane reactor option for sustainable hydrogen production from ammonia. *J. Power Sources* **2018**, *400*, 518–526. [CrossRef]
68. Jo, Y.S.; Lee, C.H.; Kong, S.Y.; Lee, K.-Y.; Yoon, C.W.; Nam, S.W.; Han, J. Characterization of a Pd/Ta composite membrane and its application to a large scale high-purity hydrogen separation from mixed gas. *Sep. Purif. Technol.* **2018**, *200*, 221–229. [CrossRef]
69. Budhi, Y.W.; Suganda, W.; Irawan, H.K.; Restiawaty, E.; Miyamoto, M.; Uemiya, S.; Nishiyama, N.; Annaland, M.V.S. Hydrogen separation from mixed gas (H2, N2) using Pd/Al2O3 membrane under forced unsteady state operations. *Int. J. Hydrogen Energy* **2020**, *45*, 9821–9835. [CrossRef]
70. Yáñez, M.; Ortiz, A.; Gorri, D.; Ortiz, I. Comparative performance of commercial polymeric membranes in the recovery of industrial hydrogen waste gas streams. *Int. J. Hydrogen Energy* **2020**. [CrossRef]
71. Pal, N.; Agarwal, M.; Maheshwari, K.; Solanki, Y.S. A review on types, fabrication and support material of hydrogen separation membrane. *Mater. Today Proc.* **2020**, *28*, 1386–1391. [CrossRef]
72. Strugova, D.; Zadorozhnyy, M.Y.; Berdonosova, E.; Yablokova, M.Y.; Konik, P.; Zheleznyi, M.; Semenov, D.; Milovzorov, G.; Padaki, M.; Kaloshkin, S.; et al. Novel process for preparation of metal-polymer composite membranes for hydrogen separation. *Int. J. Hydrogen Energy* **2018**, *43*, 12146–12152. [CrossRef]
73. Rezakazemi, M.; Shahidi, K.; Mohammadi, T. Hydrogen separation and purification using crosslinkable PDMS/zeolite A nanoparticles mixed matrix membranes. *Int. J. Hydrogen Energy* **2012**, *37*, 14576–14589. [CrossRef]
74. Peydayesh, M.; Mohammadi, T.; Bakhtiari, O. Effective hydrogen purification from methane via polyimide Matrimid® 5218-Deca-dodecasil 3R type zeolite mixed matrix membrane. *Energy* **2017**, *141*, 2100–2107. [CrossRef]
75. Hamid, M.A.A.; Chung, Y.T.; Rohani, R.; Junaidi, M.U.M. Miscible-blend polysulfone/polyimide membrane for hydrogen purification from palm oil mill effluent fermentation. *Sep. Purif. Technol.* **2019**, *209*, 598–607. [CrossRef]
76. Naderi, A.; Chung, T.-S.; Weber, M.; Maletzko, C. High performance dual-layer hollow fiber membrane of sulfonated polyphenylsulfone/Polybenzimidazole for hydrogen purification. *J. Membr. Sci.* **2019**, *591*, 117292. [CrossRef]
77. Chen, F.; Dong, S.; Wang, Z.; Xu, J.; Xu, R.; Wang, J. Preparation of mixed matrix composite membrane for hydrogen purification by incorporating ZIF-8 nanoparticles modified with tannic acid. *Int. J. Hydrogen Energy* **2020**, *45*, 7444–7454. [CrossRef]
78. Abedini, R.; Omidkhah, M.; Dorosti, F. Hydrogen separation and purification with poly (4-methyl-1-pentyne)/MIL 53 mixed matrix membrane based on reverse selectivity. *Int. J. Hydrogen Energy* **2014**, *39*, 7897–7909. [CrossRef]
79. Cao, X.; Qiao, Z.; Wang, Z.; Zhao, S.; Li, P.; Wang, J.; Wang, S. Enhanced performance of mixed matrix membrane by incorporating a highly compatible covalent organic framework into poly(vinylamine) for hydrogen purification. *Int. J. Hydrogen Energy* **2016**, *41*, 9167–9174. [CrossRef]
80. Salim, W.; Vakharia, V.; Chen, K.K.; Gasda, M.; Ho, W.W. Oxidatively stable borate-containing membranes for H2 purification for fuel cells. *J. Membr. Sci.* **2018**, *562*, 9–17. [CrossRef]
81. Nigiz, F.U.; Hilmioglu, N.D. Enhanced hydrogen purification by graphene—Poly(Dimethyl siloxane) membrane. *Int. J. Hydrogen Energy* **2020**, *45*, 3549–3557. [CrossRef]
82. Tanco, M.A.L.; Medrano, J.A.; Cechetto, V.; Gallucci, F.; Tanaka, D.A.P. Hydrogen permeation studies of composite supported alumina-carbon molecular sieves membranes: Separation of diluted hydrogen from mixtures with methane. *Int. J. Hydrogen Energy* **2020**. [CrossRef]

83. Xu, R.; He, L.; Li, L.; Hou, M.; Wang, Y.; Zhang, B.; Liang, C.; Wang, T. Ultraselective carbon molecular sieve membrane for hydrogen purification. *J. Energy Chem.* **2020**, *50*, 16–24. [CrossRef]
84. Sun, C.; Zheng, X.; Bai, B. Hydrogen purification using nanoporous graphene membranes and its economic analysis. *Chem. Eng. Sci.* **2019**, *208*, 115141. [CrossRef]
85. Yang, E.; Alayande, A.B.; Goh, K.; Kim, C.-M.; Chu, K.-H.; Hwang, M.-H.; Ahn, J.-H.; Chae, K.-J. 2D materials-based membranes for hydrogen purification: Current status and future prospects. *Int. J. Hydrogen Energy* **2021**, *46*, 11389–11410. [CrossRef]
86. Sang, P.; Zhao, L.; Xu, J.; Shi, Z.; Guo, S.; Yu, Y.; Zhu, H.; Yan, Z.; Guo, W. Excellent membranes for hydrogen purification: Dumbbell-shaped porous γ-graphynes. *Int. J. Hydrogen Energy* **2017**, *42*, 5168–5176. [CrossRef]
87. De Silva, S.; Du, A.; Senadeera, W.; Gu, Y. Strained graphitic carbon nitride for hydrogen purification. *J. Membr. Sci.* **2017**, *528*, 201–205. [CrossRef]
88. Wei, S.; Zhou, S.; Wu, Z.; Wang, M.; Wang, Z.; Guo, W.; Lu, X. Mechanistic insights into porous graphene membranes for helium separation and hydrogen purification. *Appl. Surf. Sci.* **2018**, *441*, 631–638. [CrossRef]
89. Zeynali, R.; Ghasemzadeh, K.; Sarand, A.B.; Kheiri, F.; Basile, A. Experimental study on graphene-based nanocomposite membrane for hydrogen purification: Effect of temperature and pressure. *Catal. Today* **2019**, *330*, 16–23. [CrossRef]
90. Liu, Y.; Liu, W.; Hou, J.; Dai, Y.; Yang, J. Coronoid nanographene C216 as hydrogen purification membrane: A density functional theory study. *Carbon* **2018**, *135*, 112–117. [CrossRef]
91. Wu, T.; Prasetya, N.; Li, K. Recent advances in aluminium-based metal-organic frameworks (MOF) and its membrane applications. *J. Membr. Sci.* **2020**, *615*, 118493. [CrossRef]
92. Wang, N.; Mundstock, A.; Liu, Y.; Huang, A.; Caro, J. Amine-modified Mg-MOF-74/CPO-27-Mg membrane with enhanced H_2 /CO_2 separation. *Chem. Eng. Sci.* **2015**, *124*, 27–36. [CrossRef]
93. Jin, H.; Wollbrink, A.; Yao, R.; Li, Y.; Caro, J.; Yang, W. A novel CAU-10-H MOF membrane for hydrogen separation under hydrothermal conditions. *J. Membr. Sci.* **2016**, *513*, 40–46. [CrossRef]
94. Liu, J.; Wang, Y.; Guo, H.; Fan, S. A novel heterogeneous MOF membrane MIL-121/118 with selectivity towards hydrogen. *Inorg. Chem. Commun.* **2020**, *111*, 107637. [CrossRef]
95. Xiao, J.; Tong, L.; Yang, T.; Bénard, P.; Chahine, R. Lumped parameter simulation of hydrogen storage and purification systems using metal hydrides. *Int. J. Hydrogen Energy* **2017**, *42*, 3698–3707. [CrossRef]
96. Dunikov, D.; Borzenko, V.; Blinov, D.; Kazakov, A.; Lin, C.-Y.; Wu, S.-Y.; Chu, C.-Y. Biohydrogen purification using metal hydride technologies. *Int. J. Hydrogen Energy* **2016**, *41*, 21787–21794. [CrossRef]
97. Yang, F.; Chen, X.; Wu, Z.; Wang, S.; Wang, G.; Zhang, Z.; Wang, Y. Experimental studies on the poisoning properties of a low-plateau hydrogen storage alloy LaNi 4.3 Al 0.7 against CO impurities. *Int. J. Hydrogen Energy* **2017**, *42*, 16225–16234. [CrossRef]
98. Hanada, N.; Asada, H.; Nakagawa, T.; Higa, H.; Ishida, M.; Heshiki, D.; Toki, T.; Saita, I.; Asano, K.; Nakamura, Y.; et al. Effect of CO2 on hydrogen absorption in Ti-Zr-Mn-Cr based AB2 type alloys. *J. Alloys Compd.* **2017**, *705*, 507–516. [CrossRef]
99. Zhou, C.; Fang, Z.Z.; Sun, P.; Xu, L.; Liu, Y. Capturing low-pressure hydrogen using V Ti Cr catalyzed magnesium hydride. *J. Power Sources* **2019**, *413*, 139–147. [CrossRef]
100. Song, C.; Liu, Q.; Deng, S.; Li, H.; Kitamura, Y. Cryogenic-based CO2 capture technologies: State-of-the-art developments and current challenges. *Renew. Sustain. Energy Rev.* **2019**, *101*, 265–278. [CrossRef]

 *catalysts* 

Review

Bioalcohol Reforming: An Overview of the Recent Advances for the Enhancement of Catalyst Stability

Vincenzo Palma , Concetta Ruocco * , Marta Cortese and Marco Martino

Department of Industrial Engineering, University of Salerno, Via Giovanni Paolo II 132, 84084 Fisciano (SA), Italy; vpalma@unisa.it (V.P.); mcortese@unisa.it (M.C.); mamartino@unisa.it (M.M.)
* Correspondence: cruocco@unisa.it; Tel.: +39-089-964027

Received: 13 May 2020; Accepted: 9 June 2020; Published: 12 June 2020

Abstract: The growing demand for energy production highlights the shortage of traditional resources and the related environmental issues. The adoption of bioalcohols (i.e., alcohols produced from biomass or biological routes) is progressively becoming an interesting approach that is used to restrict the consumption of fossil fuels. Bioethanol, biomethanol, bioglycerol, and other bioalcohols (propanol and butanol) represent attractive feedstocks for catalytic reforming and production of hydrogen, which is considered the fuel of the future. Different processes are already available, including steam reforming, oxidative reforming, dry reforming, and aqueous-phase reforming. Achieving the desired hydrogen selectivity is one of the main challenges, due to the occurrence of side reactions that cause coke formation and catalyst deactivation. The aims of this review are related to the critical identification of the formation of carbon roots and the deactivation of catalysts in bioalcohol reforming reactions. Furthermore, attention is focused on the strategies used to improve the durability and stability of the catalysts, with particular attention paid to the innovative formulations developed over the last 5 years.

Keywords: hydrogen; bioalcohol; reforming; coke; catalyst stability; active phase; support; promoter

---

Summary

1. Introduction

The search for clean technology approaches able to assure safe and sustainable energy production is increasingly gaining ground due to the heavy impacts of fossil fuels on the world economy (oil price fluctuation), global warming, and human health [1,2]. An effective solution proposed to reduce the consumption of conventional feedstocks involves the use of hydrogen as an energetic vector, which leads to no or very low carbon emissions, as well as the release of atmospheric pollutants [3]. Steam reforming of natural gas is the most widespread technology for hydrogen production, and applying the same technology to new-generation (biomass-derived) fuels could offer significant energy and environmental advantages [4]. Biomass is abundantly available in different forms. The use of biomass for energy generation results in a neutral carbon balance; only trace amounts of sulphur and heavy metals are present in biomass compared to fossil fuels, thus limiting the formation of harmful substances [5]. Among the available fuels produced from biomass, bioalcohols are emerging as competitive sources for hydrogen production via reforming [6–8]. Bioalcohols (i.e., alcohols generated from biomass or biological routes) can be produced from different feedstocks, including crops, agricultural and forestry waste, and food waste [9]. First-generation biofuels are produced from sugar, starch, oil-bearing crops, or animal fats. Bioethanol is mainly obtained via fermentation of sugar cane or starches, while butanol and propanol are formed as co-products via well-established technologies [10]. Wood, agricultural residues, forestry residues (cellulosic, hemicelluloses, or lignin), organic waste, food waste, and specific biomass crops are the feedstocks used for second-generation bioalcohols. Bioethanol and biobutanol can be produced via fermentation of lignocellulosic sugars via different microorganisms; for the latter, the process is more difficult and not commercialized yet (only pilot plants are available) [11]. Ethanol and butanol can also be derived from algae (third-generation bioalcohols) [12]. Bioglycerol is produced during biodiesel generation via transesterification of triglycerides, using vegetable oil as the feedstock [13]. In addition, glycerol can also be generated as a by-product, along with bioethanol (consuming up to 10% of the weight of the employed sugar) [14]. As an alternative to the conventional approaches for methanol production (which employ natural gas or coal as feedstocks), other routes are available, mainly involving biomass gasification [15] (including the conversion of municipal solid waste [16], animal waste, and agriculture wastes [17]) and CO_2 hydrogenation [18].

Bioalcohol conversion (denoted as X) to hydrogen can follow different routes, including steam reforming (Equation (1)), oxidative steam reforming (Equation (2)), dry reforming (Equation (3)), and aqueous-phase reforming (Equation (4)) [8]. Among these processes, according to the stoichiometry of the reaction, steam reforming gives the highest hydrogen yields. However, several side reactions may occur during reforming, which besides affecting the H_2 selectivity, may also be responsible for carbon formation and catalyst deactivation. In this regard, the addition of oxygen via the oxidative process has been investigated as a viable route to mitigate carbon deposition [19,20]. The main pathways favoring coke formation during reforming include the Boudouard reaction, decomposition of carbon-containing intermediates (i.e., CH_4), dehydration, and subsequent polymerization reactions [21]. The same intermediates participating in the main reactions, in fact, may also be involved in coking pathways, while the contribution of side reactions depends on the operating conditions and the nature of both the active metal and the selected support [22]. In particular, a high water content and oxygen co-feeding disadvantage carbon formation [23,24], while the effect of the temperature on coke selectivity depends on the substrate, which influences the nature of the carbon formed and the effects of the coke gasification reactions [25,26]. In fact, it is clear that the product distribution and contributions of coke formation reactions also depend on the chosen molecule; it was found that the higher the number of the hydroxyl groups in a molecule, the lower the formation of CH_4, while CO selectivity was enhanced. In addition, species with longer carbon chains promote the formation of carbonaceous deposits. In fact, the number of hydroxyl groups influences the oxygen content in the reaction intermediates, along with their contributions to dehydration reactions [27]. The characteristics of the coke formed (amorphous or filamentous, with possible whiskers) also change depending on the starting substrate and the acidic–basic properties of the catalyst [28]. In particular, alcohols were shown to be precursors of encapsulating coke, while CO and CH_4 were responsible for the formation of filaments, whose contribution to catalyst deactivation was more pronounced [29].

$$C_nH_{2n+1}OH_{(g)} + xH_2O_{(g)} \leftrightarrow nCO_{2(g)} + (x+n+1)H_{2(g)} \tag{1}$$

$$C_nH_{2n+1}OH_{(g)} + (x-2y)H_2O_{(g)} + yO_{2(g)} \leftrightarrow nCO_{2(g)} + (x+n-2y+1)H_{2(g)} \tag{2}$$

$$C_nH_{2n+1}OH_{(g)} + zCO_{2(g)} \leftrightarrow (n+z)CO_{(g)} + (n+1)H_{2(g)} \tag{3}$$

$$C_nH_{2n+1}OH_{(l)} + xH_2O_{(l)} \leftrightarrow nCO_{2(g)} + (x+n+1)H_{2(g)} \tag{4}$$

The reforming of bioalcohols has attracted the attention of several scientists, who have proposed different approaches to modulate catalysts' selectivity and improve their durability, including the use of additives and promoters (i.e., alkaline or transition metals and acidic or basic oxides) [30–32], the choice of high surface area supports or perovskites [33,34] and core–shell catalysts [35], the development of innovative preparation methods allowing enhanced dispersion of active species and improved metal–support interactions [36,37], and the synthesis of solid solutions for the confinement of active metals particles [38]. The deposition of catalysts on structured carriers (i.e., foams and monoliths) with high thermal conductivity or on microchannel walls was also shown to prevent the deterioration of catalytic performance due to the reduction of hot and cold spots [39–41]. In addition, the use of fluidized bed reactors was proposed as an effective route to separate filamentous coke from catalyst particles [42].

The present review focuses on the most recent advances to improve the stability and durability of catalysts for the reforming processes of five alcohols (ethanol, methanol, glycerol, butanol, and propanol), presenting a critical analysis of the main carbon formation roots for each substrate. Particular attention is devoted to innovative formulations developed in the last 5 years, analysing the contributions of the active phase, the support, and eventual promoters to catalyst stability. novel reactor configurations developed to improve the performances of the considered reforming processes are also assessed.

This review article is divided into three main sections based on the alcohol used in the catalytic process: ethanol, methanol, and glycerol. Finally, a further section focusing on butanol and propanol reforming is presented. Each of the main sections is divided into paragraphs to highlight the effects of the active phases, supports, and promoters. At the end of every section, a summary table is provided to compare the catalytic performance in terms of carbon formation rates.

2. Bioethanol Reforming

During ethanol steam reforming (ESR), the main pathways responsible for coke formation are ethanol cracking, dehydration reactions to ethylene (Equation (5)) and subsequent polymerization (Equation (6)), and aldol condensation of acetone (Equation (7)), followed by dehydration and oligomerization of mesityl oxide, CO disproportion (Boudouard reaction, Equation (8)), and methane dehydrogenation (Equation (9)); ethane can also be formed and subsequently dehydrogenated [43–45].

$$C_2H_5OH \rightarrow C_2H_4 + H_2O \tag{5}$$

$$C_2H_4 \rightarrow (C_2H_4)_n \rightarrow C \tag{6}$$

$$2C_2H_5OH \leftrightarrow C_3H_6O + CO + 3H_2O \tag{7}$$

$$2CO \leftrightarrow CO_2 + C \tag{8}$$

$$CH_4 \rightarrow C + 2H_2 \tag{9}$$

Based on the involved mechanism, different types of carbon can be formed (monoatomic adsorbed carbon, amorphous polymeric films, vermicular filaments, and graphitic crystalline platelets); while the monoatomic adsorbed carbon and polymeric coke are derived from the thermal decomposition of hydrocarbons, metallic sites are directly involved in the formation of filamentous and graphitic coke [46]. The crystallization of carbon coming from ethylene polymerization is favored at high temperatures, while carbide species mainly derivate from CH_4 decomposition, which can polymerize and form amorphous carbon or carbon whiskers; finally, carbon growth during Boudouard reactions may involve island formation [47,48]. For several transition metals (e.g., Ni, Co, Rh, Ru, Pd, Pt) [49], discrete Fourier transform (DFT) calculations reveal that before C-C bond breaking, an intermediate CH-CO bond is formed, which based on the reaction conditions and the nature of the metal surface can either dehydrogenate to form solid carbon or be hydrogenated to CH_4 during the C-C rupture. For example, in the case of Ni, it was shown that elevated temperatures favor intermediate decomposition over hydrogenation.

Ethanol reforming has been widely investigated under simulated bioethanol feeding (only containing water and ethanol); some studies were also performed with crude bioethanol feeding or a model mixture containing typical impurities (i.e., methanol, acetaldehyde, isopropyl alcohol, isobutyl alcohol, isoamyl alcohol). However, in the latter case, a faster catalyst deactivation was described. Glycerol and acetic acid are known as the major coke precursors; moreover, longer and heavier alcohols are not easily reformed but can be dehydrated to the corresponding olefins [50].

Liu et al. [51] described the mechanism of carbon formation on Ni/CeO_2 catalysts and the role of the hydroxyls groups in suppressing carbon formation (Figure 1). During ethanol steam reforming on the above catalysts, coke formation was mainly ascribed to the dehydrogenation of the surface methyl groups. Ni^0, in fact, was the active phase leading to both the C-C and C-H cleavages of ethanol. In general, such methyl groups are not stable on the Ni(111) surfaces and are mostly dehydrogenated to surface carbon and hydrogen. However, in this study, surface carbon was formed through the generation of nickel carbides, which can also be originated from small amounts of ethoxy species decomposing on the Ni sites to form Ni_3C. Water can easily dissociate on the Ni-CeO_x to form hydroxyls groups, which together with the lattice oxygen on the surface, promote the oxidation of deposited carbon. However, the oxygen transfer from the ceria lattice to Ni particles is an endothermic reaction and only occurs above 470 °C. Thus, two competitive processes were described on the surface (carbon deposition and its oxidation by hydroxyls); the redox nature of the support as well as the Ni-ceria interactions are crucial to improve the oxygen transport and coke removal.

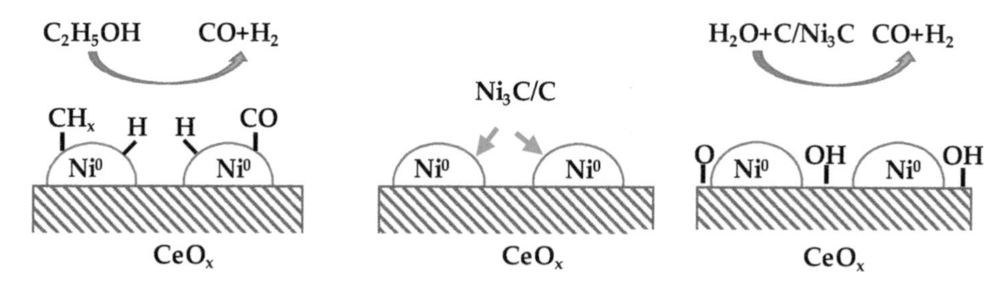

Figure 1. Mechanism of coke formation and removal on a Ni/CeO_x catalyst during ethanol steam reforming (ESR) [51].

The rate of ethylene formation throughout ESR is related to the support acidity and the amount of acidic sites on the catalysts' surfaces. Figure 2 displays the positive effects of acidic surface tailoring by the addition if Ti for a nanostructured Ni-Al catalyst [52]. Ti doping resulted in an expansion of pore sizes, thus improving the mass transfer and the contact between catalytic sites and reactants; moreover, the redox properties of the Ti^{4+}/Ti^{3+} couple strongly influenced the catalyst selectivity.

In fact, due to the titania effect on the surface basicity, the ethylene formed via ethanol dehydration was more stable and coke generated from the C_2H_4 by polymerization and carbonization was thermodynamically less favoured and easier to decompose compared to the Ni sample containing only alumina.

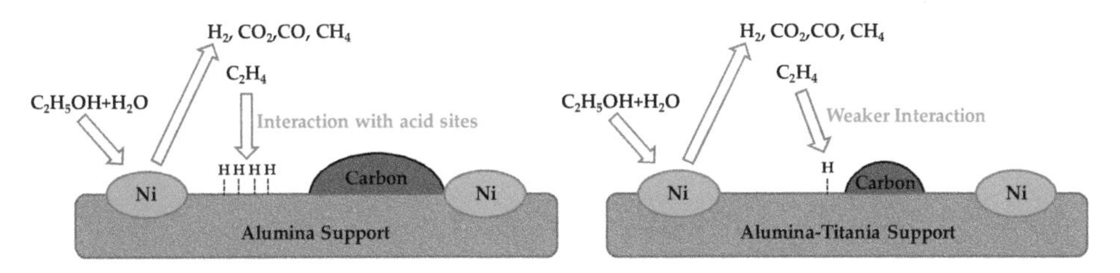

Figure 2. Mechanism of carbon formation of Ni-Al and Ni-Al-Ti catalysts during ethanol steam reforming [52].

For alumina-based catalysts, the beneficial effects of support modification by the addition of CeO_2 or La_2O_3 were shown, as well as for noble metals as active species (i.e., rhodium) [53]. The presence of low ceria loadings adds a Lewis acidity of medium strength to the CeO_2-Al_2O_3 mixed oxide, while the reduced carbon deposition on this support is related to the oxygen storage capacity and mobility of the ceria. In fact, the reversible release of oxygen makes O_x (lattice oxygen) available on the oxide surface. Carbon monoxide adsorbs on the surface, reacting with O_x and producing CO_2. Moreover, the growth of carbon fibres is hindered by CeO_2, which provides extra oxygen for gasification; solid carbon can react with oxygen lattices, further improving the yield to carbon oxides. Because CO is consumed to generate CO_2, less carbon monoxide is available to be converted to coke via the Boudouard reaction. A different mechanism for the prevention of carbon build-up was reported for La_2O_3. In the presence of lantania, carbon dioxide is subtracted to the equilibrium via the formation of $La_2O_2CO_3$, which can react with the carbon metal species in its vicinity and generate CO.

The origin and nature of coke formed over Ni/La_2O_3-αAl_2O_3 catalysts via ethanol steam reforming were reported in a work by Montero et al. [22]. The spent catalysts were characterized by scanning electron microscopy (SEM) analysis, which allowed the morphology of the carbonaceous deposits to be determined (encapsulating, filamentous, and graphitic carbon). Acetaldehyde, ethylene, and non-reacted ethanol were identified as the main coke precursors responsible for the deposition of encapsulating coke via cracking and polymerization, while filamentous and partially graphitic coke were derived from CH_4 and CO by decomposition and Boudouard reaction. As filamentous coke is located far from the metallic centers and is not able to cover the active sites (Figure 3), its influence on catalyst deactivation is reduced and the main impact on the catalyst activity loss comes from the encapsulation of active metal sites (in fact, the formation of filamentous coke caused a separation of metal nanoparticles from the support, without covering the active metal sites).

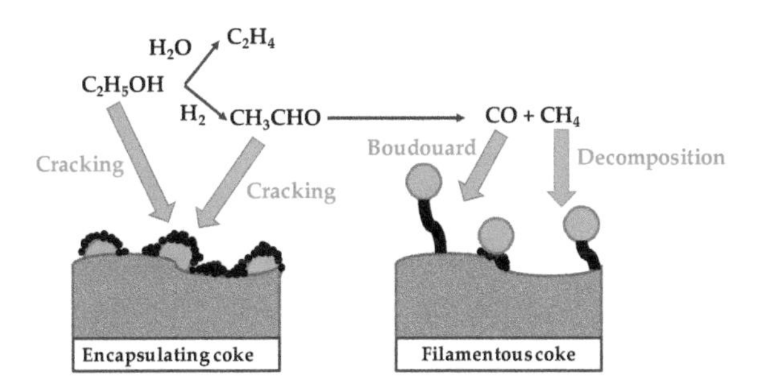

Figure 3. Mechanism and nature of coke formed over Ni/La_2O_3-αAl_2O_3 [22].

The first mechanism is predominant at high temperatures and with low space velocity, while the second one is favoured at intermediate contact times, at which methane and carbon monoxide concentrations reach a maximum; the increases of the temperature and steam-to-ethanol ratio enhance carbon gasification, reducing the concentration of coke precursors on the surface.

For nickel core–shell-structured catalysts, it was found that during ESR reaction, the ethanol molecule is first dehydrogenated to acetaldehyde, which decomposes to CH_4 and CO, with a subsequent formation of carbon lumps. Conversely, if the ethanol dehydration occurs, ethylene is formed as an intermediate, which can easily polymerise into carbon nanotubes (CNT) [54].

Co@CoO_x core–shell-structured catalysts exposed to mild oxidants (i.e., H_2O and CO_2) during ESR reaction suffer from stressing and collapse of the structure, with an oxidation of the CoO_x core by means of the oxygen-free radical species which are, therefore, no longer available for the gasification of coke; for this system, carbon is mainly formed via dehydrogenation or condensation reactions [55]. Thus, sintering and formation of amorphous coke occurred. The promoting effect of ceria related to the Ce^{3+}/Ce^{4+} redox cycle assured easier mobility of the active oxygen species, prevented re-oxidation of metallic Co particles, and at the same time allowed rapid consumption of carbonaceous species. For sol–gel alumina-supported cobalt catalysts, it was reported that the catalytic decomposition of ethylene on the Co sites was responsible for the formation of graphitic carbon, while amorphous coke was derived from the polymerization of carbon coming from methane decomposition. In addition, the influence of the CoO/Co^0 ratio on the carbon growth on the surfaces of the active sites was discussed. This ratio ruled the contribution of ethanol molecule activation and oxidation of adsorbed coke, with an optimal value of 1:3 selected to assure adequate activity and stability of the final catalyst [56].

Oxidative steam reforming of ethanol (OSRE) has also received considerable attention due to the role of gas-phase oxygen species, together with those components that are eventually presented as mobile lattice oxygen, in the oxidation of carbon deposited on the catalysts' surfaces [57,58].

For Ni-Cu bimetallic catalysts derived from hydrotalcite-like compounds, the effect of oxygen co-feeding during ethanol steam reforming was investigated at 500 °C under a molar ratio of water/ethanol/oxygen equal to 6:1:0.3. As depicted in Figure 4, O_2 addition mitigated the formation of filamentous-like coke deposits (which were observed under ESR conditions and caused the separation of metals nanoparticles from the support); oxygen promoted water production, limiting, the adsorption of hydrogen (thus hindering the reduction of NiO) and CO or CH_4 products on nickel active sites (which are converted to CO_2 and CO). Moreover, the dissolution of gasified carbon in nickel particles with further diffusion though the metal was mitigated, thus hampering the growth of filamentous coke [59].

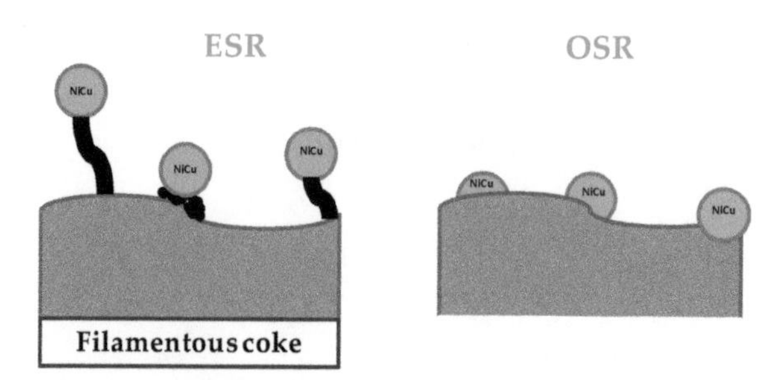

Figure 4. Effects of oxygen addition on carbon formation during ESR for Ni-Cu catalysts derived from hydrotalcite-like compounds [59].

The Pt-Ni/CeO_2-SiO_2 catalysts, tested under a reacting stream of $C_2H_5OH/H_2O/O_2$ = 1/4/0.5 at 500 °C, displayed better stability compared to the ESR measurements. The oxygen co-feeding improved coke gasification, thus resulting in diminished accumulation of carbon on the catalysts' surfaces; moreover, oxygen addition in the presence of the catalytic support having remarkable ionic conductivity encouraged the oxidation of carbon-containing species, without affecting the hydrogen production rate, which was almost the same for ESR and OSRE tests [60].

During stability tests in a fluidized bed reactor under OSRE conditions, the Pt-Ni/CeO_2-SiO_2 catalyst displayed only partial deactivation, and after 300 h of time-on-stream (TOS), the system reached a new stationary condition with no more activity decay. When a plateau was observed in the gas product distribution, carbon deposition was measured at different time-on-stream values and no change in coke selectivity was observed. This phenomenon, depicted in Figure 5, can be explained by the fact that carbon deposition was not constant during the test. On the contrary, coke progressively accumulated on the most reactive catalytic sites (involved in both reforming reactions and carbon precursors generation). Therefore, the partial deactivation of the latter sites gradually reduced the carbon formation rate (CFR), and under stationary conditions the net rate of the carbon formation (i.e., the difference between coke formation and gasification) becomes equal to zero [61].

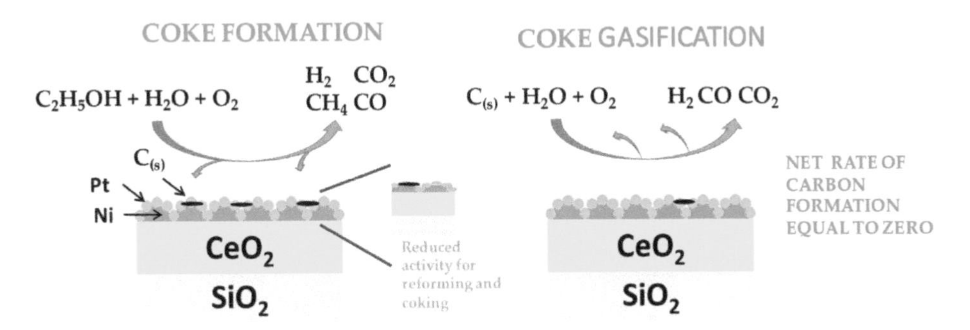

Figure 5. Description of the partial activity loss observed for Pt-Ni/CeO_2-SiO_2 catalysts during oxidative steam reforming of ethanol (OSRE) at 500 °C, where $C_2H_5OH/H_2O/O_2$ = 1:4:0.5 [61].

For ethanol dry reforming (EDR), the development of stable catalysts is a critical issue, especially when stoichiometric feeds are selected, implying that no excess of CO_2 is available to remove carbon deposits; ethylene polymerization, Boudouard, and methane decomposition reactions are the main routes for carbon deposition and for the reaction of ethanol with carbon dioxide [62]. In the presence of supports with redox properties, the removal of coke accumulated through the above pathways is favoured. CO_2, in fact, can re-fill the oxygen vacancy of the reduced support (i.e., ceria) to form CO and O; this atomic oxygen can be transferred towards the metal surface or the metal–support interface, thus favouring the gasification of carbonaceous deposits [63].

Compared to gas-phase processes, aqueous-phase reforming of ethanol has rarely been investigated in the last five years. Aqueous-phase reforming of ethanol (APRE) is carried out at temperatures below 300 °C; such mild operative conditions minimize the contribution of undesirable decomposition reactions and tend to mitigate coking phenomena (carbon formation is thermodynamically unfavourable) [64,65]. Thus, only a few works provide information about the accumulation of solid carbon on the catalysts' surfaces. Zhao et al. [66] proposed a mechanism for carbon formation over Ru-Pt/TiO_2 catalysts. Ethanol conversion via reforming passes through the cleavage of the C-C bond and the metal particle size influences the C1/C2 product ratio (larger particles assure a higher C1/C2 ratio). The energy barriers for C-C bond breaking are reduced on sites with low coordination at the corners and at step edges on metal surfaces. In fact, the most reactive sites (low coordination) can be strongly linked to the reaction intermediates (CO and CH_x), with a consequent hard regeneration of the catalytic site. Carbon formation can also be linked to Boudouard reaction, methane decomposition, ethylene polymerization, and cracking of ethane; however, CO disproportion is more likely to occur. Despite the mechanism of APR of ethanol and the carbon formation pathways not being completely understood, for noble metals catalysts supported on titania, it was shown that the by-products selectivity decrease in the order of Ru > Rh > Pt > Ir [64] and that the metal particle size strongly affects the formation of intermediate species acting as coke precursors [67].

Accordingly, in order to minimize the problem of carbon deposition, different strategies have been proposed in the literature, including the control of the metal particle sizes and the addition of a second metal or promoter able to change the electronic properties of the active species, as well as modify their interactions with the support [68–70]. In this section, the results of selected studies concerning steam reforming, oxidative steam reforming, dry reforming, and aqueous-phase reforming of ethanol are discussed. In the following paragraphs, bioethanol reforming with a focus on carbon formation will be discussed based on the nature of the active phase (Section 2.1), the chosen supports (Section 2.2), and the effects of promoters and additives (Section 2.3).

Ethanol reforming has been investigated with different noble and non-noble catalysts. Figure 6 groups the papers investigated in the present review based on the kind of active species selected.

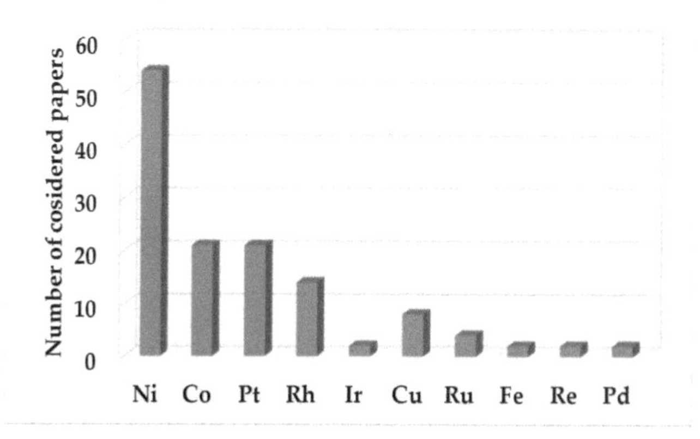

Figure 6. Number of considered papers as a function of the active phase analyzed for the ethanol reforming.

Nickel is the most widely selected metal for ethanol reforming, followed by cobalt. Moreover, among the noble metals, platinum and rhodium are most commonly selected.

2.1. The Influence of the Active Phase

In the following paragraph, the role of active species selection on catalyst stability for bioethanol reforming is discussed. In particular, various examples related to cobalt- and nickel-based catalysts are reported for the investigated processes, followed by the description of the performances of noble-metal-containing catalysts, as well as copper-based catalysts.

At 500 °C and under stoichiometric feeding conditions, CH_4 decomposition and Boudouard reaction were shown to be responsible for the formation of filamentous coke over Co_3O_4/CeO_2 catalysts during ESR [71]; carbon dissolves and subsequently precipitates on the rear of the metal, warding off the Co^0 active sites from the support and leading to a loss of the Co^0 active centers. However, by modulating the cobalt loading and particle dispersion, it is possible to vary the thickness of the filamentous coke, thus resulting in a good catalytic stability. In fact, active species dispersion and metal–support interactions are key factors affecting catalyst resistance towards deactivation. Thus, the improved stability of Co/CeO_2 catalysts during ESR at different temperatures (400, 600, and 800 °C) under a steam-to-carbon ratio of 3 compared to Ni-, Fe-, and Cu-based samples was ascribed to the strong metal–support interactions, which contributed to maintaining the Co species in an oxidized state; moreover, the gas–solid reactions between the hydrocarbon species adsorbed on the surface and the oxygen groups assured an easier gasification of such species, preventing their decomposition to coke [72]. For $PtKCo/CeO_2$ catalysts tested during oxidative steam reforming of ethanol, only a partial activity loss was observed and the reduced catalyst performance was ascribed to both carbon deposition and oxidation of cobalt metallic sites [73]. Carbon filaments were bonded to the support and contain cobalt particles at the top with the unblocked surface, not necessarily causing catalyst deactivation;

thus, only a few metallic particles were encapsulated by carbon and no longer participated in the reaction. In addition, non-filamentous coke (formed as a result of condensation and graphitization of carbon-containing species chemisorbed on the catalysts' surfaces) only covered the external surface of the active particles, while the middle part was still accessible to the reactants. Pt addition had a negligible effect on carbon deposition. It was also shown that the oxygen from the feeding can oxidize the surface atoms of cobalt particles, with a rate that increases for small particle sizes (lower than 4 nm).

The effect of the preparation method (mechanical mixing or sonication) on the stability performance of Co nanoparticles mixed with α-Al_2O_3 was reported by Riani et al. [74]. Well-dispersed Co nanoparticles displaying a good interaction with alumina were observed in the sonicated sample, while for the mechanically mixed catalyst only aggregate nanoparticles were detected. The results of stability tests carried out at 500 °C and under a steam-to-ethanol ratio of 6 proved that Co aggregates produced long carbon nanotubes, which embedded the active particles, while the tendency to form graphitic carbon is very low for the nanoparticles interacting with alumina.

For NiCo-MgAl catalysts prepared by microwave- or sonication-assisted co-precipitation, it was found that the prereduction of the catalyst is crucial to assure proper acid–base properties of the material, which affected the catalyst stability during OSRE. For the unreduced sample, in fact, a high basicity was observed, which favored certain unwanted pathways (including acetone generation), along with the formation of coke precursors [58].

The preparation method was also shown to affect the durability of Ni/SiO_2 catalysts for ESR [75]. Compared to the impregnated samples and those prepared by deposition–precipitation, the catalysts synthetized via strong electrostatic adsorption displayed improved interactions between the active phase and the support, assuring good control of the Ni particle size, which helped reduce the carbon deposition at 600 °C under a H_2O/C_2H_5OH ratio of 9. The same authors demonstrated the benefits of using silica fibers instead of the commonly used porous SiO_2. For porous catalysts, in fact, coke accumulates inside the pores and its gasification via the reaction with water is more difficult; conversely, carbon is deposited on the external surfaces of the fibers and its removal via gasification is easier.

The ultrasound-assisted synthesis of Ni/$MgAl_2O_4$ catalysts allowed sufficiently dispersed Ni particles to be obtained, even at high Ni loading [70]; the limited surface acidity, mainly due to the presence of Ni, promoted ethanol dehydration to ethylene, which for a Ni loading as high as 10% was further subjected to reforming, with very limited coking phenomena occurring.

Ni/$Ce_{0.9}Sm_{0.1}O_{2-\delta}$ nanowires were prepared via a simple two-step hydrothermal method followed by Ni deposited via wet impregnation. Such catalysts displayed high concentrations of oxygen vacancies on their surfaces and very low dimensions for Ni crystallites, assuring higher metallic surface areas compared to their analogue catalysts prepared from commercial doped ceria. Due to these improved properties, the Ni/$Ce_{0.9}Sm_{0.1}O_{2-\delta}$ sample displayed an exceptional stability with no loss of activity after 192 h of reaction at 550 °C under stoichiometric feeding conditions (no carbon deposits or filaments were detected for the spent catalyst) during ESR [76]. Ni/CeO_2 catalysts prepared via the sol–gel route displayed improved performance for APRE between 140 and 180 °C with 10 wt.% ethanol in water stream compared to the same catalyst synthetized via the solution combustion method. During the reaction, Ce_2O_3 reduced Ni^{2+} to Ni^0 and was itself oxidized to CeO_2; the Ni sites can also be substituted for Ce in the CeO_2 lattice, thus increasing oxygen mobility in the structure. The enhanced oxygen mobility of the catalyst prepared by sol–gel method resulted in an improved coking resistance [77].

For Co/CeO_2 catalysts, the use of nitrates as inorganic salt precursors was shown to increase acetaldehyde and ethylene selectivity, with severe carbon deposition [78]; conversely, the choice of cobalt acetyl acetonate strongly enhanced the catalyst stability during ESR at 450 °C under a steam-to-ethanol ratio of 10, the organic ligands linked to the metal sites decreased the size of cobalt particles, improving their dispersion and assuring them a sort of imprinting effect, increasing the sites' accessibility to ethanol and water. Similarly, He et al. [36] prepared Ni/SBA-15 starting from two different nickel precursors (nitrate and citrate) and tested them for ethanol reforming at 500 °C and a H_2O/C_2H_5OH ratio of 3. In the first case, NiC_x species were detected, which were easily removed from catalysts' surfaces. On the contrary, the choice of Ni nitrate as a precursor led to the formation of carbon nanofibers (CNF) with a regular graphite structure, which covered the Ni active sites and were more difficult to remove. Similarly, Pt-Ni/CeO_2-SiO_2 was prepared by employing different cerium precursors (nitrate, ammonium nitrate, and acetyl acetonate) and tested at 500 °C $C_2H_5OH/H_2O/O_2$ = 1:4:0.5. It was found that the organic precursor assured a templating effect on the ceria crystallites, protecting them and improving their dispersion; as a result, the carbon formation rate was reduced by half compared to the samples synthetized from inorganic precursors [37].

Ni catalysts prepared by using a smectite-derived material as the starting precursor compared to conventionally impregnated samples displayed reduced crystallite sizes of Ni and a consequent higher resistance towards carbon formation [79]. In fact, during synthesis, Ni species migrated in the smectite framework, and despite moving towards the surface during the reaction, a strong interaction with the framework was maintained, inhibiting sintering. Since this interaction limited the growth of particles, it was possible to reach considerable nickel contents (up to 35 wt.%), thus resulting in high catalytic performance for ESR.

Noble-metal-based catalysts display lower coke formation and enhanced stability with time-on-stream testing compared to Ni or Co [80,81]. For Ir/CeO_2 catalysts tested during ethanol dry reforming at 500 °C under a C_2H_5OH/CO_2 ratio of 1:1 [82], a very low amount of amorphous carbon was detected, proving that the strong interaction between the Ir particles and the ceria support efficiently prevented the sintering of Ir metal particles, as well as coke formation. Moraes et al. [83] investigated the effect on carbon build-up of Pt addition to a Ni/CeO_2 nanocube catalyst for ethanol steam reforming at 300 °C and a steam-to-ethanol ratio of 3. The formation of a catalytically inactive nickel carbide phase (solid solution on carbon in Ni) was observed over the monometallic sample, which together with the accumulation of adsorbed acetates on the surface led to significant catalyst deactivation. A depicted in Figure 7, Pt inhibited the formation of NiC_x species, enhancing the hydrogenation rate of carbonaceous species. Moreover, the noble metal is characterized by slower carbon diffusion kinetics; thus, a decrease of the carbon diffusion rate throughout the Ni lattice was assured by the platinum particles (Ni carburization during the reaction was mitigated by a discontinuous Pt surface outer shell) [84]. As a result, the CFR during 28 h of time-on-stream tests was one-fourth of the value recorded for the monometallic catalyst.

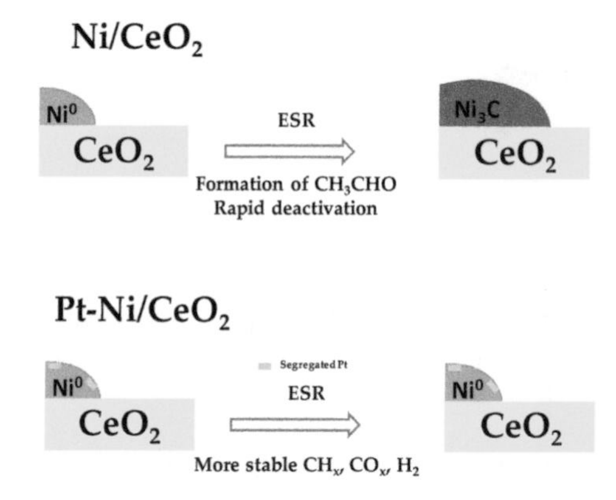

Figure 7. Effect of Pt addition on carbon formation for Pt/CeO_2 catalysts [84].

For noble-metal-based catalysts, due to the reduced solubility of carbon on the surface, it was shown that carbon nucleation is energetically unfavourable; the noble metal lattice is too large compared to graphene lattice, thus inhibiting its formation. As a consequence, the formation of superficial carbon instead of more structured deposits is promoted in the presence of supported noble metal catalysts (the solubility of carbon in the latter material is very low) [85].

Similar results were also reported by Palma et al. [86]. The addition of Rh to Ni/CeO_2-La_2O_3 samples enhanced the catalyst stability due to multiple effects, involving the reduced accumulation of acetate as well as the higher capacity for gasification of the methyl groups produced via decomposition of intermediates. In fact, acetates (derived from acetaldehyde dehydrogenation and subsequent oxidation) can be converted to CH_x species or may accumulate on the surface of the Rh-free catalyst, leading to acetone formation and subsequent carbon deposition. As a result, the Rh-Ni catalyst displayed almost no carbon formation for 24 h of reaction at 500 °C under a steam-to-ethanol ratio of 3 [87]. The excellent stability of Rh-Pt/CeO_2 catalysts during ESR under stoichiometric feeding conditions at 700 °C was ascribed to the formation of a Rh-Pt alloy containing both Rh and Pt oxidized species, which also led to very small metal particles (almost 5 nm). The control of the carbon accumulation was related to the availability of surface oxygen, which travels from the support, crossing the active metals and reaching the CH_x coke precursors [85].

The basic properties of the $Rh_{0.4}Pt_{0.4}$/CeO_2 catalyst assured stable performance during steam reforming of a bioethanol feed (having a steam-to-ethanol ratio of 3 and containing ethyl acetate 0.5 mol.%, 1,1-diethoxyethane 0.2 mol%; trace constituents: propanol, 2-methyl-1-propanol, 3-methyl-1-butanol, and 2-methyl-1-butanol) at 700 °C, where alcohol dehydration was disadvantaged, along with olefin production and further carbonaceous deposition on the catalysts' surfaces [88].

Due to the differing affinity for carbon deposit formation (different adsorption energies of the C-containing intermediate species were reported), Pt/Al_2O_3 and Rh/Al_2O_3 catalysts displayed dissimilar behaviour during steam reforming of crude bioethanol at 500 °C under a steam-to-ethanol ratio of 5 and with 1% impurities [89]. The carbon formation rate of the platinum-based catalyst was affected by the type of impurity selected and decreased in the order of IPA (Isopropyl alcohol) > 1-propanol > pure ethanol > propanal > propylamine > acetone, while the Rh/Al_2O_3 showed a higher durability, with a negligible trend related to the contaminants. A Pt-Co/CeO_2-ZrO_2-Al_2O_3 catalyst was successfully employed for the steam reforming of a real bioethanol feed coming from the industry (having a steam-to-ethanol ratio of approximately 6 and also containing sulphur/phosphorous compounds); during stability measurements at 400 °C, coke deposition did not occur [90].

In some cases, the addition of a second non-noble metal (i.e., Co or Cu) had a mitigating effect in terms of carbon deposition [32,91]. The improved stability of Ni-Co/SBA-15 catalysts upon cobalt deposition was ascribed to the formation of a carbide phase (thermodynamically stable against decomposition to metallic Co and graphitic carbon), which assures a decrease in cobalt particle sizes [92]. Reduced C_2H_4 formation was observed during ESR for NiCu ex-hydrotalcite catalysts compared to the copper-free sample, which limited the deposition of graphitic carbon, as well as the formation of carbon filaments having small metallic particles at their growing tips.

For copper–nickel oxide catalysts [93], it was also reported that the metal–oxide interface facilitates the transformation of CH_x-adsorbed species resulting from acetaldehyde decomposition into methoxy-like adsorbed species, which are readily reformed to produce H_2 instead of being decomposed to solid carbon.

Braga et al. [94] investigated the effect of Co addition to Ni/$MgAl_2O_4$ on its stability during OSRE at 500 °C under H_2O/EtOH/O_2 = 3:1:0.5. Filamentous carbon was deposited on bimetallic Ni-Cu samples and the rate of carbon deposition increased for lower Co contents. The carbon gasification rate depends on the presence of surface oxygen for the oxidation of the *C species; the small dimension of cobalt particles (lower than 5 nm) interacting with Ni enhanced the fraction of oxidized Co available to oxidize the carbon species on metal sites. Thus, cobalt deposition on Ni/$MgAl_2O_4$ effectively mitigated coke accumulation, with the best results recorded for the 4Co-4Ni catalyst.

2.2. The Role of the Support

The present paragraph describes the crucial role played by the catalytic support in preventing coking phenomena, mainly linked to its acidic and basic features, as well as its structural properties. Thus, the identification of the acidic and basic sites of the support is fundamental for the understanding of coke formation mechanisms.

For Rh/CeO_2 catalysts tested during ESR [95], NH_3 temperature-programmed desorption (TPD) results indicate the presence of strong basic sites; however, a slightly acid behaviour was also observed. Different paths of ethanol molecule activation are observed based on the site features. Acidic sites favour dehydration reactions and ethylene formation, which is a well-known coke precursor; while the dehydrogenation pathway is promoted in basic sites, leading to the formation of $C_2H_5O^*$ intermediates, which can decompose to CH_x species and further be oxidized to CO and CO_2.

$La_2Ce_{2-x}Ni_xO_{7-\delta}$ catalysts displayed very stable behaviour during oxidative steam reforming of ethanol. Ni incorporation into the pyrochlore structure, as well as the synergistic effect of nickel and cerium ions, induced the formation of Ce^{3+} ions, thus creating oxygen vacancies. This effect, together with the good dispersion of Ni particles and the presence of NiO as an impurity (sintered nickel oxides promote coke accumulation), mitigates coke deposition [96]. The same authors synthesized pyrochlore catalysts substituted by Li and Ru in A and B sites and supported them on Al_2O_3 or $La_2Zr_2O_7$. Substitution of La^{3+} by Li^+ metal cations modified the relative compositions of active metal ions Ru^{n+}/Ru^{4+} and Ce^{4+}/Ce^{3+}, thus enhancing formation of vacancies and increasing the carbon oxidation rate. Moreover, during tests performed at 350 °C under stoichiometric feeding conditions, disordered carbonaceous and ordered graphitic species were formed over the alumina-supported catalyst, while no coke accumulation was observed for the $La_2Zr_2O_7$-based sample. The basic properties of this support, in fact, suppressed ethylene formation and its successive polymerization [97].

For Cu catalysts supported on CeO_2, ZrO_2, and CeO_2-ZrO_2, and tested during ethanol dry reforming, it was found that the concentration of the oxygen vacancies decreased in the order of Cu/CeO_2-ZrO_2 > Cu/CeO_2 > Cu/ZrO_2. Thus, coke formation was inhibited and its degree of graphitization reduced with the Cu/CeO_2-ZrO_2 catalysts, which displayed the highest durability at 700 °C under a C_2H_5OH/CO_2 ratio of 1:1 [98]. $Cu/Ce_{0.8}Zr_{0.2}O_2$ catalysts displayed negligible deactivation during dry reforming of ethanol at 700 °C and a C_2H_5OH/CO_2 ratio of 1:1. In fact, the incorporation of Cu prevented the crystallite growth of the Ce-Zr-O solid solution, which may occur during preparation; thus, the mean crystallite size of the CuCeZr catalyst was approximately 3.1 nm (lower than the value of 5 nm reported for the pure ceria-zirconia support); smaller particles are reported to reduce the formation enthalpy of oxygen vacancies. As a result, coke accumulation and sintering were mitigated [99] and stable behaviour with complete ethanol conversion for 90 h of time-on-stream was recorded [100]. Dang et al. [101] investigated the dry reforming of an ethanol/glycerol mixture (4:1) over a $Ni/CaCO_3$ catalyst: $CaCO_3$ reacted with the alcohols to yield CO, H_2, and CaO, while CO_2 allowed for $CaCO_3$ to be regenerated. The presence of carbonates was beneficial for suppressing coke accumulation; the highly reactive carbonaceous species can interact with the CO_2 coming from the carbonate decomposition and can be easily converted to carbon monoxide.

The metal particle size and the dispersion of the active phase are also affected by the structure of the support. Da Costa-Serra and Chica [102] studied ethanol steam reforming for two Co-based catalysts supported on structured manganese oxides (Birnessite and Todorokite). In the latter case, the special microporous structure of the support provided high-quality positions for the stabilization of the cobalt particles (i.e., the microporous structure assures a good interaction of the metal particles with the support) during the preparation, thus leading to smaller metal particles compared to the birnessite-based sample (6 vs. 12 nm). As a result, carbon accumulated during a 24 h ESR test at 500 °C under a H_2O/C_2H_5OH ratio of 13 was almost one-half of the todorokite-supported Co catalyst. The superior stability of a Ni/KIT-6 catalyst (ordered silica) for ethanol dry reforming at 550 °C under a C_2H_5OH/CO_2 ratio of 1 compared to a nickel sample supported on mesoporous silica was ascribed to the stabilization of the Ni species via the formation of Si-O-Ni-O-Si bonds, with a consequent confinement within the pores of the ordered structure; such confinement resulted in a stronger metal–support interaction and smaller Ni particles compared to the Ni/SiO_2 catalyst, which enhanced the resistance towards coke formation [103]. By depositing a Pt-Ni/CeO_2 catalyst on high surface area mesoporous silica, a better dispersion of both metal and ceria particles was observed compared to the silica-free sample, thus enhancing the catalyst stability during ESR. The bimetallic catalysts were also prepared at different CeO_2/SiO_2 ratios (0.25, 0.30, and 0.40), and the relationship between the CFR and average Ni crystallites calculated for the spent catalysts was investigated. The smallest crystallite size, which was linked to the minimum coke selectivity, was recorded for a ratio of 30% [60].

Catalyst deposition on specific structured carriers may also affect their stability performance for ethanol reforming. The enhanced mass transfer rate and the improved gas–solid contact of structured carriers (i.e., foams, honeycombs, and monoliths) are expected to increase the contribution of coke gasification reactions, with a consequent mitigation in the deactivation of the catalyst [40,104]. Rh/CeSiO_2 catalysts supported on a ceramic monolith were subjected to a water/ethanol mixture of 3.5 at 755 °C, and no coke formation was observed during 96 h of operation [105]. Cobalt deposition on a Y_2O_3-doped ZrO_2 monolith eliminated the generation of undesired by-products (C_2H_2 and C_2H_4O), whose formation in the presence of the corresponding powder material was responsible for carbon deposition. Moreover, the stability of such structured catalysts was enhanced by infiltrating La into the monolith structure [106]. For the steam reforming of a simulated bioethanol for (containing 1 mol% of acetaldehyde, isopropyl alcohol, isobutyl alcohol, and isoamyl alcohol) over a Pt-Ni/CeO_2-ZrO_2 catalyst supported on a SiC foam, it was observed that the binder employed during the deposition of the catalytic wash coat on the structured carrier may affect catalyst stability [107]. The choice of boehmite as a binder reduced the carbon formation rate during tests at 450 °C at a steam-to-ethanol ratio of 3 compared to both the corresponding powder sample and the structured carrier containing silica as a ligand. In the latter case, in fact, very fast deactivation was observed, which was described as follows: during reforming reactions, steam can attack silica-structured Si atoms, with consequent structural changes for silica. Thus, active species particles can be wrapped or buried within the structure, losing their activity.

2.3. *The Effect of the Addition of Promoters*

The addition of promoters is a common strategy to mitigate carbon deposition on the catalysts' surfaces during ethanol reforming, and in the following paragraph, the influence of various promoters (mainly alkali metals and rare earths elements) on catalyst stability for bioethanol reforming is discussed.

For Co/CeO_2 catalysts tested during ESR, it was reported that the Co particle size influences the type of carbonaceous deposits; large cobalt particles caused the formation of carbon filaments with encapsulated cobalt (carbon diffuses inside the crystallites, destroying them and the causing growth of nanofibers), while few atomic layers of carbon, mainly located at the boundary of the Co and CeO_2 particles, were detected over small cobalt particles. The same authors found that the concentration of hydroxyls groups on the catalysts' surfaces (modulated on the basis of the steam-to-ethanol ratio) is crucial to changing coke selectivity. In this regard, a high concentration of OH species and K-O sites was assured upon potassium addition. Potassium, in fact, is able to favor the conversion of coke precursors to CH_x instead of fully dehydrogenated C = C species (mainly graphitic whiskers and layers) [108]. Moreover, the formation of carbonaceous deposits was also prevented over Co/ZrO_2 catalysts promoted by potassium, mainly due to the decreased contribution of the CO disproportion reaction [109]. La_2O_3 promotion improved the active species dispersion and the metal–support interactions for Co/CeO_2 (and Ni/CeO_2) catalysts; the lower carbon formation was attributed to the formation of $La_2O_2CO_3$ as an intermediate phase, which contributes to eliminating carbon deposits, leading to the formation of carbon monoxide [110]. Cerdà-Moreno et al. [111] investigated the effect of La addition as a promoter on the stability of CoZn hydrotalcites at 600 °C for the steam reforming of a raw bioethanol stream containing several impurities (acetaldehyde, methanol, propanol, and SO_2) and having a steam-to-ethanol ratio of 10; the improved performance of the lanthanum-containing catalyst is mainly related to the reduced C_2H_4 selectivity.

For Ni/SBA-15 catalysts, the introduction of Mn as a promoter and the subsequent change in the redox properties of the catalytic support favored the formation of a lower amount of coke lumps (carbon donuts surrounding Ni particles), with less activity loss during ESR compared to the Mn-free catalysts [54]. A similar reduction in the extent of deactivation and the same differences in the types of coke formed were observed by adding Mo to Ni/SBA-15 catalysts [112] and Mn to Co/SBA-15 catalysts [113].

The synthesis of a Ni-Ce-W oxide catalyst via reverse microemulsion allowed W and part of the Ni to be incorporated into the ceria lattice, while the remaining Ni formed highly-dispersed nano-NiO (almost 2 nm) outside the Ni–W–Ce oxide structure. Due to the synergy between Ni and W inside the ceria lattice, an increase of the oxygen vacancies with respect to the W-free sample was observed, which assured considerable resistance towards coke formation during ESR [114].

Compared to the ceria-free samples, CeO_2-promoted Ni/SBA-15 catalysts displayed well-dispersed nickel particles confined in the mesoporous channels of SBA-15 [115]; the metal–support interactions led to a homogeneous distribution of Ni and Ce with a large Ni-CeO_2 interface, which contributed to controlling the size of Ni particles. Ni's ability to break CO bonds responsible for CO disproportion and carbon deposit formation was mitigated thanks to the strong electronic perturbation induced by ceria. In addition, the OH groups present on the ceria surface, together with those derived from steam, can easily react with methyl groups to generate CO, CO_2, and H_2. The high oxygen mobility in the ceria lattice enhanced the carbon gasification rate and assured a lower rate of carbon deposition. Thus, ceria can effectively minimize sintering and coking over Ni/SBA-15 catalysts during ethanol steam reforming at 700 °C and a H_2O/C_2H5OH ratio of 4, reducing the accumulated amount of coke deposited by almost 70%.

In an attempt to improve the stability of NiMgAl catalysts, Du et al. [116] investigated the addition of various promoters, which modulated the electronic properties of the final catalyst. NiCoMgAl catalysts, compared to the Mn- or Zr-promoted samples, displayed higher durability, ascribable to the good metal dispersion and small metal size, which hindered carbon deposition and sintering (small particles of metals such as Ni have been found to block the mechanism of carbon filament growth and to decrease carbon accumulation). Conversely, an enhanced rate of carbon formation (mainly in a filamentous form) was observed for the NiMnMgAl and NiZrMgAl samples. Ce addition to NiMgAl reduced the dimensions of metal particles, while the presence of lanthanum increased the surface Ni content and the number of basic sites; in both cases, a reduced carbon formation rate was measured. Similarly, the benefits of supporting $LaNiO_3$ catalysts on $CeSiO_2$ were reported [117]. The presence of SiO_2 blocked the sintering of ceria particles and the very low carbon formation rate recorded during ESR at 700 °C, while a H_2O/C_2H_5OH ratio of 3 over the $LaNiO_3/CeSiO_2$ catalyst was ascribed to the metal–support interactions, which contributed to maintaining small dimensions for Ni crystallites. The size of metal particles, in fact, affects the nucleation rate of carbon; a critical ensemble size (ensemble of 6–7 atoms) was proposed, below which carbon formation does not occur. For these catalysts, it was also reported that after initial deactivation, new stability conditions were reached, despite the presence of carbon on the catalyst's surface. The initial loss of activity was observed when the rate of formation of CH_x species was higher than their rate of desorption to form CH_4. Such species can be further dehydrogenated to C as well as H, and the highly-reactive carbon species can both encapsulate metallic particles, leading to complete deactivation, or diffuse through the Ni crystallites and nucleate the growth of carbon filaments. In the latter case, the top surface is still in contact with the reacting mixture and the catalyst is able to maintain its activity for the ESR. Dan et al. [118] investigated the steam reforming of a fir wood crude bioethanol over Ni/Al_2O_3 catalysts (containing methanol, acetic acid, higher alcohols, esters, aldehydes, organic acids, and dimethyl furan). However, the bioethanol displayed rapid deactivation during tests at 350 °C, due to the deposition of graphitic carbon. A significant improvement in its durability was observed with the addition of rare earth oxides, with La_2O_3 decreasing the alumina surface acidity and CeO_2 promoting water activation for the gasification of coke deposits. Thus, stable performance was recorded for 4 h at 350 °C.

Likewise, the incorporation of Au into Ni/SBA-5 catalysts improved the dispersion of the NiO phase to form smaller nickel oxide particles. In addition, strengthened interactions between the SBA-15 support and NiO phase were assured, thus efficiently reducing the coke deposition on active sites [119].

Mondal et al. [120] described the performance of unpromoted and Rh-promoted Ni/CeO_2-ZrO_2 catalysts for oxidative steam reforming of a crude bioethanol stream (containing ethanol, butanediol, butandioic acid, acetic acid, and glycerol). Stability tests were performed at 600 °C and EtOH/H_2O/O_2 = 1:13:0.35. The addition of Rh led to smaller dimensions for Ni crystallites, and despite both amorphous and rod-shaped filamentous carbon nanotubes being observed for the two spent catalysts, the Rh-Ni/CeO_2-ZrO_2 lessened the deposition of encapsulating amorphous coke, thus exhibiting a significantly lower deactivation rate.

The addition of Ce or La to Ni/Al_2O_3 acted as a spacer, preventing NiO particles from aggregating, and hence increasing metal dispersion on the catalyst's surface; thus, improved stability was recorded for ethanol dry reforming. In particular, despite multiwalled carbon nanofibers being deposited on both the promoted and unpromoted catalysts, the stable performance of rare-earth-containing samples was related to the absence of encapsulated Ni particles located on the tip of the filamentous nanocarbon [121,122]. In addition, in the presence of CO_2, the formation of the $La_2O_2CO_3$ intermediate discussed above was enhanced, along with its ability to oxidize surface C_xH_y species and to preserve active metal sites. Similar results were also reported for Ce-doped Co/Al_2O_3 catalysts [123], La-doped Co/Al_2O_3 [124], and La-doped Cu/Al_2O_3 catalysts [125].

In some cases, the addition of a promoter did not lead to the desired performance improvement. For the Ni/CeO_2 catalyst, the addition of K caused faster deactivation due to the deposition of a very large amount of graphitic fibres, which were longer and thicker than those observed over the unpromoted catalyst; these graphitic fibres encapsulated the nickel crystallites, resulting in an activity loss [126]. For Pt-Ni catalysts, the addition of a third metal (K, Cs, or Rh) as a promoter was investigated, finding that caesium and rhodium are able to reduce the carbon formation rate during ESR at 450 °C and under stoichiometric feeding conditions, while in the presence of potassium, this addition was detrimental for catalyst stability [23].

Table 1 summarizes the stability performance of different catalysts proposed over the last 5 years for steam reforming, oxidative steam reforming, and dry reforming of ethanol. As discussed above, various parameters affect the catalyst durability and carbon formation rate, including the catalyst composition (active species, support, promoters) and operative conditions (temperature, feeding, and space velocity). A high ethanol concertation, low contact time, and high steam/ethanol ratio increase carbon accumulation on the catalyst's surface. Moreover, noble-metal-based catalysts displayed lower coke formation compared to Ni- or Co-based catalysts, with the addition of promoters lessen the coke selectivity. In fact, as can be seen from the data reported in Table 1, when ESR was performed over Ir- and Rh-based catalysts, the CFR was among the lowest reported in the present review. Very low carbon formation rates were also measured in the presence of Pd, as well as for bimetallic Rh-Ni catalysts. Carbon formation rates range between 10^{-2} to 10^{-7}; overall, during oxidative steam reforming of ethanol, reduced coke selectivity values were recorded compared to the steam reforming and dry reforming cases, due to the enhanced contribution of gasification reactions promoted by O_2. In this regard, the highest stability was assured over the Pt-Ni/CeO_2-SiO_2 catalysts, as well as over Ru-based perovskites containing La as a promoter. During dry reforming, the extent of catalyst deactivation was enhanced due to the absence of water in the feed, which is crucial for the promotion of carbon gasification; thus, CFR, defined as the ratio between the mass of carbon, measured in g, and the product of the catalytic mass (measured in g), the mass of carbon feed during the test (in g) and the time on stream (in hours), was of the order of 10^{-3} $g_{coke} \cdot g_{catalyst}^{-1} \cdot g_{carbon,fed}^{-1} \cdot h^{-1}$. However, a strong improvement in catalyst stability was observed by selecting ceria and zirconia instead of alumina as the catalytic support. The results reported in Table 1 also show the effects of the catalyst preparation method and the salt precursor selection on the stability of the final catalyst during ethanol reforming. In particular, the lowest values for carbon formation rates were recorded for the catalysts prepared by sonication (which assured lower dimensions for metal crystallites and a better active species–support interaction) instead of impregnation; moreover, the choice of organic salts as precursors improved the dispersion of active species, thus resulting in a more stable catalyst.

Table 1. Carbon formation rates of various catalysts employed for steam reforming, oxidative steam reforming, and dry reforming of ethanol.

Catalyst [a]	Operative Conditions [b]	WHSV [c] (h^{-1})	X EtOH (%)	Carbon Formation Rate [d] (Multiplied for 1000)	Ref.
		Ethanol steam reforming			
$1Pt\text{-}3Ni/CeO_2\text{-}SiO_2$	T = 450 °C S/E = 4 %C_2H_5OH = 10%	4.1	95% after 310 min	3	[23]
$0.5Rh\text{-}1Pt\text{-}3Ni/CeO_2\text{-}SiO_2$	T = 450 °C S/E = 4 %C_2H_5OH = 10%	4.1	93% after 1300 min	0.84	
$1Rh\text{-}3Ni/CeO_2\text{-}SiO_2$	T = 450 °C S/E = 4 %C_2H_5OH = 10%	4.1	91% after 4900 min	0.065	
$1Pt\text{-}3Ni\text{-}0.5K/CeO_2\text{-}SiO_2$	T = 450 °C S/E = 4 %C_2H_5OH = 10%	4.1	92% after 200 min	17	
$1Pt\text{-}3Ni\text{-}0.5Cs/CeO_2\text{-}SiO_2$	T = 450 °C S/E = 4 %C_2H_5OH = 10%	4.1	94% after 1600 min	0.39	
Ni-Co/mesoporous carbon (MC)	T = 375 °C S/E = 12 %C_2H_5OH = 4%	1.2	0% after 700 min	14	[32]
Ni-Co/2Zr-MC	T = 375 °C S/E = 12 %C_2H_5OH = 4%	1.2	77% after 700 min	7.1	
Ni-Co/2Y-MC	T = 375 °C S/E = 12 %C_2H_5OH = 4%	1.2	90% after 700 min	8.1	
$LaNi_{0.85}Zn_{0.15}O_{3\text{-}\delta}$	T = 700 °C S/E = 3 %C_2H_5OH = 18.6%	8.2	100% after 8 h	2.7	[34]
$10Ni/9La_2O_3\text{-}\alpha Al_2O_3$	T = 500 °C S/E = 3 %C_2H_5OH = 8 %	11.5	32 % after 20 h	0.19	[42]
$5Co/CeO_2$	T = 500 °C S/E = 3 %C_2H_5OH = 25%	6.8	94% after 6 h	25	[71]
$10Co/CeO_2$	T = 500 °C S/E = 3 %C_2H_5OH = 25%	6.8	98% after 6 h	75	
$20Co/CeO_2$	T = 500 °C S/E = 3 %C_2H_5OH = 25%	6.8	98% after 6 h	58	
$10Ni/CeO_2$	T = 300 °C S/E = 3 %C_2H_5OH = 2.5%	2.1	10% after 30 h	0.15	[84]
$1Pt10Ni/CeO_2$	T = 300 °C S/E = 3 %C_2H_5OH = 2.5%	2.1	33% after 30 h	0.19	[84]
$1Rh\text{-}10Ni/15La_2O_3\text{-}10CeO_2\text{-}Al_2O_3$	T = 500 °C S/E = 3 %C_2H_5OH = 18.8%	42.2	100% after 24 h	0.00031	[87]
Co-La/CeO_2 La/Co mol ratio 0.1	T = 420 °C S/E = 12 %C_2H_5OH = 7.7%	9.5	60% after 21 h	0.17	[110]
Ni-La/CeO_2 La/Ni molar ratio of 0.1	T = 420 °C S/E = 12 %C_2H_5OH = 7.7%	9.5	99% after 21 h	19	
3Ni/SBA-15	T = 650 °C S/E = 4 %C_2H_5OH = 4.5%	25.7	70% after 50 h	0.19	[115]
3NiCe/SBA-15 Ce/Ni molar ratio of 1:1	T = 650 °C S/E = 4 %C_2H_5OH = 4.5%	25.7	90% after 50 h	0.045	[115]
$0.4Pt\text{-}0.4Rh/CeO_2\text{-}SiO_2$ Si/Ce molar ratio of 1:2	T = 680 °C S/E = 3 %C_2H_5OH = 1.8%	14.3	100% for 72 h	0.16	[127]
$2Ir/CeO_2$ nanoparticles	T = 650 °C S/E = 3 %C_2H_5OH = 25%	9.23	80% after 45 h	0.0083	[128]
$2Ir/CeO_2$ nanoroads	T = 650 °C S/E = 3 %C_2H_5OH = 25%	9.23	55% after 45 h	0.0096	[128]
$1Rh/Al_2O_3$	T = 500 °C S/E = 3 %C_2H_5OH = 14%	40.7	80% after 45 h	0.028	[129]
$1Rh\text{-}15\%La_2O_3\text{-}Al_2O_3$	T = 500 °C S/E = 3 %C_2H_5OH = 14%	40.7	90% after 45 h	0.0029	[129]
$1Rh\text{-}15\%La_2O_3\text{-}5\%CeO_2\text{-}Al_2O_3$	T = 500 °C S/E = 3 %C_2H_5OH = 14%	40.7	97% after 45 h	0.0016	[129]

Table 1. *Cont.*

Catalyst [a]	Operative Conditions [b]	WHSV [c] (h^{-1})	X EtOH (%)	Carbon Formation Rate [d] (Multiplied for 1000)	Ref.
Co-Mg@mesoporous Al_2O_3 Co/Al molar ratio of 0.1:1 Mg/Al mol ratio 0.25:1	T = 550 °C S/E = 5 $\%C_2H_5OH$ = 6.7%	4.8	100% after 4 h	9	[130]
$LaNiO_3/ZrO_2$	T = 650 °C S/E = 3 $\%C_2H_5OH$ = 5%	27.1	80% after 50 h	0.57	[131]
$LaNi_{0.7}Co_{0.3}O_3/ZrO_2$	T = 650 °C S/E = 3 $\%C_2H_5OH$ = 5%	27.1	96% after 50 h	0.36	
$LaCoO_3/ZrO_2$	T = 650 °C S/E = 3 $\%C_2H_5OH$ = 5%	27.1	70% after 50 h	0.68	
10Ce/Ni-Mg-Al	T = 540 °C S/E = 6 $\%C_2H_5OH$ = 14.3%	2.1	83% after 10 h	0.51	[132]
5Ni/CNTs-SiO_2 fibers	T = 450 °C S/E = 9 $\%C_2H_5OH$ = 10%	2.6	87% after 22 h	1.2	[133]
10Ni/CNTs-SiO_2 fibers	T = 450 °C S/E = 9 $\%C_2H_5OH$ = 10%	2.6	100% after 22 h	1.5	
Pt@HBZ (HB zeolite)	T = 350 °C S/E = 4 $\%C_2H_5OH$ = 4%	3.4	100% after 15 h	0.23	[134]
Pt-B (B zeolite)	T = 350 °C S/E = 4 $\%C_2H_5OH$ = 4%	3.4	60% after 15 h	0.46	
2.5Co/hydroxyapatite	T = 500 °C S/E = 6 $\%C_2H_5OH$ = 4.1%	2.2	60% after 5 h	17	[135]
5Co/hydroxyapatite	T = 500 °C S/E = 6 $\%C_2H_5OH$ = 4.1%	2.2	40% after 5 h	19	
7.5Co/hydroxyapatite	T = 500 °C S/E = 6 $\%C_2H_5OH$ = 4.1%	2.2	30% after 5 h	20	
20Ni/Attapulginte	T = 700 °C S/E = 3 $\%C_2H_5OH$ = 12.6%	5.1	75% after 50 h	0.25	[136]
20Ni/5Mg-Attapulgite	T = 700 °C S/E = 3 $\%C_2H_5OH$ = 12.6%	5.1	85% after 50 h	0.23	
20Ni/10Mg-Attapulgite	T = 700 °C S/E = 3 $\%C_2H_5OH$ = 12.6%	5.1	98% after 50 h	0.047	
20Ni/20Mg-Attapulgite	T = 700 °C S/E = 3 $\%C_2H_5OH$ = 12.6%	5.1	87% after 50 h	0.097	
10Ni/20Pr-CeO_2	T = 600 °C S/E = 5 $\%C_2H_5OH$ = 15.7%	18.2	100% after 120 h	0.0016	[137]
$Pd_{0.01}Zn_{0.291}Mg_{0.7}Al_2O_4$	T = 450 °C S/E = 3 -	3.1	100% after 30 h	0.00017	[138]
$La_{0.7}Ce_{0.3}Ni_{0.7}Fe_{0.3}O_3$	T = 500 °C S/E = 4 -	1.2	98% after 50 h	0.07	[139]
2.5Pt-1Cu@SiO_2	T = 400 °C S/E = 4 -	2.9	100% after 30 h	0.33	[140]
2.5Pt@SiO_2	T = 400 °C S/E = 4 -	2.9	70% after 30 h	0.60	
2.5Pt-1Cu/SiO_2	T = 400 °C S/E = 4 -	2.9	80% after 30 h	0.79	
$10Ni/17CeO_2ZrO_25La_2O_3$	T = 500 °C S/E = 3 $\%C_2H_5OH$ = 5%	10.1	86% after 4 h	12	[141]
$10Ni/ZrO_25La_2O_3$	T = 500 °C S/E = 3 $\%C_2H_5OH$ = 5%	10.1	57% after 4 h	37	
$1Rh/17CeO_2ZrO_25La_2O_3$	T = 500 °C S/E = 3 $\%C_2H_5OH$ = 5%	10.1	92% after 4 h	0.58	
$1Rh/ZrO_25La_2O_3$	T = 500 °C S/E = 3 $\%C_2H_5OH$ = 5%	10.1	68% after 4 h	5.2	
Oxidative steam reforming of ethanol					
1Pt3Ni/CeO_2-SiO_2 Cerium precursor: nitrate	T = 500 °C S/E = 4 O_2/E = 0.5 $\%C_2H_5OH$ = 10%	12.3	59% after 100 h	0.0030	[37]
1Pt3Ni/CeO_2-SiO_2 Cerium precursor: ammonium nitrate	T = 500 °C S/E = 4 O_2/E = 0.5 $\%C_2H_5OH$ = 10%	12.3	60% after 100 h	0.0029	
1Pt3Ni/CeO_2-SiO_2 Cerium precursor: acetyl acetonate	T = 500 °C S/E = 4 O_2/E = 0.5 $\%C_2H_5OH$ = 10%	12.3	73% after 100 h	0.0014	
30Ni/CeO_2-ZrO_2	T = 600 °C S/E = 9 O_2/E = 0.35	5.1	95% after 36 h	0.92	[43]

Table 1. *Cont.*

Catalyst [a]	Operative Conditions [b]	WHSV [c] (h^{-1})	X EtOH (%)	Carbon Formation Rate [d] (Multiplied for 1000)	Ref.
1Rh-30Ni/CeO_2-ZrO_2	T = 600 °C S/E = 9 O_2/E = 0.35	5.1	85% after 36 h	0.45	
NiCo-MgAl (Ni+Co = 20 wt.%) Conventional synthesis	T = 550 °C S/E = 3 O_2/E = 0.5 %C_2H_5OH = 12.8%	91.6	100% after 100 h	7.7	[58]
NiCo-MgAl (Ni+Co = 20 wt.%) Microwave-assisted co-precipitation	T = 550 °C S/E = 3 O_2/E = 0.5 %C_2H_5OH = 12.8%	91.6	100% after 100 h	7	
NiCo-MgAl (Ni+Co = 20 wt.%) Sonication-assisted co-precipitation	T = 550 °C S/E = 3 O_2/E = 0.5 %C_2H_5OH = 12.8%	91.6	100% after 100 h	5.4	
1Pt-3Ni/CeO_2-SiO_2 CeO_2/SiO_2 ratio = 25	T = 500 °C S/E = 4 O_2/E = 0.5 %C_2H_5OH = 10%	4.1	100% after 100 h	0.0076	[60]
1Pt-3Ni/CeO_2-SiO_2 CeO_2/SiO_2 ratio = 30	T = 500 °C S/E = 4 O_2/E = 0.5 %C_2H_5OH = 10%	4.1	100% after 135 h	0.0012	
1Pt-3Ni/CeO_2-SiO_2 CeO_2/SiO_2 ratio = 40	T = 500 °C S/E = 4 O_2/E = 0.5 %C_2H_5OH = 10%	4.1	100% after 120 h	0.0065	
15Ni/$MgAl_2O_4$	T = 500 °C S/E = 3 O_2/E = 0.5 %C_2H_5OH = 2.5%	9.2	80% after 28 h	22	[94]
4Co11Ni/$MgAl_2O_4$	T = 500 °C S/E = 3 O_2/E = 0.5 %C_2H_5OH = 2.5%	9.2	70% after 28 h	21	
7.5Co7.5Ni/$MgAl_2O_4$	T = 500 °C S/E = 3 O_2/E = 0.5 %C_2H_5OH = 2.5%	9.2	70% after 28 h	7.1	
11Co4Ni/$MgAl_2O_4$	T = 500 °C S/E = 3 O_2/E = 0.5 %C_2H_5OH = 2.5%	9.2	60% after 28 h	6.7	
15Co/$MgAl_2O_4$	T = 500 °C S/E = 3 O_2/E = 0.5 %C_2H_5OH = 2.5%	9.2	60% after 28 h	1.9	
4Co4Ni/$MgAl_2O_4$	T = 500 °C S/E = 3 O_2/E = 0.5 %C_2H_5OH = 2.5%	9.2	60% after 28 h	0.11	
$La_{0.6}Sr_{0.4}CoO_{3-\delta}$	T = 600 °C S/E = 3 O_2/E = 0.5 %C_2H_5OH = 4.4%	3.6	96% after 5 h	2.1	[142]
$Mg_2AlNi_3H_zO_y$	T = 260 °C S/E = 3 O_2/E = 1.6 %C_2H_5OH = 14.4%	81.9	100% after 75 h	0.31	[143]
NiCo-MgAl (Ni+Co = 20 wt.%) Conventional synthesis	T = 550 °C S/E = 3 O_2/E = 0.5 %C_2H_5OH = 12.8%	47.3	100% after 100 h	0.072	[144]
NiCo-5PrMgAl	T = 550 °C S/E = 3 O_2/E = 0.5 %C_2H_5OH = 12.8%	47.3	100% after 100 h	0.044	
NiCo-5CeMgAl	T = 550 °C S/E = 3 O_2/E = 0.5 %C_2H_5OH = 12.8%	47.3	100% after 100 h	0.049	
$La_2Ce_{1.8}Ru_{0.2}O_7$/$La_2Zr_2O_7$	T = 400 °C S/E = 3 O_2/E = 0.6 %C_2H_5OH = 14.6%	28.1	100% after 100 h	0.0013	[145]
$Mg_xLa_{2-x}Ce_{1.8}Ru_{0.2}O_7$/$La_2Zr_2O_{7-\delta}$	T = 400 °C S/E = 3 O_2/E = 0.6 %C_2H_5OH = 14.6%	28.1	100% after 100 h	0.00021	
$Ca_xLa_{2-x}Ce_{1.8}Ru_{0.2}O_7$/$La_2Zr_2O_{7-\delta}$	T = 400 °C S/E = 3 O_2/E = 0.6 %C_2H_5OH = 14.6%	28.1	100% after 100 h	0.0011	
	Ethanol dry reforming				
1Rh/CeO_2	T = 700 °C CO_2/E = 1 -	18.5	88% after 65 h	0.0089	[63]
1Rh/CeO_2	T = 700 °C CO_2/E = 3 -	18.5	100% after 65 h	0.0033	
2Rh/CeO_2	T = 700 °C CO_2/E = 1 -	4.6	100% after 70 h	0.035	[82]
15Cu/CeO_2	T = 700 °C CO_2/E = 1 %C_2H_5OH = 30%	6.2	75% after 90 h	0.0061	[98]
15Cu/ZrO_2	T = 700 °C CO_2/E = 1 %C_2H_5OH = 30%	6.2	64% after 90 h	0.0093	
15Cu/CeO_2-ZrO_2 Ce/Zr mol ratio = 1	T = 700 °C CO_2/E = 1 %C_2H_5OH = 30%	6.2	100% after 90 h	0.0045	
10Co/Al_2O_3	T = 700 °C CO_2/E = 1 %C_2H_5OH = 20%	17.2	20% after 8 h	8.9	[123]
2Ce-10Co/Al_2O_3	T = 700 °C CO_2/E = 1 %C_2H_5OH = 20%	17.2	38% after 8 h	6.6	
3Ce-10Co/Al_2O_3	T = 700 °C CO_2/E = 1 %C_2H_5OH = 20%	17.2	50% after 8 h	4.7	
4Ce-10Co/Al_2O_3	T = 700 °C CO_2/E = 1 %C_2H_5OH = 20%	17.2	37% after 8 h	5.1	

Table 1. *Cont.*

Catalyst [a]	Operative Conditions [b]	WHSV [c] (h^{-1})	X EtOH (%)	Carbon Formation Rate [d] (Multiplied for 1000)	Ref.
5Ce-10Co/Al_2O_3	T = 700 °C CO_2/E = 1 %C_2H_5OH = 20%	17.2	34% after 8 h	5.2	[124]
10Co/Al_2O_3	T = 700 °C CO_2/E = 1 %C_2H_5OH = 20%	17.2	50% after 72 h	0.12	
3La10Co/Al_2O_3	T = 700 °C CO_2/E = 1 %C_2H_5OH = 20%	17.2	30% after 72 h	0.078	
10Ni/SiO_2-Al_2O_3	T = 750 °C CO_2/E = 1.4 -	1.8	97% after 10 h	2.8	[146]
10Ni/ Al_2O_3 calcined at 500 °C	T = 700 °C CO_2/E = 3 -	36.9	100% after 4 h	5.5	[147]
10Ni/ Al_2O_3 calcined at 600 °C	T = 700 °C CO_2/E = 3 -	36.9	100% after 4 h	6.7	
10Ni/ Al_2O_3 calcined at 700 °C	T = 700 °C CO_2/E = 3 -	36.9	100% after 4 h	9.7	

Note: [a] The metal or oxides loadings are intended for the weight of the catalyst; [b] all the tests were performed at atmospheric pressure; [c] refers to the ethanol mass flow-rate; [d] $g_{coke} \cdot g_{catalyst}^{-1} \cdot g_{carbon,fed}^{-1} \cdot h^{-1}$.

3. Oxidative Biomethanol Steam Reforming

Methanol is actually an attractive feedstock for reforming processes. It can be obtained from renewable and fossil sources [148], including biomass and CO_2, thus offering a pathway to a sustainable carbon-neutral cycle [149]. It can be easily converted to hydrogen via the (oxy-) reforming processes, and the resulting gas mixture, without a water–gas shift unit [150], can be directly used in high-temperature proton exchange membrane fuel cells (Figure 8), or [151] after purification steps in low-temperature polymer electrolyte membrane (PEM) fuel cells [152].

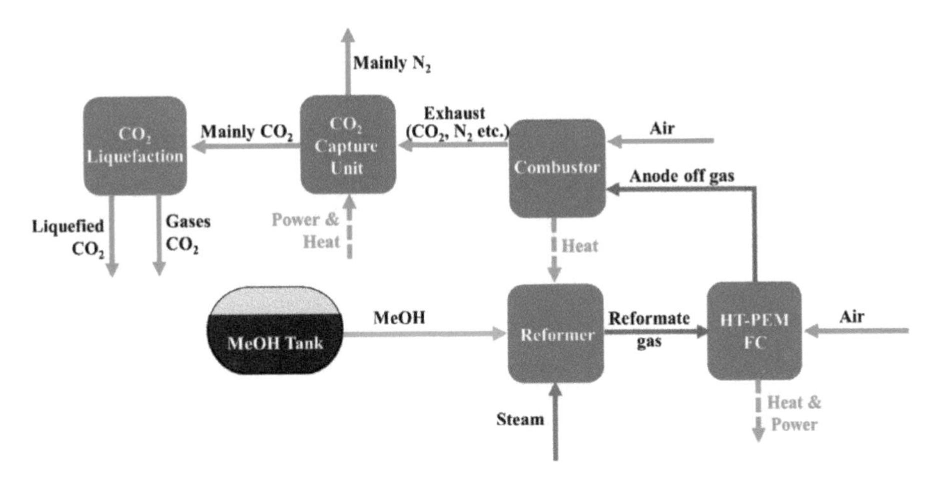

Figure 8. Block diagram of MSR-based system [150].

Many advantages were identified for using methanol for reforming processes. The molecule contains only one C atom, thus the absence of the C-C bond prevents the formation of a series of by-products. During the reforming processes [153], its reforming temperatures are relatively low; moreover, its tendency to form coke is lessened due to the high H/C ratio [154]. Methanol steam reforming (MSR) can be described by the following chemical reactions [155]:

$$CH_3OH + H_2O \leftrightarrow CO_2 + 3H_2 \ \Delta H^{\circ}_{298k} = +49.7 \text{ kJ}\cdot\text{mol}^{-1} \quad (10)$$

$$CO + H_2O \leftrightarrow CO_2 + H_2 \ \Delta H^{\circ}_{298k} = -41.2 \text{ kJ}\cdot\text{mol}^{-1} \quad (11)$$

$$CH_3OH \leftrightarrow CO + H_2 \ \Delta H^{\circ}_{298k} = +90.7 \text{ kJ·mol}^{-1} \quad (12)$$

The MSR reaction is endothermic and takes place with an increase in the number of moles; on the contrary, the water–gas shift reaction is exothermic and proceeds without a variation in the number of moles.

In this section, the results of selected studies published in the last five years on methanol (oxy-) steam reforming are reported. In the first paragraph, recent studies on the most-diffused active phases in MSR will be discussed. In the second section, comparative studies on the effects of the support will be reported. In the third section, the effects of promoters will be discussed, while theoretical and simulation studies, as well as non-conventional reactor configurations, will be reported in the fourth section. Finally, a selection of recent articles devoted to the oxidative steam reforming of methanol will be proposed. The paper distribution as a function of the active phase in the bibliographic survey reveals that copper is the preferred metal in methanol reforming; however, noble metals such as gold, platinum, and palladium are also used (Figure 9).

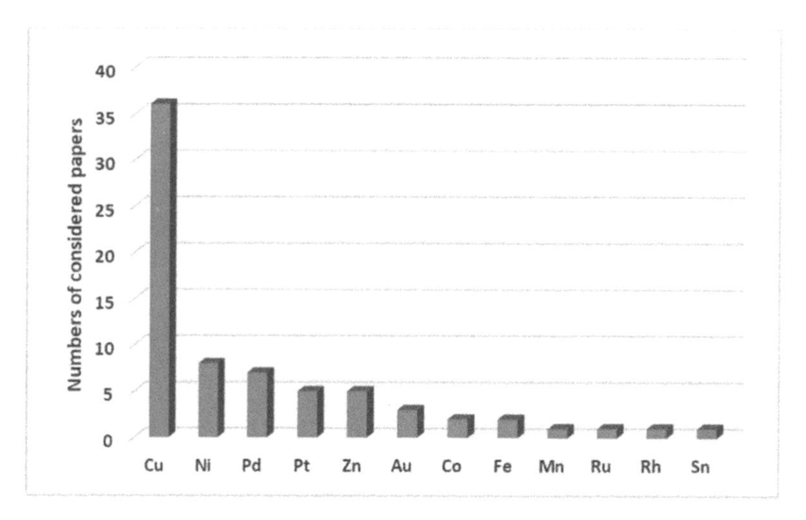

Figure 9. Number of considered papers as a function of the active phase analyzed for the methanol reforming.

3.1. *The Influence of the Active Phase*

As previously stated, in this section, the main results of a selection of articles on the effects of the active phase are reported. Most of them study copper-based catalysts, while only two cases are based on noble metal catalysts, palladium, or platinum.

Tonelli et al. investigated the stability of copper supported on ceria in MSR [156]. The ceria support was prepared by precipitation method with ammonium hydroxide from a solution of cerium (III) nitrate, while the catalyst was obtained by incipient wetness impregnation of the support with a copper (II) nitrate solution. The MSR tests were carried out on the catalyst, without reduction, at atmospheric pressure, at three temperatures (260, 280, and 300 °C), with a gaseous mixture of methanol, water, and nitrogen. The total flow rate was 82 mL·min^{-1}, with a methanol molar percentage of 5% and a H_2O/CH_3OH molar ratio equal to 1.2, while the catalyst weight was 300 mg. The activity was evaluated under steady-state conditions during a time-on-stream of 60 h and under shutdown–start-up operation, under inert conditions. In all experiments, the initial methanol conversion was higher than 80% and the selectivity to hydrogen was 100%; however, the activity decreased with the time-on-stream. In the case of steady-state conditions, the conversion losses were 76% at 300 °C, 78% at 280 °C, and 67% at 260 °C after 3000 min in reaction, while the catalyst was regenerated by air treatment at 400 °C.

For the discontinuous mode, the initial conversion and conversion until 500 min were the same as that obtained under the continuous regime. Moreover, after the stop period, the activity was recovered and self-activation occurred. The X-ray photoelectron spectroscopy (XPS) spectra suggested that the deactivation was mainly attributed to the adsorption of carbonate or formate species, which can be desorbed under inert flow or air burning. The Cu 2p XPS spectra sample after reduction and after inert treatments showed a decreased Cu surface content, which could be due to the Cu migration out of the ceria lattice or to the redispersion of the Cu particles; moreover, an overreduction of ceria may have also occurred. The effect of Cu loading (Cu = 7, 10, or 15 at%) on $Ce_XZr_{1-X}O_2$ (x = 0.4, 0.5, 0.6, 0.7, and 0.8) solid solutions was investigated by Das et al. [157]. The catalytic activity tests were performed in the temperature range of 200–330 °C, at atmospheric pressure, with a steam/methanol ratio of 1.1 (mol/mol), using nitrogen (23.5 mL min^{-1}) as the carrier and internal standard at a gas hourly space velocity (GHsV) of 40,000 h^{-1}. The results showed that the most-active catalysts had a Ce/Zr ratio equal to 0.6 and Cu loading of 10 and 15 at%; the MSR behaviour was found to be sensitive to the pretreatment, as evident in the case of the 10 at%Cu/$Ce_{0.6}Zr_{0.4}O_2$ and 15 at%Cu/$Ce_{0.6}Zr_{0.4}O_2$ catalysts, which perform better without pretreatment for the former and in the regenerated form for the latter. The time-on-stream activity tests showed a constant decrease of the CH_3OH conversion over 50 h for both 10 at%Cu/$Ce_{0.6}Zr_{0.4}O_2$ and 15 at%Cu/$Ce_{0.6}Zr_{0.4}O_2$ catalysts; the XPS analysis showed that the surface atomic ratio of copper decreased from 5.9% before the activity tests to 1.4% after the activity test, demonstrating that the formation of reduced Cu is accompanied by sintering; moreover, carbon formation also occurred. These results suggested that the formation of large aggregates of copper covered with coke could be responsible for the loss in activity. After the regeneration, these aggregates break into a mixture of oxidized (Cu^{2+}) and reduced (Cu^0 and Cu^+) copper species and the catalysts recover the activity, suggesting a correlation between the activity and different proportions of copper components in the various forms of these catalysts. Deshmane et al. studied the effect of Cu loading (Cu = 5–20 wt.%) on high surface area MCM-41 prepared by one-pot synthesis [158]. The Cu-MCM-41 catalysts were reduced before the catalytic activity tests with 4% hydrogen in argon at 550 °C for 5 h. The tests were carried out at atmospheric pressure with a mixture CH_3OH/H_2O (molar ratio = 1:3) at a GHSV of 2838 h^{-1} in the temperature range of 200–350 °C. The results showed that the methanol conversion increased with the increase of the Cu loading until 15 wt.% for Cu-MCM-41, which showed ≈89% conversion, 100% hydrogen selectivity, and 0.8% CO selectivity at 300 °C. A further increase of copper loading was detrimental; methanol conversion decreased to about 77% when the Cu loading was 20 wt.%. This result was attributed to the decrease in catalysts' surfaces area leading to a decrease in Cu dispersion. The time-on-stream test was carried out on 15wt.% Cu-MCM-41 and 20wt.% Cu-MCM-41, with the latter showing strong resistance to deactivation at 48 h of reaction, suggesting that the use of a high surface area MCM-41 support significantly enhanced the stability of Cu-based catalysts. The thermogravimetric analysis of the spent catalysts showed the presence of ~1.3% and ~2.1% carbon for 15% Cu-MCM-41 and 20% Cu-MCM-41 catalysts, respectively. Xu et al. studied the effect of the preparation method on the catalytic activity of Cu-based composite oxide catalysts for MSR [159]. Two catalyst were prepared, the first (Cu-ZnO-Al_2O_3-ZrO_2-Ga_2O_3 with Cu/Zn/Al/Zr/Ga = 14.9:30.9:3.9:10.8:1.9 mass ratio) by urea co-precipitation and active carbon co-nanocasting technique, and the other (Cu-ZnO-Al_2O_3-ZrO_2-Ga_2O_3 with Cu/Zn/Al/Zr/Ga = 13.3:28.2:3.9:10.0:1.8 mass ratio) by conventional co-precipitation method. The catalytic activity tests were performed at atmospheric pressure, with a methanol solution (CH_3OH/H_2O = 1:1 molar ratio), a liquid flow rate of 0.03 mL·min^{-1}, and a $F_{reactant}\cdot W_{cat}{}^{-1}$ = 6000 mL*$g_{cat}{}^{-1}$ h^{-1}. The results highlighted the better performance of the catalyst obtained with the active carbon co-nanocasting technique in terms of methanol conversion, hydrogen selectivity, and turnover frequency (TOF) values; moreover, no CO formation was reported. This trend was also confirmed in the stability tests carried out at 275 °C for 70 h, which showed a slight deactivation for the catalysts prepared with the active carbon co-nanocasting technique, while a continuous deactivation occurred in the other case. This result was attributed to the smaller Cu particle size distribution and higher metal–support interaction obtained with the active carbon co-nanocasting technique. Moreover, the temperature-programmed oxidation (TPO) profiles of this catalyst after the stability test showed a peak at 252 °C and a shoulder at 451 °C, which were assigned to the deposition of amorphous carbon. The TPO of the catalysts obtained by conventional co-precipitation method after the stability test showed two peaks at 411 and 657 °C; the second one was attributed to the formation of graphitic coke, which is more stable and oxidizable only at high temperatures. Thattarathody and Sheintuch investigated the kinetic and dynamic behaviour of MSR on a $CuO/ZnO/Al_2O_3$ commercial catalyst at various steam-to-carbon ratios (S/C = 0, 0.5, 1), in a steady state and with temperature ramping conditions (140–300 °C) [160]. High activity was observed above 200 °C; moreover, rate oscillation was observed under isothermal conditions at a steam-to-carbon ratio equal to 1, as evident in the volumetric product compositions shown in Figure 10.

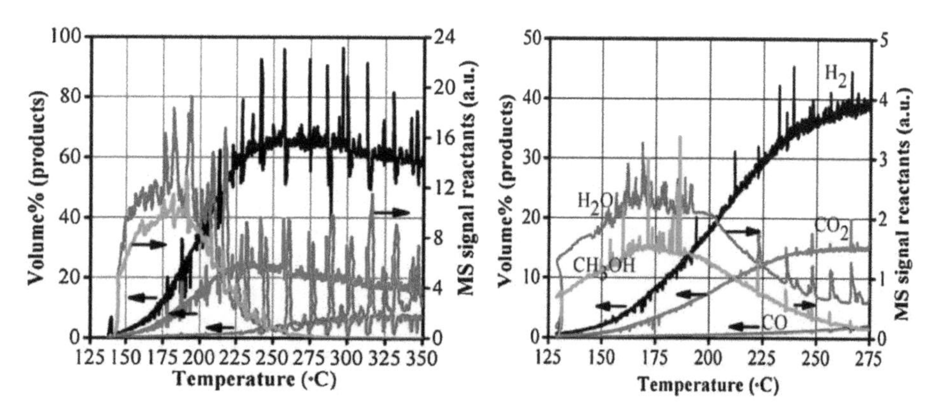

Figure 10. Comparison of product compositions during methanol steam reforming (S/C = 1) with an Ar flow of 10 mL/min (**left**) and 50 mL/min (**right**) [160].

After the tests, scanning TPO experiments were carried out to evaluate the amount of coke produced with the three different steam-to-carbon ratios (0, 0.5, and 1.0), obtaining 0.0187, 0.017, and 0.018 g of carbon per g of catalyst, respectively. Bagherzadeh et al. studied the effect of the exposition of $CuO/ZnO/Al_2O_3$ catalysts prepared by hydrothermal and co-precipitation methods for glow-discharge plasma for 45 min at 1000 V in MSR [161]. The catalytic activity tests were carried out in the temperature range of 180–300 °C, at a GHSV of 10,000 $cm^3 \cdot g_{cat}^{-1} . h^{-1}$; the volumetric flow rate of the argon carrier gas was set on 66.7 $cm^3 \cdot min^{-1}$ and the H_2O/CH_3OH molar ratio was approximately equal to 1.5. The characterization results demonstrated the benefits of coupling the co-precipitation method with glow-discharge plasma. X-ray diffraction (XRD) patterns highlighted the better dispersion of CuO (111) and another crystallite facet. Moreover, the SEM micrographs showed a more uniform distribution of isomorph particles, which had a smaller particles size and better surficial morphology, and the BET analysis demonstrated a higher specific surface area for the sample obtained by co-precipitation and following treatment with plasma. Accordingly, the results of the catalytic activity tests showed higher methanol conversion and better selectivity for the catalyst prepared by plasma-assisted co-precipitation method. For example, the methanol conversion and CO selectivity at 240 °C were 95% and 0.24%, respectively; in addition, the time-on-stream test showed no significant deactivation over 900 min of reaction. Ajamein et al. studied the effect of fuel type (sorbitol, propylene glycol, glycerol, diethylene glycol, or ethylene glycol) in the preparation of $CuO/ZnO/Al_2O_3$ nanocatalysts obtained by microwave-enhanced combustion method for MSR [162]. The catalytic activity tests were carried out in the temperature range of 160–300 °C, with a methanol/water ratio equal to 1.5. The results showed that the best performance was related to the use of ethylene glycol as fuel in the preparation of the catalyst. The characterization results demonstrated that sorbitol led to the formation of copper oxide species that were more crystalline in structure and assured a lower dispersion of crystallite active sites, especially the Cu(111) facet. The use of ethylene glycol resulted in a homogeneous morphology and narrow particles size distribution (the average surface particle size was about 265 nm). The catalytic activity tests highlighted the better performance of the catalyst prepared with the assistance of ethylene glycol, which showed total methanol conversion at 260 °C, with a negligible CO selectivity. The stability tests of the catalysts obtained by sorbitol and propylene glycol showed total conversion during the first 600 min and a considerable drop over a further 800 min. On the other hand, no deactivation occurred with the catalyst prepared by ethylene glycol. Among the polyol used, sorbitol had the highest polarity, facilitating the growth of Zn (002) crystals as the polar facet of zinc crystallites, improving the carbon monoxide formation. A comparative study on the use of different active metals supported on M-MCM-41 (M: Cu, Co, Ni, Pd, Zn, and Sn), prepared by one-pot hydrothermal procedure, was proposed by Abrokwah et al. [163]. The activity tests were carried out under the same conditions previously reported [158], highlighting the best performance of the Cu-based catalyst, which showed methanol conversion of 68%, hydrogen selectivity of 100%, and CO selectivity of 6% at 250 °C. On the other hand, PdMCM-41 and Ni-MCM-41 catalysts showed reduced activity for the water–gas shift reaction, resulting in higher CO selectivity. The stability tests were carried out at 300 °C for 40 h; the Cu-MCM-41 catalyst displayed an initial increase in the conversion with time due to the unsteady state; thereafter, the conversion stabilized at ≈74% and no apparent deactivation occurred with a further 30 h of reaction. Except for Cu-MCM-41 and Co-MCM-41, all the other catalysts showed a decreasing trend for methanol conversion, which was attributed to coking and sintering. The thermogravimetric analysis differential scanning calorimeter (TGA-DSC) thermograms showed a nominal (absolute) 0.25%–0.96% coke formation between 400 and 570 °C that was attributed to the formation of amorphous and graphitic carbon. TPD studies on cobalt-manganese oxides demonstrated that the reaction paths of adsorbed methanol lead to decomposition to CO and H_2, as well as formation of stable surface formates, which decompose at higher temperatures to CO_2 and H_2 [164]. Li et al. investigated the catalytic activity of Mn-, Fe-, Co-, Ni-, Cu-, and Zn-based catalysts in the methanol, acetic acid, and acetone steam reforming [165]. The activity tests for MSR were performed with a steam-to-carbon ratio equal to 5 and a liquid hourly space velocity (LHSV) of 12.7 h^{-1} at atmospheric pressure, in the temperature range of 200–500 °C. The results showed that Mn-, Fe-, and Zn-based catalysts were not active in MSR due to the low ability to break the chemical bonds or to activate the steam. On the contrary, Cu- and Co-based catalysts were both active; however, Co promoted methanol decomposition, showing higher CO selectivity. The supported metals showed different catalytic behaviour with respect to the unsupported one; for example, in MSR, the unsupported Cu catalysts showed lower stability than the supported one, which rapidly deactivated (Figure 11).

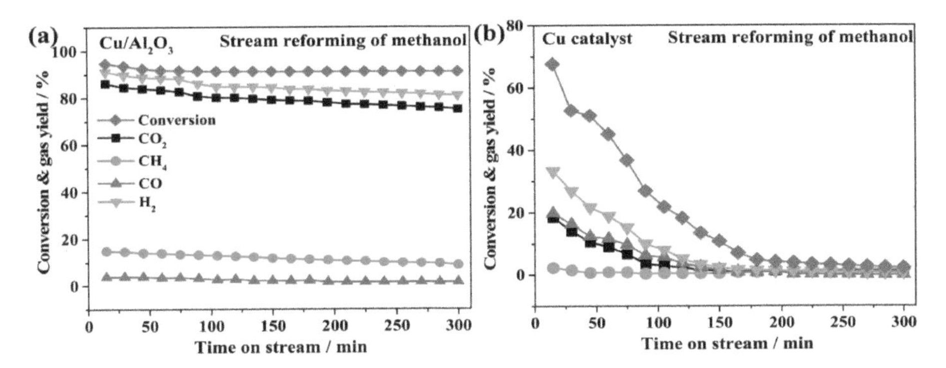

Figure 11. Time-on-stream tests (MSR) for Cu/Al_2O_3 (**a**) and Cu (**b**) catalysts. Reaction conditions: S/C = 1.5; T = 500 °C; LHSV = 12.7 h^{-1}; P = 1 atm [165].

It is worthwhile noting that the TPO profiles and the thermogravimetric analysis of the catalyst used in acetic acid and acetone steam reforming demonstrated the formation of a large amount of coke for all the catalysts. Conversely, carbon formation was really low in the case of MSR. Maiti et al. investigated the catalytic activity towards the MSR of a series of copper-ion-substituted $CuMAl_2O_4$ (M = Mg, Mn, Fe and Zn) spinels prepared via single-step solution combustion synthesis [166]. The catalytic activity tests were carried out in the temperature range of 200–330 °C, under atmospheric pressure, with a steam/methanol ratio of 1.1 (molar basis), at a GHSV of 30,000 h^{-1}. Among the studied catalysts, the $Cu_{0.1}Fe_{0.9}Al_2O_4$ was the most active, showing methanol conversion of ≈98% and CO selectivity of ≈5% at 300 °C. On the other hand, the analogous impregnated catalyst, 10 at%$CuO/FeAl_2O_4$, showed less catalytic activity. The time-on-stream tests showed a decrease of the methanol conversion from 50% to 35% over 20 h of reaction at a temperature of 250 °C. The evaluation of the XRD and high-resolution transmission electron microscopy (TEM) analysis showed the formation of a stable spinel phase containing substitutional copper ions, which stayed intact after 20 h of reaction. The decrease in the activity after the time-on-stream tests was attributed to the sintering of the catalyst, which also caused lowering of the copper surface concentration. Cu-Ni-Al spinel catalysts prepared by solid-phase method using copper hydroxide, nickel acetate, and pseudoboehmite as starting materials were studied by Qing et al. [167]. The results showed that Cu-Ni-Al spinels with molar ratios of 1:0.05:2 and 1:0.05:3 can be prepared at a calcination temperature ranging from 900 to 1100 °C; the spinel content increased with the calcination temperature. The catalytic performance was related to the calcination temperature of the catalyst, so the catalyst obtained at 1000 °C showed the best catalytic performance. The stability tests carried out at 255 °C and a weight hourly space velocity (WHSV) of 2.18 h^{-1} showed no significant deactivation after 300 h; however, the diffractograms of the spent catalysts showed a shift of the peaks to high angles, confirming a gradual release of Cu during the reaction process. Luo et al. developed nano-Ni_XMg_YO solid-solution oxides prepared by impregnation, hydrothermal treatment, and co-precipitation as catalysts for MSR [168]. The catalytic activity tests were carried out under atmospheric pressure with a mixture of water and methanol (molar ratios of 1 or 3) and a GHSV of 92,000 $mL \cdot g_{cat}^{-1} \cdot h^{-1}$ or 114,000 $mL \cdot g_{cat}^{-1} \cdot h^{-1}$, at four different reaction temperatures (400, 500, 600, and 700 °C). The best performance was obtained with the catalyst prepared by hydrothermal method, with methanol conversion of 97.4% and a hydrogen yield of 58.5%, which was maintained for 20 h at a steam-to-carbon ratio of 3. This superior catalytic activity was attributed to the nanoscale active phase and to the high micropore volume; moreover, the excellent anticarbon deposition capability was attributed to the formation of the solid solution, which was able to prevent the agglomeration of nickel particles, and to the high basicity of the magnesium oxide supports, which supplied oxygen from the adsorbed CO_2 and H_2O to burn off the amorphous carbon. Zeng et al. reported a study on the preparation and use of Pd/ZnO-based catalysts in MSR, which were synthesized by reduction with $NaHB_4$ from Pd ions supported on a zeolitic imidazolate framework-8 (ZIF-8) [169]. The activity tests were carried out with various methanol/water molar ratios ranging from 1 to 6, in the temperature range 250–380 °C, with a flow rate from 0.01 to 0.2 $mL \cdot min^{-1}$. The activity of two catalysts that were synthesized by reduction and calcined at 400 and 450 °C, respectively, was compared to that of two catalysts obtained by wet impregnation method and to that of a commercial $CuO/ZnO/Al_2O_3$ catalyst. The best performance in terms of methanol conversion (97%), CO_2 selectivity (86.3%), and stability after a time-on-stream test of 50 h was obtained with the catalyst prepared by reduction and calcined at 450 °C. This result was attributed to the larger surface area, the evenly distributed PdZn alloy activity sites, and abundant oxygen vacancies. A reaction mechanism was also suggested, in which the PdZn surface stabilizes the intermediate methyl formate and could provide a benefit to the adsorption of hydrogenated O-anchored species, such as HCOOH, H_2CO, and CH_3OH. Defects are able to alter the adsorption characteristics of these molecules, promoting dissociation; moreover, surface oxygen can enable water dissociation at low coverage, which is the key kinetic step in water splitting. Finally, the CO formation can be disfavored by the oxygen vacancies in the ZnO support. Claudio-Piedras et al. investigated the effect of the platinum precursor in Pt nanocatalysts supported on CeO_2 nanorods in MSR [170]. The catalysts were prepared by impregnation with $Pt(NH_3)_4(NO_3)_2$, $(CH_3\text{-}COCHCO\text{-}CH_3)_2Pt$, and $H_2PtCl_6{*}6H_2O$. The best performance was obtained with the catalysts prepared from the nitrate precursor, showing higher hydrogen yield and methanol conversion ascribable to the improved redox process, as well as to the better Pt dispersion on the surface of ceria, which promoted the water–gas shift (WGS) reaction at moderate temperatures. The time-on-stream tests highlighted the high stability of all the catalysts, with no significant deactivation observed after 24 h of reaction.

3.2. The Role of the Support

In this section, some studies on the effect of the support on the catalytic behaviour are reported, focusing on the preparation method, the effect of the precursors, and on the type of phase; in one article, a comparative study on the use of different supports is also discussed.

Barrios et al. prepared and tested two series of palladium catalysts in the MSR reaction, supported on $ZnO\text{-}CeO_2$ nanocomposites (Zn/Ce = 0.5, 1 or 2) and obtained by co-precipitation using oxalate or carbonate precursors [171]. For comparison, a series of catalysts were prepared by impregnation of mixed oxides, obtained through the incorporation of ZnO onto ceria by incipient wetness impregnation with two nominal loadings (3.5 wt.% and 11 wt.%), and by impregnation of ceria and ZnO with palladium acetate. The catalytic activity tests were carried out in the temperature range of 125–350 °C, with a gas mixture of CH_3OH/H_2O (1/1) diluted in He (16 vol.%) at a GHSV of 71,500 $cm^3h^{-1}g^{-1}$ and a W/F^0_{CH3OH} ratio, defined as the catalytic mass normalized for the methanol volumetric flow rate of 174 $g{\cdot}h{\cdot}m_{CH3OH}{}^{-3}$. XPS spectroscopy and the CO chemisorption experiments highlighted the formation of bulk and surface PdZn alloys in the ternary catalysts. The 2 wt.%Pd/CeO_2 catalyst caused CO decomposition at 250 °C and reversed the water–gas shift at higher temperatures. The MSR occurred in all catalysts in which ZnO was present; however, the catalysts prepared by impregnation of the supports obtained by carbonate co-precipitation showed the highest CO_2 selectivity due to the better dispersion of the ZnO phase. The time-on-stream tests carried out over 50 h of reaction showed that in the case of Pd/CeO_2, the methanol conversion (80%) and the CO_2 selectivity (8%) remained constant after 18 h on stream. The Pd/ZnO sample displayed a continuous decrease of methanol and water conversion; however, CO_2 selectivity remained almost constant at 90%. Both nanocomposite-supported Pd catalysts showed deactivation during the first 24–28 h, followed by stabilized methanol conversions of 53% and 40% for the $Pd/ZnO\text{–}CeO_2$ and $Pd/ZnO/CeO_2$ samples, respectively, with selectivity of 80%. The PdZn alloy formation seems to play a crucial role in preventing the methanol decomposition and in releasing hydrogen via inverse spillover, while the reforming reaction takes place mostly on the oxidized surface. Even though the nanocomposite-supported catalysts were less active and selective than the Pd/ZnO catalysts, they showed much more stability, so that the CeO_2 acted similarly to an "active dispersant of ZnO". Ajamein et al. investigated the effect of the precursor type, ultrasound irradiation, and urea/nitrate ratio on the catalytic performance of $CuO/ZnO/Al_2O_3$ nanocatalysts prepared by ultrasound-assisted urea–nitrate combustion method in methanol steam reforming reaction [172]. The results showed that the use of boehmite precursor instead of aluminium nitrate reduces the crystallite size, increases the dispersion, and enhances the specific surface area of copper and zinc species. Comparing sonication and conventional mechanical mixing, the mixing of primary gel provided nanocatalysts with improved homogeneity. Moreover, the CuO and ZnO crystallite sizes and the specific surface areas increased with the urea/nitrate ratio. The best performance in the catalytic activity tests in term of methanol conversion and hydrogen yield were obtained with the catalyst prepared from boehmite with a urea/nitrate ratio equal to 1 obtained by ultrasound irradiation, due to the smaller crystallite sizes and to the highly dispersed particles. This catalyst was also subjected to a time-on-stream test at 240 °C for 1200 min, with a H_2/CH_3HO ratio of 1.5 at a GHSV of 10,000 $cm^3g^{-1}h^{-1}$, showing a decrease in methanol conversion from 100% to 90%. Lin et al. reported the use of platinum atomically dispersed on α-molybdenum carbide for low-temperature (150–190 °C) aqueous-phase methanol reforming, with an average TOF of

18,046 moles of hydrogen per mole of platinum per hour [173]. The exceptional hydrogen production was attributed to the outstanding ability of α-MoC to induce water dissociation, and to the synergy between platinum and α-MoC in activating methanol with a 0.2 wt.%Pt/α-MoC catalyst. DFT studies demonstrated that the α-MoC support is able to provide highly active sites for water dissociation, with an activation energy of 0.56 eV, thus offering abundant surface hydroxyls and accelerating the methanol-reforming reaction at the interface between Pt_1 and α-MoC. Moreover, the geometry of the well-dispersed Pt_1 maximizes the exposed active interface of Pt_1/α-MoC and increases the density of active sites for the reforming reaction. In a more recent study, Cai et al. investigated Zn-modified Pt/MoC catalysts (Zn loading = 0–9.8%) prepared by temperature-programmed reaction method in low-temperature methane steam reforming [174]. The activity tests were performed with H_2O and CH_3OH at a molar ratio of 3:1, in the temperature range of 120–200 °C. The Zn doping favored the formation of α-MoC1-x phase, enhancing the Pt dispersion and the interactions between α-MoC1-x and Pt active sites. The 0.5Zn-Pt/MoC catalyst exhibited good performance in terms of hydrogen production, low CO selectivity, and good stability at 120 °C. However, at temperatures higher than 140 °C, the catalytic activity of this catalyst decreased during the initial stage of reaction, due to the sintering of Pt particles and to the change of α-MoC1-x phase. A treatment with a 15 vol.% CH_4/H_2 gas at 590 °C for 2 h was able to increase the catalytic activity of the spent 0.5Zn-Pt/MoC catalyst; however, the deactivation of the catalyst was inevitable. Liu et al. investigated the effects of supports in $Pt/In_2O_3/MOx$ catalysts (MOx = γ-Al_2O_3, MgO, Fe_2O_3, La_2O_3, or CeO_2) with Pt loading of 1 wt.% and In_2O_3 loading of 3 wt.% in MSR in the temperature range of 250–400 °C [175]. The catalytic activity tests were carried out in a flowing-type quartz tube (I.D. = 6.0 mm), at a GHSV of between 12,870 h^{-1} and 38,610 h^{-1}, with a steam-to-carbon ratio of between 0.6 to 1.8. The activity tests showed that the prereduced $1Pt/3In_2O_3/CeO_2$ catalyst exhibited the highest activity among the studied catalysts, with methanol conversion of 98.7%, hydrogen selectivity of 100%, and CO selectivity of 2.6% at 325 °C, with a steam-to-carbon ratio of 1.4 and a GHSV of 12,870 h^{-1}. These results were related to the active metal dispersion and enhanced redox properties associated with the strong interactions among Pt, In_2O_3, and CeO_2. Moreover, the $1Pt/3In_2O_3/CeO_2$ catalyst showed good stability in the time-on-stream test over 32 h. Díaz-Pérez et al. studied the Cu-based catalysts supported on SiO_2, Al_2O_3–SiO_2, TiO_2 rutile, and TiO_2 anatase metal oxides for MSR [176]. The catalysts were prepared by wet impregnation with a loading of 20 wt.%, while the activity tests were carried out with a steam-to-methanol ratio of 1:1.5. The results showed that on highly acidic supports such as Al_2O_3–SiO_2, the methanol conversion decreased with the TOS due to carbon formation. On TiO_2 anatase, the catalytic activity and stability was significantly lower than that on TiO_2 rutile, probably due to the differences in adsorbate–surface binding on rutile and anatase. The catalyst supported on nanosized SiO_2 showed the highest catalytic activity and selectivity. Time-on-stream tests were performed for over 80 h of reaction at low and high pressures, and no deactivation occurred; however, metal sintering was observed by means of HRTEM and XRD. The high activity and selectivity of Cu/SiO_2 was attributed to the low acid site concentration. Tahay et al. compared the performance of cubic and hexagonal phases of $ZnTiO_3$ with TiO_2 and ZnO as catalyst supports in MSR [177]. The $ZnTiO_3$ phases were synthesized by sol–gel method, while copper was used as the active phase. The tests were carried out in the temperature range of 150–300 °C at a WHSV 1 h^{-1} under atmospheric pressure, with a mixture of N_2/H_2O/methanol at a ratio of 6:2:1. The results of the activity test showed that the Cu/cubic-$ZnTiO_3$ catalyst exhibited high activity, high hydrogen selectivity. and low coke formation due to low–moderate acid sites in the cubic sample; moreover, the trend of methanol conversion at 300 °C is Cu/cubic-$ZnTiO_3$ > Cu/TiO_2 > Cu/hexagonal-$ZnTiO_3$. Time-on-stream tests showed a higher decrease of the methanol conversion of the $ZnTiO_3$ hexagonal-based catalyst compared to the cubic-based one at 42 h of reaction. The thermogravimetric analysis revealed the presence of three weight losses: the first one between 25 and 200 °C was attributed to the water elimination, the second between 200 and 400 °C to the low-temperature oxidation of the copper, while the weight loss after 400 °C was attributed to the coke burning. The amount of deposited carbon in the case of Cu/cubic-$ZnTiO_3$ was negligible; thus, the deactivation was only attributed to the Cu sintering process. The coke resistance of the Cu/cubic-$ZnTiO_3$ catalyst could be related to the high CO_2/CO ratio, which reduces the CO disproportionation on the catalysts' surfaces.

3.3. The Effect of the Addition of Promoters

In this section, a selection of recently published articles on the effect of promoters on the catalytic activity in MSR is reported. The sequence is settled based on the kind of promoter.

Talkhoncheh et al. studied the effect of the preparation method and the CeO_2 promotion effect on CuO/ZnO/Al_2O_3/ZrO_2-based nanocatalysts in the MSR reaction [178]. The catalysts were prepared by means of urea–nitrate combustion synthesis and homogeneous precipitation method; the catalytic activity tests were carried out at a temperature range of 200–300 °C under atmospheric pressure at a GHSV of 10,000 $cm^3 g_{cat}^{-1} \cdot h^{-1}$, feeding a mixture of H_2O/MeOH with a ratio equal to 1.5. The XRD analysis and FESEM images showed that the homogeneous precipitation method and the CeO_2 addition improves the dispersion, decreases the particle size, decreases the relative crystallinity of CuO and ZnO species, and enhances the surface homogeneity. The activity tests showed that the catalyst obtained by homogeneous precipitation method was more active in terms of high methanol conversion and low CO selectivity. Moreover, the CeO_2 addition decreased the CO selectivity and reduced the methanol conversion. The stability test carried out on the ceria-promoted catalyst prepared by urea precipitation method showed no deactivation after 1200 min of time-on-stream. In fact, CeO_2 was able to oxidize the carbon deposited on the nanocatalyst's surface. Taghizadeh et al. investigated the activity of cerium-promoted copper-based catalysts synthesized via conventional and surfactant-assisted impregnation methods using KIT-6 as support in MSR [179]. The activity tests were carried out at atmospheric pressure, with a methanol-to-water molar ratio of $\frac{1}{2}$ and at a WHSV of 2 h^{-1}. The results demonstrated that the incorporation of cerium oxide improved the performance of the Cu/KIT-6-based catalysts due to the higher dispersion and to the smaller size of Cu particles; moreover, the surfactant-assisted impregnation enhanced the physiochemical properties of the resulting Ce-promoted catalysts. The catalyst prepared with this method showed methanol conversion of ≈92%, hydrogen selectivity of 99%, and negligible CO selectivity (0.9%) at 300 °C. These results were attributed to the presence of the CTAB surfactant, which hindered the metal species migration during the drying and decreased the sintering during the calcination. The time-on-stream tests highlighted the stability of the Ce-promoted catalysts, with no significant deactivation observed over 24 h of reaction time. Phongboonchoo et al. investigated the catalytic activities of Ce-Mg-promoted Cu/Al_2O_3 catalysts prepared by co-precipitation in MSR [180]. The activity tests were carried out with a methanol/steam ratio of 1.5, 1.75, or 2, diluted in He at a temperature range of 200–300 °C. The results showed that the methanol conversion and hydrogen yield were higher with the monopromoted catalysts ($Cu_{0.3}Mg_{0.3}/Al_2O_3$ and $Cu_{0.3}Ce_{0.3}/Al_2O_3$) than those without a promoter, probably due to the higher dispersion of copper species and to the strong interaction between copper and ceria, which lessened the catalyst reduction. Moreover, the performance of the bipromoter catalyst showed higher methanol conversion, higher hydrogen yield, and lower CO selectivity than the monopromoted catalyst. The increase in the catalytic activity was attributed to the formation of smaller Cu crystallites, improved copper dispersion, and lower reduction temperature. The effect of the steam-to-carbon ratio was also investigated, demonstrating that a lower CO selectivity was obtained with a ratio of 2. Moreover, in order to optimize the reaction conditions, a theoretical study was also performed via a face-centered central composite design response surface model (FCCCD-RSM). The analysis was carried out on four main factors (temperature, steam-to-carbon ratio, Cu weight percentage, and magnesium weight fraction: Mg/(Ce+Mg)) by building a matrix and varying each factor within the level of the other factors. The results were analyzed using the Design-Expert 7.0 software package (Stat Ease Inc., Minneapolis, MN, USA), using analysis of variance and the percentage contribution of each factor to the responses. At a 95% confidence interval, the optimal operating region for maximal methanol conversion (100%) and hydrogen yield in the range of 28.9–29.4%, as well as CO selectivity of 0.16–0.18%, provided a copper level of 46–50 wt.%, a Mg/(Ce+Mg) yield of 16.2–18.0 wt.%, a temperature of 245–250 °C, and a steam-to-carbon ratio of 1.74–1.80 (Figure 12).

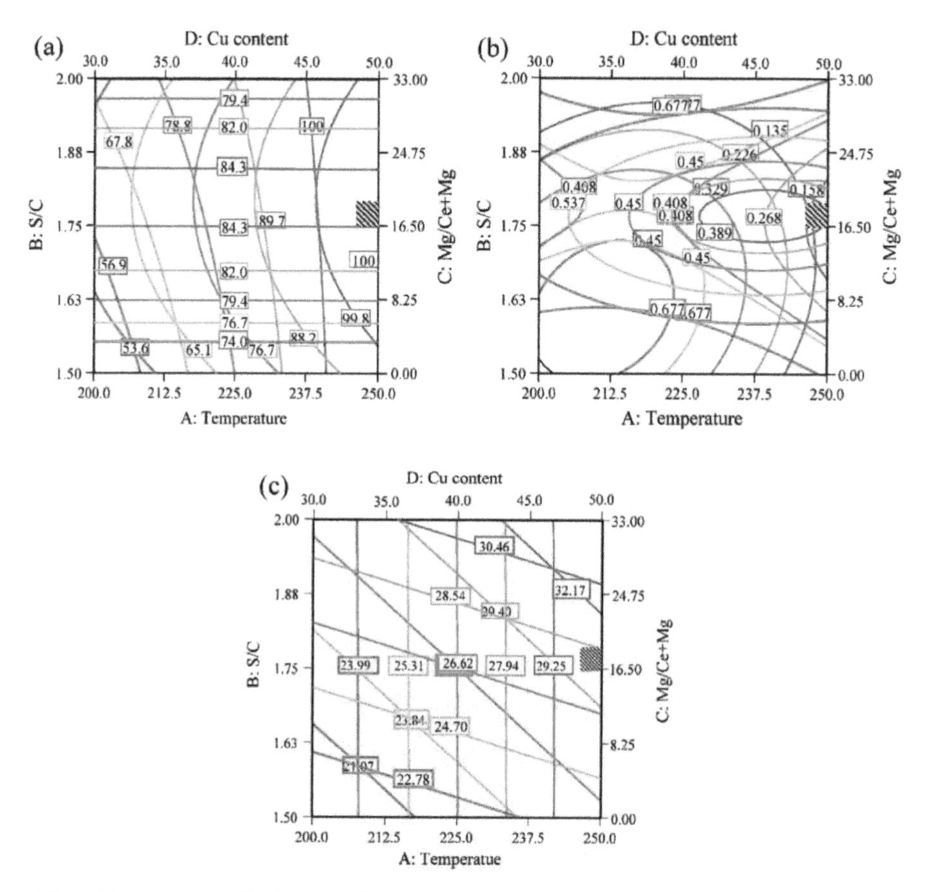

Figure 12. Contour plots for the optimal conditions for the methanol conversion (**a**); CO selectivity (**b**); H_2 yield (**c**) [180].

These results were compared to the catalytic activity of $Cu_{0.5}Ce_{0.25}Mg_{0.05}/Al_2O_3$ in a time-on-stream test of 72 h under the optimal operating parameters, showing complete methanol conversion and a CO selectivity of 0.14–0.16% with a hydrogen yield of 24–25%; furthermore, no deactivation occurred.

Hou et al. proposed Cu-Al spinel oxide as a sustained release catalyst for MSR, in which the surface modification of the spinel was performed with MgO [181]. The Cu-Al spinel with a Cu/Al atomic ratio equal to 1:3 was prepared by solid-phase method, where MgO was loaded by incipient wetness impregnation (loading = 0.9, 1.7, 3.6, and 9.5%); a reference catalyst 9.5%Mg/Al_2O_3 was also prepared. The catalytic tests were carried out with a methanol-to-water molar ratio of 0.46, the methanol WHSV was 0.98 h^{-1}, and the reaction temperature was 255 °C. The characterization results showed that the MgO strongly interacts with the Cu-Al spinel by changing the surface microstructure. Mg^{2+} cations were incorporated into the spinel structure, substituting a portion of Cu^{2+} in the lattice. The doped catalysts showed a higher reduction temperature, lower copper releasing rate, and smaller copper particle size. The addition of suitable amounts of magnesium oxide to the Cu-Al catalyst enhanced the catalytic activity and stability; the best performance was obtained with 1.7%Mg/CuAl.

The evaluation of coke formation on the spent catalysts demonstrated a correlation with the amount of the MgO loading. In the CO_2-TPD profiles, two oxidation peaks were found; the peak at 150-300 °C was attributed to the decomposition of formate, while the peak at 300–500 °C was attributed to the decomposition of carbonate. For the catalysts with low loading (≤1.7 wt.%), the low temperature peak was dominant; on the contrary, for the catalyst with high loading (9.5 wt.%), the high temperature peak was dominant. A correlation between the catalytic activity and the high temperature desorption peak was also found; thus, the lower activity of the catalysts with MgO loading higher than 1.7% was attributed to a coverage effect. The active copper was blocked by the dynamic formation of carbon deposits during the catalytic reaction. Moreover, lower Cu release rates were observed with the increase of MgO loadings and the Cu released was not able to sustain the catalytic activity; thus, fast deactivation was found for the 9.5%Mg/CuAl catalyst. Liu et al. investigated the use of $CuOZnOxGa_2O_3$-Al_2O_3 and $CuOZnOxGa_2O_3$-ZrO_2 catalysts (Ga wt.% = 0–12.4) prepared by sol–gel method in the MSR reaction [182]. The catalytic activity tests were performed at 250 and 275 °C at a GHSV of 2,200 h^{-1}, with a mixture $CH_3OH/H_2O/N_2$ in a 1:1:1.3 molar ratio. The characterization results showed that the addition of alumina and zirconia increased the specific surface area and modified the reduction temperature of the $CuOZnOGa_2O_3$ catalytic system. Zirconia promotion provided the highest reducibility and best performance in MSR for both studied temperatures. The introduction of Ga_2O_3 enhanced the hydrogen production rate but did not improve the stability of the CuZn system. The $CuZn_3Ga_2O_3ZrO_2$ catalysts showed higher stability. The hydrogen production rate at 275 °C after 44 h was 312 $mLg_{cat}^{-1}min^{-1}$, with methanol conversion of 75%. The TPO and TG mass spectrometry analyses demonstrated the presence of negligible amounts of carbon deposits below 250 °C. Mohtashami and Taghizadeh studied ZrO_2-promoted Cu-ZnO/MCM-41 catalysts prepared by sol–gel impregnation and modified impregnation methods in the MSR reaction [183]. The modified impregnation method consisted of treating the MCM-41 support with acetic acid for 5 h at room temperature before drying and calcining. The catalytic tests were performed in the temperature range of 280–320 °C; at a WHSV of 1.08, 1.62, or 2.16 h^{-1}; and with a methanol-to-water molar ratio of 1:2. The inclusion of 2% ZrO_2 to the Cu-ZnO/MCM-41 catalyst increased the methanol conversion from 90.6 to 94.0% at 300 °C. The promoted catalyst prepared through the modified impregnation method showed the best performance, with methanol conversion of 97.8%, a hydrogen selectivity of 99.0%, CO selectivity of 0.4%, and also good stability. Reduced deactivation was observed within 81 h in the time-on-stream test. The evaluation of the amount of coke deposited on the catalyst's surface after the time-on-stream tests by thermogravimetric analysis showed a significant reduction of the coke formation in the zirconia-promoted catalysts. This performance was attributed to the positive effect of the acetic acid pretreatment on the support, which modified the MCM-41 surface through the generation of oxyl groups, which were able to prevent metal particle aggregation, decrease the metal particles size, and improve the dispersion and reduction behaviours of the CuO particles. Lu et al. investigated the effect of the lanthanum addition and the effect of nickel loading on different supported Ni-based catalysts [184]. The catalysts were prepared by co-precipitation wetness impregnation method; the activity tests were carried out at a GHSV of 10,920 h^{-1}, at atmospheric pressure, with a mixture of $MeOH/H_2O$ (molar ratio = 3:1). The results demonstrated that the catalytic performance at low temperature for Ni-based catalysts can be enhanced by the addition of lanthanum, due to the formation of smaller and highly-dispersed NiO particles. The characterization results showed that lanthanum species are able to interact with nickel oxide and aluminum oxide to generate La-Ni or La-Ni-Al mixed oxides. The comparative study on the use of different supports showed that a weak interaction occurred between SiO_2 and NiO species, while with Al_2O_3, the $NiAl_2O_4$ spinel and separated NiO particles located on the outer surface were formed. The use of the MgO support generated the NiO-MgO solid solution, which decreased the amount of active nickel species. Azhena et al. studied the effect of Cu promotion in Pd/ZrO_2-based catalysts, along with the influence of the zirconia structure, in low-temperature MSR [185]. The catalysts were prepared by wet impregnation and the catalytic activity tests were carried out at atmospheric pressure with a steam-to-methanol ratio of 1.5, a contact time W_{cat}/F_{CH3OH}^0 = 83 Kg_{cat}·$mol^{-1}s^{-1}$, and in the temperature range of 180–260 °C. The results showed that the use of monoclinic zirconia provides benefits both in terms of activity and selectivity; at the same time, the selectivity is also improved by the addition of Cu. These improvements were attributed to the enhanced dispersion of the metal phase on monoclinic zirconia and to the strong interaction between Pd and Cu. Liu et al. studied the catalytic behaviour of 1wt.%Pt/xIn_2O_3/Al_2O_3 catalysts (x = 0–45 wt.%) prepared by incipient wetness impregnation in the MSR reaction [186]. The catalytic activity tests were carried out under atmospheric pressure in the temperature range of 200–500 °C. The optimal performance was obtained with a reacting mixture of H_2O/CH_3OH (mole ratio = 1.4; flow rate = 1.2 cm^3·h^{-1}) diluted in N_2 (flow rate = 30 cm^3·min^{-1}), at a GHSV of 14,040 h^{-1} and at 350 °C, with In_2O_3 loading of 30 wt.%. Under these conditions, the 1Pt/30In_2O_3/Al_2O_3 catalyst exhibited complete methanol conversion, high hydrogen selectivity (99.6%), and low CO selectivity (3.2%). The high activity was related to the intimate contact of Pt with partly reduced In_2O_3, which was hypothesized to be the active site of the reforming reaction. The performance of the 1Pt/30In_2O_3/Al_2O_3 catalyst was also compared to that of the 1Pt/30ZnO/Al_2O_3 catalyst in a time-on-stream test of 17 h; the higher activity and stability of 1Pt/30In_2O_3/Al_2O_3 was attributed to the enhanced dispersion of metallic Pt, and to the synergistic effect and strong interaction between Pt and In_2O_3, which facilitated the water activation, thus promoting the methanol reforming. Martinelli et al. investigated the effect of the sodium doping on supported Pt-based catalysts in methanol steam reforming [187]. The support was yttria-stabilized zirconia YSZ (Y/Z = 0.11) prepared by co-precipitation, while the catalysts were obtained by sequential incipient wet impregnation with a platinum salt precursor (Pt loading 2 wt.%) and sodium nitrate (Na loading 0.25, 0.5, 1, or 2.5 wt.%). The catalytic activity tests were carried out in a steady state under a feed stream containing 2.9% CH_3OH, 26.1% H_2O, 29.9% H_2, and 4.3% N_2 (balance He) at atmospheric pressure, at a GHSV of 381,000 h^{-1}, in the temperature range of 275–350 °C. The catalysts were activated in hydrogen (100 cm^3·min^{-1}) at 350 °C for 1 h (ramp rate = 4 °C·min^{-1}). The results showed that with a 2.5 wt.% of Na loading, the CO_2 selectivity was higher than 90%; the product distribution was attributed to different reaction pathways for methanol decomposition. Methanol decarbonylation was favored in the absence of sodium, while formate decarboxylation was promoted in the presence of 2.5 wt.% of Na (Figure 13). These conclusions were supported from the observed weakening of the C-H bond of formate in in situ diffuse reflectance infrared Fourier transform spectroscopy (DRIFT) studies and kinetic isotope effect experiments. The formate exhibited a ν(CH) stretching band at a low wavenumber, consistent with C–H bond weakening, thus favoring the dehydrogenation that is directly related to the decarboxylation. The hypothesis is that formate is similar to an intermediate; moreover, Na is able to favor the dehydrogenation and the selectivity can be tuned between decarbonylation and decarboxylation based on the Na dopant level.

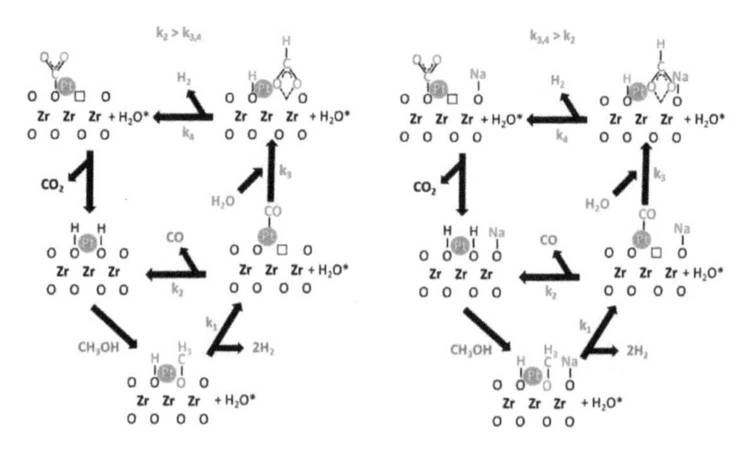

Figure 13. Proposed **MSR** pathway on undoped Pt/YSZ (**right**) and Na-doped Pt/YSZ (**left**) [187].

Zhang et al. investigated the activity of Au-based catalysts supported on modified montmorillonites in MSR [188]. The tests were carried out at atmospheric pressure, in the temperature range of 250–500 °C, with a CH_3OH/H_2O molar ratio of 1. Among the catalysts examined, Au-Ti-Ce/Na-bentonite showed the best performance, with a methanol conversion of 72% and a hydrogen selectivity of 99% at 350 °C. This result was attributed to the formation of the Au-Ce-Ti solid solution into the interlayer space of the bentonite, leading to a high surface area, small Au particle size, and a large average pore volume and diameter. The time-on-stream test showed no significant deactivation during 120 h of reaction (Figure 14).

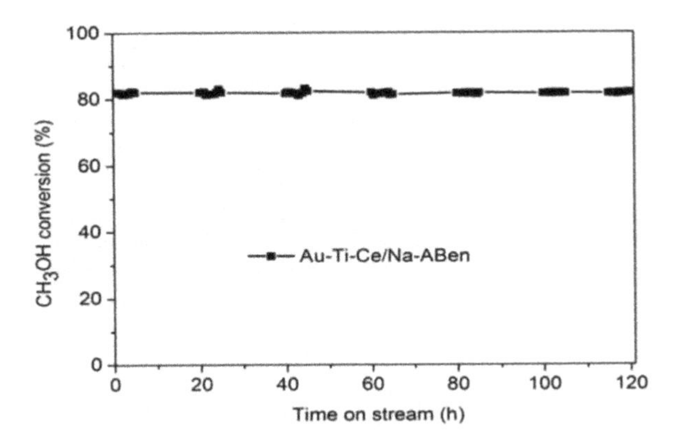

Figure 14. Time-on-stream test. T = 400 °C, CH_3OH/H_2O = 1, 0.175 mL·h^{-1} [188].

Lytkina et al. studied the influence of the composition and structure of metal-oxide-stabilized zirconia support ($M_XZr_{1-X}O_{2-\delta}$, M = Y, La, Ce) prepared by co-precipitation method in bimetallic Cu-Ni and Ru-Rh catalysts for MSR [189]. The activity tests were carried out at atmospheric pressure in the temperature range of 200–400 °C, with a mixture of methanol and water with a molar ratio of 1/1, at a GHSV of 172 h^{-1}. In this study, a bifunctional mechanism for MSR over the ZrO_2-based catalysts was proposed, in which methanol conversion proceeds on the metal atoms, while the support provides the active sites for the water activation. In cerium-doped catalysts, the fraction of trivalent cerium cations decreases with the increase of cerium amount; thus, the lower catalytic activity for the catalysts with higher cerium loading can be ascribed to the interaction of defects, clustering, or sintering. The lanthanum-doped catalysts showed higher activity than the yttrium ones; however, the selectivity was lower. The Ru-Rh-based catalysts displayed higher activity in both the MSR and the methanol decomposition reaction, which caused a drop in the selectivity. The best performance was obtained with the $Ni_{0.2}$-$Cu_{0.8}/Ce_{0.1}Zr_{0.9}O_{2-\delta}$ catalyst in terms of the hydrogen yield, selectivity, and stability. Lu et al. studied $CuZnAlO_X$ catalysts prepared by co-precipitation method and boron-modified/$CuZnAlO_X$ catalysts with various boron loadings (0.28%, 0.38%, 0.73%, 0.89%, or 4.10%) prepared by impregnation method in MSR [190]. The catalytic tests were performed at atmospheric pressure, at a GHSV of 9000 mL·g·$^{-1}$·h^{-1}, in the temperature range 160–310 °C, with a H_2O/CH_3OH molar ratio of 3. The characterization results showed that the introduction of boron affected the Cu dispersion and reducibility. The best performance was obtained with a boron loading of 0.38%; the methanol conversion reached 93% due to the presence of higher specific surface area, lower reduction temperature, and strong interactions between the boron and copper species, which suppressed the Cu particle migration. Time-on-stream tests showed no deactivation of this catalyst over a period of 102 h of reaction, except for a negligible change for the 0.38B/$CuZnAlO_X$. Maiti et al. compared the catalytic activity behaviour of sol–gel-synthesized nanostructured $Cu_XFe_{1-X}Al_2O_4$ ($0.3 \leq x \leq 0.8$; where n = 30, 40, 50, 60, 70, and 80) hercynites with the corresponding catalysts prepared by solution combustion synthesis in MSR [191]. The activity tests were performed at atmospheric pressure, in the temperature range 200–300 °C, at a GHSV of 30,000 h^{-1}, with a gas stream molar ratio for methanol/water/nitrogen of 1:1.1:8.4. The catalysts prepared by sol–gel method were more active than those prepared by solution combustion synthesis; moreover, Cu doping enhanced the catalytic activity towards methane steam reforming. The $Cu_{0.8}Fe_{0.2}Al_2O_4$ catalyst showed a methanol conversion of ~80% and low CO selectivity of 2%, even after 50 h of time-on-stream testing; this behaviour was explained as the result of a partial breakdown of the spinel lattice during the reforming reaction, with the formation of CuO followed by reduction to metallic copper, leading to a stable ratio between reduced and oxidized copper (Cu^0, Cu^+)/Cu^{2+}. XPS studies showed the presence of adventitious carbon; however, the difference in the C1s signal from the fresh to the aged samples indicated a moderate carbon accumulation, also demonstrating that the carbon accumulation was not severe. Song et al. studied the effects of ZnO content on the performance of $Zn_yCe_1Zr_9O_x$ (y = 0, 0.5, 1, 5) in MSR [192]. The catalysts were prepared by using conductive carbon black T100 as a hard template; the catalytic tests were conducted at atmospheric pressure, in the temperature range of 200–500 °C, at a steam-to-carbon of 1.4, at a GHSV of 5151 h^{-1}. The best performance was obtained with the $Zn_1Ce_1Zr_9O_x$ catalyst, which showed full methanol conversion and an improved hydrogen production rate of 0.31 mol·$h^{-1}g_{cat}^{-1}$ at 400 °C. Moreover, the stability tests, which were carried out for 24 h, showed no deactivation and a CO selectivity below 8%. The characterization results revealed the formation of a solid solution with the incorporation of Zn^{2+} into the $Ce_1Zr_9O_x$ lattice, which modulated the O_{Latt}/O_{Ads} surface ratio and generated a new Zn-O-Zr interfacial structure, increasing the CO_2 selectivity. The CO_2/CO intensity ratio in the desorption peaks observed during CH_3OH-TPD increased with the zinc molar ratio in the catalysts until y = 1 ($Zn_1Ce_1Zr_9Ox$ catalyst) and decreased from y = 1 to y = 5 ($Zn_5Ce_1Zr_9Ox$ catalyst); thus, the highest abundance of lattice oxygen in the $Zn_1Ce_1Zr_9Ox$ catalyst suppressed the undesired CO formation.

3.4. Unconventional Reactor Configuration, Simulation, and Theoretical Studies

This section deals with recently published articles on the MSR, in which the reactor configuration, in addition to being innovative compared to the conventional one (such as membrane or structured reactors), assumes a dominant role in the catalytic activity. In addition, simulations and theoretical studies have also been included.

Mateos-Pedrero et al. studied the effect of the surface area and polarity ratio of the ZnO support on the catalytic activity of CuO/ZnO in MSR [193]. The surface area of ZnO was tuned by changing of the calcination temperature, while the polarity ratio was modified by using different salt precursors. The supports were prepared by a modified hydrothermal method, using two different salt precursors, zinc acetate or zinc nitrate, and calcined at four different temperatures: 300, 350, 375, and 400 °C. The catalysts were prepared by impregnation of the support with copper nitrate at pH = 6 using ammonium hydroxide, with a metal loading of 15 wt.% and calcined at 360 °C. The results showed that the copper dispersion and surface area increased with the surface area of the support and that the reducibility of the copper species increased with the polarity ratio in the ZnO support. The activity tests showed a dependence from the surface area of the support, and therefore from the Cu dispersion; moreover, the selectivity increased with the polarity ratio. The best performance was obtained with the catalyst whose support was obtained from zinc acetate and was calcined at 375 °C. This catalyst was tested in a Pd-membrane reactor in two different experiments to evaluate the methanol conversion and gas selectivity stability in one case, and to investigate the hydrogen recovery and hydrogen permeate purity under various operating conditions in the other case. The first set of experiments showed that it is possible to reach 97% of methanol conversion at 300 °C, 2.0 bar, and WHSV = 2.73 h^{-1}, with good stability and selectivity. The second set of experiments showed that it is possible to achieve a hydrogen recovery rate of ca. 75% and a hydrogen permeate purity rate higher than 90% at 330 °C, 2.5 bar, and WHSV = 1.37 h^{-1}. Kim et al. carried out process simulation and design, as well as economic analysis, to evaluate the technical and economic feasibility of MSR in a membrane reactor for ultrapure hydrogen production [194].

The simulation was performed with Aspen HYSYS®; certain operating conditions were investigated, such as the effects of the hydrogen permeance ($1 \times 10^{-5} - 6 \times 10^{-5}$ mol·m^{-2}·s^{-1}·Pa^{-1}), H_2O sweep gas flow rate (1-20 kmol·h^{-1}), and reaction temperature (448–493 K) in a conventional packed-bed reactor, using previously reported reaction kinetics. Improved performances regarding the methanol conversions and hydrogen yield were observed for all the studied cases in the membrane reactor configuration compared to the packed-bed configuration. Higher hydrogen permeance and H_2O sweep gas flow rates were beneficial for methanol conversion, but no further improvement was obtained for H_2O sweep gas flow rates over 10 kmol·h^{-1}. A unit hydrogen production cost was also calculated, showing a higher cost for the packed-bed reactor (9.37 $ kg·H_2^{-1}) than for the membrane reactor (7.24 $ kg·H_2^{-1}). Köpfle et al. studied the activation and the catalytic performance in MSR of two Zr-containing intermetallic systems, Cu-Zr and Pd-Zr [195]. Metal mixtures of Cu and Zr were prepared with different stoichiometry ratios (Cu/Zr = 9:2, 2:1, 1:2). Moreover, two Pd-Zr systems were prepared by ALD/CVD (atomic layer deposition/chemical vapour deposition) of zirconium(IV)tert-butoxide on Pd and Zr foils (Pd/Zr ratio 2/1). The preparation of the corresponding Cu-Zr intermetallic catalyst was carried out using the Cu foil. The transitions of the initial metal and intermetallic compound structures in the active and CO_2-selective states were monitored in MSR by an inverse surface science and bulk model approach. The CO_2 selectivity and the catalytic performance of the Cu-Zr system were promising due to the formation of a beneficial Cu–ZrO_2 interface. The two Pd-Zr systems showed a low-temperature coking tendency, high water activation temperature, and low CO_2 selectivity. Zhou et al. reported a benchmark study in which the performance of porous Cu-Al fiber sintered felt, constructed using the solid-phase sintering method, was compared to that of Cu fiber sintered felt and Al fiber sintered felt. The fibers were impregnated with Cu/Zn/Al/Zr catalysts and used in a cylindrical MSR microreactor [196]. The results showed that the Cu-Al fiber based catalyst gave higher methanol conversion and hydrogen flow rates than the Cu- and Al-fiber-based catalysts. The rough-fiber-based catalyst showed a much higher methanol conversion and hydrogen flow rate than the smooth one. Moreover, the methanol conversion and hydrogen flow rate increased with the decrease of the Cu fiber weight and the increase of the Al fiber weight. The best performance in terms of hydrogen production was obtained with a three-layer Cu-Al fiber with 80% porosity and a 1.12 g Cu fiber/1.02 g Al fiber. The time-on-stream tests showed a significant deactivation of the catalysts after 32 h due to the coke formation. In further studies, a laser micromilling technique was reported for the fabrication of surface microchannels on porous copper-fiber-sintered felts [197]. The effects of the surface microchannel shape and catalyst loading on the activity in the MSR microreactor was investigated. The results showed that the rectangular shape provided a lower pressure drop, higher average velocity, and higher permeability compared to the stepped and polyline microchannels, as well as the highest methanol conversion and hydrogen flow rate. Furthermore, in the latter case, the deactivation of the catalyst observed during time-on-stream tests was attributed to the carbon deposition and catalyst loss. Tajrishi et al. studied the use of Cu/SBA-15-based nanocatalysts in a parallel-type microchannel reactor for MSR reaction [198]. SBA-15 was prepared by hydrothermal method, while the catalysts were prepared by wetness impregnation method. The catalytic activity tests were carried out at atmospheric pressure; the microchannel reactor walls were coated with the nanocatalysts and a mixture of methanol and distilled water was injected into the microreactor by a syringe pump at various flow rates (1.8, 2.4 and 3 mL·h^{-1}) through a vaporizer at 150 °C. A series of reaction conditions was evaluated on the 5%Cu/SBA-15 catalyst, such as the effects of the reaction temperature (260, 280, 300, 320, and 340 °C), the WHSV (32.76, 43.68, and 54.56 h^{-1}), and the steam-to-carbon molar ratio (1, 2, and 3) on the methanol conversion; hydrogen yield; and H_2, CO, and CO_2 selectivity. Moreover, the effects of promoters were investigated by performing a series of experiments at 300 °C at a weight GHSV of 43.68 h^{-1} and at a steam-to-carbon molar ratio of 2 on the following catalysts: xCu/SBA-15 (x = 5, 10 and 15 %), 10%Cu/yZnO/SBA-15 (y = 5, 10 and 15 %), 10%Cu/5%ZnO/2%CeO_2/SBA-15, 10%Cu/5%ZnO/2%ZrO_2/SBA-15, and 10%Cu/5%ZnO/2%CeO_2/2%ZrO_2/SBA-15. Specific surface area measurements and field emission SEM images demonstrated that the addition of CeO_2 and ZrO_2 to the

10%Cu/5%ZnO/SBA-15 catalyst led to a reduction in the agglomeration of crystallites, thus increasing the specific surface area and lowering the pore diameter. The methanol conversion and hydrogen selectivity were improved by ZrO_2 promotion, while ZnO and CeO_2 promoters reduced the CO selectivity. Moreover, CeO_2 and ZrO_2 promoted the stability of the Cu/ZnO/SBA-15-based catalysts, due to the better reducibility of CuO particles and less coke deposition. The 10%Cu/5%ZnO/2%CeO_2/2%ZrO_2/SBA-15 catalyst showed the best performance, displaying optimal methanol conversion of 95.2%, low CO selectivity of 1.4%, high H_2 yield of 90%, and good stability in the time-on-stream test over 60 h of reaction at 300 °C, at a weight space hourly velocity of 43.68 h^{-1}, and at a steam-to-carbon molar ratio of 2, due to smaller size of the copper and zinc crystallites, higher copper dispersion, and greater specific surface area. Liu et al. studied copper foams with different types of hole arrays as catalyst supports for cylindrical laminated MSR microreactors [199]. The copper foams were fabricated by laser processing method; the catalytic formulation (Cu/Zn/Al/Zr) was loaded by impregnation, while macroscopic numerical analysis was used to analyse the reactant distribution on the foams. The optimal hole array distribution was obtained on the basis of the experimental results by varying the reactant flow rate, reaction temperature, and catalyst loading. The simulation results showed that the radial distribution uniformity was improved and the axial flow velocity was increased from the copper foams with hole arrays. The copper foams whose hole size decreased in the arrays from the center to the radial direction provided the best catalytic performance; by feeding a flow rate of 10 mL·h^{-1} at 300 °C, the initial methanol conversion was 95% and the initial hydrogen production flow rate was 0.52 mol·h^{-1}; after time-on-stream tests for 24 h, the methanol conversion decreased to 70% and the hydrogen production flow rate decreased to 0.35 mol·h^{-1}. Sarafraz et al. investigated the use of the Cu-SiO_2 porous catalyst coated on the internal wall of a microreactor with parallel micropassages in MSR [200]. The catalyst was prepared by coating with copper and silica nanoparticles, using convective flow boiling heat transfer followed by calcination. The catalytic activity tests were carried out in the temperature range of 250–400 °C, with a reactant flow rate of 0.1-0.9 dm^3·min^{-1}, a catalyst loading of 0.25-1.25 g, and at a heat flux of 500 kW·m^{-2}. The highest methanol conversion was obtained at GHSV of 24,000 mL·g^{-1}·h^{-1}, at a temperature of 500 °C, and with a methanol-to-water molar ratio of 0.1. The increase in the GHSV of the reactants generated a decrease in the methanol conversion, which was attributed to the suppression of the reactant diffusion into the pores of the catalyst and to the decrease in the average film temperature of the reactor. Moreover, for the low methanol-to-water molar ratio the reaction was complete, requiring more thermal energy; therefore, greater heat flux was necessary to compensate for the temperature drop. Shanmugam et al. investigated the effect of supports (CeO_2, Al_2O_3, and ZrO_2) and an In_2O_3 co-support in Pt-based catalysts for MSR in a microchannel reactor [201]. CeO_2, Al_2O_3, and ZrO_2 were prepared by sol–gel method, the co-support was obtained by impregnation with In_2O_3, while the catalysts were obtained by impregnation. Two microchannel platelets were coated with the catalysts and sealed face-to-face by laser welding with the inlet and outlet capillaries. The activity tests were carried out in a microchannel reactor at atmospheric pressure in the temperature range of 300–375 °C, with a flow rate for the water/methanol mixture of 30 mL/h and a steam-to-carbon ratio of 1.4. Among the studied catalysts, the best performance was obtained with 15Pt/CeO_2, which exhibited complete methanol conversion, high hydrogen selectivity, and high CO formation (7 vol.%) at 350 °C. The addition of In_2O_3 reduced the CO formation to 1.9 vol.% due to the enhanced dispersion of metallic Pt nanoparticles and to the boost given to the water activation, which reacted with methanol on the Pt surface, leading to the selective formation of CO_2 and H_2 and suppressing the CO formation. The time-on-stream tests showed only slight deactivation in 100 h under reaction conditions at WHSV of 88 L·h^{-1}·g^{-1}. Zhuang et al. developed a multichannel microreactor with a bifurcation inlet manifold and rectangular outlet manifold for MSR, in which the commercial CuO/ZnO/Al_2O_3 catalyst was directly packed [202]. The catalytic tests were performed with a variable steam-to-carbon ratio (1.1–1.5), a variable WHSV (0.4–6.7 h^{-1}) in the temperature range of 225–325 °C, with two catalyst particle sizes of 50–150 mesh and 150–200 mesh. A computation fluid dynamics (CFD) simulation was also performed to study the flow distribution in the multichannel reactor. The results showed that the methanol conversion was enhanced by increasing the steam-to-carbon ratio and the temperature, as well as by decreasing the WHSV and catalyst particle size. On the other hand, the CO concentration decreased with a growth in the steam-to-carbon ratio and the WHSV, as well as decreased with the decrease of the temperature and catalyst particle size. A time-on-stream test was also carried out for 36 h, achieving a methanol conversion of 94.04% and a CO concentration of 1.05%, with a steam-to-carbon ratio of 1.3, a temperature of 275 °C, a WHSV of 0.67 h^{-1}, and a catalyst particle size of 150–200 mesh. Zhu et al. reported a modelling and design study for a multitubular packed-bed reactor for MSR on a $Cu/ZnO/Al_2O_3$ catalyst [203]. A multitubular packed-bed reformer pseudo-homogenous model was developed based on Langmuir–Hinshelwood kinetics of the MSR process to investigate the impacts of operating conditions and geometric parameters on the performance. Moreover, pressure drop, heat, and mass transfer phenomena, as well as diffusion inside the catalyst particles, were also investigated. A comparative study was also performed to evaluate the performance of a current–current reactor with respect a co-current reactor.

The simulation results showed that a co-current heat exchanger provided a lower CO concentration and better heat transfer efficacy; moreover, the lower liquid fuel flow rate and higher thermal air inlet temperature gave higher residual methanol and CO concentrations (Figure 15). At a fixed catalyst loading, the increase in the tube numbers, the growth of the baffle plate number, and the decrease in the tube diameter increased both the methanol conversion and the CO concentration. Ke and Lin performed density functional theory computations and transition state theory analyses on the intrinsic mechanism of Ni-catalysed MSR by considering 54 elementary reaction steps [204]. The microkinetic model was obtained by combining the quantum chemical results with a continuous stirring tank reactor. The microkinetic simulations showed that O*, CO*, OH*, and H* are the only surface species with non-negligible surface coverage. The main reaction pathway is described in Equations (13) and (14).

$$CH_3OH \rightarrow CH_3OH^* \rightarrow CH_3O^* \rightarrow CH_2O^* \rightarrow CHO^* \rightarrow CO^* \rightarrow CO_2^* \rightarrow CO_2 \tag{13}$$

and the rate determining step is:

$$CH_3O^* \rightarrow CH_2O^* + H^* \tag{14}$$

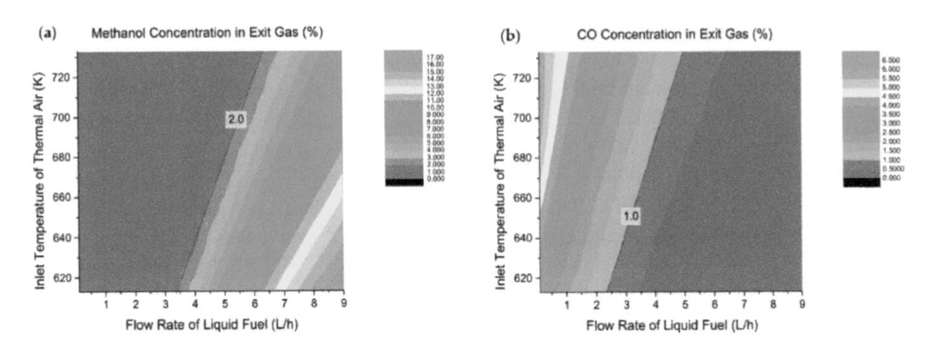

Figure 15. Profiles of the (**a**) methanol concentration and (**b**) CO concentration in the exit gas of the reformer with different inlet flow rates for the fuel mixture and inlet temperature of the thermal air [203].

Wang et al. studied a MSR rib microreactor heated by automobile exhaust [205]. The effect of the inlet exhaust and methanol-to-steam ratio on the performance of the reactor were numerically analyzed with a computational fluid dynamic study. The results showed that the methanol conversion increased with the inlet exhaust velocity and the inlet exhaust temperature; moreover, the axial temperature increased along the axis and decreased with the reactant inlet velocity. By fixing the inlet reactant velocity to 0.1 m/s and the inlet temperature to 220 °C, the best performance was obtained with a water/methanol ratio of 1.3, an exhaust inlet velocity of 1.1 m·s^{-1}, and an exhaust inlet temperature of

500 °C, obtaining methanol conversion of 99.4%, hydrogen content of 69.9%, and thermal efficiency of 28%.

3.5. Oxidative Steam Reforming of Methanol

The OSRM is a combination of an exothermic reaction between O_2 and methanol and the endothermic reaction of steam reforming. The exothermic reaction has frequently been assumed to be the partial oxidation (PO) of methanol [206]; however, more recently it was found that with the commercial $Cu/ZnO/Al_2O_3$ catalyst, the combustion of methanol is the main reaction between oxygen and methanol [207]. This process presents some advantages, such as the possibility to produce hydrogen with a very low concentration of CO; moreover, it is suitable for power variation by varying the methanol/oxygen ratio [208]. A selection of articles on this topic follows below.

Pojanavaraphan et al. investigated the catalytic performance of a series of $Au/CeO_2–Fe_2O_3$ catalysts prepared by deposition–precipitation for the OSRM [209]. The catalysts were prepared with the aim of investigating the Ce/(Ce+Fe) atomic ratio (1, 0.75, 0.5, 0.25, 0), the gold loading (1, 3, and 5 wt.%), and the calcination temperature (200, 300, and 400 °C). The catalytic activity tests were carried out at a GHSV of 30,000 mL g-cat^{-1} h^{-1}, in the temperature range of 200–400 °C, at atmospheric pressure. A mixture of distilled water and methanol was continuously injected by a syringe into a vaporizer chamber at 150 °C, at a rate of 1.5 mL h^{-1}. The H_2O/CH_3OH and O_2/CH_3OH molar ratios were varied from 1:1 to 4:1 and from 0 to 2.5, respectively. The best performance was obtained with a gold loading of 3 wt.% and Ce/(Ce+Fe) = 0.25 calcined at 300 °C; this catalyst showed an optimal Au particle size; coexistence of the CeO_2-Fe_2O_3 solid solution and free F_2O_3 phase; and strong interaction sites, such as Au-Au, Au^0-free Fe^{3+}, and Ce^{4+}-Fe^{3+}; which were indicated as the main factors for successfully improving the catalytic activity. The presence of high steam content negatively affected the formation of hydroxyl, carbonate, and formate species; the best reaction condition was $O_2/H_2O/CH_3OH$ = 0.6:2:1. Stability tests were carried out, highlighting the effect of a pretreatment with O_2 on the catalytic activity of the catalyst. The results showed that the activity of the unpretreated catalyst decreased in the methanol conversion from 95.7% to 91% and in the hydrogen yield from 85.87% to 82%, while that of the pretreated catalyst followed the same trend, but with higher conversion and hydrogen yield. The TPO profile of the unpretreated spent catalyst showed two peaks attributed to the different types of coke formed, with a total coke percentage of 0.629 wt.%. The pretreated catalyst showed the first oxidation peak at 84 °C with low intensity due to carbonaceous species, before being converted to the coke, revealing that the O_2 pretreatment is able to reduce or retard the coke formation rate during the stability test via coke gasification; the coke percentage was only 0.053 wt.%. These data demonstrated that the coke formation did not affected the stability of the pretreated catalyst, suggesting that the O_2 pretreatment could reduce or retard the coke formation rate. Pérez-Hernández et al. investigated the effect of the bimetallic Ni/Cu loading on ZrO_2 support in the autothermal steam reforming of methanol (ASRM) for hydrogen production [210]. The support was prepared by sol–gel method while the catalysts were prepared by sequential impregnation to obtain three different total metallic loading values of 3 wt.%, 15 wt.%, and 30 wt.%, with a Cu/Ni ratio equal to 4:1. TEM-EDX images highlighted the core–shell structure of the Cu/Ni nanoparticles, but in the case of the 30wt.%$Ni/Cu/ZrO_2$ sample the shell was constituted by the Ni-Cu alloy. The activity tests showed that the methanol conversion increased with the Ni/Cu loading, thus the highest conversion was obtained with the 30wt.%$Ni/Cu/ZrO_2$ catalyst; however, the 15wt.%$Ni/Cu/ZrO_2$ catalyst exhibited the highest hydrogen selectivity. A time-on-stream test was also performed on the 30wt.%$Ni/Cu/ZrO_2$ catalyst, which exhibited stable performance during a 46 h time-on-stream test at 400 °C, without apparent deactivation. Mierczynski et al. studied the performance of copper- and gold-doped copper catalysts supported on a multiwalled carbon nanotube (MWCNT) prepared by wet impregnation and deposition–precipitation methods in OSRM [211]. The tests were carried out at 200 and 300 °C on 100 mg of catalyst and with a stream composition of $H_2O/CH_3OH/O_2$ (molar ratio = 1:1:0.4), a GHSV of 26700 h^{-1}, under atmospheric pressure, with a total flowrate of 31.5 mL·min^{-1} and methanol concentration of 6% (Argon was used as the diluent gas). The in situ XRD analysis showed the occurrence of Au-Cu alloy in the bimetallic catalyst during the reduction process at 300 °C. The highest activity and hydrogen yield were obtained with the 0.5%Au–20%Cu/MWCNT bimetallic catalyst; the performance was related to the reducibility and to the highest total acidity of the catalytic system, which is able to stabilize the intermediate formed during the reaction. Jampa et al. investigated the use of Cu-loaded mesoporous ceria and Cu-loaded mesoporous ceria-zirconia catalysts synthesized by nanocasting process, using MCM-48 as the template (% of Cu loading varied from 1 to 12 wt.% in OSRM [212]. The catalytic activity tests were performed over a temperature range of 200–400 °C at atmospheric pressure, with a mixture of H_2O/CH_3OH (1.5 mL·h^{-1}, molar ratio varied from 1:1 to 3:1), He (45 mL·min^{-1}), and oxygen (45 mL·min^{-1}). The results showed that the best catalytic performance was obtained with 9 wt.% Cu loading, which assured methanol conversion of 100% and hydrogen yield of 60%; however, these results were obtained at 350 °C when mesoporous CeO_2 was used as the support and at 300 °C when CeO_2-ZrO_2 was used as the support by feeding O_2 at 5 mL·min^{-1} and a H_2O/CH_3OH molar ratio of 2:1. The time-on-stream stability tests over 169 h of reaction resulted in a continuous decrease of the methanol conversion and hydrogen yields for both the 9%Cu/CeO_2 and 9%Cu/$CeZrO_4$. The TPO profiles of the spent catalyst 9%Cu/CeO_2 showed two peaks at 207 °C and 296 °C, whereas 9%Cu/$CeZrO_4$ showed the oxidation peaks at 266 and 484 °C, indicating the presence of two different types of coke or carbonaceous species. The oxidation peaks at 207, 266, and 296 °C were assigned to the poorly polymerized coke deposited on the catalyst particles, while the peak at 484 °C was ascribed to highly polymerized coke deposited near the catalyst–support interface. Moreover, the significant decrease in the Branauer–Emmett–Teller (BET) specific surface area of the spent catalysts suggested that sintering had occurred. Based on these data, the deactivation was attributed to coke formation and agglomeration; however, the ceria/zirconia-based catalyst showed reduced CO selectivity. Thus, the main conclusion was that the addition of zirconium into the support improved the redox property, the thermal stability, and the oxygen storage capacity of the catalyst, resulting in a better performance in terms of CO oxidation reaction, and thus in a low CO level during autothermal steam reforming of methanol. Pu et al. investigated the Cu/ZnO-based catalysts promoted by Sc_2O_3 in ASRM [213]. The Cu/Sc_2O_3-ZnO catalysts were prepared by the reverse precipitation method, with Cu metal loading of 15 wt.% and Sc/Zn molar ratios of 0, 0.03, 0.05, and 0.07. The catalytic activities were evaluated in the temperature range of 220–600 °C at atmospheric pressure, while an aqueous solution of methanol at a rate of 0.05 mL/min was fed together with a mixed gas flow (50 mL·min^{-1}) of N_2 and O_2. The characterization results showed that the Sc promotion assured a reduction of the particle size and an increase of the metal dispersion of Cu. In addition, the Sc doping enhanced the interaction between the metal and support in the ZnO lattice, improving the metal dispersion and sintering resistance of the catalysts. The best performances in terms of catalytic activity and stability were obtained with the catalyst characterized by the Sc/ZnO molar ratio of 0.05 mol·mol^{-1}. The decrease of the methanol conversion in the time-on-stream tests was attributed to sintering phenomena: in fact, coke formation was suggested by the performed analysis.

To summarize, the results shown in Table 2 highlight significant differences in the performance of the catalysts studied in the reviewed articles on the methanol reforming in terms of the carbon formation rate. The highest reported rate (10^{-2} g_{coke}·$g_{catalyst}{}^{-1}$·$g_{carbon,fed}{}^{-1}$·h^{-1}) was for Cu/Al_2O_3 [165], while the lowest one ($6.3 \cdot 10^{-7}$ g_{coke}·$g_{catalyst}{}^{-1}$·$g_{carbon,fed}{}^{-1}$·h^{-1}) was for 10%Cu-10%Zn-2%Zr/MCM-41 [183]. Although in the two cases the reaction conditions were very different, both in terms of temperature and steam-to-carbon ratio (500 °C and S/C = 1.5 in the first case; 300 °C and S/C = 2 in the second case), the effect of the zirconium promoter seemed to play a major role. Zirconium, in fact, can stabilize the active species and decrease the growth of metal oxides during the synthesis process, thus allowing the size of the crystallites to be reduced and allowing a better dispersion of the active phases. On the other hand, the use of noble metals such as gold, as well as the addition of oxygen in the reforming reaction, does not seem to suppress the formation of coke; however, further studies are necessary.

Table 2. Carbon formation rates for various catalysts employed for steam reforming and oxidative steam reforming of methanol.

Catalyst [a]	Operative Conditions [b]	T (°C)	WHSV [c] (h^{-1})	X MeOH (%)	Carbon Formation Rate [d] (MULTIPLIED for 1000)	Ref.
			Methanol Steam Reforming			
15%Cu-MCM-41	H_2O/CH_3OH = 3/1	250	1.0	≈73 after 48 h	0.015	[158]
20%Cu-MCM-41	H_2O/CH_3OH = 3/1	250	1.0	≈60 after 48 h	0.024	[158]
Cu-ZnO-Al_2O_3-ZrO_2-Ga_2O_3 Cu/Zn/Al/Zr/Ga = 14.9:30.9:3.9:10.8:1.9 mass ratio	H_2O/CH_3OH = 1/1	275	4.3	≈85 after 70 h	0.010	[159]
Cu-ZnO-Al_2O_3-ZrO_2-Ga_2O_3 Cu/Zn/Al/Zr/Ga = 13.3:28.2:3.9:10.0:1.8 mass ratio	H_2O/CH_3OH = 1/1	275	4.3	≈70 after 70 h	0.014	[159]
Pd-MCM-41	H_2O/CH_3OH = 3/1	300	1.0	≈32 after 40 h	0.041	[163]
Zn-MCM-41	H_2O/CH_3OH = 3/1	300	1.0	≈5 after 40 h	0.15	[163]
Ni-MCM-41	H_2O/CH_3OH = 3/1	300	1.0	≈15 after 40 h	0.16	[163]
Cu-MCM-41	H_2O/CH_3OH = 3/1	300	3.0	≈75 after 40 h	0.049	[163]
Cu/Al_2O_3	H_2O/CH_3OH = 3/2	500	0.8	≈91 after 5 h	10	[165]
NixMgyO Impregnation	H_2O/CH_3OH = 1/1	600	65.7	51.4 after 20 h	0.0079	[168]
NixMgyO Hydrothermal method	H_2O/CH_3OH = 1/1	600	65.7	58.3 after 20 h	0.0030	[168]
NixMgyO Co-precipitation	H_2O/CH_3OH = 1/1	600	65.7	57.3 after 20 h	0.085	[168]
Cu/cubic-$ZnTiO_3$	$N_2/H_2O/CH_3OH$ = 1/2/1	250	1	≈63 after 42 h	1.1	[177]
Cu/hexagonal-$ZnTiO_3$	$N_2/H_2O/CH_3OH$ = 1/2/1	250	1	≈5 after 42 h	6.7	[177]
10%Cu-10%Zn/MCM-41	H_2O/CH_3OH = 2/1	300	1.62	≈75 after 60 h	0.0011	[183]
10%Cu-10%Zn-2%Zr/MCM-41 Impregnated	H_2O/CH_3OH = 2/1	300	1.62	≈83 after 60 h	0.00095	[183]
10%Cu-10%Zn-2%Zr/MCM-41 Sol–gel method	H_2O/CH_3OH = 2/1	300	1.62	90.2 after 60 h	0.00079	[183]
10%Cu-10%Zn-2%Zr/MCM-41 MCM-41 pretreated with acetic acid	H_2O/CH_3OH = 2/1	300	1.62	92.8 after 60 h	0.00063	[183]
10%Cu/SBA-15	H_2O/CH_3OH = 2/1	300	43.7	≈64 after 60 h	0.0031	[198]
10%Cu/5%ZnO/SBA-15	H_2O/CH_3OH = 2/1	300	43.7	≈74 after 60 h	0.0025	[198]
10%Cu/5%ZnO/2%CeO_2/SBA-15	H_2O/CH_3OH = 2/1	300	43.7	≈85 after 60 h	0.0013	[198]
10%Cu/5%ZnO/2%ZrO_2/SBA-15	H_2O/CH_3OH = 2/1	300	43.7	≈84 after 60 h	0.0031	[198]
10%Cu/5%ZnO/2%CeO_2/2%ZrO_2/SBA-15	H_2O/CH_3OH = 2/1	300	43.7	≈86 after 60 h	0.0031	[198]
			Oxidative Steam Reforming of Methanol			
xAu/CeO_2–Fe_2O_3 x = 3 wt.%, Ce/(Ce+Fe) = 0.25.	$O_2/H_2O/CH_3OH$ = 0.6/2/1.	350	11.9	91 after 12 h	0.098	[209]
xAu/CeO_2–Fe_2O_3 x = 3 wt.%, Ce/(Ce+Fe) = 0.25.	$O_2/H_2O/CH_3OH$ = 0.6/2/1. O_2 pretreatment	350	11.9	92 after 12 h	0.0082	[209]
9 wt.%Cu/CeO_2	H_2O/CH_3OH = 2/1; O_2 = 5 mL·min^{-1}	350	0.9	65 after 2.8 h	0.0042	[212]
9 wt.%Cu/$CeZrO_4$	H_2O/CH_3OH = 2/1; O_2 = 5 mL·min^{-1}	300	0.9	65 after 2.8 h	0.0032	[212]

Note: [a] The metal or oxide loadings are intended for the weight of the catalyst; [b] all the tests were performed at atmospheric pressure; [c] refers to the methanol mass flow rate; [d] $g_{coke} \cdot g_{catalyst}^{-1} \cdot g_{carbon,fed}^{-1} \cdot h^{-1}$.

4. Bioglycerol Reforming

An extensive research study concerning the applications of bioglycerol as the main by-product of biodiesel production processes (~10% by weight) [214] was carried out in recent years. Indeed, with the constant growth of the biofuel market and approximately 36 million tons of biodiesel having been produced [215], the supply of bioglycerol exceeds the global demand (~$3{\cdot}10^6$ ton produced by 2020, with a demand below ~$5{\cdot}10^5$ ton/year) [216]. The consequent drop in bioglycerol prices and the difficulties in the disposal of the produced surplus constitute a threat for the biodiesel production process, which is already fails to be competitive in terms of price [215,217,218].

Crude glycerol (i.e., glycerol derived from the biodiesel production process) is a highly viscous liquid characterized by a dark color, which includes variable quantities of soap, catalyst, alcohol, monoglycerides, diglycerides, polymer, water, unreacted triacylglycerols, and biodiesel; glycerol concentrations are in the range of 40–85%, depending on the efficiency of the biodiesel production process. Due to the presence of such impurities, it is not possible to use crude glycerol in most typical applications (e.g., cosmetics, food, alkyd resins, tobacco, pharmaceutical, polyurethane [217]), which are depicted in Figure 16.

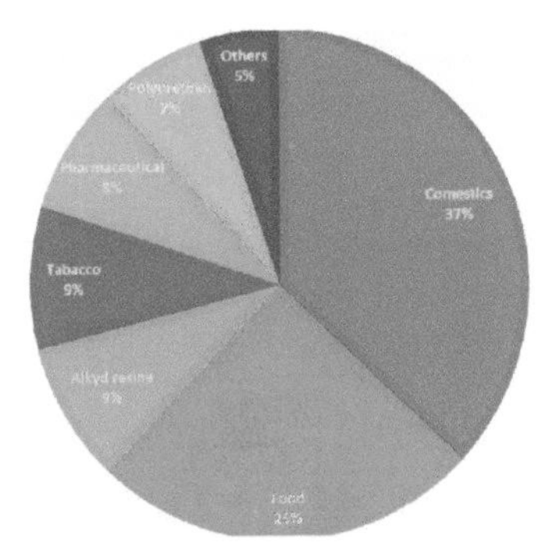

Figure 16. Crude glycerol's most typical applications [217].

Moreover, crude glycerol purification through distillation is an expensive process and most biodiesel plants use this feedstock for direct burning process; however, in this way, the recovery efficiency is low due to the combustion difficulties caused by the high viscosity and the formation of highly toxic substances (e.g., acrolein and aldehyde). Hence, the employment of surplus-produced crude glycerol to obtain other value-added chemicals could represent the best route in view of improving the biodiesel industry's economic feasibility [215,219].

An interesting pathway for crude glycerol revalorization is to use it as a renewable biomass source for H_2 production; thereby, hydrogen, as a promising fuel for moving vehicles and providing power and heat for industries, would be produced, involving less energy consumption in comparison with the traditional routes for its obtainment (water electrolysis and thermochemical processes) [219]. Hydrogen can be produced by glycerol via several processes, including steam reforming, partial oxidation, autothermal reforming, aqueous-phase reforming, dry reforming, and photocatalytic reforming. Among them, steam reforming has drawn major attention. In fact, since the steam reforming of methane is presently the most widely-used process in the industry to obtain hydrogen, the adoption of glycerol would not require many changes in the already existent systems [218,220]. However, glycerol steam reforming (GSR) is an endothermic catalytic process that requires large energy consumption and suffers from catalyst deactivation caused by coke formation. Additionally, the other routes for H_2 generation from glycerol are characterized by several disadvantages. The autothermal reforming (the process in which the partial oxidation is coupled with steam reforming) leads to a lower energy consumption compared to the SR. In fact, the steam reforming step absorbs the energy produced by the PO reaction; however, as a drawback, the presence of several hotspots caused by the heat produced during the PO can result in negative effects on the catalyst activity. The liquid-phase reaction (APR) operating at pressures of 30–80 bar and temperatures of 200–250 °C, reduces the amount of necessary heat, especially due to the liquid form of the reactant, which does not require energy for vaporization; however, this process usually requires more time compared to the SR and requires noble metals to be adopted as catalysts (generally Pt), increasing costs. Glycerol dry reforming (GDR) appears to be an attractive path for H_2 production, as it is a carbon capture and recycling (CCR) approach; indeed, with the adoption of this process, CO_2 would be converted and recycled back into fuels and added value chemicals. Regardless, the major obstacles in the application of this technology concern the sintering of catalysts particles and the coke accumulation on the catalysts' surfaces, which are even higher compared to other reforming processes. A novel route could be represented by the photocatalytic reforming of glycerol, in which H_2 would be produced under ambient conditions, in the presence of an appropriate catalyst and solar light [62,219,220].

Catalysts activity and stability are essential aspects in the transformation of glycerol into hydrogen through reforming reactions, as they allow the activation energy to be lowered, improving the kinetics of the involved chemical reactions, thus reducing the energy requirements and the correspondent costs [219]. In recent years, an extensive effort has been made towards improving the activity of catalysts. Studies have been performed on both noble and non-noble metals, different kinds of supports, and the influence of promoters. Moreover, intensified investigations on the selective removal of reaction products have also been performed in order to avoid the thermodynamic limitations (glycerol conversion and hydrogen yield) of the process [218]. Since the positive results gained have been in terms of catalyst activity, the main critical issue in glycerol reforming reactions remains the stability of the employed catalysts.

Catalyst stability, the fundamental challenge in the industrial development of glycerol reforming processes, is mainly compromised by metal sintering and coke deposits. Coke formation in glycerol reforming reactions is a truly complex aspect, since a multitude of carbon sources may be responsible for coke deposits, thus making it difficult to define the catalyst deactivation pathway [221]. Numerous side reactions involved in glycerol conversion lead to the formation of coke, such as the Boudouard reaction (Equation (8)), olefin polymerization (Equation (6)), glycerol dehydration, and rearrangement and condensation reactions [222]. Papageridis et al. [223], having gained a deepened understanding of carbon deposition, suggested that atomic carbon or C_yH_{1-y} species (C_α species) are generated by dissociative chemisorption of the hydrocarbon on the catalysts surface. Thus, the production of an amorphous film (C_β) by dehydropolymerization occurred, which can subsequently lead to the formation of graphitic carbon. Otherwise, these carbon atoms may also be dissolved and diffused through the metal to active growth areas, resulting in the precipitation of amorphous vermicular carbon (C_γ) or whiskers (Figure 17). Moreover, under harsh and protracted reaction conditions, the amorphous carbon can also be converted into graphitic form (C_δ) [223]. An accurate analysis of the nature of the formed coke is a powerful tool in the improvement of catalyst stability, as the total amount of generated coke has less impact in catalytic deactivation in comparison to the degree of graphitization of the deposits [224,225]. Temperature-programmed oxidation tests (TPO) performed in numerous research studies have pointed out the easier oxidation of amorphous carbon (T below 550 °C) compared to filamentous and graphitic carbon deposits, such as carbon nanofibers or carbon nanotubes (T over 600 °C) [33,226].

The extent of coke deposition in glycerol reforming reactions is mainly dependent on the nature of the active phase, the extent of dispersion, and the chemical characteristics of the supports or promoters.

Glycerol reforming has been investigated for various metals and the below chart summarizes the papers considered in this review, dividing them on the basis of the active phase analyzed during the study. As it is possible to see from the graph (Figure 18), in recent years, due to their good catalytic activity, low cost, and ease of availability, Ni-based catalysts have drawn major attention among the other active phases. Regarding the active phase, the most investigated materials have been Ni-, Co-, Rh-, Pt-, and Ru-based catalysts; attention has also been given to Ir, Cu, and Pd catalysts.

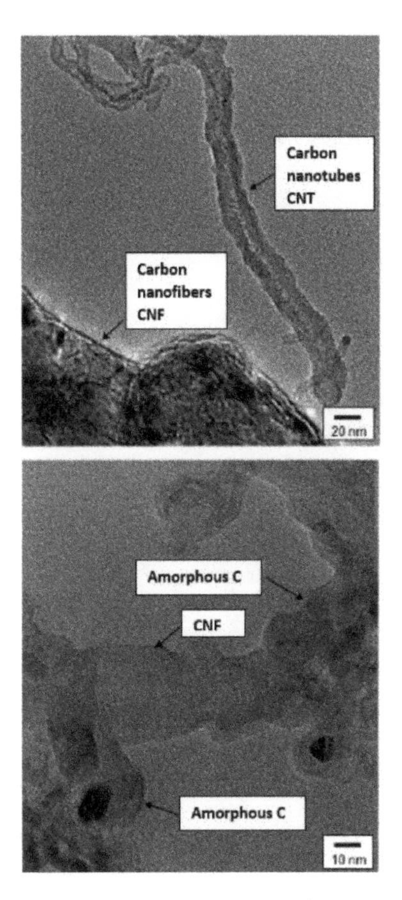

Figure 17. TEM images of amorphous and graphitic carbon formed during glycerol reforming.

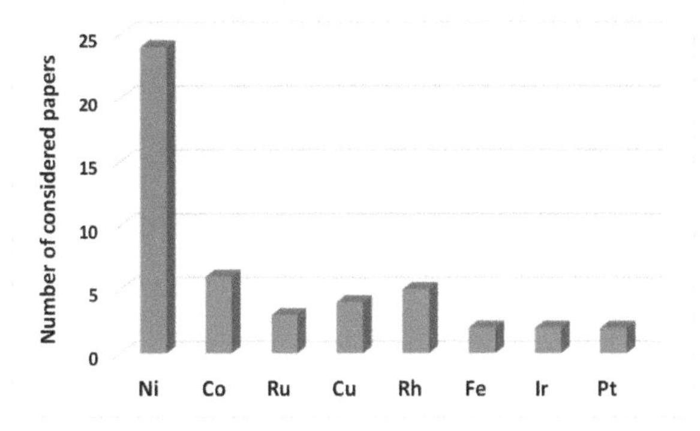

Figure 18. Number of considered papers as a function of the active phases analyzed for glycerol reforming.

4.1. *The Influence of the Active Phase*

In the following paragraph, selected papers highlighting the roles of active species on catalyst stability for glycerol reforming are discussed.

As anticipated above, nickel is widely used, mainly due to its low cost and satisfactory activity; in addition, Ni has been found to promote the water–gas shift reaction, leading to an increase of H_2 production [215]. Papageridis et al. [223] made a comparison between Ni-, Co-, and Cu/γ-alumina-supported catalysts for the steam reforming of glycerol, observing the higher ability of the Ni/Al sample to convert glycerol compared to Co/Al and Cu/Al catalysts. On the other hand, the Ni-based catalyst showed worse stability, characterized by a drastic drop in activity within the first 7 h. Rh, Ru, Pt, and Ir have been tested in GSR by Sensei et al. by preparing precious metal nanocatalysts over the promoted Al_2O_3 with MgO. In the stability test performed at 600 °C (GHSV = 35,000 mL g^{-1} h^{-1} and H_2O:$C_3H_8O_3$ = 9), Rh showed a glycerol conversion of ~98% almost constantly over 20 h, while the other metals were characterized by a drastic decrease in the conversion within the first 5 h. Moreover, from the TPO analysis of Rh/$MgAl_2O_4$, two oxidation peaks were detected, corresponding to the formation of amorphous and graphitic carbon, whereas for Ru, Pt, and Ir, only the presence of graphitic carbon has been evaluated [226]. Rh-alumina catalyst stability was also studied in GSR by Silva et al. [227]. A decrease in the glycerol conversion from 99% to 92% was achieved after 13 h of time-on-stream testing. Moreover, the characterization of the spent catalyst carried out by Raman spectroscopy, SEM/EDS, and TPO showed the presence of amorphous carbon species. According to the results of the TPO analysis, the authors regenerated the catalyst in air at 500 °C, obtaining a complete recovery of the catalytic performance. Montmorillonite (MMT)-supported nickel nanoparticle (Ni/MMT) catalysts were prepared by Jiang et al. In order to optimize the Ni loading and the calcination temperature in GSR. Stability tests were performed on the three catalysts containing 20% Ni (Ni/MMT molar ratio 1:1) and calcination was performed at different temperatures (1Ni-1MMT-800, 1Ni-1MMT-700, and 1Ni-1MMT-600, calcined at 800, 700, and 600 °C, respectively). The 1Ni-1MMT-700 catalyst showed the best stability among all the tested catalysts, probably due to the better dispersion of the metal obtained with the calcination at 700 °C. Furthermore, comparing the stability of 1Ni-1MMT-700 and 1Ni-1Al-700 under 30 h of TOS, a rapid deactivation of the alumina-based catalyst was found. This result, coupled with the characterization of the spent catalysts, pointed out the different reaction pathways for the two catalysts; indeed, with Al_2O_3 acid, dehydration reactions that lead to coke precursors are more prevalent than dehydrogenation reactions [228]. Ni behavior in the steam reforming of glycerol has also been studied over a novel fly-ash-based catalyst. In the stability evaluation, which was performed at 550 °C with a steam/glycerol ratio of 10 and three different space periods of 3 h for time-on-stream testing, glycerol conversions higher than 90% were obtained [229]. Mono- and bimetallic Ni-based catalysts (Ni-Co, Ni-Cu, and Ni-Zn) supported on attapulgite were submitted in a GSR stability investigation at 600 °C, with a steam/glycerol ratio of 9 and GHSV of 9619 h^{-1}. After 30 h of time-on-stream testing, among all the catalysts, Ni-Zn/ATP exhibited the best performance, with glycerol conversion and H_2 selectivity remaining almost stable over the time period [230]. Carrero et al. [231] evaluated the performance of bimetallic Ni-(Cu, Co, Cr)/SBA-15-supported catalysts; the addition of the second metal favored the dispersion of the Ni phase and the metal–support interactions, thus leading to better performances in GSR, particularly for th2e Ni-Cr/SBA-15 catalyst. This catalyst also showed a lower amount of deposited coke compared with the other bimetallic catalysts and a relatively good stability within 60 h at 600 °C. Bimetallic (Co, Cu, and Fe)-Ni/CNTs catalysts were analyzed in low-temperature GSR (375 °C) tests; a glycerol conversion of ~91% was achieved for Co-Ni/CNTs, which was also characterized by a lower amount of deposited coke (0.05 $mg_C \cdot mg_{cat}^{-1}$). Moreover, the thermogravimetric analysis performed on Co-Ni/CNTs revealed a shift of the coke peak to lower temperatures, pointing out that the Co addition may have increased the formation of amorphous carbon [232]. The influence of the active phase on the stability of the catalysts and the nature and extent of coke deposits was also analyzed in relation to the dry reforming of glycerol (GDR) by Tavanarad et al. [233], who in 2018 prepared and proved various

Ni/γ-alumina catalysts by varying the Ni loading from 5 to 20 wt.%. The first activity screening on the Ni content showed that the best performances were obtained with a content of nickel of 15 wt.%, while a Ni loading of 20 wt.% resulted in worse nickel dispersion and bigger crystalline structures. The 15Ni/γ-Al_2O_3 was subsequently tested at 700 °C under 20 h of time-on-stream; a drastic drop in the catalyst activity was observed in the first 5 h, followed by an almost constant glycerol conversion of ~55% for the next 15 h. TPO and SEM analyses of the catalyst after the stability test evidenced the presence of whisker-type carbonaceous species on its surface. A comparison between the stability of catalysts with different metals as the active phase and the same support specie was investigated by Bac et al. [234] via 72-h time-on-stream stability tests at 750 °C, carried out on Rh, Ni, and Co supported on Al_2O_3-ZrO_2-TiO_2 (AZT) catalysts. Better performances were obtained for Rh/AZT and Ni/AZT. Indeed, for both catalysts the CO_2 conversion was above 91% of the thermodynamic limit, characterized by decreases of only 13% and 8%, respectively. The stability of Rh/AZT was attributed to the lack of sintering and carbon deposits, while Ni/AZT exhibited the ability to post-gasify surface carbon species that were deposited on the catalyst during the first few hours. Moreover, the Co/ATZ catalyst showed a drastic drop in CO_2 conversion, suggesting the presence of severe irreversible carbon formation.

Larimi and co-workers [235] investigated the influence of the Ce/Zr ratio on the performance of a $Pt_{0.05}Ce_xZr_{0.95-x}O_2$ catalyst for the aqueous-phase reforming of glycerol (GAPR), finding a dependence of the reactivity of the catalyst by the metal ratio, which influenced the Pt oxidation state, the active metal dispersion, the surface area, and the particle size. The catalyst showing the best results in term of activity ($Pt_{0.05}Ce_{0.475}Zr_{0.475}O_2$) was subsequently tested for 50 h using TOS at 250 °C and 50 bar, during which no deactivation was detected. A comparison between monometallic Ir and bimetallic Ir-Ni and Ir-Co catalysts was evaluated by Espinosa-Moreno et al. [236], who found that IrNi bimetallic catalysts excelled, probably as a result of the high tendency of Ni to cleave to the C-C bond. Moreover, the IrNi/La_2O_3 sample exhibited improved resistance to deactivation, ascribable to the carbon species removal on Ni sites, which is promoted by the oxygen species formed upon La addition. Bimetallic Ni catalysts also showed less CH_4 and CO selectivity, pointing out that the addition of Ni to Ir reduces the methanation reaction and increases the metal–support interactions, thus leading to higher H_2 concentrations.

4.2. The Role of the Support

The role of the support is also fundamental in improving the stability of the adopted catalysts; the present paragraph discusses the benefits and drawbacks of the choices of various supports for bioglycerol reforming.

As a matter of fact, high dispersion and large surfaces are the mainly factors that enhance reactivity in glycerol reforming reactions nonetheless, support acidity and basicity properties also play key roles on the formation of coke deposits. Other authors [237,238] employed Langmuir–Hinshelwood and Eley–Rideal kinetic models coupled with statistical discrimination and thermodynamic evaluation, and identified a molecular adsorption mechanism for glycerol and steam on both Brønsted acidic and basic sites on the catalysts. Thus, they pointed out the necessity of having acidic sites on the surface to improve the process efficiency. On the other hand, acid supports (Al_2O_3) tend to improve dehydration, dehydrogenation, hydrogenolysis, and condensation reactions, which cause coke deposits, leading to rapid deactivation of catalysts [30]. Charisiou and co-authors, in order to evaluate the influence of the support on the glycerol steam reforming (GSR), performed an investigation on nickel catalysts (8 wt.%) supported on Al_2O_3, ZrO_2, and SiO_2. The results of the 20 h stability test at 600 °C pointed out the behavior of different catalysts related to the acidity or basicity of the three supports. In particular, the catalyst prepared on the acidic alumina (Ni/Al_2O_3) showed drastic deactivation, causing the blockage of the active sites by coke precursors formed on the acidic surface sites. Conversely, the more basic Ni/ZrO_2 catalyst appeared to be less affected by deactivation, which may be attributed to the stronger metal–support interactions observed during the TPR analysis, as well as to the capability of

ZrO_2 to enhance the water adsorption and activation. Lastly, the Ni/SiO_2 catalyst ensured the best outcomes in terms of both glycerol conversion and H_2 selectivity, thus leading to the advantages gained from the neutral nature of the support, exploiting both acidic and basic surface sites. Moreover, in the characterization of the spent catalysts, Ni/Al_2O_3 showed the lowest amount of deposited coke (~40%), while for the other two samples the value was almost equal or higher (~50%). On the other hand, the carbon degree of graphitization has also been estimated, which followed the order Ni/Al_2O_3 > Ni/ZrO_2 > Ni/SiO_2, thus leading to the interconnection between the nature of the deposited coke and the deactivation catalysts [224]. The influence of the support towards Ni-based catalysts in the GSR has also studied by Zamzuri and co-authors [239] through the examination of Ni/Al_2O_3, Ni/La_2O_3, Ni/ZrO_2, Ni/ SiO_2, and Ni/MgO. The TOS stability test carried out at 650 °C for 5 h showed that the catalysts' overall activity increased following the order Ni/Al_2O_3 > Ni/La_2O_3 > Ni/ZrO_2 > Ni/MgO > Ni/ SiO_2. Conversely, while Ni/La_2O_3, Ni/ZrO_2, and Ni/ SiO_2 were found to be almost stable within the time-on-stream tests, Ni/Al_2O_3 and Ni/MgO showed decreased activity. Since Al_2O_3 has attracted major attention as a support for catalysts in GSR, by virtue of its high surface area and mechanical and chemical resistance under reaction conditions, numerous strategies have been investigated in order to limit the coke deposition, which is probably caused by the acidic properties of Al_2O_3. A comparison between Ni catalysts supported on mixed lanthana-alumina and Ni catalysts supported on commercial alumina pointed out that addition of La favored the dispersion of the active phase, increasing the basicity of the support and increasing the metal–support interactions. Moreover, after the stability tests, the comparison between the spent catalysts also showed the different nature of the carbon deposits on the catalysts surface. In fact, even though the amount of deposited carbon was approximately equal (0.41 $g_{coke}{\cdot}g_{catalyst}{}^{-1}$ for Ni/Al and 0.44 $g_{coke}{\cdot}g_{catalyst}{}^{-1}$ for Ni/LaAl), TPO and Raman characterizations showed the presence of more graphitic carbon species on the Ni/Al samples than Ni/LaAl, for which almost amorphous carbon deposits were detected. This result was in agreement with the trends of the stability tests, from which it was possible to highlight the good stability in the case of Ni/LaAl and a drastic drop in the activity of Ni/Al [221]. La-modified Al_2O_3 was also investigated by Sanchez et al. as a support for Ni catalysts in GSR reactions. The stability tests were performed on the bare supports and on the catalysts, involving 4 h time-on-stream tests for Ni/Al_2O_3 and Ni/La_2O_3-Al_2O_3 at 700 °C and WHSV = 35 h^{-1}. The addition of La to the support enhanced the H_2 production, giving a result that was close to that obtained with the Ni catalyst; moreover, while Ni/Al_2O_3 showed a drop in the H_2 production over time, Ni/La_2O_3-Al_2O_3 gave an almost stable result after 90 min of reaction. The TGA analysis carried out on the spent catalysts also showed that the amount of coke deposited on Ni/Al_2O_3 was almost twice that deposited on Ni/La_2O_3-Al_2O_3. Moreover, the SEM analysis revealed that the surface of Ni/Al_2O_3 was covered by carbon filaments, which could be the reason for the catalyst deactivation [30]. Additionally, modifications of the alumina support were investigated by Bobadilla et al., who prepared NiSn bimetallic catalysts supported on Al_2O_3, evaluating the influence of the addition of Mg and Ce on the support. NiSn/Al, NiSn/AlMg, NiSn/AlCe, and NiSn/AlMgCe were submitted to stability tests at 650 °C, P = 1 atm, and with a water/glycerol molar ratio of 12 for 6 h TOS tests. The best results were obtained for the catalyst modified with both CeO_2 and MgO. In fact, the addition of these oxides resulted in a synergic effect that decreased the coke formation and favored the WGS reaction [240]. Furthermore, Charisiou and co-authors [241] performed a comparative analysis between Ni/CaO-MgO-Al_2O_3 and Ni/Al_2O_3 catalysts, which showed how the addition of CaO-MgO led to smaller Ni particles, increased the basicity and the surface amount of the Ni^0 phase, thus resulting in better GSR results. The activity and stability tests carried on the two catalysts showed a different distribution of products, as the modified support favored H_2 and CO_2 production (enhancement of the WGS reaction). Moreover, this proved that the deactivation can be prevented with the addition of CaO and MgO to the support. In addition, the modified support also led to a lower amount of coke being deposited on the catalyst's surface, altering its nature, which based on the Raman analysis appeared to be less graphitic than that deposited on the unmodified Ni/Al_2O_3 catalyst. A study concerning the support roles for Pt-based catalysts in the GSR was carried out by de Rezende et al. [28]. Several layered double-hydroxides containing Mg and Al (Mg/Al ratios of 3 and 5) as supports were tested. Specifically, four catalysts were prepared, two containing hydrated layered supports, PtMg5Al-H and PtMg3Al-H; and two containing mixed calcined oxides, PtMg5Al-O and PtMg3Al-O. Subsequently, the catalysts were submitted to GSR time-on-stream stability tests at 600 °C for 250 min, during which PtMg5Al-H showed a stable conversion rate of ~85%, while all the other catalysts exhibited losses in activity. Moreover, the authors presented a correlation between the amount of carbon deposited on the catalysts and the degree of deactivation, thus pointing out the minor contribution of carbonaceous materials derived from intermediate organic liquid compounds on the deactivation processes. This study is in contrast with what is usually found in the literature, as reaction intermediates adsorbed on the catalysts are generally considered responsible for deactivation of catalysts. An interesting study evaluated the effect of the support on Co-based catalysts adopting alumina, niobia, and 10 wt.% niobia/alumina. The three prepared catalysts containing 20 wt.% CoO were tested under 30 h of GSR reaction at 500 °C with GHSV = 200,000 h^{-1}. In the first 8 h of reactions, complete glycerol conversion was obtained for CoNb, while the rate was ~90% for CoNbAl and ~80% for CoAl. Subsequently, after 24–26 h, a strong deactivation trend was detected for all catalysts, especially for CoNb. Indeed, from the analysis of the product distribution as a function of time, it was evaluated that even though the conversion was higher for CoNb, the gas production was low, thus indicating the probability of coke formation during the process. The catalyst that showed the best stability during the reaction was CoAl, with a decrease observed only after 26 h; these results were probably associated with the higher cobalt dispersion obtained for this catalyst. However, the catalyst supported by both Nb and Al presented the best values for H_2 production during the first 8 h of reaction. On the contrary, CoAl showed reduced hydrogen formation and was also characterized by the lowest amount of deposited coke, thus suggesting a coke formation mechanism based on the hydrogenation reaction of CO and CO_2. Conversely, looking at the H_2 production results, the authors assumed that the coke formed on CoNb and CoNb catalysts was mainly caused by the CO disproportionation, thus excluding hydrogen. A further analysis of the nature of the formed carbon revealed more amorphous coke deposited on CoNb and CoNbAl, while the deposits on the CoAl catalyst were predominantly graphitic [242]. The support effect on the stability of the catalysts was also investigated for glycerol dry reforming (GDR). In particular, a comparison between Ni catalysts supported on CaO and ZrO_2 pointed out the differences in the two support species. With this purpose, several catalysts were prepared by wet impregnation on both CaO and ZrO_2, increasing the amount of Ni (5%, 10% and 15%); moreover, the bare supports and the catalysts were also characterized by means of BET surface area determination tests, SEM analysis, XRD technique tests, and TGA and TPR analysis. Afterwards, the prepared catalysts were tested by the authors in stability experiments, which were carried out in a fixed-bed reactor at 700 °C and at atmospheric pressure over 3 h in time-on-streams. This resulted in better performances for the 15%Ni/CaO catalyst, characterized by a glycerol conversion rate of 30.52% and a hydrogen yield rate of 23.06%. For the other tested catalysts, the performances decreased in the order of 10% Ni/ ZrO_2 > 5% Ni/ CaO > 10% Ni/ CaO > 15% Ni/ ZrO_2 > 5% Ni/ ZrO_2. The better outcomes achieved for the 15%Ni/CaO were found to be in agreement with the characterization results, which proved better interactions between Ni and CaO, better metal dispersion, and greater surface area results in comparison with Ni/ZrO_2, also resulting in smaller crystallite sizes for NiO species [243]. Lee et al. [244], through a comparison of Ni-supported catalysts (Ni/$LaAlO_3$, Ni/CeO_2, Ni/MgO, and Ni/MgAl) regarding glycerol aqueous-phase reforming (GAPR), highlighted the strong dependency of the glycerol conversion and hydrogen yield on the type of support and the Ni loading. The support influence was investigated by the authors on 15 wt.% Ni catalysts supported on the four abovementioned species. In the activity tests at 250 °C and 20 bar, with LHSV = 5 h^{-1}, the catalyst 15%Ni/$LaAlO_3$ showed the best outcomes, with a glycerol conversion rate of 35.8% and a H_2 selectivity rate of 67.2%. Next, under the same conditions, four $LaAlO_3$-supported catalysts with 5, 10, 15, and 20 wt.% Ni were tested, confirming that the best performances were obtained with the 15% Ni. 15%Ni/ $LaAlO_3$ product. The selectivity was also investigated with a 20 h TOS test, during which high CO_2 and H_2 yields were obtained, while low CO selectivity was achieved due to the WGS reaction. Moreover, the analysis of the spent catalysts revealed the presence of some agglomerated carbon particles on the surfaces of Ni/MgAl, Ni/CeO_2, and Ni/MgO. Conversely, the amount of deposited carbon was small and it was in a fibrous form in the case of the Ni/ $LaAlO_3$ catalyst, thus highlighting the high resistance to coking of the catalyst, which was probably related to the migration of mobile oxygen from the support ($LaAlO_3$) to the metallic Ni particles [244]. In the study by Espinosa-Moreno and co-authors on Ir-based catalysts for H_2 production through GAPR, a comparison was carried out between La_2O_3 and CeO_2 as supports. The catalytic tests performed at 270 °C and 58 bar showed better values for H_2 production with La_2O_3 catalysts compared to CeO_2-supported catalysts due to the higher pore volume and metal–support interactions obtained using lantana [236]. As can be appreciated in Figure 19, comparing the performances of IrNi/La_2O_3 and IrNi/CeO_2, much higher carbon-to-gas conversion rates were obtained when suing lantana as the support.

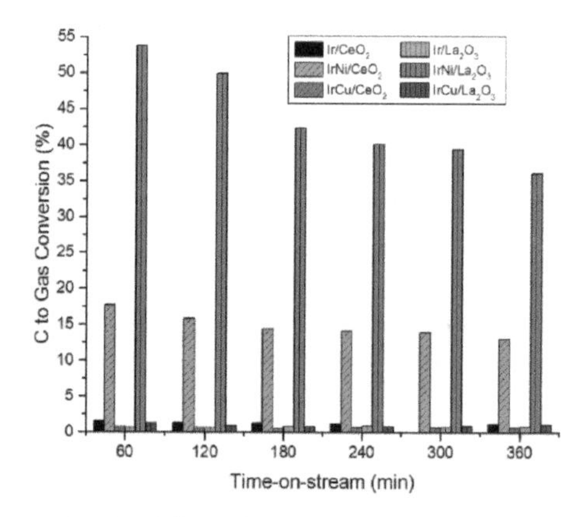

Figure 19. Carbon-to-gas conversion (%) as a function of the time-on-stream (min) stability test carried out at 270 °C and 58 bar [236].

4.3. *The Effect of the Addition of Promoters*

In this paragraph, the effects of promoters on the stability of various catalysts for glycerol reforming is discussed. In fact, a further strategy to inhibit catalyst deactivation is the adoption of additives and promoters, in order to create specific surface centers, which are mainly acidic or basic, bimetallic, or redox sites [245]. The effect of the addition of promoters to Ni/Al_2O_3 for the GSR has been widely studied in literature. Lima et al. investigated the consequences of the addition of Mg, finding an increase in the Ni dispersion and in its resistance to sintering and a decrease of the catalysts' acidity, thus promoting the stability of the catalysts. The addition of Mg also resulted in a lower amount of deposited coke, which was more graphitic [246]. A supplementary investigation on the effect of the addition of Mg to Ni/Al_2O_3 confirmed that its addition promotes the basicity of the catalyst and results in higher Ni dispersion, thus favoring the catalytic performance and lowering the carbon formation. Indeed, Dieuzeide et al. [247] pointed out that with an optimal loading of Mg of 3 wt.%, the amount of coke obtained was approximately half of the amount obtained without Mg. In addition, correlations between the structure of the deposited carbon, the Ni particle size, and the Mg presence were evaluated, since it was previously observed that a decrease in the Ni particle size and an increase in the Mg loading resulted in an increase of the degree of graphitization of the formed coke. Demsash et al. [248] evaluated the effect of ceria-promoted Ni-alumina catalysts in GSR by varying the amounts of nickel and ceria (5, 10, and 15 wt.% Ni, and 5 and 10 wt.% Ce). Among the prepared catalysts, 10Ni/Al_2O_3/5CeO_2 exhibited the best results in terms of activity and stability, with product distribution remaining almost stable over the 16 h TOS test at 650 °C. The Ce-promoted catalyst also showed the lowest coking activity, as a result of the inhibition effect of ceria to the formation of coke deposits; indeed, CeO_2 promotes the hydration of the support. Carrero et al. [231] investigated the addition of Cu, Co, and Cr to a Ni/SBA-15 catalyst for GSR applications at 600 °C, involving a steam/carbon ratio of 2 and a WHSV of 7.7 h^{-1} for a 5 h TOS test. During the stability tests, all the catalysts showed glycerol conversion rates above 85 mol%. In particular, glycerol conversion decreased following the order of Ni-Cr/SBA-15 > Ni-Co/SBA-15 > Ni/SBA-15 > Ni-Cu/SBA-15 along the reaction time, as can been seen in Figure 20.

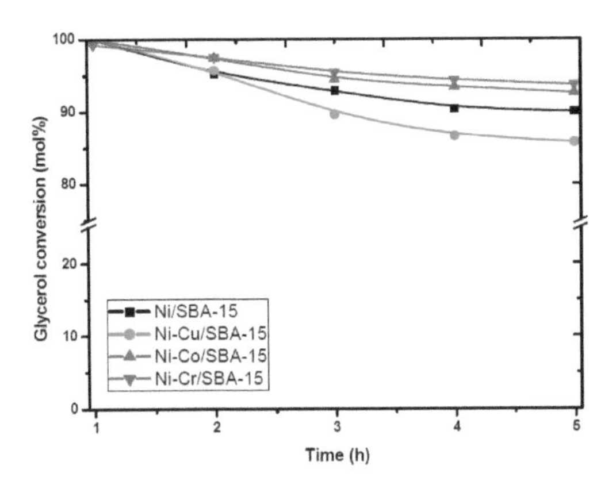

Figure 20. Conversion of glycerol over the Ni-(Cu,Co,Cr)/SBA-15 catalysts (T = 600 °C, P = 1 bar, $WHSV_{glycerol}$ = 7.7 h^{-1}, water/glycerol = 6 mol/mol) [231].

Catalyst deactivation has been shown to be correlated with the amount of coke deposited, increasing in the order of Ni-Cr/SBA-15 < Ni-Co/SBA-15 < Ni/SBA-15 < Ni-Cu/SBA-15. Focusing attention on the product distribution during the reaction with the Ni-Cu/SBA-15 samples, the catalyst characterized by the highest deactivation and largest amount of coke deposition also enhanced the formation of methane and carbon monoxide, while lower amounts of carbon dioxide and hydrogen were produced. Moreover, for the Ni-Cr/SBA-15 samples, the highest hydrogen concentrations were detected in the gas stream. Regarding Ni-Co/SBA-15, despite the high glycerol conversion values obtained, low H_2 production was achieved—even lower than Ni/SBA-15, possibly due to the lower ability of Co than Ni to break the C-C bond. SEM images of the catalysts pointed out the presence of carbon nanofibers on Ni-Cu/SBA-15 and Ni-Co/SBA-15, while no formation of carbon nanofibers was detected by the SEM analysis on Ni-Cr/SBA-15. Subsequently, Ni-Cr/SBA-15 has been tested with a 60 h TOS protocol, which revealed good stability, with a glycerol conversion rate above 93% throughout the time period. Prior to investigating the promotion of Ni catalysts, the same authors evaluated the influence of additives on Co-catalysts in GSR. The study was focused on the comparison between Co/SBA-15 and promoted Co-M/SBA-15 with M/Zr, Ce, or La catalysts. The tests, which were carried out at 500 and 600 °C, with a water/glycerol ratio of 6 and a WHSV of 7.7 h^{-1}, showed that the addition of promoters significantly improved the stability of Co/SBA-15, especially at 500 °C, where the non-promoted catalyst suffered a drastic deactivation (from 98% to 75%) in the first 2 h. Among the promoted catalysts, Co/Ce/SBA-15 ensured the best results, with a glycerol conversion rate of ~100% during the 5 h time-on-stream test; moreover, despite the higher activity of this catalyst, the amount of coke deposited on its surface was lower compared to that observed for Co/SBA-15 (0.025 $g_{coke} \cdot g_{cat} \cdot h^{-1}$ for Co/Ce/SBA-15 and 0.028 $g_{coke} \cdot g_{cat} \cdot h^{-1}$ for Co/SBA-15). The products distribution analysis also pointed out that the use of Co/Ce/SBA-15 led to higher CO_2 and H_2 concentrations, while the CO formation was partially suppressed, highlighting the tendency of ceria to improve the oxygen mobility and water reactivity, enhancing the water–gas shift reaction. On the other hand, the addition of La and Zr resulted in lower hydrogen production; furthermore, the presence of ethylene was detected on Co/Zr/SBA-15, possibly caused by the enhancement of the dehydration reactions on the zirconium acid sites. Co/Ce/SBA-15 was then tested with a 50 h TOS test, during which high stability was reached (glycerol conversion rate of ~100%). However, the amount produced hydrogen decreased, while the methane concentration increased. Moreover, the amount of coke was found to be higher and the authors suggested the possibility of a secondary reaction, such as $C+2H_2 \leftrightarrow CH_4$ [249]. The advantages gained by the addition of Mg to Ni catalysts were verified by Veiga and co-workers, who studied glycerol steam reforming with nickel supported by activated carbon. During stability tests at 650 °C, Mg-promoted catalysts, among other promoters (MgO, La_2O_3, Y_2O_3), showed higher initial conversion and lower deactivation rates as a result of the capability of Mg to enhance the steam adsorption and stabilize Ni against sintering [250]. Other strategies used to improve the stability of catalysts for glycerol steam reforming involved the development of new types of catalysts and their subsequent doping with additives. Nickel, as a promoter for Fe/Mg-containing metallurgical waste, was prepared via solid-state impregnation of Ni into the structure of the metallurgical residue. Its stability was evaluated at 580 °C and 1 atm with a 48 h of TOS test. The developed catalyst exhibited good stability, with a glycerol conversion rate of ~90% and a hydrogen yield rate of ~80%; moreover, the amount of coke formation was low (2.7 $mg_{coke}\ g_{cat}^{-1}\ h^{-1}$) and its nature was filamentous. This study also evidenced the ability of Mg to activate steam and promote the water–gas shift reaction [251]. An easily reducible $NiAl_2O_4$ spinel was developed using a novel method from a Ni-Al mixed-metal alkoxide, which was tested for use in GSR and was shown to have a highly porous structure and surface area. Moreover, the effect of the addition of 10 wt.% CeO_2 the catalyst stability was evaluated. The TOS tests performed at 630 °C and 1 atm for 16 h showed the high potential of the catalyst, characterized by low coke formation (0.0004 $g_{coke}\ g_{cat}^{-1}\ h^{-1}$). Indeed, the addition of ceria diminished the coke deposits and favored their gasification. Additionally, the formation of a well-dispersed $CeAlO_3$ phase hindered the growth of filamentous carbon [245]. The role of CeO_2 as a promoter was also studied by Dobosz et al. [252], who evaluated its addition to a calcium hydroxyapatite (HAp)-supported cobalt catalyst. During the TOS stability test, which was carried out at 800 °C for 6 h, the catalyst doped with ceria (10Co-Ce/Hap) was stable than the 10Co/Hap catalyst. In fact, even if it a decrease in the hydrogen selectivity was observed, this effect was slower than the trend observed for the undoped catalyst. Ramesh et al. focused their attention on the influence of copper on perovskite catalysts for GSR, comparing a perovskite catalyst with two copper decorated perovskites ($LaNi_{0.9}Cu_{0.1}O_3$ and $LaNi_{0.5}Cu_{0.5}O_3$). Comparing their activity, better results were obtained for $LaNi_{0.9}Cu_{0.1}O_3$, which had a glycerol conversion rate of 73% and a hydrogen selectivity rate of 67%. The stability of the 0.1%-copper-decorated perovskite was confirmed by tests carried out at 650 °C, S/C = 3, and LHSV = 10,000 h^{-1}, during which the catalyst exhibited an almost constant product distribution over the 24 h TOS test. The characterization of the spent catalysts showed the presence of small amounts of graphitic carbon in the perovskites without copper (detected by TGA analysis, in which two oxidation peaks were observed at 550 and 600 °C). Conversely, the addition of copper resulted in only one oxidation peak, suggesting the presence of only amorphous carbon on the surface of the spent catalyst [33]. The Ni/Al_2O_3 catalyst and the improvements obtained with the addition of promoters were also studied in the development of the glycerol dry reforming (GDR) process. Harun and co-authors evaluated the effect of the Ag promotion via the preparation and testing of α-alumina-supported catalysts containing 15% Ni; in particular, four catalysts with Ag loading rates of 0 wt.%, 1 wt.%, 3 wt.%, and 5 wt.% were tested at 700 °C for 3 h, with a glycerol/carbon dioxide feed ratio of 1:1. The results showed that even if similar trends were shown by all of the catalysts (increase of the glycerol conversion in the first 0.5 h followed by a decrease after 1 h and a final stabilization), among the investigated catalysts, Ag(3)-Ni/Al_2O_3 gave the best glycerol conversion result (33.5%). Hence, this catalyst was tested with a 72 h stability test at 800 °C, during which the glycerol conversion firstly decreased from the 46% to the 33% after 10 h, and then remained almost stable for the next 60 h. After 10 h of reaction, H_2 and CO concentrations increased, while CH_4 decreased; the authors suggested that coke gasification by water may have been responsible. TGA analysis performed on the spent catalyst highlighted the presence of low- and high-temperature oxidation peaks, suggesting the presence of both amorphous and graphitic carbon deposits, as was further confirmed by the SEM images, in which it was possible to appreciate the presence of encapsulated solid carbon deposits and filamentous (whisker-like) carbon deposits [31]. The addition of lanthanum to Ni-based catalysts was also investigated in the GDR. A 3wt.% La-promoted Ni/Al_2O_3 in comparison with the unpromoted catalyst led to better metal dispersion, as evaluated by the increase in the specific surface area and the decrease of the crystallite size (La-promoted catalyst: BET = 96 m^2 g^{-1}, crystallite size = 9.1 nm; unpromoted catalyst: BET = 85 m^2 g^{-1}, crystallite size = 12.8 nm). The catalysts stability was tested with a 72-h time-on-stream test at 750 °C and 1 atm with WHSV = $3.6{\cdot}10^{-4\ mL}$ g^{-1} h^{-1}, during which no severe deactivation was encountered [253]. Regarding the aqueous-phase reforming of glycerol (GAPR), Reynoso et al. compared two $Pt/CoAl_2O_4$ catalysts (Pt loadings of 0.3 wt.% and 1 wt.%) and monometallic Pt/alumina and Co/alumina in 100 h time-on-stream experiments. The results of the tests, which were performed at 260 °C and 50 bar with WHSV = 0.68 h^{-1}, pointed out that while for the bimetallic catalysts the glycerol conversion was high (above 99%) and stable during time-on-stream testing, the monometallic catalysts suffered from deactivation. Similar outcomes were also obtained in terms of the conversion of carbon to gas, which was almost stable at 95% for the bimetallic catalysts, while CoAl and PtAl showed decreases of 36% and 30%, respectively. Raman spectroscopy and temperature-programmed hydrogenation (TPH) were used to characterize the spent catalysts. from the Raman spectra, the presence of ordered graphite-like structures on the monometallic catalysts was evaluated, while no carbonaceous deposits were encountered on the bimetallic formulations. Through the TPH analysis, the presence of a low temperature peak (T = 200 °C, ascribable to the hydrogenation of the most amorphous carbon) was detected on the Pt-containing samples, suggesting that the hydrogenation of carbon deposits could happen in the proximity of the well-dispersed PT centers as a consequence of the spillover effect. However, the CoAl catalyst showed a higher amount of deposited carbon (13.5 $\mu mol_C{\cdot}g_{cat}{}^{-1}$), while for Pt-containing samples the values were much lower (1.2–1.9 $\mu mol_C{\cdot}g_{cat}{}^{-1}$) [254]. Furthermore, Pendem et al. [255] investigated the addition of potassium to hydrotalcite (Pt-KHT) catalysts for GAPR, finding that the K promotion increased the basicity of the catalyst and improved the hydrogen production. The results of the stability test carried out at 250 °C showed a hydrogen selectivity of 67.4% after 3 h of reaction.

Table 3 provides the values for thee carbon formation rate obtained from the glycerol steam reforming stability tests performed in the various considered studies. An analysis of the tendency to form coke pointed out that both the active phase and the support play key roles in the final amount of deposited carbon. Among the metals used for active phases, Ni showed better stability; indeed, all the catalytic formulations based on Ni showed lower amounts of carbonaceous deposits. In particular, the best carbon formation rates were seen for the $10Ni/Al_2O_3$ and $10Ni/Al_2O_3/5CeO_2$ catalysts tested by Demshash et al. (0.00067 and 0.000424 $g_{coke}{\cdot}g_{catalyst}{}^{-1}{\cdot}g_{carbon,fed}{}^{-1}{\cdot}h^{-1}$, respectively) [248]. Moreover, the comparison between these two catalysts also highlighted that the support modification obtained using CeO_2 can lead to a further performance increase for the catalysts. However, the eventual use of promoters in the preparation of catalysts is also of fundamental importance. Comparing the performances of the catalysts prepared by Carrero et al. [231,249], it is possible to observe that for SBA-15, the catalyst prepared using Ni as the active phase was characterized by high rates of carbon formation; conversely, the addition of Co and Cr resulted in an improvement of the obtained results. Indeed, from the investigation of the stability tests performed on Ni-Co/SBA-15 and Co/SBA-15 (stability tests were carried out at 600 °C and WHSV = 7 h^{-1}), it seems that the Ni did not affect the formation of coke, as the carbon formation rate values for the two catalysts were equal. In addition, among all the catalysts tested by the authors, the lowest carbon formation rates were obtained with a catalyst based on Ni (Ni-Cr/SBA-15 carbon formation rate = 3.009 $g_{coke}{\cdot}g_{catalyst}{}^{-1}{\cdot}g_{carbon,fed}{}^{-1}{\cdot}h^{-1}$), thus confirming the better results obtained for Ni catalysts for glycerol steam reforming.

Table 3. Carbon formation rate for various catalysts used for glycerol steam reforming.

Catalyst [a]	Operative Conditions [b]	T (°C)	WHSV [c] (h^{-1})	X Glycerol (%)	Carbon Formation Rate [d] (Multiplied for 1000)	Ref.
Glycerol steam reforming						
$8Ni/Al_2O_3$ $8Ni/4La_2O_3$-Al_2O_3	m_{cat} = 200 mg 31 v.v. % $C_3H_8O_3$ and H_2O (63% H_2O, 7% $C_3H_8O_3$ and 30% He) TOS = 4 h	700	GHSV = 50,000 mL $g^{-1}h^{-1}$	~75 ~80 After 4 h	0.141 0.152	[221]
Rh/alumina	0.1 mL min^{-1} of aqueous glycerol, P = 4.5 bar m_{cat} = 800 mg TOS = 13 h	400	7.8	~92 After 13 h	0.045	[227]
14.5Ni/SBA-15 14.5Ni-4Co/SBA-15 14.3Ni-3.6Cr/SBA-15 15Ni-4Cu/SBA-15	S/C = 2 Water/glycerol = 6 mol/mol m_{cat} = 300 mg TOS = 5 h	600	7.7	~92 ~94 ~95 ~88 After 5 h	27.879 5.092 3.009 51.157	[231]
$8Ni/Al_2O_3$ 8Ni/CaO-MgO-Al_2O_3	m_{cat} = 200 mg 31 v.v. % $C_3H_8O_3$ and H_2O (63% H_2O, 7% $C_3H_8O_3$ and 30% He) TOS = 20 h	600	GHSV = 50,000 mL $g^{-1}h^{-1}$	~70 ~80 After 20 h	0.057 0.048	[241]
10% CeO_2 addition to $NiAl_2O_4$ spinel	m_{cat} = 500 mg Water/glycerol = 9 (glycerol solution)/Ar = 1 TOS = 16 h	630	GHSV = 19600 $cm^3g_{cat}{}^{-1}h^{-1}$	90 After 16 h	0.16	[245]
$10Ni/Al_2O_3$ $10Ni/Al_2O_3/5CeO_2$	m_{cat} = 1 g 30 wt.% glycerol feed TOS = 16 h	650	12	-	0.00067 0.000424	[248]
7Co/SBA-15 7Co-8.5Zr/SBA-15 7Co-8.5Ce/SBA-15 7Co-8.5La/SBA-15	S/C = 2 Water/glycerol = 6 mol/mol m_{cat} = 300 mg TOS = 5 h	600	7.7	~75 >90 >90 >90 After 5 h	5.092 5.555 5.555 4.629	[249]
12.5Ni-UGS	m_{cat} = 500 mg S/C = 3 Water/glycerol = 9 (Water+glycerol)/Ar = 1:4 TOS = 48 h	580	GHSV = 20600 $cm^3g_{cat}{}^{-1}h^{-1}$	90 After 48 h	0.17	[251]

Note: [a] The metal or oxides loadings are intended for the weight of the catalyst; [b] refers to the glycerol mass flow rate; [c] $g_{coke}\cdot g_{catalyst}{}^{-1}\cdot g_{carbon,fed}{}^{-1}\cdot h^{-1}$.

5. Other Bioalcohol Reforming

In recent years, hydrogen production through reforming processes has been widely studied. In addition to the most investigated biosources, other bioalcohols have been explored as models for hydroxyl-bearing oxygenates. Indeed, research towards the reforming processes of model compounds is a powerful tool that can be used to approach the more complex implementation of bio-oil for H_2 production [256]. The paucity of available literature concerning reforming of butanol (C_4H_9OH) and propanol (C_3H_7OH), which are higher molecular weight alcohols and minor constituents of bio-oil in comparison with methanol or ethanol, makes this an interesting topic. Indeed, as reported in Figure 21, numerous advantages are gained through the use of butanol as a biohydrogen source [257,258]. Along with the other alcohols, reforming of butanol and propanol are processes in which the use of catalysts plays a key role in reactivity regarding the complete conversion, hydrogen yield, and stability. Indeed, different catalysts may induce different reaction pathways; therefore, the selection of a proper catalytic formulation is of prime importance in order to reduce the promotion of undesired by-products and to inhibit coke formation. The mechanisms that lead to coke formation in the reforming processes of the two investigated alcohols are different. Butanol reforming proceeds through numerous pathways; direct reforming and the formation of 1-butene, butyraldehyde, and coke deposits are mainly linked to butyraldehyde rather than 1-butene [259,260]. Instead, regarding the coke formation in the reforming process of propanol, the dehydration and dehydrogenation reactions lead to the formation of propene and propanal, which can subsequently decompose to CO and ethane or condense to form heavier compounds, leading to catalyst deactivation [256].

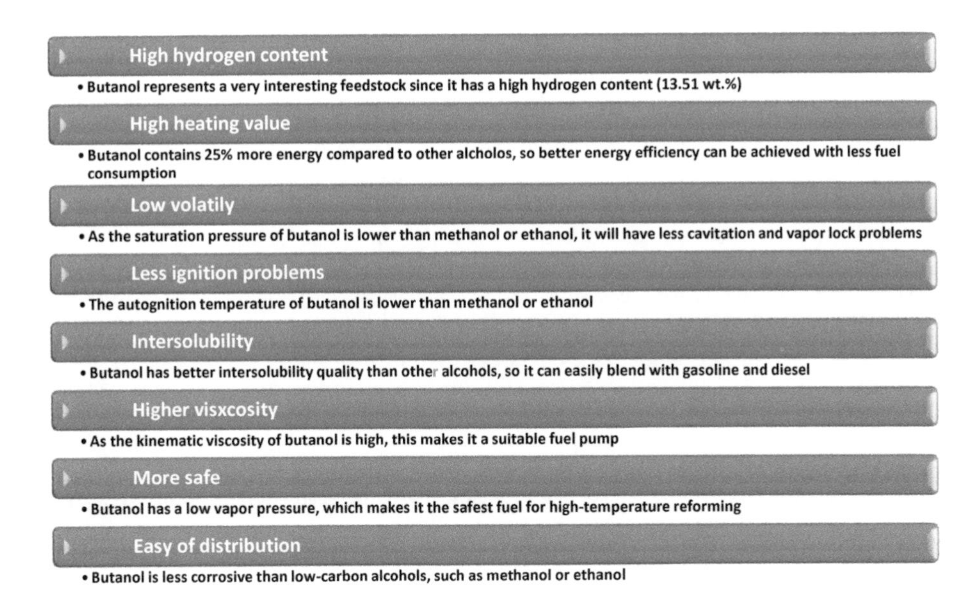

Figure 21. Advantages of using butanol [259].

The metal phase ability to activate the alcohol, the role of the support in the dispersion of the active phase and the eventual addition of promoters are crucial aspects to be taken into account in order to disadvantage the formation of coke, thus promoting catalytic stability [259].

5.1. The Influence of the Active Phase

This section focuses on the roles of active species selection in catalytic performance for butanol reforming; the stability of nickel-, ruthenium-, and platinum-based catalysts is discussed below.

The activity of nickel, as the most-used metal for reforming catalysts, has also been investigated in the steam reforming of butanol. Dhanala et al. [261] performed a comparative study of steam reforming (SR) and oxidative steam reforming (OSRB) of isobutanol on Ni/γ-Al_2O_3 catalysts with different nickel loadings. Stability tests were carried out by the authors on the 30NiAl sample for the OSRB at 600 °C with WHSV = 7.6 h^{-1} for 12 h of time-on-stream testing, during which 100% carbon conversion to gaseous products persisted over time. The products distributions after the first 2 h of testing reached steady-state values, with a H_2 yield rate of ~60%, confirming the stability of the tested catalyst. A further investigation on the spent catalysts showed the presence of complex criss-crossed carbon nanofiber networks on Ni/γ-Al_2O_3 catalysts for both SR and OSRB, probably caused by the decomposition of carbonaceous compounds. The authors also tested catalysts with different Ni loadings (in particular 30NiAl and 20NiAl) and the XRD analysis revealed that 30NiAl has a larger average crystallite size than 20NiAl. Moreover, the further characterization of the two spent catalysts pointed out a correlation between the extent of carbon deposits and the nickel crystallite sizes, while the carbon nanofiber networks over 30NiAl were found to be denser and bigger than 20NiAl. In addition, the O_2 addition to the common steam reforming led to less and smaller carbon nanofibers for the OSRB compared to SR. An investigation of Ni catalysts with various particle sizes (3.6–8.8 nm) was also carried out by Wang et al. [262] through the preparation and testing of mesoporous MgO-supported Ni catalysts for the butanol steam reforming (BSR). The different nickel particle sizes were achieved by the addition of a specific amount of nickel precursor. The as-prepared catalysts with different Ni loadings were denoted as Ni_xO/MgO, with x = 0.04, 0.08, 0.12, 0.15, and 0.20. From the activity tests, $Ni_{0.12}O$/MgO (Ni size of 5.62 nm, as assessed by TEM analysis) showed better BSR results; thus, the catalyst was also submitted to a stability test for 20 h at 450 °C (S/C = 20, liquid flow rate = 5.5mL·h^{-1}). During the first 10 h of reaction, butanol conversion declined from 95.2% to 92.1% and the H_2 selectivity declined from 68.1% to 64.9%. Then, in the following 10 h, their values remained almost unchanged; moreover, the tests on the spent catalyst showed no severe coke deposition. A comparison between the activity of Ni and Co was carried out by Yadav and co-authors through the investigation of two multiwalled carbon-nanotube-supported catalysts. From the activity tests, Ni/CNT showed higher butanol conversion and hydrogen yield values for all the investigated temperatures (500, 450, 400, and 350 °C); consequently, the catalyst was tested with 20 h of time-on-stream testing at 500 °C (W/F_{A0} = 4.2 g h·mol^{-1}). Only slight decreases in results were observed (butanol conversion decreased from 63.6% to 59.6% and H_2 yield decreased from 0.63 mol/mol to 0.57 mol/mol) [263]. A comprehensive study on the deactivation mechanism of a 0.5 wt.% Rh/ZrO_2 catalyst involved in steam reforming and autothermal steam reforming (ATRB) of butanol was performed by Harju et al. [264]. Butanol conversion on the catalyst and the bare support was investigated by the authors for both processes at 500, 600, and 700 °C (S/C molar ratio of 4 and O_2/C molar ratio of 0.1). The results showed the highest performance at 700 °C, with an initial complete conversion for both SR and ATR on the Rh/ZrO_2 catalyst and the bare support. However, while the conversion of the catalyst remained almost constant for 23 h, butanol conversion with ZrO_2 collapsed within 15 min. The authors suggested that the deactivation of Rh/ZrO_2 was mainly caused by the coke deposits on and near the Rh nanoparticles as a result of butyraldehyde and condensation product formation. Moreover, the deactivation of the catalyst follows a different route depending on the operative temperature; while at 700 °C the formation of coke precursors is mainly caused by gas-phase reactions, at lower temperatures the formation of coke deposits takes place mostly through reactions on the catalysts' surfaces. Furthermore, Harju et al. [265] performed a study on the same type of catalyst (Rh/ZrO_2) used in the aqueous-phase reforming of n-butanol. Their attention was focused on the effect of the variation of the support particle sizes. The ZrO_2 support was crushed and sieved in order to obtain three catalysts with particle sizes in the range of 40–60, 60–100, and 250–420 μm. The stability tests, carried out at 220 °C and 35 bar with LHSV = 150 h^{-1} for 4 h TOS, pointed out that the catalyst deactivation was faster for large support particles, due to the high concentrations of butyraldehyde inside the catalyst particles, which favored the formation of deposits caused by aldol condensation reactions. However, the catalysts with smaller support particle sizes (4–60 μm and 60–100 μm) also showed decreases in the butanol conversion, albeit with lower values. Regarding the selectivity towards gaseous products, the values were high and almost constant for 40–60-μm and 60–100-μm Rh/ZrO_2 catalysts (from 99% to 97% in 2.5 h), whereas for the catalyst with the larger particle size this dropped off severely over time. Moreover, the elemental analysis of the spent catalyst revealed that the formation of coke deposits increased as the particle sizes of the catalysts increased. Yadav et al. [266] compared the activity and stability of two noble metal catalysts, Pt/Al_2O_3 and Pd/ Al_2O_3, for butanol steam reforming. The 20-h time-on-stream investigations conducted at 500 °C exhibited a decrease from 82.5% to 74.1% for butanol conversion with Pt/Al, while for the Pd/Al catalyst the conversion decreased from 80.5% to 74.1%. The H_2 yield decreased from 0.37 to 0.28 mol/mol for Pt/Al and from 0.3 to 0.24 mol/mol for the PdAl catalyst, thus pointing out the good stability of both noble-metal-based catalysts. Wang et al. [256] performed a study on n-propanol steam reforming with ruthenium and ruthenium-nickel bimetallic catalysts supported on ceria-alumina oxides, analyzing the influence of the variation of the loading of nickel, loading of ceria, and the calcination of the ruthenium precursor. The catalysts were denoted as AxCyN3Rc or AxCyN3Rnc, where "nc" was used if the ruthenium precursor not calcined and "c" was used if it was calcined; moreover, a stood for the alumina carrier, x was the wt.% loading of the ceria, and y was the weight percent loading of nickel. In this section, the contributions of the calcination of Ru and Ni loading will be discussed, while the effect of the addition of ceria will be discussed in the following section. The investigation of the calcination of the Ru precursor pointed out that poor ruthenium dispersion is achieved after the calcination of the precursor salt, worsening the steam reforming results. On the other hand, from the tests carried out on catalysts with different Ni loading values, it was seen that the increase of Ni from the 2.5% to 10% resulted in a better n-propanol SR activity due to the capability of Ni to activate the n-propanol molecules. Moreover, from the comparison between the catalysts with and without Ni, a lower H_2 formation rate and CO selectivity were found in the absence of Ni, while CH_4 was higher, suggesting that Ru is less active than Ni in the conversion of methane.

5.2. The Role of the Support

Various supports were selected for butanol reforming. This section describes the effect of the support choice on catalyst stability for butanol reforming.

Due to its high surface area, alumina has been widely adopted as a support for reforming catalysts; however, it has also been documented that the acidic nature of such a support may result in an enhancement of coke deposits, mainly caused by dehydration reactions [259]. Bikzarra et al. [267] investigated the effect of Al_2O_3 modifications by CeO_2, La_2O_3, and MgO for Ni-based catalysts, since their basicity could moderate the acid properties of alumina. Moreover, CeO_2 and La_2O_3 were also adopted to improve the metal dispersion and prevent sintering, while MgO improved the catalysts' strength and enhanced the steam adsorption. Hence, five catalysts were prepared (Ni/Al_2O_3, Ni/CeO_2-Al_2O_3, Ni/La_2O_3-Al_2O_3, Ni/MgO-Al_2O_3) and tested in the n-butanol SR at three temperatures (600, 700, and 800 °C), with a S/C ratio of 5. Among all the catalysts, Ni/CeO_2-Al_2O_3 showed the highest performance for all the investigated temperatures, reaching equilibrium hydrogen yields at 700 and 800 °C, while Ni/MgO-Al_2O_3 exhibited the lowest hydrogen yield in all the performed tests. The adoption of Ni/ La_2O_3-Al_2O_3 resulted in experimental results that were much lower than those predicted by thermodynamic equilibrium. Additionally, the CeO_2-supported catalysts also led to high CO_2 selectivity values being obtained, suggesting the enhancement of the WGS reaction. On the other hand, reduced CO_2 selectivity values were obtained with the adoption of Ni/MgO-Al_2O_3, along with high C_2, C_3, and C_4 selectivity. Unfortunately, both CeO_2- and MgO-supported catalysts showed deactivation; for Ni/CeO_2-Al_2O_3, decreases in the H_2 yield rate and CO_2 selectivity were observed by the authors. XPS analysis on the spent catalysts showed different carbon species for investigated samples, covering 62% to 93% of the total area. Lobo and co-authors [268] evaluated the support modification effects for the aqueous steam reforming of 1-propanol. Three Pt/alumina catalysts were prepared through atomic layer deposition, obtaining ~1-nm layer of Al_2O_3, TiO_2, or CeO_2 (Pt-Al, Pt-Ti, Pt-Ce), which was tested at 230–260 °C and 69 bar in the presence of liquid water. All the catalysts demonstrated good activity for 1-propanol conversion; the results decreased in the order of Pt-Ti > Pt-Al > Pt-Ce. The authors suggested that the TiO_2 coating enhanced the rate of the Pt clusters, while the presence of CeO_2 had the opposite effect. Moreover, the characterization of catalysts showed that the coating of Al_2O_3 with both TiO_2 and CeO_2 led to a better Pt dispersion than that observed on the bare alumina, and since Pt-Al performed better than Pt-Ce, this suggested that the dispersion of the Pt particles was not the most critical aspect in determining the catalytic activity. Li et al. [269] investigated the impacts of the acidic sites on the coke formation in the steam reforming of 1-propanol through the comparison of Ni/Al_2O_3 and Ni/SiO_2; indeed, while on Ni/Al_2O_3 the presence of both Lewis and Brønsted acidic sites was observed, on Ni/SiO_2 only Lewis acidic sites were detected by the authors. The 4-h time-on-stream stability tests performed at 600 °C with a S/C ratio of 1.5 and a liquid flow rate of 0.12 $mL \cdot min^{-1}$ showed that the 1-propanol conversion was high in both catalysts (~100% on Ni/Al_2O_3 and ~98% on Ni/SiO_2); however, the product distribution differed depending on the adopted support. For Ni/Al_2O_3, the yields of H_2 and CO_2 were only 40 and 30%, respectively, and there were significant amounts of CH_4 and CO. Additionally, for Ni/SiO_2, the methane formation was suppressed, resulting in increases of H_2, CO_2, and CO yields. The increased activity of Ni/SiO_2 compared to Ni/Al_2O_3 could be attributed to the presence of less-basic sites or the absence of Brønsted acidic sites on the silica-supported catalyst. Moreover, neither catalyst showed signs of deactivation over the time period. The TEM analysis carried out on the spent catalysts highlighted the presence of amorphous coke and carbon nanotubes on both the samples; however, the amounts and the characteristics of the deposits differed depending on the support. Indeed, Ni/Al_2O_3 displayed more amorphous coke deposits, while coke deposits were mostly carbon-nanotube-shaped on Ni/SiO_2. Moreover, the diameter and length of the carbon nanotubes were higher on Ni/SiO_2 compared to Ni/Al_2O_3. The Ni/SiO_2 catalyst was found to promote the formation of carbon nanotubes; in fact, nanofiber-shaped carbon deposits were also detected on its surface. On the other hand, finer carbon nanotubes that were mostly hollow were observed for the Ni/SiO_2 sample, while the thicker ones were mainly solid. The reason for this result could be the different growth mechanisms of carbon nanotubes at different growth stages; in the early growth stage, coke was mainly deposited from the inside, while in the latter stages of growth, coke was deposited from the outside, thus increasing the diameter of the carbon nanotubes.

5.3. The Effect of the Addition of Promoters

In this section, the influence of the addition of promoters on the stability of the catalysts for butanol reforming is discussed. In particular, the addition of noble metals to reforming catalysts is a valuable feature that improves their coke resistance and stability. Indeed, despite their high cost, the addition of small percentages of noble metals (e.g., Pt, Pd, and Ru) could result in an enhancement of the metal dispersion and reducibility; moreover, they also possess high reforming activity, due to the enhancement of the steam adsorption on the catalysts' surfaces.

A study concerning the Pt, Pd, and Ru doping of Ni-Cu catalysts supported on a La-Mg mixed-metal oxide for butanol SR pointed out the benefits of the addition of noble metals, which resulted in the absence of coke deposits, thus increasing the stability of the catalysts. In particular, the stability of the catalysts, which was tested at 500 °C and 1 atm with GHSV = 3120 h^{-1} for 30 h TOS, showed the better performance of the Pt-promoted catalyst, characterized by improved butanol conversion and hydrogen yield. Subsequently, in order to test the long-term stability of the investigated catalytic formulations, up to 10 cycles were carried out, during which the stability rates of the catalysts were in the following sequence: Ru > Pt > Pd. In fact, the Ru-promoted catalyst showed better stability, despite producing lower H_2 concentrations [270]. Furthermore, Sharma et al. [271] observed that promotion of Ru on Ni xerogel catalysts for the autothermal reforming of isobutanol resulted in improved hydrogen production and reduced coke formation. Indeed, the catalyst tested for 25 h TOS at 700 °C with SV = 217,000 h^{-1}, H_2O/C = 2, and O/C = 0.1 showed a decrease in hydrogen yield from 1.53 to 1.4, showing a slow decrease of the reforming extent. Moreover, on the catalysts' surfaces, very low carbon deposits were detected by the authors. Lei et al. [272] evaluated the advantages of the ZnO promotion on the performance of Pt/Al_2O_3 in the aqueous-phase reforming of 1-propanol. An improvement in the sintering resistance of the Pt nanoparticles was observed upon addition of Zn. In particular, three catalysts (Pt/ZnO/Al_2O_3, ZnO/Pt/Al_2O_3, and Pt/Al_2O_3) were prepared and tested for APR at 250 °C and 64 bar; the hydrogen selectivity decreased in the order of ZnO/Pt/Al_2O_3 > Pt/ZnO/Al_2O_3 > Pt/Al_2O_3. Indeed, ZnO/Pt/Al_2O_3 showed a H_2 formation rate of 50.5%. Moreover, both ZnO-promoted catalysts exhibited greater stability under APR conditions.

6. Conclusions

Bioalcohols reforming is a very active research area for clean hydrogen production. During the last five years, the issue of catalyst deactivation has been widely investigated for ethanol steam reforming and oxidative reforming, while only a few studies are available for dry reforming and aqueous-phase reforming. The strategies for minimizing carbon deposition and particle sintering involve the optimization of the dispersion of active species and the improvement of metal–support interactions. In this regard, numerous catalytic formulations have been proposed (mainly based on Ni,Co and Pt as the active phases), including bimetallic catalysts and samples containing additives, both of which were prepared using different routes and employing various salt precursors. The addition of promoters was also investigated as a valuable strategy to reduce the coke selectivity; at that end, different criteria have been proposed and the authors have studied the effects of promoter selection, promoter loading, preparation method, acidic–basic properties, as well as structural properties of the support. The operative conditions (temperature, space velocity, feeding composition, and ethanol concentration in the reactant stream) also affect the amounts of carbon deposits on the catalysts' surfaces; the lowest carbon formation rates (of the order of 10^{-6}–10^{-7} $g_{coke} \cdot g_{catalyst}^{-1} \cdot g_{carbon,fed}^{-1} \cdot h^{-1}$) were recorded for noble-metal-based and promoted catalysts. In particular, when ESR was performed over Ir- and Rh-based catalysts, the CFR value was among the lowest reported in the present review. Very low carbon formation rates were also measured in the presence of Pd, as well as over bimetallic Rh-Ni catalysts. Regarding the tests performed in the presence of oxygen, the highest stability was assured over the Pt-Ni/CeO_2-SiO_2 catalysts, as well as over Ru-based perovskites containing La as the promoter. During dry reforming of ethanol, the choice of ceria and zirconia as catalytic supports instead of alumina reduced the extent of the catalyst deactivation.

MSR is a very attractive alternative for hydrogen production due to the very low tendency for coke formation. Most of the published work on this topic has been focused on Cu-based catalytic systems; the low costs of these materials make them particularly attractive for industrial applications. More sophisticated catalytic formulations, including noble-metal-based catalysts, could provide highly efficient and stable systems, with improved performance in terms of resistance towards deactivation. However, few articles have been published on these catalysts, meaning it is currently hard to establish the real potential of these materials. Although there is high potential for the O_2-assisted process in terms of enhanced hydrogen productivity and the possibility to tune the methanol/oxygen ratio in the power generation, such systems do not seem to provide a valuable alternative to the MSR in terms of reducing the carbon formation rate. The use of alternative reactor configurations is very intriguing; membrane reactor studies have demonstrated the competitiveness of these systems, both in terms of hydrogen production and operating costs.

Bioglycerol reforming processes represent a promising route to improve the economic feasibility of the biodiesel industry. The catalytic transformation of bioglycerol into hydrogen is mainly limited by the undesired formation of coke deposits on the catalysts' surfaces, thus affecting the catalysts' stability, making this an unfeasible industrial process. Therefore, many research studies have been devoted to the deep understanding of the deactivation mechanisms and the possible ways to avoid this problem. In this regard, feasible modifications of Ni-based catalysts have been reported as promising methods to improve the hydrogen production. Moreover, further studies have been performed on the support modifications devoted to improve the basicity of catalysts (e.g., the addition of more basic oxides to high surface area supports, such as Al_2O_3) and on the effects resulting from the addition of promoters.

Among all the catalysts presented in this review, Ni-based catalytic formulations have shown better stability, being characterized by lower carbon formation rates, with reduced amounts of carbon deposits shown by 10Ni/Al_2O_3 and 10Ni/Al_2O_3/5CeO_2. Moreover, analysis on the influences of supports and promoters on the stability of catalysts has shown that both of these aspects contribute to reducing the amount of carbon formed. Particular attention has been given to the modification of supports that include more basic compounds (e.g., 10Ni/Al_2O_3/5CeO_2 presented fewer carbon deposits than 10Ni/Al_2O_3), as well as the use of metals as promoters (e.g., Cr, La, Ce). Reforming processes for other bioalcohols have also been investigated in this review, as powerful tools for approaching the more complex implementation of bio-oil in the field of H_2 production. In particular, butanol and propanol reforming processes have been analyzed, focusing on the influences of the active phase, supports, and eventual promoters on the stability of the catalysts and the coke formation. Indeed, for both bioalcohols, coke formation represents the main threat in the development process, and several actions can be taken to improve the stability of the catalysts.

Author Contributions: All authors have read and agreed to the published version of the manuscript. Conceptualization, V.P. and C.R.; methodology, M.M.; formal analysis, M.C. and M.M.; investigation, C.R. and M.M.; resources, V.P. and M.C. data curation, C.R.; writing—review and editing, C.R., M.C. and M.M.; visualization, V.P.; supervision, V.P.

Funding: This work has received funding from the European Union's Horizon 2020 research and innovation programme under the Marie Skłodowska-Curie grant agreement No 734561.

Conflicts of Interest: The authors declare no conflict of interest.

Abbreviations

APR	Aqueous-phase reforming
APRE	Aqueous-phase reforming of ethanol
ASRM	Autothermal steam reforming of methanol
ATR	Autothermal reforming
BET	Brunauer–Emmett–Teller surface area measurements
BSR	Butanol steam reforming
CCR	Carbon capture and recycling
CFD	Computational fluid dynamics
CFR	Carbon formation rate
CNF	Carbon nanofibers
CNT	Carbon nanotubes
DFT	Discrete Fourier transform
DRIFT	Diffuse reflectance infrared Fourier transform spectroscopy
DSC	Differential scanning calorimeter
EDR	Ethanol dry reforming
EDS	Energy-dispersive X-ray spectrometry
EDX	Energy-dispersive X-ray analysis
GAPR	Aqueous-phase reforming of glycerol
GDR	Glycerol dry reforming
GHSV	Gas hourly space velocity
GSR	Glycerol steam reforming
LHSV	Liquid hourly space velocity
MSR	Methanol steam reforming
MMT	Montmorillonite
OSRE	Oxidative steam reforming of ethanol
OSRM	Oxidative steam reforming of methanol
OSRB	Oxidative steam reforming of butanol
PEM	Polymer electrolyte membrane
PO	Partial oxidation
SEM	Scanning electron microscopy
TEM	Transmission electron microscopy
TGA	Thermogravimetric analysis
TOF	Turnover frequencies
TOS	Time-on-stream
TPD	Temperature-programmed desorption
TPH	Temperature-programmed hydrogenation
TPO	Temperature-programmed oxidation
TPR	Temperature-programmed reduction
WGS	Water–gas shift
WHSV	Weight hourly space velocity
X	Alcohol conversion
XPS	X-ray photoelectron spectroscopy
XRD	X-ray diffraction

References

1. Susmozas, A.; Iribarren, D.; Dufour, J. Assessing the life-cycle performance of hydrogen production via biofuel reforming in Europe. *Resources* **2015**, *4*, 398–411. [CrossRef]
2. Al-Maamary, H.M.S.; Kazem, H.A.; Chaichan, M.T. The impact of oil price fluctuations on common renewable energies in GCC countries. *Renew. Sustain. Energy Rev.* **2017**, *75*, 989–1007. [CrossRef]
3. Calles, J.A.; Carrero, A.; Vizcaíno, A.J.; García-Moreno, L.; Megía, P.J. Steam reforming of model Bio-Oil aqueous fraction using Ni-(Cu, Co, Cr)/SBA-15 catalysts. *Int. J. Mol. Sci.* **2019**, *20*, 512. [CrossRef] [PubMed]

4. Colmenares, J.C.; Colmenares Quintero, R.F.; Pieta, I.S. Catalytic dry reforming for biomass-based fuels processing: Progress and future perspectives. *Energy Technol.* **2016**, *4*, 881–890. [CrossRef]
5. Lisý, M.; Lisá, H.; Jecha, D.; Baláš, M.; Križan, P. Characteristic properties of alternative biomass fuels. *Energies* **2020**, *13*, 1448. [CrossRef]
6. Palma, V.; Barba, D.; Meloni, E.; Ruocco, C.; Martino, M. Chapter 2—Membrane reactors for H_2 and energy production. In *Current Trends and Future Developments on (Bio-) Membranes*; Basile, A., Nappom, T.W., Eds.; Elsevier: Amsterdam, The Netherlands, 2020; pp. 33–56. [CrossRef]
7. Alique, D.; Bruni, G.; Sanz, R.; Calles, J.A.; Tosti, S. Ultra-Pure hydrogen via co-valorization of olive mill wastewater and bioethanol in pd-membrane reactors. *Processes* **2020**, *8*, 219. [CrossRef]
8. Frusteri, F.; Bonura, G. 5-Hydrogen production by reforming of bio-alcohols. In *Compendium of Hydrogen Energy*; Subramani, V., Basile, A., Veziroğlu, T.N., Eds.; Woodhead Publishing: Oxford, UK, 2015; pp. 109–136.
9. Melikoglu, M.; Singh, V.; Leu, S.Y.; Webb, C.; Lin, C.S.K. Biochemical production of bioalcohols. In *Handbook of Biofuels Production*, 2nd ed.; Luque, R., Lin, C.S.K., Wilson, K., Clark, J., Eds.; Woodhead Publishing: Oxford, UK, 2016; pp. 237–258.
10. Datta, A.; Hossain, A.; Roy, S. An overview on biofuels and their advantages and disadvantages. *Asian J. Chem.* **2019**, *31*, 1851–1858. [CrossRef]
11. Ail, S.S.; Dasappa, S. Biomass to liquid transportation fuel via Fischer Tropsch synthesis—Technology review and current scenario. *Renew. Sustain. Energy Rev.* **2016**, *58*, 267–286. [CrossRef]
12. Behera, S.; Singh, R.; Arora, R.; Sharma, N.K.; Shukla, M.; Kumar, S. Scope of algae as third generation biofuels. *Front. Bioeng. Biotechnol.* **2015**, *2*. [CrossRef]
13. Seretis, A.; Tsiakaras, P. Crude bio-glycerol aqueous phase reforming and hydrogenolysis over commercial $SiO_2Al_2O_3$ nickel catalyst. *Renew. Energy* **2016**, *97*, 373–379. [CrossRef]
14. Coronado, C.J.R. Glycerol: Production, consumption, prices, characterization and new trends in combustion. *Renew. Sustain. Energy Rev.* **2013**, *27*, 475–493. [CrossRef]
15. Liu, Z.; Peng, W.; Motahari-Nezhad, M.; Shahraki, S.; Beheshti, M. Circulating fluidized bed gasification of biomass for flexible end-use of syngas: A micro and nano scale study for production of bio-methanol. *J. Clean. Prod.* **2016**, *129*, 249–255. [CrossRef]
16. Iaquaniello, G.; Centi, G.; Salladini, A.; Palo, E.; Perathoner, S.; Spadaccini, L. Waste-to-methanol: Process and economics assessment. *Bioresour. Technol.* **2017**, *243*, 611–619. [CrossRef]
17. Anitha, M.; Kamarudin, S.K.; Shamsul, N.S.; Kofli, N.T. Determination of bio-methanol as intermediate product of anaerobic co-digestion in animal and agriculture wastes. *Int. J. Hydrogen Energy* **2015**, *40*, 11791–11799. [CrossRef]
18. Bonfim-Rocha, L.; Gimenes, M.L.; Bernardo de Faria, S.H.; Silva, R.O.; Esteller, L.J. Multi-objective design of a new sustainable scenario for bio-methanol production in Brazil. *J. Clean. Prod.* **2018**, *187*, 1043–1056. [CrossRef]
19. Hartley, U.W.; Amornraksa, S.; Kim-Lohsoontorn, P.; Laosiripojana, N. Thermodynamic analysis and experimental study of hydrogen production from oxidative reforming of n-butanol. *Chem. Eng. J.* **2015**, *278*, 2–12. [CrossRef]
20. Palma, V.; Ruocco, C.; Meloni, E.; Ricca, A. Highly active and stable Pt-Ni/CeO_2-SiO_2 catalysts for ethanol reforming. *J. Clean. Prod.* **2017**, *166*, 263–272. [CrossRef]
21. Bepari, S.; Kuila, D. Steam reforming of methanol, ethanol and glycerol over nickel-based catalysts-A review. *Int. J. Hydrogen Energy* **2019**. [CrossRef]
22. Montero, C.; Remiro, A.; Valle, B.; Oar-Arteta, L.; Bilbao, J.; Gayubo, A.G. Origin and nature of coke in ethanol steam reforming and its role in deactivation of Ni/La_2O_3–αAl_2O_3 catalyst. *Ind. Eng. Chem. Res.* **2019**, *58*, 14736–14751. [CrossRef]
23. Palma, V.; Ruocco, C.; Meloni, E.; Ricca, A. Influence of Catalytic Formulation and Operative Conditions on Coke Deposition over CeO_2-SiO_2 Based Catalysts for Ethanol Reforming. *Energies* **2017**, *10*, 1030. [CrossRef]
24. Horng, R.-F.; Lai, M.-P.; Chiu, W.-C.; Huang, W.-C. Thermodynamic analysis of syngas production and carbon formation on oxidative steam reforming of butanol. *Int. J. Hydrogen Energy* **2016**, *41*, 889–896. [CrossRef]
25. Valle, B.; Aramburu, B.; Olazar, M.; Bilbao, J.; Gayubo, A.G. Steam reforming of raw bio-oil over Ni/La_2O_3-αAl_2O_3: Influence of temperature on product yields and catalyst deactivation. *Fuel* **2018**, *216*, 463–474. [CrossRef]

26. Coll, R.; Salvadó, J.; Farriol, X.; Montane, D. Steam reforming model compounds of biomass gasification tars: Conversion at different operating conditions and tendency towards coke formation. *Fuel Process. Technol.* **2001**, *74*, 19–31. [CrossRef]
27. Larmier, K.; Chizallet, C.; Cadran, N.; Maury, S.; Abboud, J.; Lamic-Humblot, A.-F.; Marceau, E.; Lauron-Pernot, H. Mechanistic investigation of isopropanol conversion on alumina catalysts: Location of active sites for alkene/ether production. *ACS Catal.* **2015**, *5*, 4423–4437. [CrossRef]
28. De Rezende, S.M.; Franchini, C.A.; Dieuzeide, M.L.; De Farias, A.M.D.; Amadeo, N.; Fraga, M. Glycerol steam reforming over layered double hydroxide-supported Pt catalysts. *Chem. Eng. J.* **2015**, *272*, 108–118. [CrossRef]
29. Ochoa, A.; Aramburu, B.; Valle, B.; Resasco, D.E.; Bilbao, J.; Gayubo, A.G.; Castaño, P. Role of oxygenates and effect of operating conditions in the deactivation of a Ni supported catalyst during the steam reforming of bio-oil. *Green Chem.* **2017**, *19*, 4315–4333. [CrossRef]
30. Sánchez, N.; Encinar, J.M.; Nogales, S.; González, J.F. Lanthanum Effect on Ni/Al_2O_3 as a Catalyst Applied in Steam Reforming of Glycerol for Hydrogen Production. *Processes* **2019**, *7*, 449. [CrossRef]
31. Harun, N.; Abidin, S.Z.; Osazuwa, O.; Taufiq-Yap, Y.H.; Azizan, M.T. Hydrogen production from glycerol dry reforming over Ag-promoted Ni/Al_2O_3. *Int. J. Hydrogen Energy* **2019**, *44*, 213–225. [CrossRef]
32. Gharahshiran, V.S.; Yousefpour, M.; Amini, V. A comparative study of zirconia and yttria promoted mesoporous carbon-nickel-cobalt catalysts in steam reforming of ethanol for hydrogen production. *Mol. Catal.* **2020**, *484*, 110767. [CrossRef]
33. Ramesh, S.; Yang, E.-H.; Jung, J.-S.; Moon, D.J. Copper decorated perovskite an efficient catalyst for low temperature hydrogen production by steam reforming of glycerol. *Int. J. Hydrogen Energy* **2015**, *40*, 11428–11435. [CrossRef]
34. Shao, J.; Zeng, G.; Li, Y. Effect of Zn substitution to a $LaNiO_{3-\delta}$ perovskite structured catalyst in ethanol steam reforming. *Int. J. Hydrogen Energy* **2017**, *42*, 17362–17375. [CrossRef]
35. Sun, C.; Zheng, Z.; Wang, S.; Li, X.; Wu, X.; An, X.; Xie, X. Yolk-shell structured Pt-CeO_2 @Ni-SiO_2 as an efficient catalyst for enhanced hydrogen production from ethanol steam reforming. *Ceram. Int.* **2018**, *44*, 1438–1442. [CrossRef]
36. He, S.; Mei, Z.; Liu, N.; Zhang, L.; Lu, J.; Li, X.; Wang, J.; He, D.; Luo, Y. Ni/SBA-15 catalysts for hydrogen production by ethanol steam reforming: Effect of nickel precursor. *Int. J. Hydrogen Energy* **2017**, *42*, 14429–14438. [CrossRef]
37. Ruocco, C.; Palma, V.; Ricca, A. Hydrogen production by oxidative reforming of ethanol in a fluidized bed reactor using a Pt Ni/CeO_2SiO_2 catalyst. *Int. J. Hydrogen Energy* **2019**, *44*, 12661–12670. [CrossRef]
38. Wang, F.; Zhang, L.; Deng, J.; Zhang, J.; Han, B.; Wang, Y.; Li, Z.; Yu, H.; Cai, W.; Deng, Z. Embedded Ni catalysts in Ni-O-Ce solid solution for stable hydrogen production from ethanol steam reforming reaction. *Fuel Process. Technol.* **2019**, *193*, 94–101. [CrossRef]
39. Ricca, A.; Palma, V.; Martino, M.; Meloni, E. Innovative catalyst design for methane steam reforming intensification. *Fuel* **2017**, *198*, 175–182. [CrossRef]
40. Palma, V.; Ruocco, C.; Castaldo, F.; Ricca, A.; Boettge, D. Ethanol steam reforming over bimetallic coated ceramic foams: Effect of reactor configuration and catalytic support. *Int. J. Hydrogen Energy* **2015**, *40*, 12650–12662. [CrossRef]
41. Koc, S.; Avci, A.K. Reforming of glycerol to hydrogen over Ni-based catalysts in a microchannel reactor. *Fuel Process. Technol.* **2017**, *156*, 357–365. [CrossRef]
42. Montero, C.; Ochoa, A.; Castaño, P.; Bilbao, J.; Gayubo, A.G. Monitoring Ni 0 and coke evolution during the deactivation of a $Ni/La_2O_3{-}\alpha Al_2O_3$ catalyst in ethanol steam reforming in a fluidized bed. *J. Catal.* **2015**, *331*, 181–192. [CrossRef]
43. Mondal, T.; Pant, K.K.; Dalai, A.K. Catalytic oxidative steam reforming of bio-ethanol for hydrogen production over Rh promoted $Ni/CeO_2{-}ZrO_2$ catalyst. *Int. J. Hydrogen Energy* **2015**, *40*, 2529–2544. [CrossRef]
44. Li, M.-R.; Wang, G.-C. The mechanism of ethanol steam reforming on the Co^0 and Co^{2+} sites: A DFT study. *J. Catal.* **2018**, *365*, 391–404. [CrossRef]
45. Grewal, T.P.K.; Roy, S. Modeling the effect of coke deposition in a heat integrated ethanol reformer. *Int. J. Hydrogen Energy* **2016**, *41*, 19863–19880. [CrossRef]
46. Argyle, M.D.; Bartholomew, C.H. Heterogeneous catalyst deactivation and regeneration: A review. *Catalysts* **2015**, *5*, 145–269. [CrossRef]

47. Trevisanut, C.; Mari, M.; Millet, J.-M.M.; Cavani, F. Chemical-loop reforming of ethanol over metal ferrites: An analysis of structural features affecting reactivity. *Int. J. Hydrogen Energy* **2015**, *40*, 5264–5271. [CrossRef]
48. Sharma, Y.C.; Kumar, A.; Prasad, R.; Upadhyay, S.N. Ethanol steam reforming for hydrogen production: Latest and effective catalyst modification strategies to minimize carbonaceous deactivation. *Renew. Sustain. Energy Rev.* **2017**, *74*, 89–103. [CrossRef]
49. Li, D.; Li, X.; Gong, J. Catalytic reforming of oxygenates: State of the art and future prospects. *Chem. Rev.* **2016**, *116*, 11529–11653. [CrossRef] [PubMed]
50. Lazar, M.D.; Senila, L.; Dan, M.; Mihet, M. Chapter 10—Crude bioethanol reforming process: The advantage of a biosource exploitation. In *Ethanol*; Basile, A., Iulianelli, A., Dalena, F., Veziroğlu, T.N., Eds.; Elsevier: Amsterdam, The Netherlands, 2019; pp. 257–288.
51. Liu, Z.; Duchoň, T.; Wang, H.; Peterson, E.W.; Zhou, Y.; Luo, S.; Zhou, J.; Matolin, V.; Stacchiola, D.J.; Rodríguez, J.A.; et al. Mechanistic Insights of Ethanol Steam Reforming over Ni–CeOx(111): The Importance of Hydroxyl Groups for Suppressing Coke Formation. *J. Phys. Chem. C* **2015**, *119*, 18248–18256. [CrossRef]
52. Gonçalves, A.A.S.; Faustino, P.B.; Assaf, J.M.; Jaroniec, M. One-Pot Synthesis of Mesoporous Ni–Ti–Al Ternary Oxides: Highly Active and Selective Catalysts for Steam Reforming of Ethanol. *ACS Appl. Mater. Interfaces* **2017**, *9*, 6079–6092. [CrossRef]
53. Osorio-Vargas, P.; Flores-González, N.A.; Navarro, R.M.; Fierro, J.L.; Campos, C.H.; Reyes, P. Improved stability of Ni/Al2O3 catalysts by effect of promoters (La_2O_3, CeO_2) for ethanol steam-reforming reaction. *Catal. Today* **2016**, *259*, 27–38. [CrossRef]
54. Jo, S.W.; Kwak, B.S.; Kim, K.M.; Do, J.Y.; Park, N.-K.; Lee, T.J.; Lee, S.-T.; Kang, M. Reasonable harmony of Ni and Mn in core@shell-structured NiMn@SiO_2 catalysts prepared for hydrogen production from ethanol steam reforming. *Chem. Eng. J.* **2016**, *288*, 858–868. [CrossRef]
55. Chen, M.; Wang, C.; Wang, Y.; Tang, Z.; Yang, Z.; Zhang, H.; Wang, J. Hydrogen production from ethanol steam reforming: Effect of Ce content on catalytic performance of Co/Sepiolite catalyst. *Fuel* **2019**, *247*, 344–355. [CrossRef]
56. Passos, A.R.; Martins, L.; Pulcinelli, S.H.; Santilli, C.; Briois, V. Correlation of Sol-Gel Alumina-Supported Cobalt Catalyst Processing to Cobalt Speciation, Ethanol Steam Reforming Activity, and Stability. *ChemCatChem* **2017**, *9*, 3918–3929. [CrossRef]
57. Spallina, V.; Matturro, G.; Ruocco, C.; Meloni, E.; Palma, V.; Fernandez, E.; Melendez, J.; Tanaka, A.P.; Sole, J.V.; Annaland, M.V.S.; et al. Direct route from ethanol to pure hydrogen through autothermal reforming in a membrane reactor: Experimental demonstration, reactor modelling and design. *Energy* **2018**, *143*, 666–681. [CrossRef]
58. Muñoz, M.; Moreno, S.; Molina, R. Oxidative steam reforming of ethanol (OSRE) over stable NiCo–MgAl catalysts by microwave or sonication assisted coprecipitation. *Int. J. Hydrogen Energy* **2017**, *42*, 12284–12294. [CrossRef]
59. Passos, A.R.; Pulcinelli, S.H.; Santilli, C.V.; Briois, V. Operando monitoring of metal sites and coke evolution during non-oxidative and oxidative ethanol steam reforming over Ni and NiCu ex-hydrotalcite catalysts. *Catal. Today* **2019**, *336*, 122–130. [CrossRef]
60. Palma, V.; Ruocco, C.; Meloni, E.; Ricca, A. Oxidative steam reforming of ethanol on mesoporous silica supported PtNi/CeO_2 catalysts. *Int. J. Hydrogen Energy* **2017**, *42*, 1598–1608. [CrossRef]
61. Palma, V.; Ruocco, C.; Meloni, E.; Ricca, A. Oxidative reforming of ethanol over CeO_2 -SiO_2 based catalysts in a fluidized bed reactor. *Chem. Eng. Process. Process. Intensif.* **2018**, *124*, 319–327. [CrossRef]
62. Yu, J.; Odriozola, J.; Reina, T.R. Dry reforming of ethanol and glycerol: Mini-review. *Catalsts* **2019**, *9*, 1015. [CrossRef]
63. Zhao, S.; Cai, W.; Li, Y.; Yu, H.; Zhang, S.; Cui, L. Syngas production from ethanol dry reforming over Rh/CeO_2 catalyst. *J. Saudi Chem. Soc.* **2018**, *22*, 58–65. [CrossRef]
64. Nozawa, T.; Yoshida, A.; Hikichi, S.; Naito, S. Effects of Re addition upon aqueous phase reforming of ethanol over TiO_2 supported Rh and Ir catalysts. *Int. J. Hydrogen Energy* **2015**, *40*, 4129–4140. [CrossRef]
65. Xiong, H.; DeLaRiva, A.; Datye, A.K.; Wang, Y. Low-temperature aqueous-phase reforming of ethanol on bimetallic PdZn catalysts. *Catal. Sci. Technol.* **2015**, *5*, 254–263. [CrossRef]
66. Zhao, Z.; Zhang, L.; Tan, Q.; Yang, F.; Faria, J.; Resasco, D.E. Synergistic bimetallic Ru-Pt catalysts for the low-temperature aqueous phase reforming of ethanol. *AIChE J.* **2018**, *65*, 151–160. [CrossRef]

67. Gogoi, P.; Nagpure, A.S.; Kandasamy, P.; Satyanarayana, C.V.V.; Raja, T.; Kandasamy, P.; Thirumalaiswamy, R. Insights into the catalytic activity of Ru/NaY catalysts for efficient H_2 production through aqueous phase reforming. *Sustain. Energy Fuels* **2020**, *4*, 678–690. [CrossRef]
68. Mulewa, W.; Tahir, M.; Amin, N.A.S. MMT-supported Ni/TiO_2 nanocomposite for low temperature ethanol steam reforming toward hydrogen production. *Chem. Eng. J.* **2017**, *326*, 956–969. [CrossRef]
69. Nejat, T.; Jalalinezhad, P.; Hormozi, F.; Bahrami, Z. Hydrogen production from steam reforming of ethanol over Ni-Co bimetallic catalysts and MCM-41 as support. *J. Taiwan Inst. Chem. Eng.* **2019**, *97*, 216–226. [CrossRef]
70. Di Michele, A.; Dell'Angelo, A.; Tripodi, A.; Bahadori, E.; Sànchez, F.; Motta, D.; Dimitratos, N.; Rossetti, I.; Ramis, G.; Sanchez, F. Steam reforming of ethanol over Ni/MgAl2O4 catalysts. *Int. J. Hydrogen Energy* **2019**, *44*, 952–964. [CrossRef]
71. Carvalho, F.L.; Asencios, Y.J.O.; Bellido, J.D.; Assaf, J.M. Bio-ethanol steam reforming for hydrogen production over Co_3O_4/CeO_2 catalysts synthesized by one-step polymerization method. *Fuel Process. Technol.* **2016**, *142*, 182–191. [CrossRef]
72. Konsolakis, M.; Ioakimidis, Z.; Kraia, T.; Marnellos, G.E. Hydrogen Production by Ethanol Steam Reforming (ESR) over CeO_2 Supported Transition Metal (Fe, Co, Ni, Cu) Catalysts: Insight into the Structure-Activity Relationship. *Catalsts* **2016**, *6*, 39. [CrossRef]
73. Greluk, M.; Słowik, G.; Rotko, M.; Machocki, A. Steam reforming and oxidative steam reforming of ethanol over PtKCo/CeO_2 catalyst. *Fuel* **2016**, *183*, 518–530. [CrossRef]
74. Riani, P.; Garbarino, G.; Canepa, F.; Busca, G. Cobalt nanoparticles mechanically deposited on α-Al_2O_3: A competitive catalyst for the production of hydrogen through ethanol steam reforming. *J. Chem. Technol. Biotechnol.* **2019**, *94*, 538–546. [CrossRef]
75. Mhadmhan, S.; Natewong, P.; Prasongthum, N.; Samart, C.; Reubroycharoen, P. Investigation of Ni/SiO_2 fiber catalysts prepared by different methods on hydrogen production from ethanol steam reforming. *Catalysts* **2018**, *8*, 319. [CrossRef]
76. Rodrigues, T.S.; De Moura, A.B.; Silva, F.; Candido, E.G.; Da Silva, A.G.; De Oliveira, D.C.; Quiroz, J.; Camargo, P.H.; Bergamaschi, V.S.; Ferreira, J.C.; et al. Ni supported $Ce_{0.9}Sm_{0.1}O_{2-\delta}$ nanowires: An efficient catalyst for ethanol steam reforming for hydrogen production. *Fuel* **2019**, *237*, 1244–1253. [CrossRef]
77. Roy, B.; Leclerc, C. Study of preparation method and oxidization/reduction effect on the performance of nickel-cerium oxide catalysts for aqueous-phase reforming of ethanol. *J. Power Sources* **2015**, *299*, 114–124. [CrossRef]
78. Sohn, H.; Ozkan, U.S. Cobalt-Based Catalysts for Ethanol Steam Reforming: An Overview. *Energy Fuels* **2016**, *30*, 5309–5322. [CrossRef]
79. Yoshida, H.; Yamaoka, R.; Arai, M. Stable Hydrogen Production from Ethanol through Steam Reforming Reaction over Nickel-Containing Smectite-Derived Catalyst. *Int. J. Mol. Sci.* **2014**, *16*, 350–362. [CrossRef]
80. Ogo, S.; Sekine, Y. Recent progress in ethanol steam reforming using non-noble transition metal catalysts: A review. *Fuel Process. Technol.* **2020**, *199*, 106238. [CrossRef]
81. Hou, T.; Yu, B.; Zhang, S.; Xu, T.; Wang, D.; Cai, W. Hydrogen production from ethanol steam reforming over Rh/CeO2 catalyst. *Catal. Commun.* **2015**, *58*, 137–140. [CrossRef]
82. Hou, T.; Lei, Y.; Zhang, S.; Zhang, J.; Cai, W. Ethanol dry reforming for syngas production over Ir/CeO_2 catalyst. *J. Rare Earths* **2015**, *33*, 42–45. [CrossRef]
83. Moraes, T.S.; Neto, R.C.R.; Ribeiro, M.; Mattos, L.V.; Kourtelesis, M.; Ladas, S.; Verykios, X.; Noronha, F.B. The study of the performance of PtNi/CeO_2–nanocube catalysts for low temperature steam reforming of ethanol. *Catal. Today* **2015**, *242*, 35–49. [CrossRef]
84. Moraes, T.S.; Neto, R.C.R.; Ribeiro, M.; Mattos, L.V.; Kourtelesis, M.; Ladas, S.; Verykios, X.; Noronha, F.B. Ethanol conversion at low temperature over CeO_2—Supported Ni-based catalysts. Effect of Pt addition to Ni catalyst. *Appl. Catal. B Environ.* **2016**, *181*, 754–768. [CrossRef]
85. Cifuentes, B.; Valero, M.F.; Conesa, J.A.; Cobo, M. Hydrogen Production by Steam Reforming of Ethanol on Rh-Pt Catalysts: Influence of CeO_2, ZrO_2, and La_2O_3 as Supports. *Catalsts* **2015**, *5*, 1872–1896. [CrossRef]
86. Palma, V.; Ruocco, C.; Meloni, E.; Ricca, A. Renewable hydrogen from ethanol reforming over CeO_2-SiO_2 based catalysts. *Catalsts* **2017**, *7*, 226. [CrossRef]

87. Campos, C.H.; Pecchi, G.; Fierro, J.L.G.; Osorio-Vargas, P. Enhanced bimetallic Rh-Ni supported catalysts on alumina doped with mixed lanthanum-cerium oxides for ethanol steam reforming. *Mol. Catal.* **2019**, *469*, 87–97. [CrossRef]
88. Sanchez, N.; Ruiz, R.; Cifuentes, B.; Cobo, M. Hydrogen from glucose: A combined study of glucose fermentation, bioethanol purification, and catalytic steam reforming. *Int. J. Hydrogen Energy* **2016**, *41*, 5640–5651. [CrossRef]
89. Bilal, M.; Jackson, S.D. Ethanol steam reforming over Pt/Al_2O_3 and Rh/Al_2O_3 catalysts: The effect of impurities on selectivity and catalyst deactivation. *Appl. Catal. A Gen.* **2017**, *529*, 98–107. [CrossRef]
90. Iulianelli, A.; Palma, V.; Bagnato, G.; Ruocco, C.; Huang, Y.; Veziroglu, N.T.; Basile, A. From bioethanol exploitation to high grade hydrogen generation: Steam reforming promoted by a Co-Pt catalyst in a Pd-based membrane reactor. *Renew. Energy* **2018**, *119*, 834–843. [CrossRef]
91. Nichele, V.; Signoretto, M.; Pinna, F.; Ghedini, E.; Compagnoni, M.; Rossetti, I.; Cruciani, G.; Di Michele, A. Bimetallic Ni–Cu catalysts for the low-temperature ethanol steam reforming: Importance of metal–support interactions. *Catal. Lett.* **2014**, *145*, 549–558. [CrossRef]
92. Rodriguez-Gomez, A.; Caballero, A. Bimetallic Ni-Co/SBA-15 catalysts for reforming of ethanol: How cobalt modifies the nickel metal phase and product distribution. *Mol. Catal.* **2018**, *449*, 122–130. [CrossRef]
93. Chen, L.-C.; Cheng, H.; Chiang, C.-W.; Lin, S.D. Sustainable hydrogen production by ethanol steam reforming using a partially reduced copper-nickel oxide catalyst. *ChemSusChem* **2015**, *8*, 1787–1793. [CrossRef]
94. Braga, A.H.; Ribeiro, M.; Noronha, F.B.; Galante, D.; Bueno, J.M.C.; Santos, J.B.O. Effects of Co addition to supported Ni catalysts on hydrogen production from oxidative steam reforming of ethanol. *Energy Fuels* **2018**, *32*, 12814–12825. [CrossRef]
95. Baruah, R.; Dixit, M.; Parejiya, A.; Basarkar, P.; Bhargav, A.; Sharma, S. Oxidative steam reforming of ethanol on rhodium catalyst—I: Spatially resolved steady-state experiments and microkinetic modeling. *Int. J. Hydrogen Energy* **2017**, *42*, 10184–10198. [CrossRef]
96. Weng, S.-F.; Hsieh, H.-C.; Lee, C.-S. Hydrogen production from oxidative steam reforming of ethanol on nickel-substituted pyrochlore phase catalysts. *Int. J. Hydrogen Energy* **2017**, *42*, 2849–2860. [CrossRef]
97. Hsieh, H.-C.; Chang, Y.-C.; Tsai, P.-W.; Lin, Y.-Y.; Chuang, Y.-C.; Sheu, H.-S.; Lee, C.-S. Metal substituted pyrochlore phase $Li_xLa_{2-x}Ce_{1.8}Ru_{0.2}O_{7-\delta}$ (x = 0.0–0.6) as an effective catalyst for oxidative and auto-thermal steam reforming of ethanol. *Catal. Sci. Technol.* **2019**, *9*, 1406–1419. [CrossRef]
98. Cao, D.; Zeng, F.; Zhao, Z.; Cai, W.; Li, Y.; Yu, H.; Zhang, S.; Qu, F. Cu based catalysts for syngas production from ethanol dry reforming: Effect of oxide supports. *Fuel* **2018**, *219*, 406–416. [CrossRef]
99. Cao, N.; Cai, W.; Li, Y.; Li, C.; Yu, H.; Zhang, S.; Qu, F. Syngas Production from Ethanol Dry Reforming over $Cu/Ce_{0.8}Zr_{0.2}O_2$ Catalyst. *Catal. Lett.* **2017**, *147*, 2929–2939. [CrossRef]
100. Chen, Q.; Cai, W.; Liu, Y.; Zhang, S.; Li, Y.; Huang, D.; Wang, T.; Li, Y. Synthesis of $Cu\text{-}Ce_{0.8}Zr_{0.2}O_2$ catalyst by ball milling for CO_2 reforming of ethanol. *J. Saudi Chem. Soc.* **2019**, *23*, 111–117. [CrossRef]
101. Dang, C.; Wu, S.; Yang, G.; Cao, Y.; Wang, H.; Peng, F.; Yu, H. Syngas production by dry reforming of the mixture of glycerol and ethanol with $CaCO_3$. *J. Energy Chem.* **2020**, *43*, 90–97. [CrossRef]
102. Da Costa-Serra, J.F.; Chica, A. Catalysts based on Co-Birnessite and Co-Todorokite for the efficient production of hydrogen by ethanol steam reforming. *Int. J. Hydrogen Energy* **2018**, *43*, 16859–16865. [CrossRef]
103. Wei, Y.; Cai, W.; Deng, S.; Li, Z.; Yu, H.; Zhang, S.; Yu, Z.; Cui, L.; Qu, F. Efficient syngas production via dry reforming of renewable ethanol over Ni/KIT-6 nanocatalysts. *Renew. Energy* **2020**, *145*, 1507–1516. [CrossRef]
104. Chai, R.; Li, Y.; Zhang, Q.; Zhao, G.; Liu, Y.; Lu, Y. Monolithic $Ni\text{-}MO_x/Ni$-foam (M = Al, Zr or Y) catalysts with enhanced heat/mass transfer for energy-efficient catalytic oxy-methane reforming. *Catal. Commun.* **2015**, *70*, 1–5. [CrossRef]
105. Moraes, T.S.; Borges, L.E.P.; Farrauto, R.; Noronha, F.B. Steam reforming of ethanol on $Rh/SiCeO_2$ washcoated monolith catalyst: Stable catalyst performance. *Int. J. Hydrogen Energy* **2018**, *43*, 115–126. [CrossRef]
106. Gaudillere, C.; González, J.J.; Chica, A.; Serra, J.M. YSZ monoliths promoted with Co as catalysts for the production of H_2 by steam reforming of ethanol. *Appl. Catal. A Gen.* **2017**, *538*, 165–173. [CrossRef]
107. Palma, V.; Ruocco, C.; Ricca, A. Ceramic foams coated with $PtNi/CeO_2ZrO_2$ for bioethanol steam reforming. *Int. J. Hydrogen Energy* **2016**, *41*, 11526–11536. [CrossRef]
108. Turczyniak, S.; Greluk, M.; Słowik, G.; Gac, W.; Zafeiratos, S.; Machocki, A. Surface state and catalytic performance of ceria-supported cobalt catalysts in the steam reforming of ethanol. *ChemCatChem* **2017**, *9*, 782–797. [CrossRef]

109. Greluk, M.; Rybak, P.; Słowik, G.; Rotko, M.; Machocki, A. Comparative study on steam and oxidative steam reforming of ethanol over 2KCo/ZrO_2 catalyst. *Catal. Today* **2015**, *242*, 50–59. [CrossRef]
110. Greluk, M.; Rotko, M.; Turczyniak-Surdacka, S. Enhanced catalytic performance of La_2O_3 promoted Co/CeO_2 and Ni/CeO_2 catalysts for effective hydrogen production by ethanol steam reforming. *Renew. Energy* **2020**, *155*, 378–395. [CrossRef]
111. Cerdá-Moreno, C.; Da Costa-Serra, J.F.; Chica, A. Co and La supported on Zn-Hydrotalcite-derived material as efficient catalyst for ethanol steam reforming. *Int. J. Hydrogen Energy* **2019**, *44*, 12685–12692. [CrossRef]
112. Kim, D.; Kwak, B.S.; Park, N.-K.; Han, G.B.; Kang, M. Dynamic hydrogen production from ethanol steam-reforming reaction on Ni_xMo_y/SBA-15 catalytic system. *Int. J. Energy Res.* **2015**, *39*, 279–292. [CrossRef]
113. Kwak, B.S.; Lee, G.; Park, S.-M.; Kang, M. Effect of MnOx in the catalytic stabilization of Co_2MnO_4 spinel during the ethanol steam reforming reaction. *Appl. Catal. A Gen.* **2015**, *503*, 165–175. [CrossRef]
114. Liu, Z.; Xu, W.; Yao, S.; Johnson-Peck, A.C.; Zhao, F.; Michorczyk, P.; Kubacka, A.; Stach, E.A.; Fernández-García, M.; Senanayake, S.D.; et al. Superior performance of Ni–W–Ce mixed-metal oxide catalysts for ethanol steam reforming: Synergistic effects of W- and Ni-dopants. *J. Catal.* **2015**, *321*, 90–99. [CrossRef]
115. Li, D.; Zeng, L.; Li, X.; Wang, X.; Ma, H.; Assabumrungrat, S.; Gong, J. Ceria-promoted Ni/SBA-15 catalysts for ethanol steam reforming with enhanced activity and resistance to deactivation. *Appl. Catal. B Environ.* **2015**, *176*, 532–541. [CrossRef]
116. Du, Y.-L.; Wu, X.; Cheng, Q.; Huang, Y.-L.; Huang, W. Development of Ni-Based Catalysts Derived from Hydrotalcite-Like Compounds Precursors for Synthesis Gas Production via Methane or Ethanol Reforming. *Catalysts* **2017**, *7*, 70. [CrossRef]
117. Marinho, A.L.A.; Rabelo-Neto, R.C.; Noronha, F.B.; Mattos, L.V. Steam reforming of ethanol over Ni-based catalysts obtained from $LaNiO_3$ and $LaNiO_3$/$CeSiO_2$ perovskite-type oxides for the production of hydrogen. *Appl. Catal. A Gen.* **2016**, *520*, 53–64. [CrossRef]
118. Dan, M.; Senila, L.; Roman, M.; Mihet, M.; Lazar, M.D. From wood wastes to hydrogen—Preparation and catalytic steam reforming of crude bio-ethanol obtained from fir wood. *Renew. Energy* **2015**, *74*, 27–36. [CrossRef]
119. He, S.; He, S.; Zhang, L.; Li, X.; Wang, J.; He, D.; Lu, J.; Luo, Y. Hydrogen production by ethanol steam reforming over Ni/SBA-15 mesoporous catalysts: Effect of Au addition. *Catal. Today* **2015**, *258*, 162–168. [CrossRef]
120. Mondal, T.; Pant, K.K.; Dalai, A.K. Oxidative and non-oxidative steam reforming of crude bio-ethanol for hydrogen production over Rh promoted Ni/CeO_2-ZrO_2 catalyst. *Appl. Catal. A Gen.* **2015**, *499*, 19–31. [CrossRef]
121. Bahari, M.B.; Phuc, N.H.H.; Abdullah, B.; Alenazey, F.; Vo, D.-V.N. Ethanol dry reforming for syngas production over Ce-promoted Ni/Al_2O_3 catalyst. *J. Environ. Chem. Eng.* **2016**, *4*, 4830–4838. [CrossRef]
122. Bahari, M.B.; Phuc, N.H.H.; Alenazey, F.; Vu, K.B.; Ainirazali, N.; Vo, D.-V.N. Catalytic performance of La-Ni/Al_2O_3 catalyst for CO_2 reforming of ethanol. *Catal. Today* **2017**, *291*, 67–75. [CrossRef]
123. Fayaz, F.; Danh, H.T.; Nguyen-Huy, C.; Vu, K.B.; Abdullah, B.; Vo, D.-V.N. Promotional Effect of Ce-dopant on Al_2O_3-supported Co Catalysts for Syngas Production via CO_2 Reforming of Ethanol. *Procedia Eng.* **2016**, *148*, 646–653. [CrossRef]
124. Fayaz, F.; Bach, L.-G.; Bahari, M.B.; Nguyen, D.T.; Vu, K.B.; Kanthasamy, R.; Samart, C.; Nguyen-Huy, C.; Vo, D.-V.N. Stability evaluation of ethanol dry reforming on Lanthania-doped cobalt-based catalysts for hydrogen-rich syngas generation. *Int. J. Energy Res.* **2018**, *43*, 405–416. [CrossRef]
125. Shafiqah, M.-N.N.; Nguyen, T.D.; Jun, L.N.; Bahari, M.B.; Phuong, P.T.; Abdullah, B.; Vo, D.-V.N. *Production of Syngas from Ethanol CO_2 Reforming on La-Doped Cu/Al2O3: Impact of Promoter Loading*; AIP Publishing LLC: Melville, NY, USA, 2019; p. 020011.
126. Słowik, G.; Greluk, M.; Rotko, M.; Machocki, A. Evolution of the structure of unpromoted and potassium-promoted ceria-supported nickel catalysts in the steam reforming of ethanol. *Appl. Catal. B Environ.* **2018**, *221*, 490–509. [CrossRef]
127. Cifuentes, B.; Hernández, M.; Monsalve, S.; Cobo, M. Hydrogen production by steam reforming of ethanol on a RhPt/CeO_2/SiO_2 catalyst: Synergistic effect of the Si:Ce ratio on the catalyst performance. *Appl. Catal. A Gen.* **2016**, *523*, 283–293. [CrossRef]

128. Wang, F.; Zhang, L.; Zhu, J.; Han, B.; Zhao, L.; Yu, H.; Deng, Z.; Shi, W. Study on different CeO_2 structure stability during ethanol steam reforming reaction over Ir/CeO_2 nanocatalysts. *Appl. Catal. A Gen.* **2018**, *564*, 226–233. [CrossRef]
129. Osorio-Vargas, P.; Campos, C.H.; Navarro, R.M.; Fierro, J.L.; Reyes, P. Rh/Al_2O_3–La_2O_3 catalysts promoted with CeO_2 for ethanol steam reforming reaction. *J. Mol. Catal. A Chem.* **2015**, *407*, 169–181. [CrossRef]
130. Gunduz, S.; Doğu, T. Hydrogen by steam reforming of ethanol over Co–Mg incorporated novel mesoporous alumina catalysts in tubular and microwave reactors. *Appl. Catal. B Environ.* **2015**, *168*, 497–508. [CrossRef]
131. Zhao, L.; Han, T.; Wang, H.; Zhang, L.; Liu, Y. Ni-Co alloy catalyst from $LaNi_{1-x}Co_xO_3$ perovskite supported on zirconia for steam reforming of ethanol. *Appl. Catal. B Environ.* **2016**, *187*, 19–29. [CrossRef]
132. Bepari, S.; Basu, S.; Pradhan, N.C.; Dalai, A.K. Steam reforming of ethanol over cerium-promoted Ni-Mg-Al hydrotalcite catalysts. *Catal. Today* **2017**, *291*, 47–57. [CrossRef]
133. Prasongthum, N.; Xiao, R.; Zhang, H.; Tsubaki, N.; Natewong, P.; Reubroycharoen, P. Highly active and stable Ni supported on CNTs-SiO_2 fiber catalysts for steam reforming of ethanol. *Fuel Process. Technol.* **2017**, *160*, 185–195. [CrossRef]
134. Dai, R.; Zheng, Z.; Sun, C.; Li, X.; Wang, S.; Wu, X.; An, X.; Xie, X. Pt nanoparticles encapsulated in a hollow zeolite microreactor as a highly active and stable catalyst for low-temperature ethanol steam reforming. *Fuel* **2018**, *214*, 88–97. [CrossRef]
135. Dobosz, J.; Małecka, M.; Zawadzki, M. Hydrogen generation via ethanol steam reforming over Co/HAp catalysts. *J. Energy Inst.* **2018**, *91*, 411–423. [CrossRef]
136. Chen, M.; Wang, Y.; Yang, Z.; Liang, T.; Liu, S.; Zhou, Z.; Li, X. Effect of Mg-modified mesoporous Ni/Attapulgite catalysts on catalytic performance and resistance to carbon deposition for ethanol steam reforming. *Fuel* **2018**, *220*, 32–46. [CrossRef]
137. Xiao, Z.; Li, Y.; Hou, F.; Wu, C.; Pan, L.; Zou, J.; Wang, L.; Zhang, X.; Liu, G.; Li, G. Engineering oxygen vacancies and nickel dispersion on CeO_2 by Pr doping for highly stable ethanol steam reforming. *Appl. Catal. B Environ.* **2019**, *258*, 117940. [CrossRef]
138. Lee, J.H.; Do, J.Y.; Park, N.-K.; Ryu, H.-J.; Seo, M.W.; Kang, M. Hydrogen production on $Pd_{0.01}Zn_{0.29}Mg_{0.7}Al_2O_4$ spinel catalyst by low temperature ethanol steam reforming reaction. *J. Energy Inst.* **2019**, *92*, 1064–1076. [CrossRef]
139. Yang, P.; Li, N.; Teng, J.; Wu, J.; Ma, H. Effect of template on catalytic performance of $La_{0.7}Ce_{0.3}Ni_{0.7}Fe_{0.3}O_3$ for ethanol steam reforming reaction. *J. Rare Earths* **2019**, *37*, 594–601. [CrossRef]
140. Dai, R.; Zheng, Z.; Yan, W.; Lian, C.; Wu, X.; An, X.; Xie, X. Dragon fruit-like Pt-Cu@mSiO_2 nanocomposite as an efficient catalyst for low-temperature ethanol steam reforming. *Chem. Eng. J.* **2020**, *379*, 122299. [CrossRef]
141. Zhurka, M.D.; Lemonidou, A.A.; Kechagiopoulos, P.N. Elucidation of metal and support effects during ethanol steam reforming over Ni and Rh based catalysts supported on (CeO_2)-ZrO_2-La_2O_3. *Catal. Today* **2020**. [CrossRef]
142. Morales, M.; Segarra, M. Steam reforming and oxidative steam reforming of ethanol over $La_{0.6}Sr_{0.4}CoO_{3-}$ perovskite as catalyst precursor for hydrogen production. *Appl. Catal. A Gen.* **2015**, *502*, 305–311. [CrossRef]
143. Fang, W.; Pirez, C.; Paul, S.; Jiménez-Ruiz, M.; Jobic, H.; Dumeignil, F.; Jalowiecki-Duhamel, L. Advanced functionalized $Mg_2AlNiXHZOY$ nano-oxyhydrides ex-hydrotalcites for hydrogen production from oxidative steam reforming of ethanol. *Int. J. Hydrogen Energy* **2016**, *41*, 15443–15452. [CrossRef]
144. Muñoz, M.; Moreno, S.; Molina, R. Promoter effect of Ce and Pr on the catalytic stability of the Ni-Co system for the oxidative steam reforming of ethanol. *Appl. Catal. A Gen.* **2016**, *526*, 84–94. [CrossRef]
145. Hsieh, H.-C.; Tsai, P.-W.; Chang, Y.-C.; Weng, S.-F.; Sheu, H.-S.; Chuang, Y.-C.; Lee, C.-S. Oxidative steam reforming of ethanol over $M_xLa_{2-x}Ce_{1.8}Ru_{0.2}O_{7-\delta}$ (M = Mg, Ca) catalysts: Effect of alkaline earth metal substitution and support on stability and activity. *RSC Adv.* **2019**, *9*, 39932–39944. [CrossRef]
146. Bej, B.; Bepari, S.; Pradhan, N.C.; Neogi, S. Production of hydrogen by dry reforming of ethanol over alumina supported nano-NiO/SiO_2 catalyst. *Catal. Today* **2017**, *291*, 58–66. [CrossRef]
147. Samsudeen, K.; Ahmed, A.F.; Yahya, M.; Ahmed, A.; Anis, F. Effect of Calcination Temperature on Hydrogen Production via Ethanol Dry Reforming Over Ni/Al_2O_3 Catalyst. *Int. J. Res. Sci.* **2018**, *4*, 5–9. [CrossRef]
148. Usman, M.; Daud, W.M.A.W. Recent advances in the methanol synthesis via methane reforming processes. *RSC Adv.* **2015**, *5*, 21945–21972. [CrossRef]

149. Kothandaraman, J.; Kar, S.; Goeppert, A.; Sen, R.; Prakash, G.K.S. Advances in Homogeneous Catalysis for Low Temperature Methanol Reforming in the Context of the Methanol Economy. *Top. Catal.* **2018**, *61*, 542–559. [CrossRef]
150. Lee, H.; Jung, I.; Roh, G.; Na, Y.; Kang, H. Comparative Analysis of On-Board Methane and Methanol Reforming Systems Combined with HT-PEM Fuel Cell and CO_2 Capture/Liquefaction System for Hydrogen Fueled Ship Application. *Energies* **2020**, *13*, 224. [CrossRef]
151. Araya, S.S.; Liso, V.; Cui, X.; Li, N.; Zhu, J.; Sahlin, S.L.; Jensen, S.; Nielsen, M.; Kær, S. A Review of The Methanol Economy: The Fuel Cell Route. *Energies* **2020**, *13*, 596. [CrossRef]
152. Iulianelli, A.; Ghasemzadeh, K.; Basile, A. Progress in Methanol Steam Reforming Modelling via Membrane Reactors Technology. *Membranes* **2018**, *8*, 65. [CrossRef]
153. Palma, V.; Ruocco, C.; Martino, M.; Meloni, E.; Ricca, A. Bimetallic supported catalysts for hydrocarbons and alcohols reforming reactions. In *Hydrogen Production, Separation and Purification for Energy*; Institution of Engineering and Technology: London, UK, 2017; pp. 39–70.
154. Xu, X.; Shuai, K.; Xu, B. Review on copper and palladium based catalysts for methanol steam reforming to produce hydrogen. *Catalysts* **2017**, *7*, 183. [CrossRef]
155. Iulianelli, A.; Ribeirinha, P.; Mendes, A.; Basile, A. Methanol steam reforming for hydrogen generation via conventional and membrane reactors: A review. *Renew. Sustain. Energy Rev.* **2014**, *29*, 355–368. [CrossRef]
156. Tonelli, F.; Gorriz, O.; Tarditi, A.M.; Cornaglia, L.; Arrúa, L.; Abello, M.C. Activity and stability of a CuO/CeO_2 catalyst for methanol steam reforming. *Int. J. Hydrogen Energy* **2015**, *40*, 13379–13387. [CrossRef]
157. Das, D.; Llorca, J.; Domínguez, M.; Colussi, S.; Trovarelli, A.; Gayen, A. Methanol steam reforming behavior of copper impregnated over CeO_2–ZrO_2 derived from a surfactant assisted coprecipitation route. *Int. J. Hydrogen Energy* **2015**, *40*, 10463–10479. [CrossRef]
158. Deshmane, V.G.; Abrokwah, R.Y.; Kuila, D. Synthesis of stable Cu-MCM-41 nanocatalysts for H_2 production with high selectivity via steam reforming of methanol. *Int. J. Hydrogen Energy* **2015**, *40*, 10439–10452. [CrossRef]
159. Xu, T.; Zou, J.; Tao, W.; Zhang, S.; Cui, L.; Zeng, F.; Wang, D.; Cai, W. Co-nanocasting synthesis of Cu based composite oxide and its promoted catalytic activity for methanol steam reforming. *Fuel* **2016**, *183*, 238–244. [CrossRef]
160. Thattarathody, R.; Sheintuch, M. Kinetics and dynamics of methanol steam reforming on CuO/ZnO/alumina catalyst. *Appl. Catal. A Gen.* **2017**, *540*, 47–56. [CrossRef]
161. Bagherzadeh, S.B.; Haghighi, M. Plasma-enhanced comparative hydrothermal and coprecipitation preparation of $CuO/ZnO/Al_2O_3$ nanocatalyst used in hydrogen production via methanol steam reforming. *Energy Convers. Manag.* **2017**, *142*, 452–465. [CrossRef]
162. Ajamein, H.; Haghighi, M.; Alaei, S. The role of various fuels on microwave-enhanced combustion synthesis of $CuO/ZnO/Al_2O_3$ nanocatalyst used in hydrogen production via methanol steam reforming. *Energy Convers. Manag.* **2017**, *137*, 61–73. [CrossRef]
163. Abrokwah, R.Y.; Deshmane, V.G.; Kuila, D. Comparative performance of M-MCM-41 (M: Cu, Co, Ni, Pd, Zn and Sn) catalysts for steam reforming of methanol. *J. Mol. Catal. A Chem.* **2016**, *425*, 10–20. [CrossRef]
164. Papadopoulou, E.; Ioannides, T. Methanol Reforming over Cobalt Catalysts Prepared from Fumarate Precursors: TPD Investigation. *Catalysts* **2016**, *6*, 33. [CrossRef]
165. Li, J.; Mei, X.; Zhang, L.; Yu, Z.; Liu, Q.; Wei, T.; Wu, W.; Dong, D.; Xu, L.; Hu, X. A comparative study of catalytic behaviors of Mn, Fe, Co, Ni, Cu and Zn–Based catalysts in steam reforming of methanol, acetic acid and acetone. *Int. J. Hydrogen Energy* **2020**, *45*, 3815–3832. [CrossRef]
166. Maiti, S.; Llorca, J.; Dominguez, M.; Colussi, S.; Trovarelli, A.; Priolkar, K.R.; Aquilanti, G.; Gayen, A. Combustion synthesized copper-ion substituted $FeAl_2O_4$ ($Cu_{0.1}Fe_{0.9}Al_2O_4$): A superior catalyst for methanol steam reforming compared to its impregnated analogue. *J. Power Sources* **2016**, *304*, 319–331. [CrossRef]
167. Qing, S.-J.; Hou, X.-N.; Liu, Y.-J.; Wang, L.; Li, L.-D.; Gao, Z.-X. Catalytic performance of Cu-Ni-Al spinel for methanol steam reforming to hydrogen. *J. Fuel Chem. Technol.* **2018**, *46*, 1210–1217. [CrossRef]
168. Luo, X.; Hong, Y.; Wang, F.; Hao, S.; Pang, C.H.; Lester, E.H.; Wu, T. Development of nano Ni_xMg_yO solid solutions with outstanding anti-carbon deposition capability for the steam reforming of methanol. *Appl. Catal. B Environ.* **2016**, *194*, 84–97. [CrossRef]

169. Zeng, Z.; Liu, G.; Geng, J.; Jing, D.; Hong, X.; Guo, L. A high-performance PdZn alloy catalyst obtained from metal-organic framework for methanol steam reforming hydrogen production. *Int. J. Hydrogen Energy* **2019**, *44*, 24387–24397. [CrossRef]
170. Claudio-Piedras, A.; Ramírez-Zamora, R.M.; Alcántar-Vázquez, B.C.; Gutiérrez-Martínez, A.; Mondragón-Galicia, G.; Morales-Anzures, F.; Pérez-Hernández, R.; Modragón-Galicia, G. One dimensional Pt/CeO_2-NR catalysts for hydrogen production by steam reforming of methanol: Effect of Pt precursor. *Catal. Today* **2019**. [CrossRef]
171. Barrios, C.; Bosco, M.V.; Baltanás, M.A.; Bonivardi, A.L. Hydrogen production by methanol steam reforming: Catalytic performance of supported-Pd on zinc–cerium oxides' nanocomposites. *Appl. Catal. B Environ.* **2015**, *179*, 262–275. [CrossRef]
172. Ajamein, H.; Haghighi, M.; Shokrani, R.; Abdollahifar, M. On the solution combustion synthesis of copper based nanocatalysts for steam methanol reforming: Effect of precursor, ultrasound irradiation and urea/nitrate ratio. *J. Mol. Catal. A Chem.* **2016**, *421*, 222–234. [CrossRef]
173. Lin, L.; Zhou, W.; Gao, R.; Yao, S.; Zhang, X.; Xu, W.; Zheng, S.; Jiang, Z.; Yu, Q.; Li, Y.-W.; et al. Low-temperature hydrogen production from water and methanol using Pt/α-MoC catalysts. *Nature* **2017**, *544*, 80–83. [CrossRef]
174. Cai, F.; Ibrahim, J.J.; Fu, Y.; Kong, W.; Zhang, J.; Sun, Y. Low-temperature hydrogen production from methanol steam reforming on Zn-modified Pt/MoC catalysts. *Appl. Catal. B Environ.* **2020**, *264*, 118500. [CrossRef]
175. Liu, X.; Men, Y.; Wang, J.; He, R.; Wang, Y. Remarkable support effect on the reactivity of Pt/In_2O_3/MO_x catalysts for methanol steam reforming. *J. Power Sources* **2017**, *364*, 341–350. [CrossRef]
176. Diaz-Perez, M.A.; Moya, J.; Serrano-Ruiz, J.C.; Faria, J. Interplay of support chemistry and reaction conditions on copper catalyzed methanol steam reforming. *Ind. Eng. Chem. Res.* **2018**, *57*, 15268–15279. [CrossRef] [PubMed]
177. Tahay, P.; Khani, Y.; Jabari, M.; Bahadoran, F.; Safari, N.; Zamanian, A. Synthesis of cubic and hexagonal $ZnTiO_3$ as catalyst support in steam reforming of methanol: Study of physical and chemical properties of copper catalysts on the H_2 and CO selectivity and coke formation. *Int. J. Hydrogen Energy* **2020**, *45*, 9484–9495. [CrossRef]
178. Talkhoncheh, S.K.; Minaei, S.; Ajamein, H.; Haghighi, M.; Abdollahifar, M. Synthesis of CuO/ZnO/Al_2O_3 /ZrO_2/CeO_2 nanocatalysts via homogeneous precipitation and combustion methods used in methanol steam reforming for fuel cell grade hydrogen production. *RSC Adv.* **2016**, *6*, 57199–57209. [CrossRef]
179. Taghizadeh, M.; Akhoundzadeh, H.; Rezayan, A.; Sadeghian, M. Excellent catalytic performance of 3D-mesoporous KIT-6 supported Cu and Ce nanoparticles in methanol steam reforming. *Int. J. Hydrogen Energy* **2018**, *43*, 10926–10937. [CrossRef]
180. Phongboonchoo, Y.; Thouchprasitchai, N.; Pongstabodee, S. Hydrogen production with a low carbon monoxide content via methanol steam reforming over $Cu_xCe_yMg_z/Al_2O_3$ catalysts: Optimization and stability. *Int. J. Hydrogen Energy* **2017**, *42*, 12220–12235. [CrossRef]
181. Hou, X.; Qing, S.; Liu, Y.; Li, L.; Gao, Z.; Qin, Y. Enhancing effect of MgO modification of Cu–Al spinel oxide catalyst for methanol steam reforming. *Int. J. Hydrogen Energy* **2020**, *45*, 477–489. [CrossRef]
182. Liu, X.; Toyir, J.; De La Piscina, P.R.; Homs, N. Hydrogen production from methanol steam reforming over Al_2O_3- and ZrO_2- modified $CuOZnOGa_2O_3$ catalysts. *Int. J. Hydrogen Energy* **2017**, *42*, 13704–13711. [CrossRef]
183. Mohtashami, Y.; Taghizadeh, M. Performance of the ZrO2 promoted Cu ZnO catalyst supported on acetic acid-treated MCM-41 in methanol steam reforming. *Int. J. Hydrogen Energy* **2019**, *44*, 5725–5738. [CrossRef]
184. Lu, J.; Li, X.; He, S.; Han, C.; Wan, G.; Lei, Y.; Chen, R.; Liu, P.; Chen, K.; Zhang, L.; et al. Hydrogen production via methanol steam reforming over Ni-based catalysts: Influences of Lanthanum (La) addition and supports. *Int. J. Hydrogen Energy* **2017**, *42*, 3647–3657. [CrossRef]
185. Azenha, C.; Mateos-Pedrero, C.; Queirós, S.; Concepción, P.; Mendes, A. Innovative ZrO_2-supported CuPd catalysts for the selective production of hydrogen from methanol steam reforming. *Appl. Catal. B Environ.* **2017**, *203*, 400–407. [CrossRef]
186. Liu, D.; Men, Y.; Wang, J.; Kolb, G.; Liu, X.; Wang, Y.; Sun, Q. Highly active and durable Pt/In_2O_3/Al_2O_3 catalysts in methanol steam reforming. *Int. J. Hydrogen Energy* **2016**, *41*, 21990–21999. [CrossRef]
187. Martinelli, M.; Jacobs, G.; Graham, U.; Crocker, M. Methanol Steam Reforming: Na Doping of Pt/YSZ Provides Fine Tuning of Selectivity. *Catalysts* **2017**, *7*, 148. [CrossRef]

188. Zhang, R.; Huang, C.; Zong, L.; Lu, K.; Wang, X.; Cai, J. Hydrogen Production from Methanol Steam Reforming over TiO_2 and CeO_2 Pillared Clay Supported Au Catalysts. *Appl. Sci.* **2018**, *8*, 176. [CrossRef]
189. Lytkina, A.A.; Orekhova, N.V.; Ermilova, M.; Yaroslavtsev, A.B. The influence of the support composition and structure ($M_XZr_{1-X}O_{2-\delta}$) of bimetallic catalysts on the activity in methanol steam reforming. *Int. J. Hydrogen Energy* **2018**, *43*, 198–207. [CrossRef]
190. Lu, P.-J.; Cai, F.-F.; Zhang, J.; Liu, Y.; Sun, Y.-H. Hydrogen production from methanol steam reforming over B-modified CuZnAlO catalysts. *J. Fuel Chem. Technol.* **2019**, *47*, 791–798. [CrossRef]
191. Maiti, S.; Das, D.; Pal, K.; Llorca, J.; Soler, L.; Colussi, S.; Trovarelli, A.; Priolkar, K.R.; Sarode, P.; Asakura, K.; et al. Methanol steam reforming behavior of sol-gel synthesized nanodimensional $Cu_xFe_{1-x}Al_2O_4$ hercynites. *Appl. Catal. A Gen.* **2019**, *570*, 73–83. [CrossRef]
192. Song, Q.; Men, Y.; Wang, J.; Liu, S.; Chai, S.; An, W.; Wang, K.; Li, Y.; Tang, Y. Methanol steam reforming for hydrogen production over ternary composite ZnyCe1Zr9Ox catalysts. *Int. J. Hydrogen Energy* **2020**, *45*, 9592–9602. [CrossRef]
193. Mateos-Pedrero, C.; Silva, H.; Tanaka, D.P.; Liguori, S.; Iulianelli, A.; Basile, A.; Mendes, A. CuO/ZnO catalysts for methanol steam reforming: The role of the support polarity ratio and surface area. *Appl. Catal. B Environ.* **2015**, *174*, 67–76. [CrossRef]
194. Kim, S.; Yun, S.-W.; Lee, B.; Heo, J.; Kim, K.; Kim, Y.-T.; Lim, H. Steam reforming of methanol for ultra-pure H_2 production in a membrane reactor: Techno-economic analysis. *Int. J. Hydrogen Energy* **2019**, *44*, 2330–2339. [CrossRef]
195. Köpfle, N.; Mayr, L.; Schmidmair, D.; Bernardi, J.; Knop-Gericke, A.; Hävecker, M.; Klötzer, B.; Penner, S. A Comparative Discussion of the Catalytic Activity and CO_2-Selectivity of Cu-Zr and Pd-Zr (Intermetallic) Compounds in Methanol Steam Reforming. *Catalysts* **2017**, *7*, 53. [CrossRef]
196. Zhou, W.; Ke, Y.; Pei, P.; Yu, W.; Chu, X.; Li, S.; Yang, K. Hydrogen production from cylindrical methanol steam reforming microreactor with porous Cu-Al fiber sintered felt. *Int. J. Hydrogen Energy* **2018**, *43*, 3643–3654. [CrossRef]
197. Ke, Y.; Zhou, W.; Chu, X.; Yuan, D.; Wan, S.; Yu, W.; Liu, Y. Porous copper fiber sintered felts with surface microchannels for methanol steam reforming microreactor for hydrogen production. *Int. J. Hydrogen Energy* **2019**, *44*, 5755–5765. [CrossRef]
198. Tajrishi, O.Z.; Taghizadeh, M.; Kiadehi, A.D. Methanol steam reforming in a microchannel reactor by Zn-, Ce- and Zr- modified mesoporous Cu/SBA-15 nanocatalyst. *Int. J. Hydrogen Energy* **2018**, *43*, 14103–14120. [CrossRef]
199. Liu, Y.; Zhou, W.; Lin, Y.; Chen, L.; Chu, X.; Zheng, T.; Wan, S.; Lin, J. Novel copper foam with ordered hole arrays as catalyst support for methanol steam reforming microreactor. *Appl. Energy* **2019**, *246*, 24–37. [CrossRef]
200. Sarafraz, M.; Safaei, M.R.; Goodarzi, M.; Arjomandi, M. Reforming of methanol with steam in a micro-reactor with Cu–SiO_2 porous catalyst. *Int. J. Hydrogen Energy* **2019**, *44*, 19628–19639. [CrossRef]
201. Shanmugam, V.; Neuberg, S.; Zapf, R.; Pennemann, H.; Kolb, G. Hydrogen production over highly active Pt based catalyst coatings by steam reforming of methanol: Effect of support and co-support. *Int. J. Hydrogen Energy* **2020**, *45*, 1658–1670. [CrossRef]
202. Zhuang, X.; Xia, X.; Xu, X.; Li, L. Experimental investigation on hydrogen production by methanol steam reforming in a novel multichannel micro packed bed reformer. *Int. J. Hydrogen Energy* **2020**, *45*, 11024–11034. [CrossRef]
203. Zhu, J.; Araya, S.S.; Cui, X.; Sahlin, S.L.; Kær, S.K. Modeling and design of a multi-tubular packed-bed reactor for methanol steam reforming over a Cu/ZnO/Al_2O_3 catalyst. *Energies* **2020**, *13*, 610. [CrossRef]
204. Ke, C.; Lin, Z. Density functional theory based micro-and macro-kinetic studies of Ni-catalyzed methanol steam reforming. *Catalysts* **2020**, *10*, 349. [CrossRef]
205. Wang, G.; Wang, F.; Chen, B. Performance study on methanol steam reforming rib micro-reactor with waste heat recovery. *Energies* **2020**, *13*, 1564. [CrossRef]
206. Udani, P.P.C.; Gunawardana, P.V.D.S.; Lee, H.C.; Kim, D.H. Steam reforming and oxidative steam reforming of methanol over CuO–CeO_2 catalysts. *Int. J. Hydrogen Energy* **2009**, *34*, 7648–7655. [CrossRef]
207. Kim, J.H.; Jang, Y.S.; Kim, D.H. Multiple steady states in the oxidative steam reforming of methanol. *Chem. Eng. J.* **2018**, *338*, 752–763. [CrossRef]

208. Turco, M.; Bagnasco, G.; Costantino, U.; Marmottini, F.; Montanari, T.; Ramis, G.; Busca, G. Production of hydrogen from oxidative steam reforming of methanol: I. Preparation and characterization of Cu/ZnO/Al_2O_3 catalysts from a hydrotalcite-like LDH precursor. *J. Catal.* **2004**, *228*, 43–55. [CrossRef]
209. Pojanavaraphan, C.; Satitthai, U.; Luengnaruemitchai, A.; Gulari, E. Activity and stability of Au/CeO_2–Fe_2O_3 catalysts for the hydrogen production via oxidative steam reforming of methanol. *J. Ind. Eng. Chem.* **2015**, *22*, 41–52. [CrossRef]
210. Pérez-Hernández, R.; Gutiérrez-Martínez, A.; Espinosa-Pesqueira, M.; Estanislao, M.L.; Palacios, J. Effect of the bimetallic Ni/Cu loading on the ZrO_2 support for H_2 production in the autothermal steam reforming of methanol. *Catal. Today* **2015**, *250*, 166–172. [CrossRef]
211. Mierczynski, P.; Vasilev, K.; Mierczynska, A.; Maniukiewicz, W.; Ciesielski, R.; Rogowski, J.; Szynkowska, I.M.; Trifonov, A.Y.; Dubkov, S.V.; Gromov, D.G. The effect of gold on modern bimetallic Au–Cu/MWCNT catalysts for the oxy-steam reforming of methanol. *Catal. Sci. Technol.* **2016**, *6*, 4168–4183. [CrossRef]
212. Jampa, S.; Jamieson, A.M.; Chaisuwan, T.; Luengnaruemitchai, A.; Wongkasemjit, S. Achievement of hydrogen production from autothermal steam reforming of methanol over Cu-loaded mesoporous CeO2 and Cu-loaded mesoporous CeO_2–ZrO_2 catalysts. *Int. J. Hydrogen Energy* **2017**, *42*, 15073–15084. [CrossRef]
213. Pu, Y.-C.; Li, S.-R.; Yan, S.; Huang, X.; Wang, D.; Ye, Y.-Y.; Liu, Y.-Q. An improved Cu/ZnO catalyst promoted by Sc_2O_3 for hydrogen production from methanol reforming. *Fuel* **2019**, *241*, 607–615. [CrossRef]
214. Adeniyi, A.G.; Ighalo, J.O. A review of steam reforming of glycerol. *Chem. Pap.* **2019**, *73*, 2619–2635. [CrossRef]
215. Roslan, N.A.; Abidin, S.Z.; Ideris, A.; Vo, D.-V.N. A review on glycerol reforming processes over Ni-based catalyst for hydrogen and syngas productions. *Int. J. Hydrogen Energy* **2019**. [CrossRef]
216. Bulutoglu, P.S.; Say, Z.; Bac, S.; Ozensoy, E.; Avci, A.K. Dry reforming of glycerol over Rh-based ceria and zirconia catalysts: New insights on catalyst activity and stability. *Appl. Catal. A Gen.* **2018**, *564*, 157–171. [CrossRef]
217. Bagnato, G.; Iulianelli, A.; Sanna, A.; Basile, A. Glycerol Production and Transformation: A Critical Review with Particular Emphasis on Glycerol Reforming Reaction for Producing Hydrogen in Conventional and Membrane Reactors. *Membranes* **2017**, *7*, 17. [CrossRef]
218. Silva, J.; Soria, M.A.; Madeira, L.M. Challenges and strategies for optimization of glycerol steam reforming process. *Renew. Sustain. Energy Rev.* **2015**, *42*, 1187–1213. [CrossRef]
219. Schwengber, C.A.; Alves, H.J.; Schaffner, R.A.; Silva, F.A.; Sequinel, R.; Bach, V.R.; Ferracin, R.J. Overview of glycerol reforming for hydrogen production. *Renew. Sustain. Energy Rev.* **2016**, *58*, 259–266. [CrossRef]
220. Seadira, T.; Sadanandam, G.; Ntho, T.A.; Lu, X.; Masuku, C.M.; Scurrell, M. Hydrogen production from glycerol reforming: Conventional and green production. *Rev. Chem. Eng.* **2018**, *34*, 695–726. [CrossRef]
221. Charisiou, N.; Siakavelas, G.; Papageridis, K.; Baklavaridis, A.; Tzounis, L.; Polychronopoulou, K.; Goula, M. Hydrogen production via the glycerol steam reforming reaction over nickel supported on alumina and lanthana-alumina catalysts. *Int. J. Hydrogen Energy* **2017**, *42*, 13039–13060. [CrossRef]
222. Gallegos-Suárez, E.; Guerrero-Ruiz, A.; Rodríguez-Ramos, I. Efficient hydrogen production from glycerol by steam reforming with carbon supported ruthenium catalysts. *Carbon* **2016**, *96*, 578–587. [CrossRef]
223. Papageridis, K.; Siakavelas, G.; Charisiou, N.D.; Avraam, D.; Tzounis, L.; Kousi, K.; Goula, M. Comparative study of Ni, Co, Cu supported on γ-alumina catalysts for hydrogen production via the glycerol steam reforming reaction. *Fuel Process. Technol.* **2016**, *152*, 156–175. [CrossRef]
224. Charisiou, N.D.; Papageridis, K.; Siakavelas, G.; Tzounis, L.; Kousi, K.; Baker, M.A.; Hinder, S.J.; Sebastian, V.; Polychronopoulou, K.; Goula, M. Glycerol Steam Reforming for Hydrogen Production over Nickel Supported on Alumina, Zirconia and Silica Catalysts. *Top. Catal.* **2017**, *60*, 1226–1250. [CrossRef]
225. Charisiou, N.; Siakavelas, G.; Tzounis, L.; Dou, B.; Sebastian, V.; Hinder, S.; Baker, M.; Polychronopoulou, K.; Goula, M. Ni/Y_2O_3–ZrO_2 catalyst for hydrogen production through the glycerol steam reforming reaction. *Int. J. Hydrogen Energy* **2020**, *45*, 10442–10460. [CrossRef]
226. Senseni, A.Z.; Rezaei, M.; Meshkani, F. Glycerol steam reforming over noble metal nanocatalysts. *Chem. Eng. Res. Des.* **2017**, *123*, 360–366. [CrossRef]
227. Silva, J.; Ribeiro, L.; Órfão, J.; Soria, M.A.; Madeira, L.M. Low temperature glycerol steam reforming over a Rh-based catalyst combined with oxidative regeneration. *Int. J. Hydrogen Energy* **2019**, *44*, 2461–2473. [CrossRef]

228. Jiang, B.; Zhang, C.; Wang, K.; Dou, B.; Song, Y.; Chen, H.; Xu, Y. Highly dispersed Ni/montmorillonite catalyst for glycerol steam reforming: Effect of Ni loading and calcination temperature. *Appl. Therm. Eng.* **2016**, *109*, 99–108. [CrossRef]
229. Bepari, S.; Pradhan, N.C.; Dalai, A.K. Selective production of hydrogen by steam reforming of glycerol over Ni/Fly ash catalyst. *Catal. Today* **2017**, *291*, 36–46. [CrossRef]
230. Wang, Y.; Chen, M.; Yang, Z.; Liang, T.; Liu, S.; Zhou, Z.; Li, X. Bimetallic Ni-M (M = Co, Cu and Zn) supported on attapulgite as catalysts for hydrogen production from glycerol steam reforming. *Appl. Catal. A Gen.* **2018**, *550*, 214–227. [CrossRef]
231. Carrero, A.; Calles, J.; García-Moreno, L.; Vizcaíno, A. Production of Renewable Hydrogen from Glycerol Steam Reforming over Bimetallic Ni-(Cu,Co,Cr) Catalysts Supported on SBA-15 Silica. *Catalysts* **2017**, *7*, 55. [CrossRef]
232. Wang, R.; Liu, S.; Liu, S.; Li, X.; Zhang, Y.; Xie, C.; Zhou, S.; Qiu, Y.; Luo, S.; Jing, F.; et al. Glycerol steam reforming for hydrogen production over bimetallic MNi/CNTs (M Co, Cu and Fe) catalysts. *Catal. Today* **2019**. [CrossRef]
233. Tavanarad, M.; Meshkani, F.; Rezaei, M. Production of syngas via glycerol dry reforming on Ni catalysts supported on mesoporous nanocrystalline Al_2O_3. *J. CO2 Util.* **2018**, *24*, 298–305. [CrossRef]
234. Bac, S.; Say, Z.; Koçak, Y.; Ercan, K.E.; Harfouche, M.; Ozensoy, E.; Avci, A.K. Exceptionally active and stable catalysts for CO_2 reforming of glycerol to syngas. *Appl. Catal. B Environ.* **2019**, *256*, 117808. [CrossRef]
235. Larimi, A.; Kazemeini, M.; Khorasheh, F. Aqueous phase reforming of glycerol using highly active and stable $Pt_{0.05}Ce_xZr_{0.95-x}O_2$ ternary solid solution catalysts. *Appl. Catal. A Gen.* **2016**, *523*, 230–240. [CrossRef]
236. Espinosa-Moreno, F.; Balla, P.; Shen, W.; Chavarría-Hernández, J.C.; Ruiz-Gómez, M.; Tlecuitl-Beristain, S. Ir-Based Bimetallic Catalysts for Hydrogen Production through Glycerol Aqueous-Phase Reforming. *Catalysts* **2018**, *8*, 613. [CrossRef]
237. Kousi, K.; Chourdakis, N.; Matralis, H.; Kontarides, D.; Papadopoulou, C.; Verykios, X. Glycerol steam reforming over modified Ni-based catalysts. *Appl. Catal. A Gen.* **2016**, *518*, 129–141. [CrossRef]
238. Cheng, C.-K.; Foo, S.Y.; Adesina, A.A. Glycerol Steam Reforming over Bimetallic Co−Ni/Al_2O_3. *Ind. Eng. Chem. Res.* **2010**, *49*, 10804–10817. [CrossRef]
239. Zamzuri, N.H.; Mat, R.; Amin, N.A.S.; Talebian-Kiakalaieh, A. Hydrogen production from catalytic steam reforming of glycerol over various supported nickel catalysts. *Int. J. Hydrogen Energy* **2017**, *42*, 9087–9098. [CrossRef]
240. Bobadilla, L.; Penkova, A.; Alvarez, A.; Domínguez, M.; Romero-Sarria, F.; Centeno, M.A.; Odriozola, J.; Leal, M.I.D. Glycerol steam reforming on bimetallic $NiSn/CeO_2$–MgO–Al_2O_3 catalysts: Influence of the support, reaction parameters and deactivation/regeneration processes. *Appl. Catal. A Gen.* **2015**, *492*, 38–47. [CrossRef]
241. Charisiou, N.; Papageridis, K.; Tzounis, L.; Sebastian, V.; Hinder, S.; Baker, M.; Alketbi, M.; Polychronopoulou, K.; Goula, M. Ni supported on CaO-MgO-Al_2O_3 as a highly selective and stable catalyst for H_2 production via the glycerol steam reforming reaction. *Int. J. Hydrogen Energy* **2019**, *44*, 256–273. [CrossRef]
242. Menezes, J.P.D.S.; Duarte, K.R.; Manfro, R.L.; Souza, M.M.V.M. Effect of niobia addition on cobalt catalysts supported on alumina for glycerol steam reforming. *Renew. Energy* **2020**, *148*, 864–875. [CrossRef]
243. Arif, N.N.M.; Vo, D.-V.N.; Azizan, M.T.; Abidin, S.Z. Carbon Dioxide Dry Reforming of Glycerol for Hydrogen Production using Ni/ZrO_2 and Ni/CaO as Catalysts. *Bull. Chem. React. Eng. Catal.* **2016**, *11*, 200. [CrossRef]
244. Lee, H.-J.; Shin, G.S.; Kim, Y.-C. Characterization of supported Ni catalysts for aqueous-phase reforming of glycerol. *Korean J. Chem. Eng.* **2015**, *32*, 1267–1272. [CrossRef]
245. Yancheshmeh, M.S.; Sahraei, O.A.; Aissaoui, M.; Iliuta, M.C. A novel synthesis of $NiAl_2O_4$ spinel from a Ni-Al mixed-metal alkoxide as a highly efficient catalyst for hydrogen production by glycerol steam reforming. *Appl. Catal. B Environ.* **2020**, *265*, 118535. [CrossRef]
246. Lima, D.S.; Calgaro, C.O.; Perez-Lopez, O.W. Hydrogen production by glycerol steam reforming over Ni based catalysts prepared by different methods. *Biomass Bioenergy* **2019**, *130*, 105358. [CrossRef]
247. Dieuzeide, M.; Laborde, M.; Amadeo, N.; Cannilla, C.; Bonura, G.; Frusteri, F. Hydrogen production by glycerol steam reforming: How Mg doping affects the catalytic behaviour of Ni/Al_2O_3 catalysts. *Int. J. Hydrogen Energy* **2016**, *41*, 157–166. [CrossRef]

248. Demsash, H.; Mohan, R. Steam reforming of glycerol to hydrogen over ceria promoted nickel–alumina catalysts. *Int. J. Hydrogen Energy* **2016**, *41*, 22732–22742. [CrossRef]
249. Carrero, A.; Vizcaíno, A.; Calles, J.; García-Moreno, L. Hydrogen production through glycerol steam reforming using Co catalysts supported on SBA-15 doped with Zr, Ce and La. *J. Energy Chem.* **2017**, *26*, 42–48. [CrossRef]
250. Veiga, S.; Bussi, J. Steam reforming of crude glycerol over nickel supported on activated carbon. *Energy Convers. Manag.* **2017**, *141*, 79–84. [CrossRef]
251. Sahraei, O.A.Z.; Larachi, F.; Abatzoglou, N.; Iliuta, M. Hydrogen production by glycerol steam reforming catalyzed by Ni-promoted Fe/Mg-bearing metallurgical wastes. *Appl. Catal. B Environ.* **2017**, *219*, 183–193. [CrossRef]
252. Dobosz, J.; Cichy, M.; Zawadzki, M.; Borowiecki, T. Glycerol steam reforming over calcium hydroxyapatite supported cobalt and cobalt-cerium catalysts. *J. Energy Chem.* **2018**, *27*, 404–412. [CrossRef]
253. Siew, K.W.; Lee, H.C.; Gimbun, J.; Chin, S.Y.; Khan, M.R.; Taufiq-Yap, Y.H.; Cheng, C.-K. Syngas production from glycerol-dry(CO_2) reforming over La-promoted Ni/Al_2O_3 catalyst. *Renew. Energy* **2015**, *74*, 441–447. [CrossRef]
254. Reynoso, A.; Iriarte-Velasco, U.; Gutiérrez-Ortiz, M.A.; Ayastuy, J. Highly stable $Pt/CoAl_2O_4$ catalysts in Aqueous-Phase Reforming of glycerol. *Catal. Today* **2020**. [CrossRef]
255. Pendem, C.; Sarkar, B.; Siddiqui, N.; Konathala, L.N.S.K.; Baskar, C.; Bal, R. K-Promoted Pt-Hydrotalcite Catalyst for Production of H_2 by Aqueous Phase Reforming of Glycerol. *ACS Sustain. Chem. Eng.* **2017**, *6*, 2122–2131. [CrossRef]
256. Wang, M.; Au, C.T.; Lai, S.Y. H2 production from catalytic steam reforming of n-propanol over ruthenium and ruthenium-nickel bimetallic catalysts supported on ceria-alumina oxides with different ceria loadings. *Int. J. Hydrogen Energy* **2015**, *40*, 13926–13935. [CrossRef]
257. Patel, R.; Patel, S. Process development for bio-butanol steam reforming for PEMFC application. *Int. J. Eng. Technol.* **2018**, *7*, 110–112. [CrossRef]
258. Patel, R.; Patel, S. Effect of operating conditions on hydrogen production in butanol reforming: A review. *Int. Conf. Multidiscip. Res. Pract.* **2015**, 145–150.
259. Patel, R.; Patel, S. Renewable hydrogen production from butanol: A review. *Clean Energy* **2017**, *1*, 90–101. [CrossRef]
260. Harju, H.; Lehtonen, J.; Lefferts, L. Steam reforming of n -butanol over Rh/ZrO_2 catalyst: Role of 1-butene and butyraldehyde. *Appl. Catal. B Environ.* **2016**, *182*, 33–46. [CrossRef]
261. Dhanala, V.; Maity, S.K.; Shee, D. Oxidative steam reforming of isobutanol over $Ni/\gamma\text{-}Al_2O_3$ catalysts: A comparison with thermodynamic equilibrium analysis. *J. Ind. Eng. Chem.* **2015**, *27*, 153–163. [CrossRef]
262. Wang, Y.; Yang, X.; Wang, Y. Catalytic performance of mesoporous MgO supported Ni catalyst in steam reforming of model compounds of biomass fermentation for hydrogen production. *Int. J. Hydrogen Energy* **2016**, *41*, 17846–17857. [CrossRef]
263. Yadav, A.K.; Vaidya, P.D. Renewable hydrogen production by steam reforming of butanol over multiwalled carbon nanotube-supported catalysts. *Int. J. Hydrogen Energy* **2019**, *44*, 30014–30023. [CrossRef]
264. Harju, H.; Lehtonen, J.; Lefferts, L. Steam- and autothermal-reforming of n-butanol over Rh/ZrO_2 catalyst. *Catal. Today* **2015**, *244*, 47–57. [CrossRef]
265. Harju, H.; Pipitone, G.; Lefferts, L. Influence of the catalyst particle size on the aqueous phase reforming of n-butanol over Rh/ZrO_2. *Front. Chem.* **2020**, *8*, 17. [CrossRef]
266. Yadav, A.K.; Vaidya, P.D. A study on the efficacy of noble metal catalysts for butanol steam reforming. *Int. J. Hydrogen Energy* **2019**, *44*, 25575–25588. [CrossRef]
267. Bizkarra, K.; Barrio, V.; Yartu, A.; Requies, J.M.; Arias, P.L.; Cambra, J.F. Hydrogen production from n-butanol over alumina and modified alumina nickel catalysts. *Int. J. Hydrogen Energy* **2015**, *40*, 5272–5280. [CrossRef]
268. Lobo, R.; Marshall, C.L.; Dietrich, P.J.; Ribeiro, F.H.; Akatay, C.; Stach, E.A.; Mane, A.; Lei, Y.; Elam, J.; Miller, J.T. Understanding the chemistry of H_2 production for 1-propanol reforming: Pathway and support modification effects. *ACS Catal.* **2012**, *2*, 2316–2326. [CrossRef]
269. Li, Y.; Zhang, L.; Zhang, Z.; Liu, Q.; Zhang, S.; Liu, Q.; Hu, G.; Wang, Y.; Hu, X. Steam reforming of the alcohols with varied structures: Impacts of acidic sites of Ni catalysts on coking. *Appl. Catal. A Gen.* **2019**, *584*, 117162. [CrossRef]

270. Shejale, A.D.; Yadav, G.D. Noble metal promoted Ni–Cu/La_2O_3–MgO catalyst for renewable and enhanced hydrogen production via steam reforming of bio-based n-butanol: Effect of promotion with Pt, Ru and Pd on catalytic activity and selectivity. *Clean Technol. Environ. Policy* **2019**, *21*, 1323–1339. [CrossRef]
271. Sharma, M.P.; Akyurtlu, J.F.; Akyurtlu, A. Autothermal reforming of isobutanol over promoted nickel xerogel catalysts for hydrogen production. *Int. J. Hydrogen Energy* **2015**, *40*, 13368–13378. [CrossRef]
272. Lei, Y.; Lee, S.; Low, K.-B.; Marshall, C.L.; Elam, J.W. Combining Electronic and Geometric Effects of ZnO-Promoted Pt Nanocatalysts for Aqueous Phase Reforming of 1-Propanol. *ACS Catal.* **2016**, *6*, 3457–3460. [CrossRef]

Article

Kinetics of the Catalytic Thermal Degradation of Sugarcane Residual Biomass Over Rh-Pt/CeO_2-SiO_2 for Syngas Production

Eliana Quiroga [1], Julia Moltó [2,3], Juan A. Conesa [2,3], Manuel F. Valero [1] and Martha Cobo [1,*]

[1] Energy, Materials and Environment Laboratory, Chemical Engineering Department, Universidad de La Sabana, Campus Universitario Puente del Común, Km. 7 Autopista Norte, 250001 Bogotá, Colombia; elianaquco@unisabana.edu.co (E.Q.); manuel.valero@unisabana.edu.co (M.F.V.)

[2] Department of Chemical Engineering, University of Alicante, P.O. Box 99, E-03080 Alicante, Spain; julia.molto@ua.es (J.M.); ja.conesa@ua.es (J.A.C.)

[3] University Institute of Engineering of Chemical Processes, University of Alicante, P.O. Box 99, E-03080 Alicante, Spain

* Correspondence: martha.cobo@unisabana.edu.co; Tel.: +57-1-861-5555 (ext. 25003); Fax: +57-1-861-5555

Received: 3 April 2020; Accepted: 23 April 2020; Published: 6 May 2020

Abstract: Thermochemical processes for biomass conversion are promising to produce renewable hydrogen-rich syngas. In the present study, model fitting methods were used to propose thermal degradation kinetics during catalytic and non-catalytic pyrolysis (in N_2) and combustion (in synthetic air) of sugarcane residual biomass. Catalytic processes were performed over a Rh-Pt/CeO_2-SiO_2 catalyst and the models were proposed based on the Thermogravimetric (TG) analysis, TG coupled to Fourier Transformed Infrared Spectrometry (TG-FTIR) and TG coupled to mass spectrometry (TG-MS). Results showed three different degradation stages and a catalyst effect on product distribution. In pyrolysis, Rh-Pt/CeO_2-SiO_2 catalyst promoted reforming reactions which increased the presence of H_2. Meanwhile, during catalytic combustion, oxidation of the carbon and hydrogen present in biomass favored the release of H_2O, CO and CO_2. Furthermore, the catalyst decreased the overall activation energies of pyrolysis and combustion from 120.9 and 154.9 kJ mol^{-1} to 107.0 and 138.0 kJ mol^{-1}, respectively. Considering the positive effect of the Rh-Pt/CeO_2-SiO_2 catalyst during pyrolysis of sugarcane residual biomass, it could be considered as a potential catalyst to improve the thermal degradation of biomass for syngas production. Moreover, the proposed kinetic parameters are useful to design an appropriate thermochemical unit for H_2-rich syngas production as a non-conventional energy technology.

Keywords: biomass conversion; hydrogen production; kinetic models; lignocellulosic residue; thermal degradation

---

1. Introduction

The increase in energy consumption due to population growth and the dependence on fossil fuels have enlarged greenhouse gases emissions (GHG) with a major impact on environment and global warming [1]. As a result, the use of renewable resources for sustainable energy production has been recently promoted [2]. Lignocellulosic biomass, which includes agricultural and agroindustrial residues [3], is considered as an interesting renewable resource since it has low cost, could be carbon neutral [4], and its conversion implies low GHG emissions [5]. Different processes have been proposed for the use of lignocellulosic biomass, such as pyrolysis [6], gasification [7], combustion [8,9], carbonization [10] and liquefaction [11]. A combination of processes has been proposed as a non-conventional energy technology to produce hydrogen (H_2) from biomass [6,12,13]. H_2 has a high calorific value and can be used in fuel cells (FC), which convert chemical energy into power and heat [14].

Colombia is the third Latin American country in biomass production [15] and generates approximately 72 million tons of agricultural waste per year with a potential energy of at least 331,000 TJ/year [16]. Sugarcane press-mud is a byproduct obtained from the clarification of sugarcane juice during the non-centrifugal sugar production [17]. This residue is obtained with a yield of 3 to 5 wt% [18], which represents about 1.36 Mton/year of sugarcane press-mud [19]. Currently, this residue is used as a raw material for organic fertilizers [17] or, more often, discarded in large quantities, generating pollution in sources of water. Sugarcane press-mud has been used to produce bioethanol through fermentation [19]; nonetheless, it contains approximately 30 wt% lignocellulosic rich solid waste that is currently discarded [19]. This solid waste will be hereinafter called sugarcane residual biomass and will be the focus of this study.

Among thermochemical processes, pyrolysis of biomass is a widely used technology to produce power or syngas in the absence of O_2 [20,21]. Moreover, it is the most studied process since it precedes other thermochemical processes as gasification and combustion [22]. Some studies have shown that, due to the lignocellulosic composition of sugarcane residual biomass, pyrolysis is an alternative to convert it into valuable products, such as H_2-rich syngas, bio-oils and biochar [3,23,24]. However, one of the main problems during thermochemical processes is the low quality of the produced gas due to the presence of higher organic and oxygenated compounds known as tars [25]. These condensable compounds decrease gas yields and process efficiency [26]. In order to improve the gas quality, some authors have proposed the integration of pyrolysis with reforming [27] or gasification [28] to reform the volatiles obtained during pyrolysis, therefore obtaining a H_2-rich gas stream [27]. Nevertheless, these processes are more complex, since each one operates under different optimal conditions [29]. Thus, costs increase because of the need of more than one thermochemical unit [13,27,30]. Hence, catalytic pyrolysis has emerged as a feasible and economic alternative due to several reactions taking place, such as catalytic cracking, reforming and deoxygenation reactions of heavy compounds that allow for organic compounds degradation [6] and carbon conversion [31]. Consequently, less tars and a H_2-rich gas can be obtained in a single step.

Several catalysts have shown to improve the formation of gases during biomass pyrolysis [22,31,32]. Among them, there is a trend in the use of Ni-based catalyst due to the higher activity and low cost [25]. However, they can present deactivation due to the formation of coke on the catalyst surface [33]. Moreover, catalysts with noble metals such as Pt/Al_2O_3 [34], Rh-perovskite [30] and Pt-Rh/MgAl(O) [28] have been tested during integrated pyrolysis processes with steam reforming to improve the quality of condensable and non-condensable gas streams from pyrolysis [35,36]. In these studies, Pt and Rh have shown great activity promoting reforming reactions; Pt has a great selectivity to H_2, and Rh has a great capacity to break O–H bonds, which deliver an increase in the H_2 and CO yields [36]. Although these catalysts have been used in steam reforming, it has been observed that during pyrolysis H_2O is present throughout the temperature range, allowing the reforming reactions to take place [31]. The above is caused by the H_2O contained in the sample and the degradation of hemicellulose and lignin [31]. Thus, studying low noble metal loading (<1% wt) catalysts in pyrolysis could improve the composition of the gas streams obtained from this step, reducing additional equipment requirements or subsequent high temperature conditions.

Besides, in order to obtain a rich gas outlet stream and to avoid catalyst deactivation, the use of multifunctional catalyst that combine different supports has been recently proposed [33]. The presence of CeO_2 in catalysts such as Ni-Ce/Al_2O_3 and Ce/HZSM-5 avoids deactivation, since its redox properties prevent coke formation [33]. Furthermore, noble metal catalysts have shown resistance to deactivation and higher gas yields during biomass pyrolysis [28,30] and other thermochemical processes such as combustion [34] and gasification [37]. CeO_2 used as catalyst support can improve the thermal stability and basicity of the catalyst, which increases CO_2 adsorption by inhibiting coke formation and reducing its deactivation [38]. Additionally, supports such as SiO_2 offer high surface area, increasing the availability of active sites in the catalytic structure [38]. In this sense, a Rh-Pt/CeO_2-SiO_2 catalyst designed by Cifuentes et al. [39] for ethanol steam reforming has shown elevated activity and selectivity to H_2. Consequently, it is proposed to evaluate this catalyst during the biomass pyrolysis for H_2-rich syngas production.

In that sense, understanding the thermal degradation of biomass under different atmospheres is an important step in the design of biomass conversion process to obtain H_2. Thereby, differences in product distribution and kinetic parameters under catalytic and non-catalytic conditions must be addressed for different degradation atmospheres. Thus, pyrolysis (N_2) is commonly compared to combustion (O_2) [40–43], because the latter is the traditional thermal degradation process employed to handle lignocellulosic solid wastes. For that purpose, TG analysis have been widely used in the characterization of thermal degradation of different types of biomass, such as nutshell, pine sawdust [1], other lignocellulosic biomass [40] and plastics [41]. TG analysis provides real-time information on the thermal degradation of the sample as a function of time and temperature [44,45].

Once thermal degradation of the sample is studied, kinetic studies could be performed. Degradation kinetics are an important tool to understand the progress of decomposition reactions [46]. Besides, kinetic study provides kinetic parameters as activation energy (E_i), pre-exponential factor (k) and reaction models that describe the thermal degradation and allow for the design of thermochemical units suitable for this type of residues [41]. Kinetic models of biomass pyrolysis are determined based on the correlation between thermal degradation analysis and information about the released products. This could be done by integrating TG analysis with FTIR (TG-FTIR) and mass spectrometry (TG-MS) [8,47].

Kinetic modelling is usually performed by numerical methods like model fitting methods, which estimate kinetic parameters of the thermal decomposition process using an integral approach, hence the correlation with experimental data is easy and precise [9,48,49]. Gangavati et al. [24] reported the kinetic parameters found through TGA of a press-mud obtained from a sugar mill in India. Parameters were calculated using different relations from literature such as Coast and Redfern, Agrawall and Sivasubramanian methods in order to compare the values obtained [24]. Meanwhile, Garrido et al. [41] studied the thermal decomposition of viscoelastic memory foam by TG Analysis under different atmospheres and proposed a model with three consecutive reactions and the kinetic parameters using integral methods that involve all the heating rates evaluated, which gives more accurate parameters [49,50]. The above agree with Anca-Couce et al. [51], who compared kinetic parameters obtained by model free methods and model fitting methods during beechwood pyrolysis and concluded that model fitting methods are more reliable and show a better fit. However, for catalytic processes, the kinetic parameters have been only obtained by model free methods. For instance, Yang et al. [22] evaluated the effect of the multifunctional Ni-CaO-Ca_2SiO_4 catalyst on the kinetics of catalytic pyrolysis of straw, sawdust and cellulose finding an increase in the intensity of H_2 and CO observed by TG-MS and the reduction of activation energies for all biomasses. Moreover, Loy et al. [5] reported a kinetic parameter during non-catalytic and catalytic pyrolysis of rice husk, using rice hull ash catalyst, obtaining E_i values in the range of 190–186 kJ mol^{-1} and 154–150 kJ mol^{-1}, respectively. The parameters were obtained by model free methods [5].

Therefore, this study aimed to evaluate the effect of the Rh-Pt/CeO_2-SiO_2 catalyst during the pyrolysis and combustion of sugarcane residual biomass. Thermal degradation kinetic models were proposed, and their parameters calculated by model fitting methods based on the released products obtained from TG, TG-FTIR and TG-MS analysis. Obtaining the accurate kinetic parameters of catalytic conditions under different atmospheres and understanding the products evolution of the biomass catalytic pyrolysis will help us with a rigorous reactor design of the thermal degradation of sugarcane residual biomass.

2. Results and Discussion

2.1. Biomass Characterization

Table 1 presents the results of the ultimate and proximal analysis carried out on the sugarcane residual biomass. The composition obtained in the ultimate analysis is comparable to that reported for other biomass such as peat [52], pinewood [53] and vegetable waste [48], among others. The sugarcane residual biomass contains a low percentage of nitrogen and does not contain sulfur, which makes it promising for its thermal conversion since it reduces the emissions of SO_2, NO_x and soot [4]. Besides, it presents a higher heating value of 22.9 MJ kg^{-1}, which is within the average of the energy contained in the traditional coal found in Colombia [54]. Moreover, it is similar to the reported for the olive peel, which has been used to obtain H_2 through pyrolysis [55]. The ashes percentage reported for different types of biomass is lower than 10% [40,45,53]. Sugarcane residual biomass has an 8.1 wt% ash, which is composed mainly of Al, K, Fe and Si, according to the ICP-MS analysis. Depending on the concentration of these metals, they can act as catalysts, modifying the products of the decomposition. However, concentrations of these metals in sugarcane residual biomass are below the minimum concentrations that affect thermal degradation products [56].

Table 1. Characterization of the sugarcane residual biomass.

Moisture (as Received) [wt%]	2
Elemental Analysis [wt%]	
C	50.0
H	7.2
N	0.9
S	N.D
O (by difference) *	33.7
Proximate Analysis [wt%]	
Ashes [1]	8.1
Volatile matter [1]	82.8
Fixed carbon [1]	9.1
Calorific Value [MJ kg^{-1}]	
HHVdb	22.9
LHVdb	21.3
ICP-MS [mg g^{-1}]	
Al	3.30
K	2.58
Fe	2.52
Si	1.07
Mg	1.04

* Free of ashes. [1] Dry basis. N.D not detected.

The IR spectrum of the sugarcane residual biomass (Figure 1) shows signals from the bands associated with the vibrations of the CH_2 and CH_3 (2935–2915 cm^{-1}) and the –OH (3370–3420 cm^{-1}) stretches, which are attributed to the functional groups present in hemicellulose, cellulose and lignin [53]. Besides, the presence of these components is confirmed by bands in 1740, 1329 and 1375 cm^{-1}, characteristic of cellulose and hemicellulose and bands in 1463 and 1240 cm^{-1}, associated with lignin [48,57]. The bands identified at 1740 and 1620 cm^{-1} are related to ketone and ester groups, associated with the fat content of biomass [57]. Bands at 2980 and 2925 cm^{-1} correspond to the stretches of the methyl (C–H) and methylene (=CH_2) groups, respectively [1,58]. Finally, bands between 1200 and 900 cm^{-1} are related to the overlap of polysaccharide and siloxane, and the peak centered at 1050 cm^{-1} is attributed to the symmetric stretching of the polysaccharides (C–O–C) [52]. Therefore, sugarcane residual biomass is composed mainly of volatile matter, which in previous studies was found to be mainly composed of cellulose, hemicellulose and lignin [23]. The content of crude fiber (hemicellulose,

cellulose and lignin) in the sugarcane residual biomass and the characterization of the complete sample was reported in previous studies [19]. Nevertheless, sugarcane residual biomass also presents approximately 18 wt% of ashes and fixed carbon that will have to be considered in the subsequent analysis. The TG analysis of the sugarcane residual biomass is analyzed as follows.

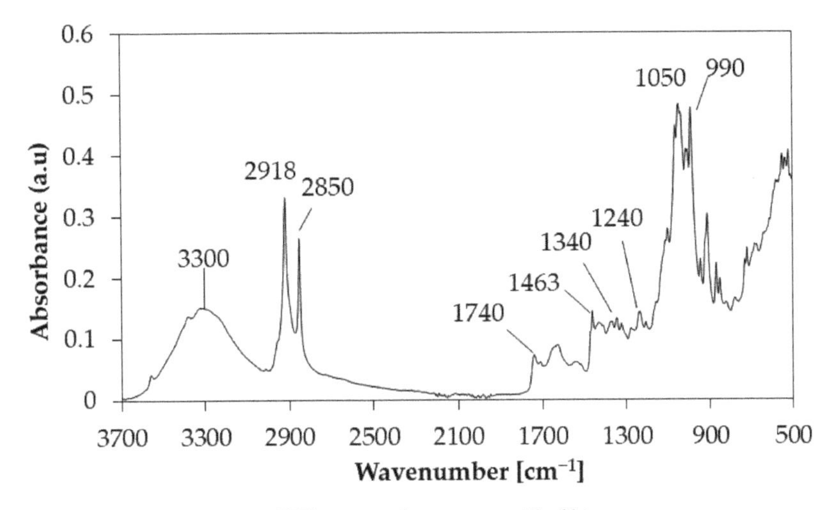

Figure 1. FTIR spectra of sugarcane residual biomass.

2.2. TG Analysis

Thermal degradation of the sugarcane residual biomass was analyzed under N_2 and air atmospheres, in the presence and in the absence of Rh-Pt/CeO_2-SiO_2 catalyst. The biomass/catalyst ratio is an important aspect in thermo-catalytic processes. Sebestyén et al. [59] analyzed two biomass/catalyst ratios, i.e., 10:1 and 1:1, during the catalytic pyrolysis of biomass in the presence of HZSM-5 and found that the effect of the catalyst is poorly visible at a ratio of 10:1. The 1:1 ratio allows to observe the effect of the catalyst in kinetic studies, but it must be varied for applications in pilot-scale reactors. Thus, following previous methodologies of catalytic pyrolysis [6,22], in this study, we used a biomass/catalyst ratio of 1:1 to clearly observe the effect of the catalyst at a lab-scale. Figure 2 shows the DTG curves obtained during the pyrolysis and combustion of both biomass and biomass/catalyst 1:1. During thermal degradation of the biomass by pyrolysis (Figure 2a,b) and combustion (Figure 2c,d), three characteristic degradation stages were identified: dehydration, devolatilization and degradation [26]. The first degradation zone (Stage I) corresponds to the dehydration phase, in which the moisture contained in the sample and some volatile compounds are released [40]. The degradation in this stage is intense in comparison with other studies; this is due to degradation of hemicellulose, cellulose and lignin that forms $H_2O_{(g)}$ from the OH-groups [59]; this is consistent with that reported for pine sawdust, salt sawdust, walnut shell [1] and wood sawdust [45]. The second zone (Stage II) corresponds to the degradation of hemicellulose and cellulose. This stage presents a greater peak for the thermal degradation during combustion, due to the presence of O_2, which allows the conversion of the carbonaceous residues, since oxidation reactions are favored [40]. During pyrolysis (Figure 2a,b), a shoulder is observed, which may be associated with the overlap between the degradation of hemicellulose and cellulose. Finally, there is the third degradation stage (Stage III), which ends at a temperature close to 500 °C and coincides with that reported for sugarcane press-mud [24] and pine sawdust [1]. The last stage corresponds to the overlapping of cellulose and lignin, since, as reported by Yang et al. [57] and Naik et al. [53], the degradation of lignin occurs over the entire range of temperature, with a maximum peak of degradation at temperatures >500 °C. However, no significant weight loss was observed at these temperatures during either pyrolysis or combustion of the sugarcane residual biomass.

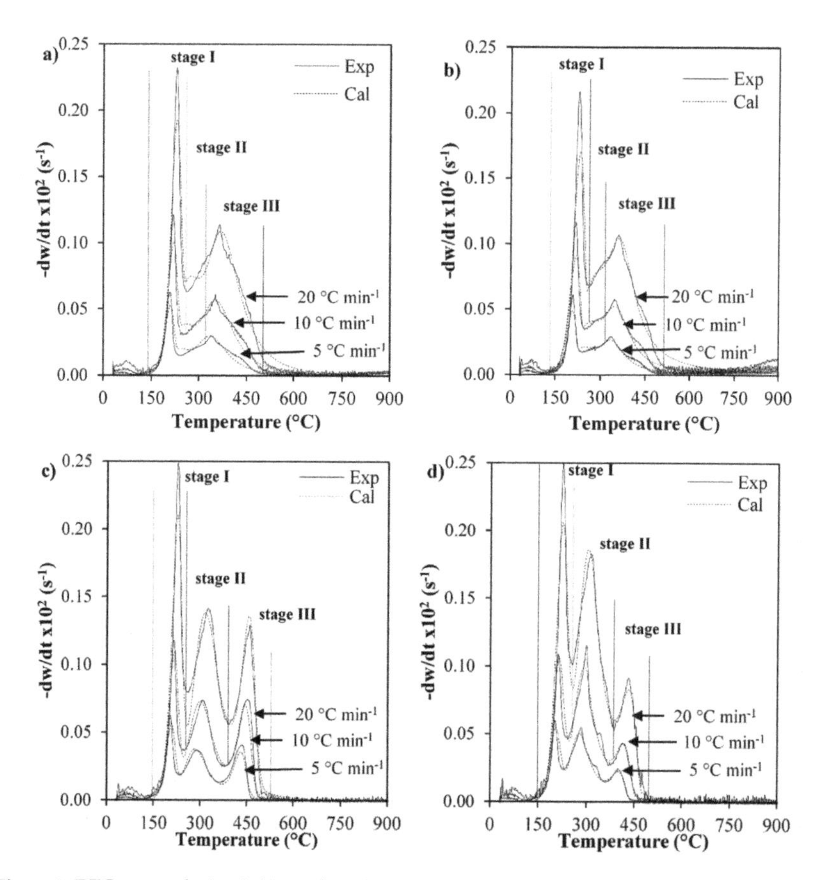

Figure 2. DTG curves during (**a**,**b**) pyrolysis (N_2) and (**c**,**d**) combustion (synthetic air) of (**a**,**c**) biomass and (**b**,**d**) biomass/catalyst 1:1. Catalyst: Rh-Pt/CeO_2-SiO_2. Continuous lines for experimental data (Exp) and dashed lines for calculated data (Cal).

Additionally, during pyrolysis and combustion, no further degradation of samples was achieved at temperatures >500 °C. The final solid obtained for pyrolysis and combustion was 20 and 8 wt%, respectively. Differences between the conversion of the sample in both atmospheres are attributed to the presence of O_2, which favors oxidation reactions including fixed carbon content [40]. Therefore, the final weight percentage of the combustion reaction represents the same value of ash present in the sample (see Table 1). On the contrary, during pyrolysis, the final weight fraction at 900 °C was higher, as carbonaceous compounds (fixed carbon) remained unreacted even at higher temperatures increasing the final biochar. Nonetheless, the final solid fraction of 20 wt% is similar to that reported for wood sawdust [45] and Chlorella vulgaris [60].

In Figure 2, it is observed that, with an increase in the heating rate, the decomposition rates increase. Additionally, a higher conversion is observed during pyrolysis of biomass with a heating rate of 5 °C min^{-1}. Similar results were reported by Mishra and Mohanty [1], who reported that at higher heating rates, biomass does not react completely, causing a greater production of carbonaceous residues. On the other hand, there is no difference between the conversions of the sample for the different reaction rates in the oxidizing atmosphere, surely caused by the presence of oxygen which accelerates decomposition.

Additionally, Figure 2b,d shows the DTG curves during pyrolysis and combustion of sugarcane residual biomass in the presence of the Rh-Pt/CeO_2-SiO_2 catalyst. For both atmospheres, a shift is observed to the left of the curves for all the heating rates. This suggests that the catalyst reduces thermal degradation temperatures [26], causing the sample to degrade faster and at lower temperature. Catalytic pyrolysis also shows a greater degradation of the sample, which suggests that the catalyst is active even at temperatures <400 °C, where different reactions are occurring compared to the sample without catalyst (Figure 2a). On the contrary, Loy et al. [61] observed a lower degradation during pyrolysis of rice husk biomass with commercial Ni powder catalysts compared to the sample without catalyst. They attributed this behavior to the possible coke deposits due to polymerization reactions that deactivates the catalyst, leading to a lower conversion of the sample [61]. This suggests that catalytic pyrolysis over Rh-Pt/CeO_2-SiO_2 allows further degradation of the biomass because it is active for the decomposition of other compounds at higher temperatures.

Current results suggest that the pyrolysis and combustion of biomass follow a different path, since there is a difference between the degradation of the three main components (i.e., hemicellulose, cellulose and lignin). Therefore, it is important to associate each degradation stage with the released products in order to propose the possible reactions that are occurring. For this, TG-FTIR and TG-MS analyses were carried out, and the results are shown in the upcoming sections.

2.3. TG-FTIR Analysis

Figure 3 shows the FTIR spectra obtained for each temperature of the degradation stages mentioned in Section 2.2. Additionally, Table 2 presents the summary of the main signals observed for functional groups such as C=O, C=C, O–H and C–O–H under both atmospheres. Moreover, characteristic bands of the released gaseous products, i.e., CO_2, CO, H_2O and CH_4, are listed. During the non-catalytic pyrolysis (Figure 3a), in stage I of degradation occurring at 243 °C, bands of CO_2 were observed between 1800 and 1400 cm^{-1} (C=O and C=C groups) [52] and close to 2400–2200 cm^{-1}. These bands decrease when the temperature increases, which indicates that C=O and C=C bonds are breaking at higher temperatures. Then, in stages II and III (338 and 375 °C), bands between 3200 and 2700 cm^{-1}, associated with the symmetrical and asymmetric vibrations of the C–H groups [62], appear and increase with temperature. In all the spectra, there is a noise zone between 3900–3300 cm^{-1}, which may be associated with the moisture of the samples. Although in DTG curves (Figure 2) a fourth stage was not identified, a less intense band between 2100–1900 cm^{-1} was observed at 842 °C. This band is associated with the formation of CO due to the breaking of the C=O and C=C bonds and possible OH bonds [42,52], which explains the decrease in the intensity of the bands at 1684, 1718 and 1509 cm^{-1}. Moreover, it confirms that the degradation of these groups leads to the formation of CH_4, due to the increase of bands at 2924 and 1440 cm^{-1} favored by the increase in temperature. The presence of other functional groups such as –CH_2OH, OCH_3, CHO and C–O–H (furans) can be identified in the region between 1900 and 1100 cm^{-1} [47,52].

Table 2. Assignment of the observed bands to the functional groups during the pyrolysis and combustion of sugarcane residual biomass [62,63].

Species	Functional Group	Wavenumber (cm^{-1})
CH_4	C–H	2924, 1440
CO_2	C=O	2361
CO	C–O	2115
Aldehydes, ketones and acids	C=O	1900–1600
Carboxylic acids	C=O	1173
Aromatics	C=C	1640
Overtones of CO_2	C=O	726–586
Hydroxyl group	O–H	3900–3600
H_2O	O–H	1509, 1757
Hydroxyl group of phenolic compounds	O–H	1336, 1450
Phenols	C–O–H	1223

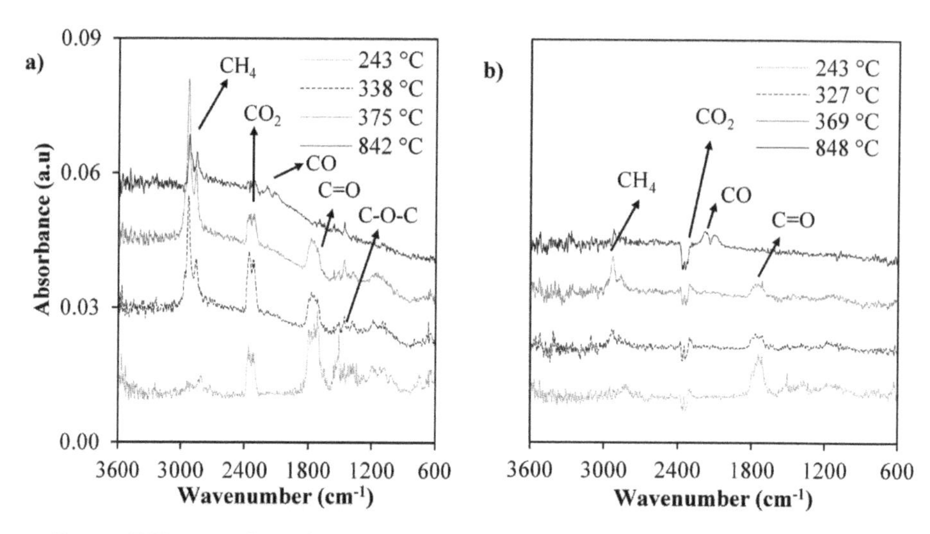

Figure 3. FTIR spectra obtained at the maximum decomposition rates during pyrolysis of (**a**) biomass and (**b**) biomass/catalyst 1:1 at 10 °C min^{-1}. Catalyst: Rh-Pt/CeO_2-SiO_2.

Concerning the spectra obtained for the catalytic pyrolysis (Figure 3b), it shows the signal of the same functional groups as the non-catalytic process. However, bands show a lower intensity and appear at lower temperatures, which coincides with the observed in Figure 2a,b, where the curves shifted to the left due to de decrease in the degradation temperature of the biomass. Besides, the CO_2 band (2330 cm^{-1}) is observed negative, which indicates that CO_2 reacts in the presence of catalyst. This may be related with the dry reforming of hydrocarbons that can be favored [61]. Consequently, a decrease in the intensity of the C=C and CH_4 bands is also observed. During pyrolysis (Figure 3a), the formation of alkanes, alkenes, other hydrogenated compounds and some carboxylic acids or esters are favored. On the contrary, during catalytic pyrolysis (Figure 3b), other reactions are favored by the presence of catalyst that possibly produce H_2 and CO. H_2 cannot be observed through TG-FTIR, but a small peak at about 2115 cm^{-1}, corresponding to CO, suggests that reforming and tar-cracking reactions are occurring [6]. Hence, it can be said that the sugarcane residual biomass follows the three stages of degradation characteristic of biomass in inert atmosphere or pyrolysis [45].

The spectra derived from the maximum temperatures of degradation in the combustion of sugarcane residual biomass are shown in Figure 4. The same bands were identified in both atmospheres (pyrolysis and combustion), but with different intensities. During combustion, the band of CO_2 at 2330 cm^{-1} shows a greater intensity than during pyrolysis. Likewise, in non-catalytic conditions (Figure 4a), CO_2 increases with temperature. Besides, the bands of CO and hydroxyl groups (–OH at 3727 and 669 cm^{-1}), which are related with water [63], appear in the second stage of degradation. Otherwise, in catalytic combustion, bands of CO and CO_2 were identified since stage II of degradation (Figure 4b). The above indicates that the formation of CO_2 through decarboxylation reactions, due to the breakdown of carboxylic acids [62,63], is occurring at lower temperatures, confirming that observed by TG analysis. In the case of pyrolysis, the intensity of the band of the C=O groups increases with temperature (Figure 5).

TG-FTIR results (Figures 3 and 4) provide useful evidence of the formation of volatile organic species during pyrolysis and combustion of the sugarcane residual biomass, in which catalytic pyrolysis promotes the conversion of carbon and the breaking of C=C bonds, releasing more volatiles. Meanwhile, during combustion, despite obtaining a higher sample conversion, the main volatiles obtained were CO_2 and CO, with higher intensities than during pyrolysis. However, due to the similarity between the functional groups of some compounds, the bands cannot be easily assigned. Consequently, the products of degradation are better detailed by TG-MS analysis, as presented in what follows.

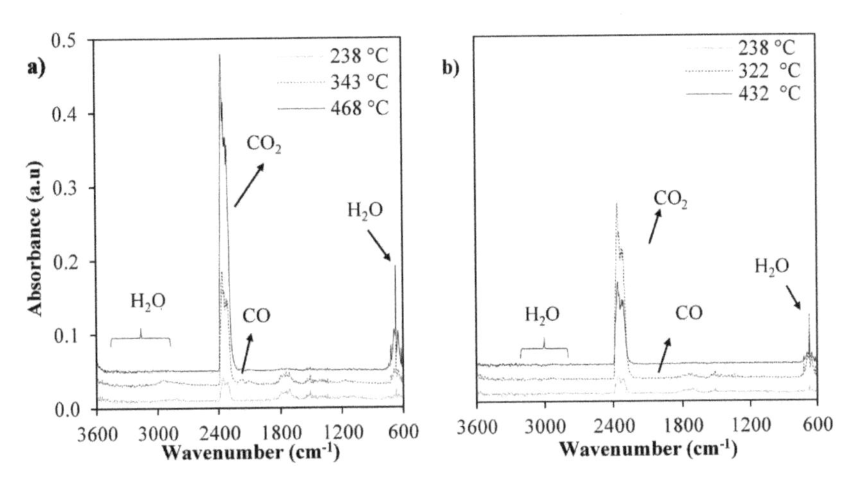

Figure 4. FTIR spectra obtained at the maximum decomposition rates during combustion of (**a**) biomass and (**b**) biomass/catalyst 1:1 at 10 °C min^{-1}. Catalyst: Rh-Pt/CeO_2-SiO_2.

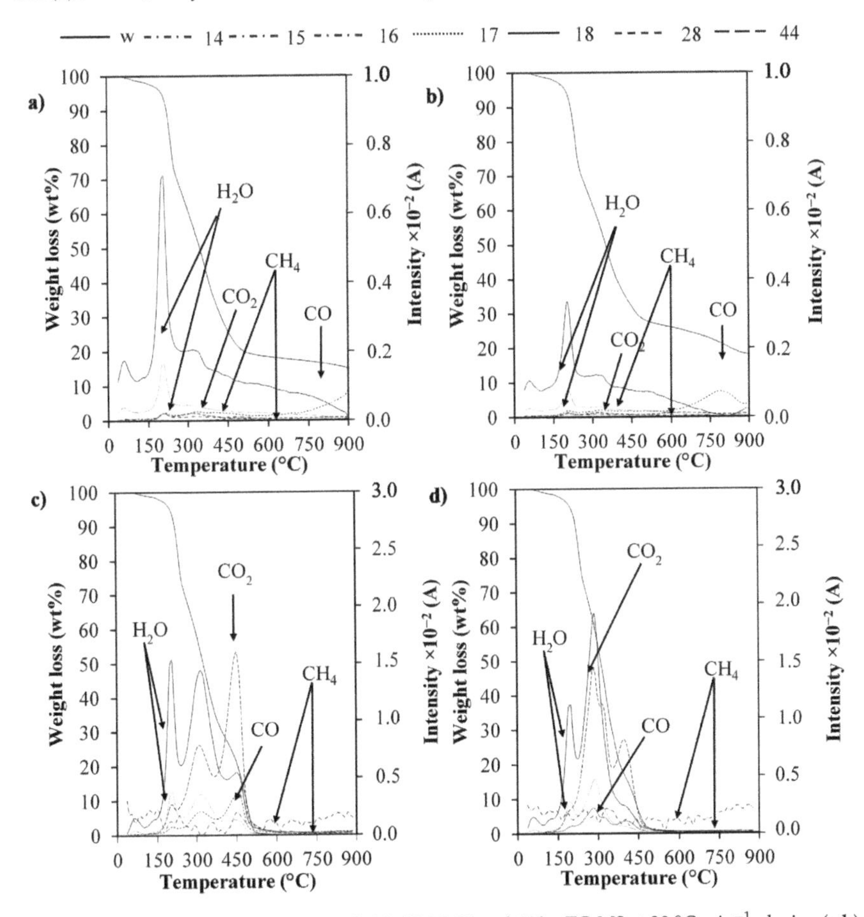

Figure 5. Profiles of the ions m/z = 14, 15, 16, 17, 18, 28 and 44 by TG-MS at 30 °C min^{-1}, during (**a**,**b**) pyrolysis and (**c**,**d**) combustion, for (**a**,**c**) biomass y (**b**,**d**) biomass/catalyst 1:1. Catalyst: Rh-Pt/CeO_2-SiO_2.

2.4. TG-MS Analysis

Figure 5 shows the ion profiles of CO (m/z = 28), CO_2 (m/z = 44), CH_4 (m/z = 14, 15 and 16) and H_2O (m/z = 17 and 18) during catalytic and non-catalytic thermal degradation in pyrolysis and combustion atmospheres. Additionally, a semi-quantitative analysis was carried out, by integrating the intensity vs. temperature data of each ion (Table 3). Moreover, non-catalytic pyrolysis was used to normalize the areas obtained for all catalytic and non-catalytic pyrolysis and combustion; this was done in order to compare the syngas composition under the different conditions. The appearance of the degradation products corresponds to each of the three stages of degradation identified for both pyrolysis and combustion, which is consistent with that observed in TGA and TG-FTIR analyses. In addition, H_2O and CH_4 profiles show changes in their intensity in both atmospheres from the first stages of degradation (Figure 5a,c). Finally, CO and CO_2 are clearly identified during combustion in the three stages of degradation (238, 343 and 468 °C).

Table 3. Relative proportion of key species with respect to non-catalytic pyrolysis.

Species	Key Ion Fragment	m/z	Non-catalytic Pyrolysis	Catalytic Pyrolysis	Non-Catalytic Combustion	Catalytic Combustion
H_2	H_2^+	2	1.0	2.4	0.5	0.6
Methane	CH_3^+	15	1.0	0.6	1.8	1.5
	CH_4^+	16	1.0	0.8	16.6	21.2
Water	OH^+	17	1.0	0.6	2.6	2.2
	H_2O^+	18	1.0	0.5	2.3	1.9
CO	CO^+	28	1.0	1.2	3.6	2.3
C_2^+ Hydrocarbons	$C_2H_2^+$	26	1.0	0.7	2.2	1.3
	$C_2H_3^+$	27	1.0	0.6	2.1	1.0
C_3^+Hydrocarbons	$C_3H_5^+$	41	1.0	0.8	2.0	1.5
	$C_3H_6^+$	42	1.0	0.8	3.1	2.8
	$C_3H_7^+$	43	1.0	0.8	7.3	5.9
Aldehydes	CHO^+	29	1.0	0.9	3.4	1.7
Formaldehyde	CH_2O^+	30	1.0	0.9	11.1	8.7
CO_2	CO_2^+	44	1.0	0.9	28.5	26.6
Alcohols	CH_2OH^+	31	1.0	0.8	56.5	52.4
	$C_2H_5OH^+$	46	1.0	1.0	14.1	13.1

The combustion of sugarcane residual biomass showed CO, CO_2 and H_2O as the main products, with intensities higher than pyrolysis (Figure 5c vs. Figure 5a and Table 3). Furthermore, in presence of catalyst, intensity of CO and CO_2 peaks increased when compared with the combustion without catalyst (Figure 5d vs. Figure 5c). This confirms that the presence of catalyst favors the formation of these products through oxidation reactions of the heavier hydrocarbons (C^+) [59], by chemisorption of O_2 at the active sites of the catalyst. These active sites break the C=C bonds and promote the formation of CO and CO_2 [42]. Therefore, C_2^+, C_3^+ and CH_2O^+ reduced their proportions. The same behavior was observed by TG-FTIR.

Figure 6 shows the ion profiles that make up some of the products of biomass pyrolysis: H_2 (m/z = 2), $C_2H_2^+$ and $C_2H_3^+$, named C_2^+ hydrocarbons (m/z = 26, 27), $C_3H_5^+$ and $C_3H_6^+$, named C_3^+ hydrocarbons (m/z = 39, 41, 42) and formaldehyde (CH_2O^+, m/z = 30) and the effect of the Rh-Pt/CeO_2-SiO_2 catalyst therein. Figure 6b,d shows that the ion profiles are displaced slightly to the left, suggesting that the catalyst promotes the reactions at lower temperatures [59]. In addition, in the atmosphere of non-catalytic pyrolysis (Figure 6a), an increase in the H_2 profile can be observed and only occurs at temperatures above 600 °C. On the other hand, for the biomass/catalyst sample, two peaks of H_2 at 300 and 550 °C are observed, accompanied by a CO peak. Yang et al. [22] studied the catalytic pyrolysis of sawdust and straw on Ni-CaO-Ca_2SiO_4 catalyst and observed only one H_2 peak at 400 °C. Therefore, our Rh-Pt/CeO_2-SiO_2 catalyst is showing itself able to promote reforming reactions between the water released in both, dehydration and degradation stages, and the C_2^+ and

C_3^+ compounds, leading to a release of CO and H_2 [33]. This is confirmed by the decrease in the intensity of C_2^+ and C_3^+ ions and the spectra obtained in Section 2.3 (Figure 3a and Table 3), and by the observed in Figure 5a, where the H_2O intensity is lower for the catalytic pyrolysis and is more revealing that the signal observed in TG-FTIR (Figure 3b). Besides, Table 3 shows that during catalytic pyrolysis, the amount of H_2 is 2.4 times higher than non-catalytic conditions. Minh Loy el al. [61] reported an increase of 1.4 times H_2 composition over non-catalytic pyrolysis of rice husk in the presence of Ni powder catalyst. Furthermore, despite this catalyst being designed for reforming processes, in which it is active at temperatures above 600 °C [39], in Figure 6b it can be seen that Rh-Pt/CeO_2-SiO_2 is active for the production of H_2 even at 300 °C.

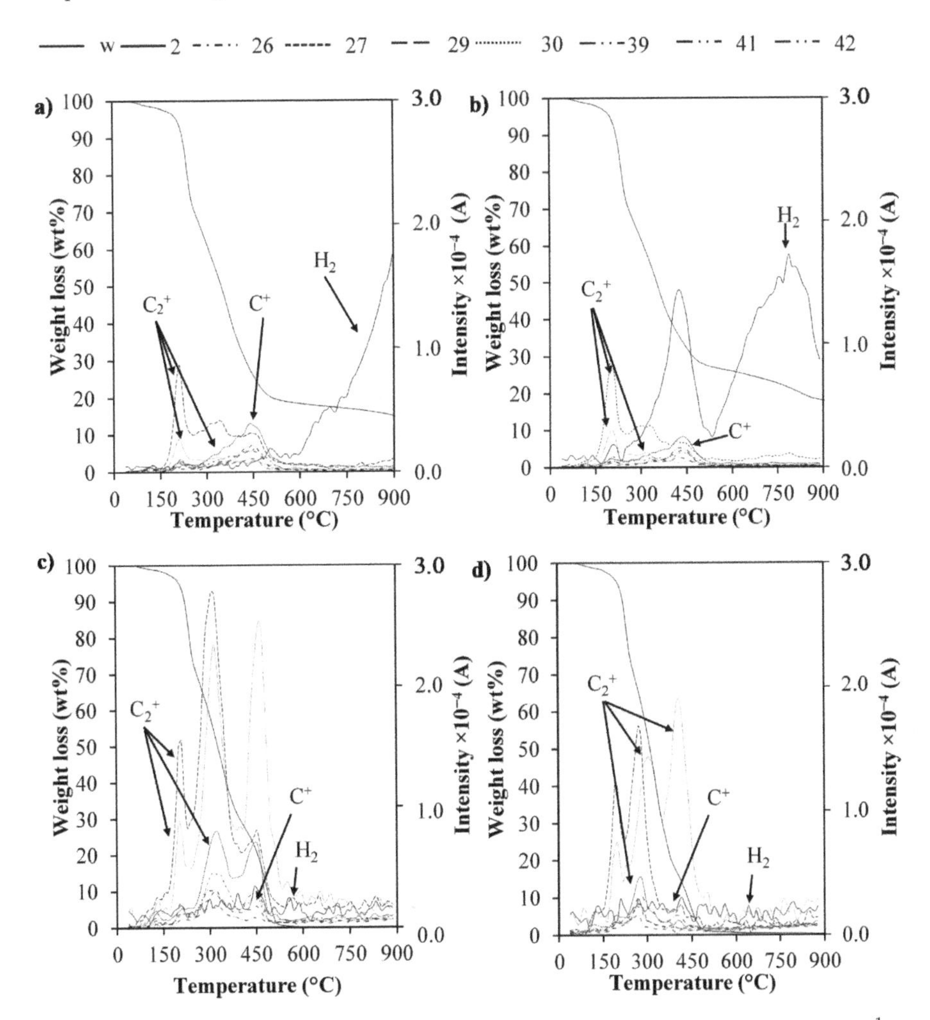

Figure 6. Profiles of the ions m/z = 2, 26, 27, 27, 29, 30, 39, 41 y 42 by TG-MS at 30 °C min^{-1}, during (**a**,**b**) pyrolysis and (**c**,**d**) combustion, for (**a**,**c**) biomass y (**b**,**d**) biomass/catalyst 1:1. Catalyst: Rh-Pt/CeO_2-SiO_2.

On the other hand, Table 3 shows that combustion produces half as much H_2 as non-catalytic pyrolysis, which explains the low intensity of the H_2 ion signal observed in Figure 6c,d. This confirms that the presence of O_2 in the medium causes the hydrocarbons preferentially convert into CO and

CO_2 (Figure 5c,d). These results are consistent with Yuan et al. [40], who reported the presence of H_2 during pyrolysis of rice husks and other wood and plant residues, while did not observe the presence of H_2 during the combustion of the same residues.

These results show in detail some of the compounds identified during the decomposition of sugarcane residual biomass in the two atmospheres, where different reactions occur. Moreover, the catalyst makes the reactions occur faster and at lower temperatures. This leads to a proposal of two different mechanisms of degradation. In the case of pyrolysis, it is observed that in stage I dehydration of the sample occurs, and in sequence, the hemicellulose and cellulose are degraded in light volatile compounds such as CH_4, ethane and ethylene. Meanwhile, during stage II, heavier compounds are obtained, including propylene and carboxylic acids. In Stage III, the higher hydrocarbons disappear as the temperature increases (> 400 °C), and this leads to the formation of H_2, CO and CO_2. During the catalytic pyrolysis, it is observed that the catalyst affects the degradation reactions of the heavy compounds, which occurs at a lower temperature and allows for a higher production of H_2 and CO.

The mechanism proposed for combustion and pyrolysis follows three main parallel reactions. Nevertheless, as noted in the results shown in Sections 2.3 and 2.4, different product distribution occurs mainly by the presence of oxygen during combustion that favors decarboxylation reactions, obtaining CO and CO_2. Therefore, in the following section, the kinetic models for pyrolysis and combustion are proposed in order to obtain the parameters that describe the thermal decomposition of this residue and the effect of the Rh-Pt/CeO_2-SiO_2 catalyst.

2.5. Kinetic Model

Despite differences between the product distribution in the two atmospheres evaluated, both present three stages of degradation, as can be seen in DTG decomposition curves (Section 2.2). The proposed model consists of three parallel reactions (according to the stages of degradation identified), each one following an independent reaction, according to Equation (1). The same models were used in the absence (pyrolysis) and in the presence (combustion) of air and were used to find the kinetic parameters of the reactions in the presence and in the absence of Rh-Pt/CeO_2-SiO_2 catalyst.

$$c_{si0}\text{Component}_i \rightarrow (c_{si0} - v_{i\infty})Residue_i + v_{i\infty}Volatiles_i \quad (1)$$

where "Component$_i$" (i = 1 to 3) refers to different fractions of the original material, "$Volatiles_i$" are the gases and condensable volatiles evolved in the corresponding reactions (i = 1 to 3), and "$Residue_i$" is the char formed in the decomposition of each Component$_i$ (i = 1 to 3). In addition, a fraction of material is introduced into the model that cannot be decomposed under the test conditions, c_{inert0}, which will be different in pyrolysis and combustion. The parameters $v_{i\infty}$ and $(c_{si0} - v_{i\infty})$ are the yield coefficients for volatiles and solid residues, respectively. Finally, c_{si0} represents the sum of initial mass fractions of Component$_i$ [41,43].

Due to the temperature interval where the first component (Component$_1$) is decomposed, it would mainly be related to the degradation of hemicellulose [64]. In the same way, Component$_2$ would be related to the decomposition of the cellulosic components of the biomass, and the third component (Component$_3$) would represent lignin fractions [64]. Nevertheles, the analysis of these different fractions does not represent exacttly the contributions of each component (c_{Si0}), as mentioned in previus works [41,64].

This is also supported by the evolution of volatiles observed in the TG-FTIR and TG-MS analyses (Sections 2.3 and 2.4). The same reaction pathway described by Equation (1) is valid in the presence of oxygen, but obviously it differs from pyrolysis in the presence of oxygen as reactant in the three reactions.

The reaction conversion is defined as the ratio between the mass fractions of the solid reacted at any time $(c_{Si0} - w_{Si})$ and the corresponding initial fraction of this component [43]:

$$\alpha_i = \frac{c_{Si0} - w_{Si}}{c_{Si0}}; i = 1,\ 2,\ 3 \quad (2)$$

In the previous expression, w_{si} is the weight fraction related to the decomposition of Component$_i$. Applying the kinetic law for the proposed solid decomposition, the kinetic expression for the decomposition of each Component$_i$ can be expressed as follows:

$$-\frac{d(w_{si}/c_{si0})}{dt} = d\alpha_i/dt = k_i(1-\alpha_i)^{n_i} \tag{3}$$

where n_i is the reaction order and k_i the kinetic constant of the corresponding reaction that follow the Arrhenius equation:

$$k_i = k_{i0}exp^{E_i/RT},\ i = 1,\ 2,\ 3 \tag{4}$$

k_{i0} is the pre-exponential factor (s^{-1}) and E_i is the apparent activation energy (kJ mol^{-1}). To solve the differential equations described in Equation (3) and find the degree of conversion α_i, the Euler method was used. With the experimental data obtained from the TGA analysis (Section 2.2), 12 parameters were assumed (n_i, k_{i0}, E_i and c_{Si0}) for each of the atmospheres (catalytic and non-catalytic pyrolysis and combustion), but these parameters were maintained for the runs performed at different heating rates. The model data calculated was the change of the mass fraction of each Component$_i$ with time ($-dw_s/dt$):

$$w_s^{cal} = 1 - c_{inert0} - \sum_i c_{Si0}\alpha_i \tag{5}$$

$$-\frac{dw_s^{cal}}{dt} = \sum_i c_{Si0}\frac{d\alpha_i}{dt} \tag{6}$$

The optimization was carried out in Microsoft Excel spreadsheet, using the Solver function. The approach of the functions to be optimized was made by minimizing the square differences between the calculated and experimental data. The objective function (O.F.) to be minimized in each case was:

$$F.O = \sum_{m=1}^{M}\sum_{j=1}^{J}\left(w_s^{exp} - w_s^{cal}\right)^2 \tag{7}$$

where M is the number of experiments, which in this case were three, one for each heating rate (5, 10 and 20 °C min^{-1}), and J is the number of points used in the optimization of each experiment. Finally, the variation coefficient (VC (%)) was calculated to validate the model obtained (Equation (8)) [41,43].

$$VC = \frac{\sqrt{F.O/(J-P)}}{\overline{w_s^{exp}}} * 100 \tag{8}$$

where J is the number of data points in each experiment (approx. 300), and P is number of parameters optimised. $\overline{w_s^{exp}}$ is the average of experimental values of mass fraction for each run.

The kinetic parameters obtained from the optimization of the experimental curves are presented in Table 4. It was observed that the presence of Rh-Pt/CeO_2-SiO_2 catalyst has a positive effect over E_i values of Component$_1$ and Component$_2$ in pyrolysis and Component$_1$ and Component$_3$ in combustion. For the pyrolysis of Components 1 and 2, the presence of catalyst resulted in a decrease of E_i from 133.6 and 108.9 kJ mol^{-1} to 104.2 and 75.1 kJ mol^{-1}, respectively. Meanwhile, in combustion the catalyst decreased the E_i values for Components 1 and 3, from 144.2 and 210.5 kJ mol^{-1} to 127.5 and 156.6 kJ mol^{-1}, respectively. This confirms that the Rh-Pt/CeO_2-SiO_2 catalyst has a positive effect by reducing the difficulty of thermal degradation [59] of sugarcane residual biomass.

The values of E_i and the kinetic constants obtained for the pyrolysis are similar to those reported for rice husk [61], considering that Component$_1$ is mainly related to the decomposition of hemicellulose, Component$_2$ to the cellulose and Component$_3$ to the lignin. Moreover, Wang et al. [32] reported E_i values of 154, 113.3 and 206 kJ mol^{-1}, for hemicellulose, cellulose and lignin, respectively. Although the presence of catalyst does not decrease the E_i for all of the degradation stages, it reduces the E_i of the overall process from 120.9 to 107.0 kJ mol^{-1} in pyrolysis and from 154.9 to 138.0 kJ mol^{-1} in combustion (Table 4). This shows that the Rh-Pt/CeO_2-SiO_2 catalyst is promising for the use in biomass thermal degradation processes and may offer lower energy requirements. Besides, those values are comparable with the values of catalytic pyrolysis reported by Fong et al. [60] over HZSM-5 zeolite/limestone for algae biomass (between 145 and 156 kJ mol^{-1}) and those of in-situ catalytic pyrolysis of rice husk over Ni powder catalyst reported by Loy et al. [61] (between 50 and 163 kJ mol^{-1}), all of them obtained by model free methods.

Table 4. Kinetic parameters obtained for the catalytic and non-catalytic pyrolysis and combustion of sugarcane residual biomass.

Component	Kinetic Parameter	Non-catalytic Pyrolysis	Catalytic Pyrolysis	Non-catalytic Combustion	Catalytic Combustion
1	T_{max}	243	243	238	238
	k_{i0} [s^{-1}]	2×10^{12}	1×10^{9}	3×10^{13}	5×10^{11}
	E_i [kJ·mol^{-1}]	133.6	104.2	144.2	127.5
	n_i	1.0	0.8	1.3	1.3
	c_{i0}	0.18	0.17	0.24	0.26
2	T_{max}	338	327	343	322
	k_{i0} [s^{-1}]	4×10^{8}	7×10^{4}	7×10^{7}	1×10^{10}
	E_i [kJ·mol^{-1}]	108.9	75.1	110.0	130.3
	n_i	4.3	4.6	2.9	3.0
	c_{i0}	0.28	0.48	0.49	0.53
3	T_{max}	375	369	468	432
	k_{i0} [s^{-1}]	5×10^{7}	4×10^{9}	2×10^{13}	5×10^{9}
	E_i [kJ·mol^{-1}]	120.2	141.7	210.5	156.6
	n_i	3.0	2.6	1.2	0.9
	c_{i0}	0.36	0.18	0.21	0.13
C_{inert0}		0.18	0.18	0.07	0.07
VC		3.4%	1.1%	2.7%	1.3%

In the absence of catalyst, c_{i0} calculated values for decomposition of Component$_1$ (Table 3) are similar because dehydration and volatile released products at lower temperatures are the same for both atmospheres [44]; as it was analized in previous Sections (2.3 and 2.4). On the contrary, calculated c_{i0} values for Components 2 and 3 vary considerably between pyrolysis and combustion. For Component$_2$, pyrolysis conversion (0.28) is lower than in combustion (0.49) due to the oxidation reactions that are taking place during combustion, obtainig the greatest intensities of CO and CO_2 ions in the TG-MS results, compared with pyrolysis (Section 2.4). Furthermore, during decomposition of Component$_3$, the degradation of heavier compounds occurs for pyrolysis [61], while in combustion such degradation happened in earlier stages due to the presence of oxygen. Therefore, c_{i0} values of pyrolysis (0.36) for Component$_3$ are higher than the one obtained for combustion (0.21).

In the presence of Rh-Pt/CeO_2-SiO_2 catalyst, c_{i0} values for pyrolysis and combustion increase in Component$_2$ compared to non-catalytic conditions. For pyrolysis, contribution of this Component$_2$ increases due to the degradation of heavier compounds (C^{2+} and C^{3+}) through refomirng reactions [59], which are favored by the catalyst, resulting in the release of H_2. In the case of combustion, the catalyst favors the oxidation reactions and the released C^{2+} and C^{3+} are oxidazed to CO and CO_2 in the second stage of degradation. Since the major conversion of the C^{2+} and C^{3+} at catalytic conditions occurs at lower temperatures, the c_{i0} values of the Component$_3$ were lower for both atmospheres (pyrolysis and combustion) compared with the non-catalytic process.

Figure 2 shows the fitting data for DTG curves. Both pyrolysis and combustion present a good agreement between experimental and calculated curves, showing a good fit using the proposed and optimized model. The variation coefficient for the data obtained to both atmospheres under catalytic and non-catalytic conditions were <5%. The lower Ei values observed, compared with non-catalytic processes, reveal the effect of the Rh-Pt/CeO_2-SiO_2 catalyst on the thermal degradation of biomass. Besides, the lower activation energies obtained in comparison with other biomass makes Rh-Pt/CeO_2-SiO_2 catalyst promising for its use in pyrolysis to syngas production. Full activity and stability tests and catalyst characterization during the catalytic pyrolysis of sugarcane residual biomass over Rh-Pt/CeO_2-SiO_2 is currently ongoing in our laboratory.

3. Materials and Methods

3.1. Biomass Recolection, Pretreatment and Characerization

The liquid sugarcane press-mud residue was collected from Tolima, Colombia. Initially, the press-mud residue was hydrolyzed at 130 °C for 1 h in an autoclave (TOMY Digital Biology, Tokyo, Japan) for subsequent fermentation. Then, the sugarcane press-mud was filtered using a sieve (70-mesh, 212 μm) to remove the solid phase. Afterwards, the solid residue containing the lignocellulosic material was dried at 60 °C for 72 h, grounded and sieved in a AS200 sieve (Retsch, Haan, Germany). Finally, the dry solid fraction (sugarcane residual biomass) with particle sizes <212 μm was the biomass used in this study.

Samples were characterized by elemental analysis using a CHNS analyzer FlashEA 1112 Series (Thermo Fisher Scientific, Waltham, MA, USA). Oxygen content was determined by difference on a dry ash basis. The proximate analysis was performed by thermogravimetry in a TGA/DSC1 (Mettler Toledo, Columbus, OH, USA), following the method described by García et al. [65] The enthalpy of combustion was measured in a calorimetric pump AC-350 (LECO, St. Joseph, MI, USA); this was used to determine the lower and higher heating value on dry basis (LHV_{db} and HHV_{db}), according to Equations (9) and (10).

$$LHV_{db}\left(\text{cal g}^{-1}\right) = \Delta H_{combustion}\left(\text{cal g}^{-1}\right) - 10.56\ (\%N) - 22.01\ (\%S) - 52.56\ (\%H) \quad (9)$$

$$HHV_{db}\left(\text{cal g}^{-1}\right) = LHV_{db}\left(\text{cal g}^{-1}\right) + 52.56\ (\%H) \quad (10)$$

where % N, S and H are the weight percentages from elemental analysis of the sample, $\Delta H_{combustion}$ is the enthalpy of combustion of the biomass, and the numbers represent the different formation enthalpies in cal g^{-1}.

Moreover, the quantitative analysis of the composition of the ashes was carried out by inductive coupling plasma mass spectrometry (ICP-MS 7700x) (Algilent Technologies, Santa Clara, CA, USA). Samples were prepared following the EPA 3051A method (acid digestion with microwaves for sediments, sludges, soils and solids). For this, the digestion of 0.1 g of biomass was performed using 4 mL of HNO_3 and 1 mL of H_2O_2, then the digestion was completed by microwave with a maximum power of 950 W. Finally, the digested sample was filtered with glass fiber and diluted into water to a volume of 25 mL and analyzed in the ICP-MS. To obtain information of the functional groups in the biomass, the sample was characterized by Attenuated Total Reflection Fourier Transform Spectroscopy (ATR-FTIR, IFS 66S) (Bruker, Billerica, MA, USA). Each IR spectrum was obtained in a scanning range of 4000 and 600 cm^{-1} with 4 cm^{-1} of resolution.

3.2. Catalyst Synthesis

The Rh-Pt/CeO_2-SiO_2 catalyst was prepared following the methodology proposed by Cifuentes et al. [38] For this, the mixed support was obtained by dissolving the $Ce(NO_3)_3 \cdot 6H_2O$ (99.9%, Merck, Darmstadt, Germany), as CeO_2 precursor, in distilled water and added slowly to the SiO_2 (Merck, Darmstadt, Germany). Subsequently, the support was dried for 24 h in an oven at 80 °C and calcined at 500 °C for 4 h. Rhodium (III) chloride hydrate ($RhCl_3 \cdot H_2O$) (Merck, Darmstadt, Germany) and hexachloroplatinic acid hexahydrate ($H_2PtCl_6 \cdot 6H_2O$) (Merck, Darmstadt, Germany) were used as precursor salts of the metals and were added by the incipient wet impregnation method [39] up to a total load of 0.4 wt% of each metal. These loads of rhodium (Rh) and platinum (Pt) were selected considering their reported activity in reforming reactions [39]. The final solid was dried at 80 °C for 24 h, then calcined at 700 °C for 2 h and reduced with 8% H_2/He at a flowrate of 200 mL min^{-1}. Finally, to ensure a particle size <177 μm, the final solid obtained was sieved on an 80-mesh sieve. The effect of the Rh-Pt/CeO_2-SiO_2 catalyst was evaluated using a 1:1 biomass/catalyst ratio. This ratio was selected based on the results reported by [6,22,59]. A complete characterization of the catalyst has been previously reported [38], with a surface area of 104 m^2 g_{cat}^{-1}.

3.3. TG Analysis

Thermal degradation of biomass was evaluated at three different heating rates (5, 10 and 20 °C min^{-1}) up to 900 °C in two reaction atmospheres, pyrolysis (N_2) and combustion (synthetic air). These conditions were applied to the sugarcane residual biomass samples with and without catalyst, for a total of 12 experiments. The analysis was carried out in a thermobalance model STA6000 (Perkin Elmer, Waltham, MA, USA). For all the experiments, 2–5 mg of dried samples was used, with a total flow rate of 100 mL min^{-1}. To ensure the reproducibility of the experiments, duplicates of experiments were carried out randomly, ensuring a difference <5%. Weight loss was defined as the ratio between the mass of the solid at any time (m) and the initial mass of the solid (m_0). Moreover, the DTG curves represent the weight change with time.

3.4. TG-FTIR Analysis

Volatile compounds obtained during the thermal degradation were analyzed by TG-FTIR analysis, using a TGA/DSC1 (Mettler Toledo, Columbus, OH, USA), coupled to a Nicolet 6700 FT-IR spectrometer (Thermo Scientific, Waltham, MA, USA). The experiments were carried out in the two reaction atmospheres, pyrolysis and combustion, with a flow rate of 100 mL min^{-1}, heating up to 900 °C at 10 °C min^{-1} with approximately 5 mg of the samples. The absorbance was measured with a resolution of 4 cm^{-1} in a range of 3600–600 cm^{-1}.

3.5. TG-MS Analysis

To identify diatomic molecules such as H_2, which cannot be identified by TG-FTIR, and to associate the identified functional groups with specific compounds, TGA-MS analysis was performed. Tests were carried out in a thermobalance TGA/SDTA851e/LF/1600 model (Mettler Toledo, Columbus, OH, USA), coupled to a mass spectrometer Thermostar GSD301T model (Pfeiffer vacuum, Asslar, Germany), which works on Square-Input Response (SIR) mode with ionization of 70 eV. In these experiments, the gases used were He (pyrolysis) and He:O_2 = 4:1 (combustion), both with a flow rate of 100 mL min^{-1} and approximately 5 mg of sample, heating up to 900 °C at 30 °C min^{-1}. To track a broad spectrum of compounds, two different runs were performed. In the first, the mass/charge ratios (m/z) were followed in the range of 2–46 and the next, in the range of 50–106. The response of the different ions was divided by the He response (m/z = 4). Finally, to obtain the proportions of the species, the areas of the followed ions were calculated integrating the TG-MS results.

3.6. Kinetic Model

A model fitting method was used for the kinetic modeling. For that purpose, a model explaining thermal decomposition in both atmospheres (pyrolysis and combustion) was proposed. This methodology has been used in kinetic models for biomass [44] and other types of materials [41]. The kinetic model proposed for the pyrolysis of biomasses could be interpreted considering the materials formed by three different fractions, as shown in Equation (1). Note that, at first, none of the components is related to a particular chemical structure; i.e., $Component_i$ do not correspond to celullose, hemicellulose or lignin fractions [64].

All raw and processed Excel data from TG, TG-FTIR, TG-MS analysis and the fitting method of the estimated kinetic parameters can be downloaded from [dataset] [66].

4. Conclusions

Catalytic and non-catalytic thermal degradation of sugarcane residual biomass under non-isothermal conditions was studied for pyrolysis and combustion. Under the oxidizing atmosphere of combustion H_2O, CO_2 and CO are mainly produced. Contrarily, H_2O, C^{+2}, C^{+3}, CO and H_2 are the main products during pyrolysis. Catalytic pyrolysis over the Rh-Pt/CeO_2-SiO_2 catalyst increases the production of H_2 at 300 and 550 °C. Moreover, a decrease in H_2O, C^{+2} and C^{+3} products indicates that the catalyst accelerates the formation of light hydrocarbons, favored by cracking and reforming reactions of the heavier compounds.

Product distributions obtained from TG-FTIR and TG-MS analysis were used to propose two kinetic models for the thermal degradation of sugarcane residual biomass. The proposed models presented a good fit (VC <5%) with the experimental data based on the parallel decomposition of three different components. The evolution of volatiles takes place in three different stages: dehydration (stage I), degradation of hemicellulose and cellulose (stage II) and degradation of cellulose and lignin (stage III). The presence of catalyst shows a positive effect on the kinetic parameter reducing the activation energies for both pyrolysis and combustion. In the case of catalytic pyrolysis, the overall activation energies decrease by about 20%–30%, compared with the non-catalytic pyrolysis. In this way, the Rh-Pt/CeO_2-SiO_2 catalyst improves the performance of the sugarcane residual biomass pyrolysis and is presented as a suitable catalyst for obtaining H_2-rich syngas. Furthermore, the kinetic parameters obtained can be used in thermochemical unit design for catalytic pyrolysis.

Author Contributions: Conceptualization, E.Q., J.A.C. and M.C.; methodology, E.Q., M.F.V. and J.A.C.; validation, E.Q., M.F.V. and J.A.C.; investigation, E.Q.; writing—original draft preparation, E.Q.; resources, J.M., J.A.C. and M.C. writing—review and editing, J.M., J.A.C. and M.C.; visualization, J.M. and M.C.; supervision, J.M., M.F.V. and M.C.; project administration, J.M. and M.C.; funding acquisition, J.M. and M.C. All authors have read and agreed to the published version of the manuscript.

Funding: This research was funded by Colciencias (Francisco Jose de Caldas Fund) and Universidad de La Sabana through the Project ING-221 (Colciencias contract 548–2019) and The International Relations Department of University of Alicante for the financial support through the program named "University Development Cooperation 2018".

Acknowledgments: E. Quiroga acknowledges the Universidad de La Sabana for the Teaching Assistant Scholarship for her master's studies.

Conflicts of Interest: There are no conflicts of interest to declare.

References

1. Mishra, R.K.; Mohanty, K. Pyrolysis kinetics and thermal behavior of waste sawdust biomass using thermogravimetric analysis. *Bioresour. Technol.* **2018**, *251*, 63–74. [CrossRef]
2. Organización de las Naciones Unidas (ONU) Combatir el Cambio Climático—Desarrollo Sostenible. Available online: http://www.un.org/sustainabledevelopment/es/combatir-el-cambio-climatico/ (accessed on 11 April 2017).
3. Ansari, K.B.; Gaikar, V.G. Pressmud as an alternate resource for hydrocarbons and chemicals by thermal pyrolysis. *Ind. Eng. Chem. Res.* **2014**, *53*, 1878–1889. [CrossRef]
4. Xiang, Z.; Liang, J.; Morgan, H.M., Jr.; Liu, Y.; Mao, H.; Bu, Q. Thermal behavior and kinetic study for co-pyrolysis of lignocellulosic biomass with polyethylene over Cobalt modified ZSM-5 catalyst by thermogravimetric analysis. *Bioresour. Technol.* **2018**, *247*, 804–811. [CrossRef] [PubMed]
5. Loy, A.C.M.; Gan, D.K.W.; Yusup, S.; Chin, B.L.F.; Lam, M.K.; Shahbaz, M.; Unrean, P.; Acda, M.N.; Rianawati, E. Thermogravimetric kinetic modelling of *in-situ* catalytic pyrolytic conversion of rice husk to bioenergy using rice hull ash catalyst. *Bioresour. Technol.* **2018**, *261*, 213–222. [CrossRef]

6. Zhao, M.; Memon, M.Z.; Ji, G.; Yang, X.; Vuppaladadiyam, A.K.; Song, Y.; Raheem, A.; Li, J.; Wang, W.; Zhou, H. Alkali metal bifunctional catalyst-sorbents enabled biomass pyrolysis for enhanced hydrogen production. *Renew. Energy* **2020**, *148*, 168–175. [CrossRef]
7. Wu, Y.; Liao, Y.; Liu, G.; Ma, X. Syngas production by chemical looping gasification of biomass with steam and CaO additive. *Int. J. Hydrog. Energy* **2018**, *43*, 19375–19383. [CrossRef]
8. Gunasee, S.D.; Carrier, M.; Gorgens, J.F.; Mohee, R. Pyrolysis and combustion of municipal solid wastes: Evaluation of synergistic effects using TGA-MS. *J. Anal. Appl. Pyrolysis* **2016**, *121*, 50–61. [CrossRef]
9. Wang, Q.; Wang, G.; Zhan, J.; Lee, J.-Y.; Wang, H.; Wang, C. Combustion behaviors and kinetics analysis of coal, biomass and plastic. *Thermochim. Acta* **2018**, *669*, 140–148. [CrossRef]
10. Sharma, H.B.; Sarmah, A.K.; Dubey, B. Hydrothermal carbonization of renewable waste biomass for solid biofuel production: A discussion on process mechanism, the influence of process parameters, environmental performance and fuel properties of hydrochar. *Renew. Sustain. Energy Rev.* **2020**, *123*, 109761. [CrossRef]
11. Jaswal, R.; Shende, A.; Nan, W.; Amar, V.; Shende, R. Hydrothermal liquefaction and photocatalytic reforming of pinewood (pinus ponderosa)-derived acid hydrolysis residue for hydrogen and bio-oil production. *Energy Fuels* **2019**, *33*, 6454–6462. [CrossRef]
12. Duman, G.; Watanabe, T.; Uddin, M.A.; Yanik, J. Steam gasification of safflower seed cake and catalytic tar decomposition over ceria modified iron oxide catalysts. *Fuel Process. Technol.* **2014**, *126*, 276–283. [CrossRef]
13. Guo, F.; Li, X.; Liu, Y.; Peng, K.; Guo, C.; Rao, Z. Catalytic cracking of biomass pyrolysis tar over char-supported catalysts. *Energy Convers. Manag.* **2018**, *167*, 81–90. [CrossRef]
14. Lucia, U. Overview on fuel cells. *Renew. Sustain. Energy Rev.* **2014**, *30*, 164–169. [CrossRef]
15. Portafolio "Colombia tiene potencial para producir energía con biomasa" | Infraestructura | Economía | Portafolio. Available online: https://www.portafolio.co/economia/infraestructura/colombia-tiene-potencial-para-producir-energia-con-biomasa-505377 (accessed on 18 February 2019).
16. Escalante, H.; Orduz, J.; Zapata, H.; Cardona, M.C.; Duarte, M. *Atlas del Potencial Energético de la Biomasa Residual en Colombia*; Universidad Industrial de Santander: Bucaramanga, Colombia, 2011.
17. García-Torres, R.; Rios-Leal, E.; Martínez-Toledo, Á.; Ramos-Morales, F.; Cruz-Sánchez, S.; Cuevas-Díaz, M.D.C. Uso de cachaza y bagazo de caña de azúcar en la remoción de hidrocarburos en el suelo contaminado. *Rev. Int. Contam. Ambient.* **2011**, *27*, 31–39.
18. Procaña Asociación Colombiana de Productores y Proveedores de caña de azucar. Available online: http://www.procana.org/new/quienes-somos/subproductos-y-derivados-de-la-ca~na.html (accessed on 13 March 2017).
19. Sanchez, N.; Ruiz, R.Y.; Cifuentes, B.; Cobo, M. Dataset for controlling sugarcane press-mud fermentation to increase bioethanol steam reforming for hydrogen production. *Waste Manag.* **2019**, *98*, 1–13. [CrossRef]
20. Shen, Y.; Wang, J.; Ge, X.; Chen, M. By-products recycling for syngas cleanup in biomass pyrolysis—An overview. *Renew. Sustain. Energy Rev.* **2016**, *59*, 1246–1268. [CrossRef]
21. Domínguez, A.; Menéndez, J.A.; Pis, J.J. Hydrogen rich fuel gas production from the pyrolysis of wet sewage sludge at high temperature. *J. Anal. Appl. Pyrolysis* **2006**, *77*, 127–132. [CrossRef]
22. Yang, H.; Ji, G.; Clough, P.T.; Xu, X.; Zhao, M. Kinetics of catalytic biomass pyrolysis using Ni-based functional materials. *Fuel Process. Technol.* **2019**, *195*, 106145. [CrossRef]
23. Gupta, N.; Tripathi, S.; Balomajumder, C. Characterization of pressmud: A sugar industry waste. *Fuel* **2011**, *90*, 389–394. [CrossRef]
24. Gangavati, P.B.; Safi, M.J.; Singh, A.; Prasad, B.; Mishra, I.M. Pyrolysis and thermal oxidation kinetics of sugar mill press mud. *Thermochim. Acta* **2005**, *428*, 63–70. [CrossRef]
25. Yu, H.; Liu, Y.; Liu, J.; Chen, D. High catalytic performance of an innovative Ni/magnesium slag catalyst for the syngas production and tar removal from biomass pyrolysis. *Fuel* **2019**, *254*, 115622. [CrossRef]
26. Garba, M.U.; Inalegwu, A.; Musa, U.; Aboje, A.A.; Kovo, A.S.; Adeniyi, D.O. Thermogravimetric characteristic and kinetic of catalytic co-pyrolysis of biomass with low- and high-density polyethylenes. *Biomass Convers. Biorefin.* **2018**, *8*, 143–150. [CrossRef]
27. Akubo, K.; Nahil, M.A.; Williams, P.T. Pyrolysis-catalytic steam reforming of agricultural biomass wastes and biomass components for production of hydrogen/syngas. *J. Energy Inst.* **2019**, *92*, 1987–1996. [CrossRef]
28. Albertazzi, S.; Basile, F.; Brandin, J.; Einvall, J.; Fornasari, G.; Hulteberg, C.; Sanati, M.; Trifirò, F.; Vaccari, A. Pt-Rh/MgAl(O) catalyst for the upgrading of biomass-generated synthesis gases. *Energy Fuels* **2009**, *23*, 573–579. [CrossRef]

29. Santamaria, L.; Arregi, A.; Alvarez, J.; Artetxe, M.; Amutio, M.; Lopez, G.; Bilbao, J.; Olazar, M. Performance of a Ni/ZrO_2 catalyst in the steam reforming of the volatiles derived from biomass pyrolysis. *J. Anal. Appl. Pyrolysis* **2018**, *136*, 222–231. [CrossRef]
30. Ammendola, P.; Lisi, L.; Piriou, B.; Ruoppolo, G. Rh-perovskite catalysts for conversion of tar from biomass pyrolysis. *Chem. Eng. J.* **2009**, *154*, 361–368. [CrossRef]
31. Lu, Q.; Li, W.; Zhang, X.; Liu, Z.; Cao, Q.; Xie, X.; Yuan, S. Experimental study on catalytic pyrolysis of biomass over a Ni/Ca-promoted Fe catalyst. *Fuel* **2020**, *263*, 116690. [CrossRef]
32. Wang, C.; Li, L.; Zeng, Z.; Xu, X.; Ma, X.; Chen, R.; Su, C. Catalytic performance of potassium in lignocellulosic biomass pyrolysis based on an optimized three-parallel distributed activation energy model. *Bioresour. Technol.* **2019**, *281*, 412–420. [CrossRef]
33. Balasundram, V.; Ibrahim, N.; Kasmani, R.M.; Hamid, M.K.A.; Isha, R.; Hasbullah, H.; Ali, R.R. Thermogravimetric catalytic pyrolysis and kinetic studies of coconut copra and rice husk for possible maximum production of pyrolysis oil. *J. Clean. Prod.* **2017**, *167*, 218–228. [CrossRef]
34. Federici, J.A.; Vlachos, D.G. Experimental studies on syngas catalytic combustion on Pt/Al_2O_3 in a microreactor. *Combust. Flame* **2011**, *158*, 2540–2543. [CrossRef]
35. Li, S.; Zheng, H.; Zheng, Y.; Tian, J.; Jing, T.; Chang, J.S.; Ho, S.H. Recent advances in hydrogen production by thermo-catalytic conversion of biomass. *Int. J. Hydrog. Energy* **2019**, *44*, 14266–14278. [CrossRef]
36. Rioche, C.; Kulkarni, S.; Meunier, F.C.; Breen, J.P.; Burch, R. Steam reforming of model compounds and fast pyrolysis bio-oil on supported noble metal catalysts. *Appl. Catal. B Environ.* **2005**, *61*, 130–139. [CrossRef]
37. Tomishige, K.; Asadullah, M.; Kunimori, K. Syngas production by biomass gasification using $Rh/CeO_2/SiO_2$ catalysts and fluidized bed reactor. *Catal. Today* **2004**, *89*, 389–403. [CrossRef]
38. Cifuentes, B.; Valero, M.F.; Conesa, J.A.J.; Cobo, M. Hydrogen Production by Steam Reforming of Ethanol on Rh-Pt Catalysts: Influence of CeO_2, ZrO_2, and La_2O_3 as Supports. *Catalysts* **2015**, *5*, 1872–1896. [CrossRef]
39. Cifuentes, B.; Hernández, M.; Monsalve, S.; Cobo, M. Hydrogen production by steam reforming of ethanol on a $RhPt/CeO_2/SiO_2$ catalyst: Synergistic effect of the Si:Ce ratio on the catalyst performance. *Appl. Catal. A Gen.* **2016**, *523*, 283–293. [CrossRef]
40. Yuan, R.; Yu, S.; Shen, Y. Pyrolysis and combustion kinetics of lignocellulosic biomass pellets with calcium-rich wastes from agro-forestry residues. *Waste Manag.* **2019**, *87*, 86–96. [CrossRef]
41. Garrido, M.A.; Font, R.; Conesa, J.A. Kinetic study and thermal decomposition behavior of viscoelastic memory foam. *Energy Convers. Manag.* **2016**, *119*, 327–337. [CrossRef]
42. Safar, M.; Lin, B.J.; Chen, W.H.; Langauer, D.; Chang, J.S.; Raclavska, H.; Pétrissans, A.; Rousset, P.; Pétrissans, M. Catalytic effects of potassium on biomass pyrolysis, combustion and torrefaction. *Appl. Energy* **2019**, *235*, 346–355. [CrossRef]
43. Conesa, J.A.; Soler, A. Decomposition kinetics of materials combining biomass and electronic waste. *J. Therm. Anal. Calorim.* **2017**, *128*, 225–233. [CrossRef]
44. Ding, Y.; Ezekoye, O.A.; Zhang, J.; Wang, C.; Lu, S. The effect of chemical reaction kinetic parameters on the bench-scale pyrolysis of lignocellulosic biomass. *Fuel* **2018**, *232*, 147–153. [CrossRef]
45. Zhang, X.; Deng, H.; Hou, X.; Qiu, R.; Chen, Z. Pyrolytic behavior and kinetic of wood sawdust at isothermal and non-isothermal conditions. *Renew. Energy* **2019**, *142*, 284–294. [CrossRef]
46. Dhyani, V.; Bhaskar, T. A comprehensive review on the pyrolysis of lignocellulosic biomass. *Renew. Energy* **2018**, *129*, 695–716. [CrossRef]
47. Gu, X.; Ma, X.; Li, L.; Liu, C.; Cheng, K.; Li, Z. Pyrolysis of poplar wood sawdust by TG-FTIR and Py–GC/MS. *J. Anal. Appl. Pyrolysis* **2013**, *102*, 16–23. [CrossRef]
48. Gogoi, M.; Konwar, K.; Bhuyan, N.; Borah, R.C.; Kalita, A.C.; Nath, H.P.; Saikia, N. Assessments of pyrolysis kinetics and mechanisms of biomass residues using thermogravimetry. *Bioresour. Technol. Rep.* **2018**, *4*, 40–49. [CrossRef]
49. Caballero, A.; Conesa, J.A. Mathematical considerations for nonisothermal kinetics in thermal decomposition. *J. Anal. Appl. Pyrolysis* **2005**, *73*, 85–100. [CrossRef]
50. Arenas, C.N.; Navarro, M.V.; Martínez, J.D. Pyrolysis kinetics of biomass wastes using isoconversional methods and the distributed activation energy model. *Bioresour. Technol.* **2019**, *288*, 121485. [CrossRef]
51. Anca-Couce, A.; Berger, A.; Zobel, N. How to determine consistent biomass pyrolysis kinetics in a parallel reaction scheme. *Fuel* **2014**, *123*, 230–240. [CrossRef]

52. Yang, J.; Chen, H.; Zhao, W.; Zhou, J. TG-FTIR-MS study of pyrolysis products evolving from peat. *J. Anal. Appl. Pyrolysis* **2016**, *117*, 296–309. [CrossRef]
53. Naik, S.; Goud, V.V.; Rout, P.K.; Jacobson, K.; Dalai, A.K. Characterization of Canadian biomass for alternative renewable biofuel. *Renew. Energy* **2010**, *35*, 1624–1631. [CrossRef]
54. Ministerio de Minas y Energía de Colombia. *El Carbón Colombiano: Fuente de Energía para el Mundo*; Ministerio de Minas y Energía de Colombia: Bogotá, Colombia, 2016.
55. Demirbaş, A. Gaseous products from biomass by pyrolysis and gasification: Effects of catalyst on hydrogen yield. *Energy Convers. Manag.* **2002**, *43*, 897–909. [CrossRef]
56. Wang, K.; Zhang, J.; Shanks, B.H.; Brown, R.C. The deleterious effect of inorganic salts on hydrocarbon yields from catalytic pyrolysis of lignocellulosic biomass and its mitigation. *Appl. Energy* **2015**, *148*, 115–120. [CrossRef]
57. Yang, H.; Yan, R.; Chen, H.; Lee, D.H.; Zheng, C. Characteristics of hemicellulose, cellulose and lignin pyrolysis. *Fuel* **2007**, *86*, 1781–1788. [CrossRef]
58. Carpio, R.B.; Zhang, Y.; Kuo, C.T.; Chen, W.T.; Schideman, L.C.; de Leon, R.L.; Charles, L.; Leon, R.L. De Characterization and thermal decomposition of demineralized wastewater algae biomass. *Algal Res.* **2019**, *38*, 101399. [CrossRef]
59. Sebestyén, Z.; Barta-Rajnai, E.; Bozi, J.; Blazsó, M.; Jakab, E.; Miskolczi, N.; Czégény, Z. Thermo-catalytic Pyrolysis of biomass and plastic mixtures using HZSM-5. *Appl. Energy* **2017**, *207*, 114–122. [CrossRef]
60. Fong, M.J.B.; Loy, A.C.M.; Chin, B.L.F.; Lam, M.K.; Yusup, S.; Jawad, Z.A. Catalytic pyrolysis of *Chlorella vulgaris*: Kinetic and thermodynamic analysis. *Bioresour. Technol.* **2019**, *289*, 121689. [CrossRef] [PubMed]
61. Minh Loy, A.C.; Yusup, S.; Fui Chin, B.L.; Wai Gan, D.K.; Shahbaz, M.; Acda, M.N.; Unrean, P.; Rianawati, E. Comparative study of *in-situ* catalytic pyrolysis of rice husk for syngas production: Kinetics modelling and product gas analysis. *J. Clean. Prod.* **2018**, *197*, 1231–1243. [CrossRef]
62. Long, Y.; Ruan, L.; Lv, X.; Lv, Y.; Su, J.; Wen, Y. TG-FTIR analysis of pyrolusite reduction by major biomass components. *Chin. J. Chem. Eng.* **2015**, *23*, 1691–1697. [CrossRef]
63. Mehmood, M.A.; Ahmad, M.S.; Liu, Q.; Liu, C.G.; Tahir, M.H.; Aloqbi, A.A.; Tarbiah, N.I.; Alsufiani, H.M.; Gull, M. Helianthus tuberosus as a promising feedstock for bioenergy and chemicals appraised through pyrolysis, kinetics, and TG-FTIR-MS based study. *Energy Convers. Manag.* **2019**, *194*, 37–45. [CrossRef]
64. Conesa, J.A.; Domene, A. Biomasses pyrolysis and combustion kinetics through n-th order parallel reactions. *Thermochim. Acta* **2011**, *523*, 176–181. [CrossRef]
65. García, R.; Pizarro, C.; Lavín, A.G.; Bueno, J.L. Biomass proximate analysis using thermogravimetry. *Bioresour. Technol.* **2013**, *139*, 1–4. [CrossRef]
66. Quiroga, E.; Moltó, J.; Conesa, J.A.; Valero, M.F.; Cobo, M. Data of Pyrolysis and Combustion of Sugarcane Residual Biomass over Rh-Pt/CeO2 -SiO2 Catalyst by tg, TG-FTIR and TG-MS. Mendeley Dataset. Available online: https://data.mendeley.com/datasets/gkysct9wjz/1 (accessed on 17 October 2019).

Article

Hydrogen Production from Steam Reforming of Acetic Acid as a Model Compound of the Aqueous Fraction of Microalgae HTL Using Co-M/SBA-15 (M: Cu, Ag, Ce, Cr) Catalysts

Pedro J. Megía, Alicia Carrero *, José A. Calles and Arturo J. Vizcaíno *

Chemical and Environmental Engineering Group, Rey Juan Carlos University, c/Tulipán s/n, 28933 Móstoles, Spain; pedro.megia@urjc.es (P.J.M.); joseantonio.calles@urjc.es (J.A.C.)

* Correspondence: alicia.carrero@urjc.es (A.C.); arturo.vizcaino@urjc.es (A.J.V.); Tel.: +34-91-488-8088 (A.C.); +34-91-488-8096 (A.J.V.)

Received: 30 October 2019; Accepted: 28 November 2019; Published: 2 December 2019

Abstract: Hydrogen production derived from thermochemical processing of biomass is becoming an interesting alternative to conventional routes using fossil fuels. In this sense, steam reforming of the aqueous fraction of microalgae hydrothermal liquefaction (HTL) is a promising option for renewable hydrogen production. Since the HTL aqueous fraction is a complex mixture, acetic acid has been chosen as model compound. This work studies the modification of Co/SBA-15 catalyst incorporating a second metal leading to Co-M/SBA-15 (M: Cu, Ag, Ce and Cr). All catalysts were characterized by N_2 physisorption, ICP-AES, XRD, TEM, H_2-TPR, H_2-TPD and Raman spectroscopy. The characterization results evidenced that Cu and Ag incorporation decreased the cobalt oxides reduction temperatures, while Cr addition led to smaller Co^0 crystallites better dispersed on the support. Catalytic tests done at 600 °C, showed that Co-Cr/SBA-15 sample gave hydrogen selectivity values above 70 mol % with a significant reduction in coke deposition.

Keywords: microalgae; acetic acid; steam reforming; hydrogen; cobalt; mesostructured materials

---

1. Introduction

An increase in global pollution has resulted in a search for alternative energy resources that can be substituted in place of widely used fossil fuels [1]. It is known that energy provided from hydrogen does not result in pollutant emissions when it is used in fuel cell applications [2–4]. In addition, hydrogen is extensively used in chemical and petroleum industries [5,6]. Nowadays, a hydrogen-based energy system must use renewable energy sources to be sustainable. In this sense, hydrogen production processes such as biomass gasification, and steam reforming (SR) of pyrolysis bio-oil have been widely described in the literature [7–10]. However, the use of microalgae hydrothermal liquefaction integrated with the steam reforming of the aqueous fraction is less known. Microalgae HTL requires temperatures between 250–350 °C and high pressures that can maintain the water coming from the microalgae crops in liquid state (40–250 bar). This process provides a great advantage when compared to the traditional biomass pyrolysis process, as it does not require a previous stage for biomass drying associated with high energy consumption [11–13]. Microalgae HTL products are a complex mixture of different compounds where carboxylic acids, ketones, phenols, aldehydes, fatty acids and nitrogen compounds [14] can be easily found along with a high water content. For this reason, they are not suitable for use as a fuel. However, this worthless aqueous fraction can be revalorized by hydrogen production through catalytic steam reforming [15,16] but the complex composition mentioned above usually forces the use of model compounds [17–20]. Among them, acetic acid is a major component, which can account even for the 56% of the water-soluble products [17]. The overall equation of the acetic acid steam reforming is:

$$C_2H_4O_2+2\,H_2O \rightarrow 2\,CO_2+4\,H_2 \quad (1)$$

Nowadays, SR catalysts are a critical point of study where activity, hydrogen selectivity and deactivation are the main concerns of the scientific community. Many papers can be found using different active phases such as Ni, Co, Pt or Ru, with Ni being the most studied [21]. Hu et al. [22] studied the performance of different transition metals supported over Al_2O_3 in acetic acid steam reforming. Their study led to the conclusion that Ni and Co were more active than the other metals tested (Fe and Cu). They attributed this behavior to the ability for cracking not only C-C bonds, but also C-H bonds. However, Co-based catalysts have been less reported despite the fact that they also provide high activity at moderate temperatures and also increases hydrogen yield [23,24].

Catalysts support selection is also an important point. For example, when Co was supported on Al_2O_3 or TiO_2 high metal dispersion was reported but cobalt aluminates or titanates were formed avoiding the reduction of some Co species [25]. On the other hand, the interaction of Co with silica has been studied leading to the conclusion that this support does not affect to its reducibility but instead promote the sintering of cobalt particles in the calcination and reduction steps [26,27]. Apart from that, there are other advanced supports such as SBA-15, which is a mesostructured material with high surface area that may allow higher metal dispersion when compared with the amorphous silica. Furthermore, SBA-15 presents an uniform distribution of mesopores that hinders the formation of Co agglomerates preventing also catalysts deactivation due to metal sintering [28].

Co-based catalysts have shown deactivation through sintering and surface cobalt oxidation [21]. Pereira et al. [29] proposed the preparation of bimetallic catalysts to stabilize Co/SiO_2 catalyst to safeguard the Co particles in a reduced state during the reforming. Combining diverse metals in the same carrier has been reported as an effective way to improve the catalyst performance by facilitating the metal reducibility [30]. As reducibility promoters noble metals, transition metals or CeO_2 among others can be used. Wang et al. [31] reported that Cu addition to Ni/attapulgite catalyst decreased the temperature for the reduction of nickel species. In line with this, Eschemann et al. [32] proved the efficiency of silver as a reduction promoter in Co/TiO_2 catalyst since Co-Ag bonds improve the reducibility of cobalt oxides [32,33]. Besides, Harun et al. [34] achieved better Ni^0 dispersion over Al_2O_3 surface when Ag was included in the catalyst formulation. Similarly, it was described that CeO_2, presents a synergistic effect with cobalt oxides since more oxygen vacancies are formed leading to higher reducibility [35]. In addition to promoting the cobalt reducibility to avoid possible crystallites oxidation, it is necessary to obtain a small crystallite size in order to increase activity and reduce the coke formation according to its growth mechanism [36]. Accordingly, Cerdá-Moreno et al. [37] found that lower Co particle size for ethanol steam reforming led to better catalytic activity. Recently, we have found that Ni-Cr/SBA-15 showed better catalytic behavior than Ni/SBA-15 in the steam reforming of pyrolysis bio-oil aqueous fraction by decreasing Ni^0 particles size [38]. Furthermore, Casanovas et al. [39,40] reported that the incorporation of Cr to Co/ZnO samples results in better catalytic performance when these catalysts were tested in ethanol steam reforming.

So far, we have not been able to find any references using the promoters described above in Co/SBA-15 catalysts to be tested in acetic acid steam reforming. Therefore, the main goal of this study is the preparation of novel cobalt catalysts incorporating a second metal leading to Co-M/SBA-15 (M: Cu, Ag, Ce and Cr) to achieve high hydrogen production rate through acetic acid steam reforming as model compound of microalgae HTL aqueous fraction.

2. Results and Discussion

2.1. Catalysts Characterization

Nitrogen physisorption profiles displayed in Figure 1 show type IV isotherms with a H1-type hysteresis loop according to the IUPAC classification, indicating the preservation of the initial mesoestructure of SBA-15 used as the support of these samples. Textural properties calculated from these analyses are summarized in Table 1 along with other physicochemical properties.

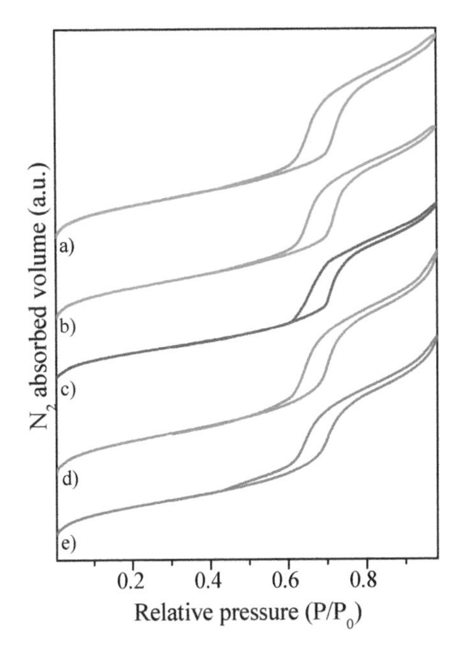

Figure 1. N_2 physisorption isotherms of calcined (**a**) Co/SBA-15; (**b**) Co-Cu/SBA-15; (**c**) Co-Ag/SBA-15; (**d**) Co-Ce/SBA-15; (**e**) Co-Cr/SBA-15 catalysts at 77K.

Table 1. Physicochemical properties of Co-M/SBA-15 (M: Cu, Ag, Ce, Cr) catalysts.

Catalyst	Co [a] (wt.%)	M [a] (wt.%)	S_{BET} ($m^2 \cdot g^{-1}$)	D_{pore} [b] (nm)	V_{pore} [c] ($cm^3 \cdot g^{-1}$)	D_{Co^0} [d] (nm)	Dispersion (%) [e]
SBA-15	-	-	550 ± 3	7.5 ± 0.1	0.97 ± 0.02	-	-
Co/SBA-15	6.4 ± 0.1	-	503 ± 4	7.2 ± 0.1	0.83 ± 0.01	9.5 ± 0.5	7.5 ± 0.2
Co-Cu/SBA-15	6.5 ± 0.1	2.0 ± 0.1	476 ± 4	7.2 ± 0.1	0.79 ± 0.03	9.7 ± 0.3	6.3 ± 0.1
Co-Ag/SBA-15	6.4 ± 0.1	1.6 ± 0.1	419 ± 4	6.9 ± 0.1	0.71 ± 0.01	12.3 ± 0.4	3.9 ± 0.6
Co-Ce/SBA-15	6.6 ± 0.1	1.7 ± 0.1	494 ± 1	7.4 ± 0.1	0.84 ± 0.01	9.6 ± 0.2	6.5 ± 0.1
Co-Cr/SBA-15	6.8 ± 0.1	1.8 ± 0.1	469 ± 1	7.1 ± 0.1	0.81 ± 0.02	7.2 ± 0.1	9.9 ± 0.3

[a] Determined by ICP-AES (M: Cu, Ag, Ce or Cr) in reduced samples, [b] BJH desorption average pore diameter, [c] Measured at $P/P_0 = 0.97$, [d] Determined from XRD of reduced catalysts by Scherrer equation from the (111) diffraction plane of Co^0, [e] Determined from H_2-TPD results using formula from Li et al. [41] assuming H/Co = 1.

The metals loading is close to the nominal value used during the catalysts preparation. Metal addition to bare SBA-15 leads to a decrease in BET surface area with Co-Ag/SBA-15 being the sample with the smallest pore size, pore volume and surface area. This phenomena has been described previously [42] and was ascribed to Ag structures growing in the mesopores of SBA-15. Similar textural properties were found in Co-(Cu, Ce or Cr)/SBA-15 samples.

Figure 2 shows the XRD patterns of the calcined samples. Peaks corresponding to cubic Co_3O_4 appear in all samples (JCDPS 01-071-4921). Attending to Co-Cu/SBA-15 sample, a small peak at

38.3° can be observed due to the formation of monoclinic CuO (JCDPS 01-089-2531). In case of Co-Ce/SBA-15, two small peaks over 28.5° and 47.5° can be seen due to the presence of cubic CeO_2 (JCDPS 01-089-8436). Ag and Cr oxides were not detected by XRD due to the overlap of the main diffraction peaks of cubic Ag_2O (JCPDS 00-012-0793), rhombohedral Cr_2O_3 (JCPDS 00-002-1362) and cubic $CoCr_2O_4$ spinel (JCPDS 00-022-1084), with the Co_3O_4 pattern. The higher Co content compared to Ag and Cr also contributes to the non-detection of Ag and Cr oxides by XRD as were observed in previous works [43,44]. XRD patterns corresponding to Co-(Cu, Ag or Ce)/SBA-15 present narrower Co_3O_4 peaks and slightly larger Co_3O_4 crystallites were obtained comparing when compared to Co/SBA-15 sample.

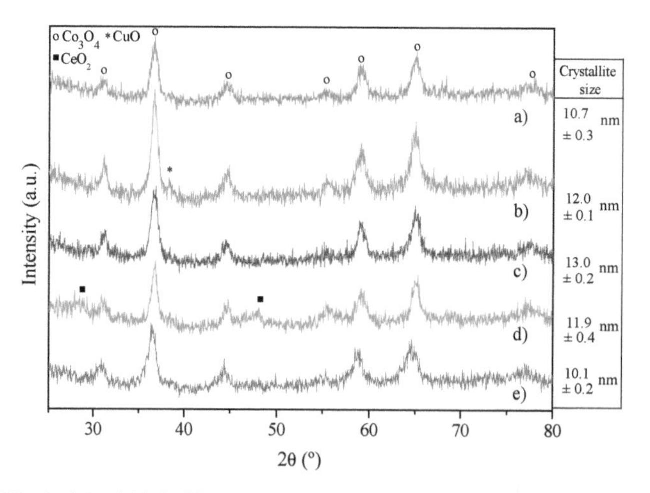

Figure 2. XRD of calcined (**a**) Co/SBA-15; (**b**) Co-Cu/SBA-15; (**c**) Co-Ag/SBA-15; (**d**) Co-Ce/SBA-15; (**e**) Co-Cr/SBA-15 catalysts. Co_3O_4 crystallites sizes calculated from the (311) diffraction plane using Scherrer equation are displayed on the right.

Figure 3 shows the TEM micrographs of calcined samples. Irregular metal oxides particles can be observed, some of them formed in the channels of SBA-15, while other particles were formed over the external surface as previously reported [43]. The presence of Co and promoters (Cu, Ag, Ce or Cr) were evaluated in the corresponding sample by EDX indicating an intimate contact between Co oxide and promoters. Co-Ag/SBA-15 catalyst has large metallic nanostructures through the SBA-15 channels and Ag_2O particles can be also observed over the support [45]. The incorporation of high Ag loadings (> 1wt. %) affects support structure and distribution of Ag_2O particles over the catalyst because the probability of Ag-Ag bond formation increases [32,33]. On the other side, it is noticeable how Co-Cr/SBA-15 sample clearly shows the highest dispersion over the support with very small metal oxide particles, which is in agreement with the lower metal diameter calculated from XRD (Table 1).

Figure 4 displays the H_2-TPR profiles of the calcined catalysts. In the case of Co/SBA-15 sample, the reduction profile shows two main reduction stages. The first one with maxima found at 248–267 °C and a shoulder around 332 °C. These peaks are attributed to the reduction of Co_3O_4 to CoO and subsequently to Co^0. The reduction stage at high temperature, with a maximum placed at 494 °C, can be attributed to the presence of Co-oxide species with stronger interaction with the support [46]. Cu addition led to a clear decrease of the reduction temperature as observed in Co-Cu/SBA-15 profile.

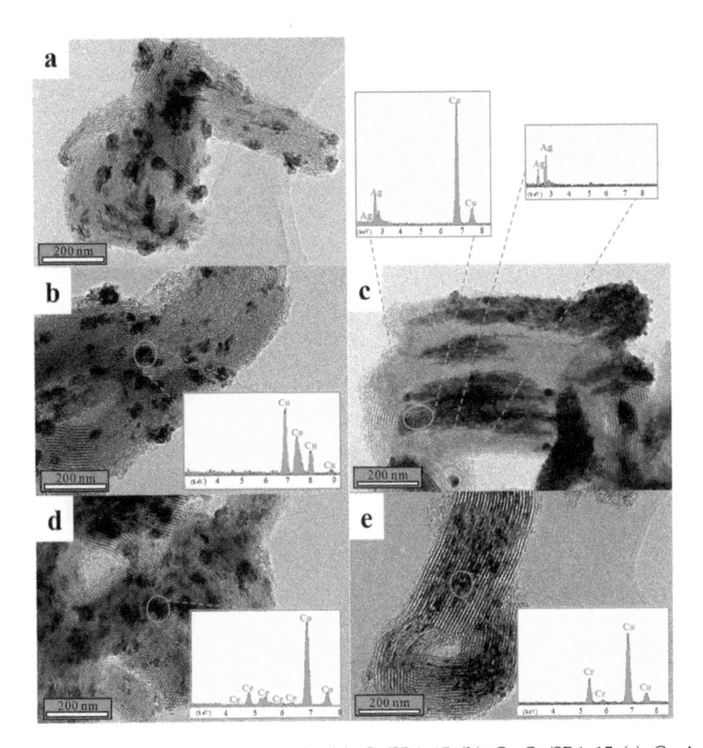

Figure 3. TEM micrographs of calcined samples (**a**): Co/SBA-15; (**b**): Co-Cu/SBA-15; (**c**): Co-Ag/SBA-15; (**d**): Co-Ce/SBA-15; (**e**): Co-Cr/SBA-15.

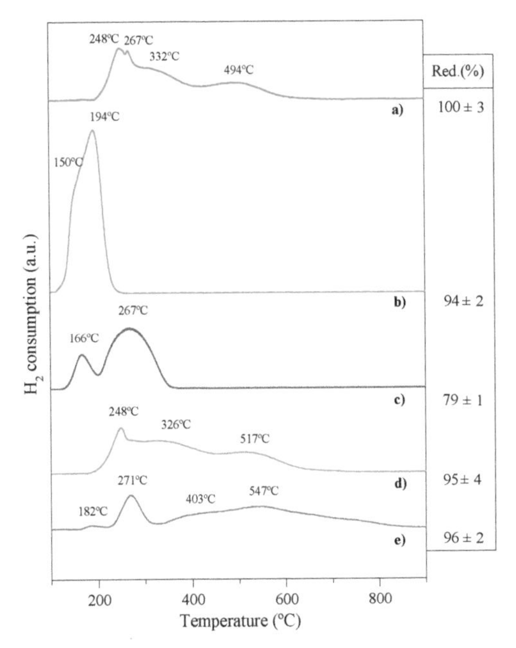

Figure 4. H_2-TPR profiles for (**a**) Co/SBA-15; (**b**) Co-Cu/SBA-15; (**c**) Co-Ag/SBA-15; (**d**) Co-Ce/SBA-15; (**e**) Co-Cr/SBA-15 samples. Red. (%) data displayed on the right correspond to reducibility.

The reduction zone is located at temperatures between 140-260 °C with two maxima at 150 and 194 °C. Whereas the lower temperature peak is ascribed to the simultaneous reduction of CuO and Co_3O_4 to Cu^0 and CoO respectively, the other one is related to the reduction of CoO to Co^0 [47]. This effect of Cu in lowering reduction temperature of metal oxides was observed in previous works for Ni-based catalysts [43]. Co-Ag/SBA-15 catalyst showed two clearly different reduction areas, also at low temperature. While the zone over 267 °C is related to the Co oxides next to Ag, the other one around 166 °C is attributed to the reduction of segregated Ag_2O particles to Ag^0 [48]. On the other hand, Co-Ce/SBA-15 sample showed a reduction profile similar to Co/SBA-15 with the peak at 494 °C shifted to higher reduction temperature due to an emerging peak assigned to superficial cerium oxide [49]. Finally, in the reduction profile of Co-Cr/SBA-15 had a new peak around 182 °C, probably due to the reduction of Cr-oxides to Cr^{3+} which can be affected by the presence of Co_3O_4 [50] although it could not be detected by XRD. The peak attributed to Co_3O_4 reduction at 271 °C remained unaltered whereas the peak of CoO reduction shifts to higher temperatures due to the presence of Cr species [51] or to the confinement of Co oxides into SBA-15 channels because of their smaller size. Based on the literature, the most likely option is the formation of a cobalt chromate mixed oxide [52], although none could be detected by XRD due to the overlap of the main diffraction lines of $CoCr_2O_4$ with those of Co_3O_4. The XRD patterns of the samples after reduction at 700 °C under pure H_2 flow are displayed in Figure 5. No peaks ascribed to Co_3O_4 pattern can be detected whereas cubic Co^0 (JCDPS 00-001-1259) peaks corresponding to (111), (200) and (220) planes showing the reflection at 2θ = 44.4°, 51.3° and 75.4° can be observed in all samples after the reduction process.

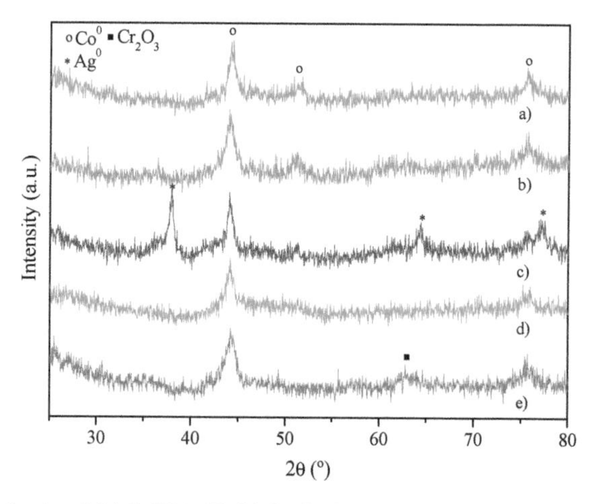

Figure 5. XRD of reduced (**a**) Co/SBA-15; (**b**) Co-Cu/SBA-15; (**c**) Co-Ag/SBA-15; (**d**) Co-Ce/SBA-15; (**e**) Co-Cr/SBA-15 catalysts at 700 °C.

Cubic Ag^0 (JCDPS 00-043-1038) diffraction peaks arose in the Co-Ag/SBA-15 sample at 2θ = 38.1°, 64.5° and 77°, ascribed to (111), (200) and (220) reflection planes, respectively. In this case, some Co-oxides could remain in this sample explaining its low reducibility (see Figure 4) but they were not detected because there is an overlapping between Ag^0 and Co_3O_4 patterns at 38.1° and 64.5°. In Co-Cr/SBA-15 catalyst a peak placed at 2θ = 63.7° was assigned to rhombohedral Cr_2O_3 (JCDPS 00-002-1362) probably coming from the release of CoO from the spinel $CoCr_2O_4$. No diffraction peaks of cubic Cu^0 (JCPDS 00-001-1241) were distinguished in Co-Cu/SBA-15 sample due to the overlapping between Cu^0 and Co^0 diffraction peaks. Co-Ce/SBA-15 reduced sample showed only the diffraction peak of metallic Co. The absence of CeO_2 diffraction peaks prompted us to think about the formation of a non-stoichiometric $CeO_{2-\sigma}$ that cannot be detected by XRD [53].

Co^0 crystallite sizes were calculated by the Scherrer equation from the diffraction plane (111). In general, whereas Co-Cu/SBA-15 and Co-Ce/SBA-15 samples present a crystallite size similar to Co/SBA-15, Co-Ag/SBA-15 had the largest crystallites (see Table 1) which differs from the literature as silver loading in Co-Ag/SBA-15 is higher than in references [31,32]. In contrast, Co-Cr/SBA-15 presented the lowest Co crystallite size because making a parallelism with the paper of Amin et al. [54] Cr-oxides can suppress the extension growth of Cu-oxides in that case, Co-oxides in our case.

H_2-TPD analysis was carried out in order to measure the dispersion of the metallic phase over the support. The results, summarized in Table 1, follow the opposite trend as Co^0 crystallite sizes calculated from the Scherrer equation. Co-Cr/SBA-15 sample reached the highest active phase dispersion over the support. This effect can be clearly observed in Figure 6, where Co^0 crystallite sizes are displayed against dispersion and it is clear that the only promoter that improves the base Co/SBA-15 catalyst is Cr. In addition, other authors have reported smaller crystallite size when Cr was incorporated to the catalyst formulation suggesting the capacity of Cr_2O_3 to act as a textural promoter preventing metallic sintering [55–57]. It should be noted that in a previous work we reported the same behavior with Ni-Cr/SBA-15 sample [38,43], in line with the results obtained by Xu et al. during the co-impregnation of Cr and Ni over char as support [58].

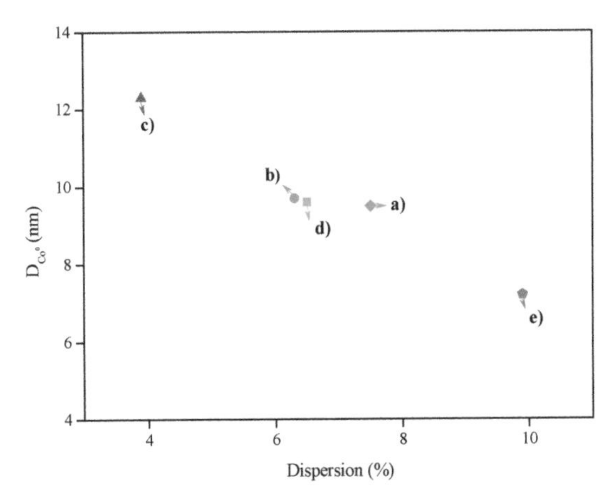

Figure 6. Comparison between Co^0 crystallites size and Dispersion over the SBA-15 material used as support for (**a**) Co/SBA-15; (**b**) Co-Cu/SBA-15; (**c**) Co-Ag/SBA-15; (**d**) Co-Ce/SBA-15; (**e**) Co-Cr/SBA-15.

2.2. Catalytic Tests

AASR (acetic acid steam reforming) reactions were carried out after the reduction of the catalysts. All experiments were performed using an aqueous solution of acetic acid with a S/C molar ratio = 2 and a WHSV = 30.1 h^{-1} at atmospheric pressure and 600 °C using N_2 as carrier gas. Conversion data are not shown because all catalysts reached complete conversion along 5 h of time-on-stream, which implies high activity for all the samples in acetic acid conversion at these reaction conditions. However, different product distributions were achieved indicating different activities in acetic acid steam reforming reaction, ascribed to the role of a second metal in secondary reactions. In this sense, hydrogen and carbon co-products distribution (dry basis) are displayed in Figure 7.

The H_2 content expected at equilibrium at the experimental conditions, predicted by means of the software GasEQ, based on the method of free Gibbs energy minimization, is also shown. Regarding products distribution, all catalysts reached high hydrogen concentration, above 53%. As known, Co-based catalysts allow the breaking of C-C bonds (only methane is produced as hydrogen-containing product) but also of C-H bonds [22]. Moreover, an effective catalyst must also be active in WGS reaction in order to eliminate CO from the metal surface during steam reforming. Over Co, methane reforming and WGS activity was presented and this clearly shown by products formation. Among them, CO_2 formation is highest and followed by CO, CH_4, thus WGS is more pronounced compared to other disproportionation and decomposition reactions. Cu, Ag and Ce addition to Co/SBA-15 decreases the hydrogen content in the gas outlet stream in line with higher co-carbon products percentages. In contrast, Co-Cr/SBA-15 reached the highest hydrogen concentration in the product stream. This behavior is related to the small Co crystallite size (see Table 1) leading to higher active sites surface area [59,60]. Therefore, Cr addition improved the catalytic performance by preventing Co agglomeration. In fact, Casanovas et al. [40] have published similar behavior adding Cr to Co/ZnO being more active and selective for ethanol steam reforming. On the other side, Co-Ag/SBA-15 achieved the lowest hydrogen concentration and therefore carbon containing products composition was higher, probably due to the pore blocking effect and the highest Co crystallite size. Co-Cu/SBA-15 and Co-Ce/SBA-15 showed higher CO_2/CO molar ratio compared to the other samples (3/2) suggesting that the activity for WGS reaction was increased [60]. If WGS reaction is favored, an increase in the hydrogen production is expected but the hydrogen content reached with these two catalysts was lower than with Co/SBA-15 (CO_2/CO ratio = 1.7) thus, it is possible to assume that the presence of a second metal hinders reactants access to Co active centers, thereby avoiding their catalytic role breaking C-H bonds. Finally, Co-Cu/SBA-15, Co-Ag/SBA-15 and Co-Ce/SBA-15 showed an increase of CH_4 from 2% to almost 5% in comparison to the Co/SBA-15 sample. CH_4 formation can be due to the decomposition of acetic acid or methanation [61]. Particularly, Co-Cu/SBA-15 and Co-Ce/SBA-15 produce more CH_4 in line with the reduction of H_2 and CO content which indicates that Cu and Ce promote the methanation reaction ($3H_2 + CO \rightarrow CH_4 + H_2O$) [62]. Instead, the increase of produced methane with Co-Ag/SBA-15 could be due to the decomposition of acetic acid since the CO content was kept constant while both CO_2 and CH_4 concentrations increase, which would be in accordance with the stoichiometry of the reaction $CH_3COOH \rightarrow CH_4 + CO_2$. However, other parallel and consecutive reactions varying the CO, CO_2 and CH_4 content can be taking place.

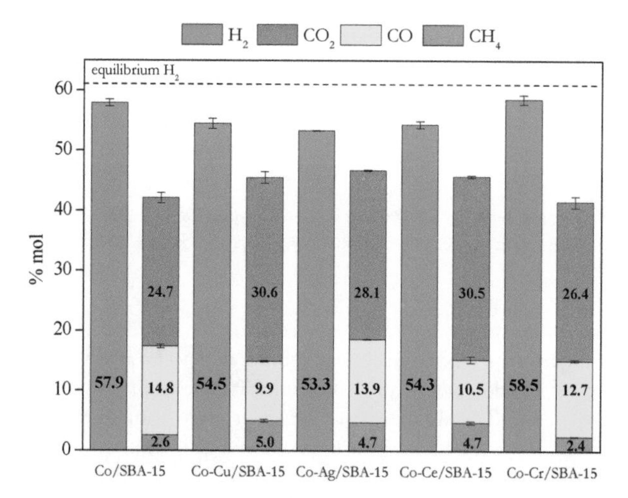

Figure 7. Products distribution in outlet gas stream produced in the acetic acid steam reforming over Co-M/SBA-15 (M: Cu, Ag, Ce, Cr) catalysts at T = 600 °C, P = 1 atm, time-on-stream = 5 h.

Regarding the evolution of H_2 selectivity, calculated as the ratio between hydrogen produced and 4 times the reacted acetic acid (stoichiometry), with reaction time showed in Figure 8 Co-Cu/SBA-15 and Co-Ag/SBA-15 samples exhibited a decrease at 2 h but after that, it remains almost constant. Regardless, the H_2 selectivity of the rest of catalysts remains almost unaltered with time-on-stream. Therefore, no deactivation was detected for Co/SBA-15, Co-Ce/SBA-15 and Co-Cr/SBA-15 samples.

In addition, it can be assessed that Co-Cr/SBA-15 sample also achieved the highest H_2 selectivity close to the thermodynamic value at the present reaction conditions. This result is promising compared to those obtained by Ni-based catalysts widely referenced in literature for acetic acid steam reforming reactions. In this sense, Thaicharoensutcharittham et al. [63] reported that $Ni/Ce_{0.75}Zr_{0.25}O_2$ catalyst with a Ni loading of 5 wt.% reached hydrogen selectivity of 33.54 mol % with a S/C = 1,n using a S/C = 3. On the other hand, Wang et al. [64] achieved hydrogen selectivity between 54.5 and 70.9 mol % for reaction tests carried out at 550 °C and 650 °C respectively, with S/C = 3 using Ni/Attapulgite catalysts. In another work, Nogueira et al. [65] published the catalytic performance of Ni catalysts supported on (MgO)-modified γ-Al_2O_3 reaching, a H_2 selectivity of 67.5 mol % at higher S/C ratio (S/C = 4). Additionally, our group tested at similar operation conditions (600 °C, GHSV: 11000 h^{-1}) Ni-based catalysts in AASR with a S/C = 4 [38]. In that work, we achieved up to 60 mol % of hydrogen content for both Ni/SBA-15 and Ni-Cr/SBA-15, which implied H_2 selectivities between 56.6–59.9 mol %. These values are lower than those achieved with Co-M/SBA-15 catalysts in the present work, even though lower S/C ratio has been used that should lead to worse catalytic results. Despite differences in reaction conditions, mainly S/C molar ratio, these H_2 selectivity values are lower than that achieved by Co-Cr/SBA-15 sample. Furthermore, we also observed the beneficial effect of adding Cr to catalysts in our recently published works [38,43], where we reached using Cr as promoter added to Ni/SBA-15 catalysts, better catalytic performance using different feedstock in steam reforming reaction.

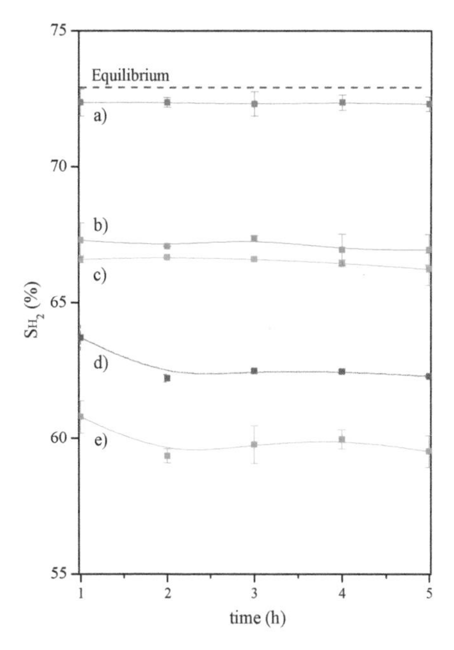

Figure 8. Hydrogen selectivity of gas stream produced in the acetic acid steam reforming over (**a**) Co-Cr/SBA-15; (**b**) Co/SBA-15; (**c**) Co-Ce/SBA-15; (**d**) Co-Ag/SBA-15; (**e**) Co-Cu/SBA-15 catalysts at T = 600 °C, P = 1 atm.

Coke formation during steam reforming has been reported as the main cause of SR catalyst deactivation [36]. It must be emphasized that catalyst deactivation is not only related to the amount of coke, but also to the nature of the coke formed, the morphology and the location over the catalyst structure [66]. In this sense, XRD patterns of used catalysts after 5 h (TOS) are shown in Figure 9.

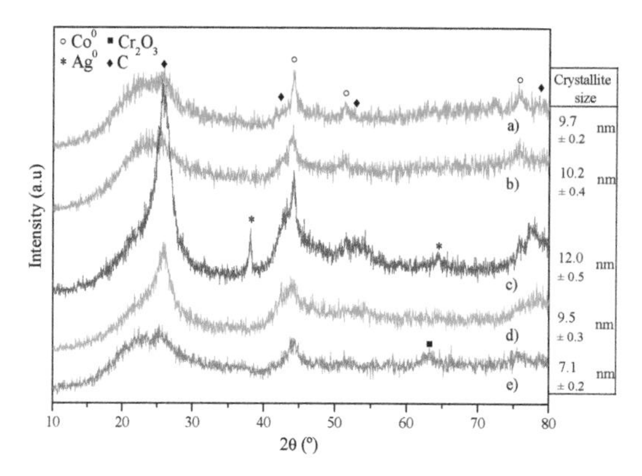

Figure 9. XRD patterns of used (**a**) Co/SBA-15; (**b**) Co-Cu/SBA-15; (**c**) Co-Ag/SBA-15; (**d**) Co-Ce/SBA-15; (**e**) Co-Cr/SBA-15 catalysts. Co^0 crystallites sizes calculated from the (111) diffraction plane using Scherrer equation are displayed on the right.

Peaks corresponding to cubic Co^0 (JCDPS 00 001 1259) at $2\theta = 44.4°$, $51.3°$ and $75.4°$ can be still distinguished. In contrast to reduced samples (Figure 5), reflection peaks corresponding to graphitic carbon (JCDPS 00-041-1487) at $2\theta = 26.5°$, $42.6°$, $53.9°$ and $78.8°$ ascribed to (002), (100), (004) and (006) reflection planes, respectively, appear as a consequence of the coke deposition along the acetic acid steam reforming being more pronounced in Co-Ag/SBA-15 sample. Cobalt crystallites sizes of used catalysts (calculated from Scherrer equation) are shown on the right side of Figure 9. Comparing these results with those found in reduced samples (Table 1), it can be concluded that cobalt crystallites sizes were very similar, which indicates no significant sintering throughout the reforming reaction.

TGA can be used for the identification of the type of coke formed during the reaction since more ordered coke will need higher temperature to be oxidized [67]. It is normally reported that amorphous carbon is more reactive than graphitic in reactions with O_2 [68] because it oxidizes at low temperatures whereas filamentous or graphitic carbon does at higher temperatures [69–71]. Figure 10 displays the derivative thermogravimetric (DTG) curves of the used catalysts along with the amount of coke formed during the reaction in terms of $mg_{coke} \cdot g_{cat}^{-1} \cdot h^{-1}$.

There are significant differences in the total coke content, in the order Co-Ag/SBA-15 > Co-Ce/SBA-15 > Co-Cu/SBA-15 > Co/SBA-15 > Co-Cr/SBA-15 which follows the reverse order of the hydrogen content in the outlet stream during AASR (see Figure 6). In general, all DTG profiles show a maximum around 500 °C and a shoulder around 550 °C, indicating the formation of some kind of carbon nanofibers with different ordering degree [69,70]. Co-Ag/SBA-15 showed a maximum around 441 °C which can be related to the formation of some defective carbon deposits. Co-Ag/SBA-15 obtained the worst catalytic results (high CH_4 concentration and the lowest H_2 concentration), in line with the highest carbon deposition. Besides, it is noteworthy that Co-Cr/SBA-15 reduced the coke production two times compared to Co/SBA-15. It is known that Cr_2O_3 has been used as an oxide catalyst with outstanding carbon deposition resistance properties [72,73]. In our case, the reduction in carbon deposition can be also ascribed to the role of chromium avoiding the formation of large Co crystallites as it could be observed by TEM and measured by the Scherrer equation, because smaller Co crystallites will prevent the initiation of carbon nucleation leading to coke formation [74]. On the other hand, Cr_2O_3 has catalytic activity in the WGS reaction, lowering the CO concentration into the gas phase surrounding the catalytic bed, thus favoring the formation of H_2 and CO_2 [75]. In this sense, the extent Boudouard reaction ($2\,CO\,(g) \rightarrow CO_2\,(g) + C\,(s)$), which is one of the main routes for coking, will be reduced.

Used catalysts were also analyzed by TEM as shown in Figure 11. In all cases, carbon nanofibers with different ordering degree can be observed. Besides, Co-Ag/SBA-15 micrograph shows some zones of defective coke deposits, in concordance with DTG results.

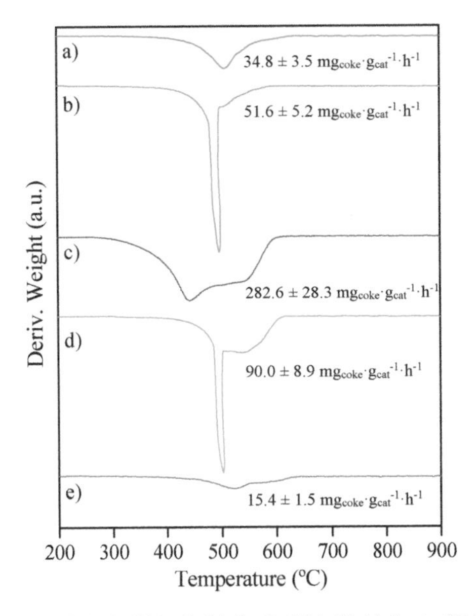

Figure 10. DTG curves of used (**a**) Co/SBA-15; (**b**) Co-Cu/SBA-15; (**c**) Co-Ag/SBA-15; (**d**) Co-Ce/SBA-15; (**e**) Co-Cr/SBA-15 samples after 5 h time-on-stream.

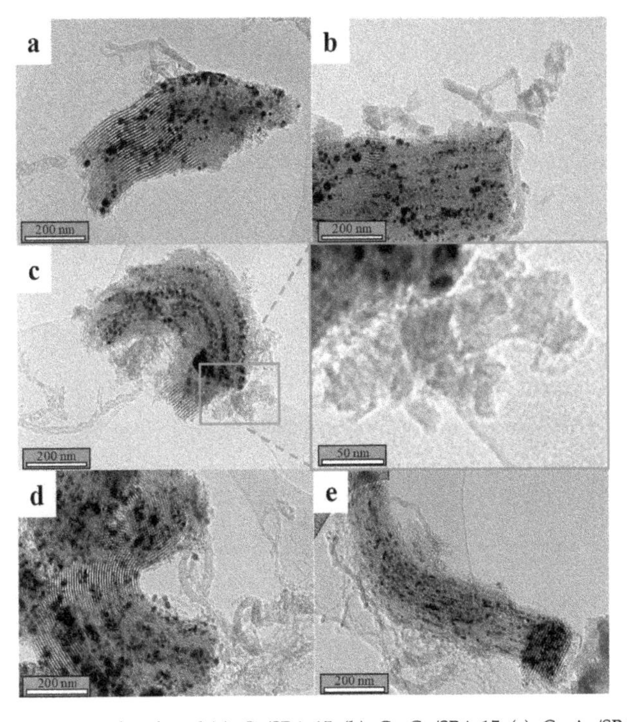

Figure 11. TEM micrographs of used (**a**): Co/SBA-15; (**b**): Co-Cu/SBA-15; (**c**): Co-Ag/SBA-15; (**d**): Co Ce/SBA 15; (**e**): Co-Cr/SBA-15 (5 h time-on-stream at 600 °C).

Finally, the Raman spectra of used catalysts in the range 1200–1700 cm^{-1} are presented in Figure 12. As it can be observed, two main bands appear in all cases, at 1330–1340 (D-band) and 1586–1591 cm^{-1} (G-band). G-band is ascribed to the stretching mode of carbon sp^2 bonds of condensed graphitic aromatic structures such as graphite layer [76], whereas D-band is related to the carbon atoms vibration of disordered aromatic structures such as amorphous or defective filamentous carbon [70,77–79]. The presence of both bands exhibits the heterogeneity of carbon species constituting the coke formed during the AASR reaction. It has been reported that the intensity of the D band relative to the G band can be used as a qualitative measure of the formation of different kinds of carbon with different degree of graphitization or disorder in the carbon structure [78–80]. Smaller I_D/I_G values indicate higher crystallinity due to higher contribution of the graphitic carbon structures formed [81,82] but it also implies more layers constituting the deposited carbon [83]. In these sense, the estimated values are summarized also in Figure 12. As can be seen, the I_D/I_G ratio decreases in the following order: Co-Cr/SBA-15 (I_D/I_G = 0.80) > Co/SBA-15 (I_D/I_G = 0.65) > Co-Ce/SBA-15 (I_D/I_G = 0.61) > Co-Ag/SBA-15 (I_D/I_G = 0.53) > Co-Cu/SBA-15 (I_D/I_G = 0.48). These results indicate that carbon deposition over the Co-Cu/SBA-15 sample occurs in larger extent on the Co surface when compared with the other samples, leading to the growth of well-ordered carbon, which may be responsible of catalyst deactivation since it act as a shell covering the active Co sites layer by layer [80]. It must be highlighted that the H_2 selectivity represented in Figure 8, decreases in the same order as I_D/I_G ratio. Therefore, the H_2 selectivity is directly related to the kind of carbon deposited on the catalyst.

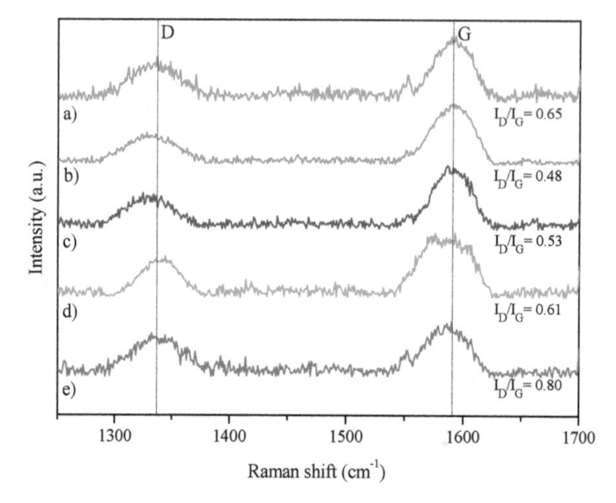

Figure 12. Raman spectra of used (**a**) Co/SBA-15; (**b**) Co-Cu/SBA-15; (**c**) Co-Ag/SBA-15; (**d**) Co-Ce/SBA-15; (**e**) Co-Cr/SBA-15 catalysts.

An AASR test done at long time-on-stream displayed in Figure 13 showed that Co-Cr/SBA-15 achieved good stability after 50 h time-on-stream. Conversion values were near 95% at the end of the reaction, while almost constant hydrogen selectivity (~72 mol %) was obtained. These results evidence that Co-Cr/SBA-15 sample is a promising option for acetic acid steam reforming, since hydrogen selectivity remains close to the equilibrium value for a long period and, in addition, this value is greater than those obtained with the Ni-based catalysts described in literature.

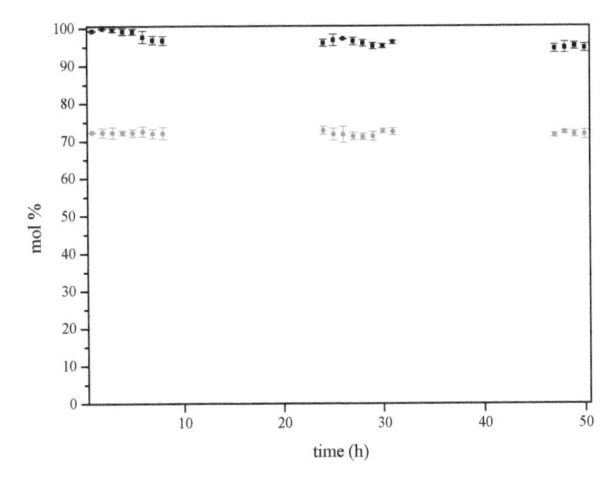

Figure 13. Acetic acid conversion (■) and hydrogen selectivity (●) during stability test of Co-Cr/SBA-15 catalyst at T = 600 °C, P = 1 atm.

3. Experimental Section

3.1. *Catalysts Synthesis*

Mesostructured SBA-15 material, synthesized using the hydrothermal method described elsewhere [84], was used as catalysts support. Pluronic 123 and TEOS were used as surfactant and silica precursor (Aldrich, St. Louis, MO, USA) respectively.

Synthesis of Co-M/SBA-15 (M: Cu, Ag, Ce or Cr) catalysts was accomplished by the incipient wetness impregnation method described in previous work [85]. Metal loading was selected as 7 wt.% of Co and 2 wt.% of promoter [86]. In this way, mixed aqueous solutions of the corresponding nitrates were used for the co-impregnation: $Co(NO_3)_2{\cdot}6H_2O$ (Acros Organics, Morris Plains, NJ, USA) and $Cu(NO_3)_2{\cdot}3H_2O$, $Cr(NO_3)_3{\cdot}9H_2O$, $Ce(NO_3)_3{\cdot}6H_2O$, $AgNO_3$ (Aldrich, St. Louis, MO, USA). Subsequently, the prepared samples were calcined under air at 550 °C.

3.2. *Catalysts Characterization*

N_2 adsorption/desorption at 77 K on a TRISTAR 3000 sorptometer (Micromeritics, Norcross, GA, USA) was used for the measurement of textural properties. Prior to the analysis samples were outgassed under vacuum at 200 °C for 4 h. To determine the chemical composition of the catalysts, ICP-AES technique was used. The equipment was a VISTA-PRO AX CCD-Simultaneous ICP-AES spectrophotometer (Varian, Palo Alto, CA, USA). Samples were previously treated by acidic digestion. XRD measurements were recorded using an X'pert PRO diffractometer (Philips, Eindhoven, The Netherlands) using Cu Kα radiation. The Scherrer equation was used to estimate the metal crystallites mean diameter. Reducibility of the samples was studied by TPR analyses. A Micromeritics (Norcross, GA, USA) AUTOCHEM 2910 system was used. The experiment is carried out flowing 35 N mL/min of gas (10% H_2/Ar) through the sample and increasing temperature up to 980 °C with a 5 °C/min heating ramp. Samples were previously outgassed under Ar flow at 110 °C for 30 min. Co dispersion of the catalysts was determined by hydrogen TPD in the same apparatus. For that, the samples were first reduced under 35 N mL/min of gas (10% H_2/Ar), then cooled to 50 °C, and saturated with H_2. After that, the physically absorbed H_2 is removed by flushing Ar and finally heated up to 700 °C at 5 °C/min in Ar flow (30 N mL/min). TEM micrographs were obtained on a 200 kV JEM 2100 microscope (JEOL, Tokyo, Japan), with a resolution of 0.25 nm at the National Centre for Electron Microscopy (CNME, Complutense University of Madrid, Madrid, Spain). It also has the possibility to achieve microanalysis results by energy dispersive X-ray spectroscopy (EDX). Samples preparation involve their suspension in acetone and subsequently deposition on a carbon-coated copper or nickel grid. Carbon deposited during catalytic tests was measured by thermogravimetric analysis (TGA), TEM and Raman spectroscopy. TGA analysis were performed in airflow with a heating rate of 5 °C/min up to 1000 °C on a SDT 2960 thermobalance (TA Instruments, New Castle, DE, USA). Raman spectra were recorded using a NRS-5000/7000 series Raman spectrometer (JASCO, Tokyo, Japan) at the IMDEA Energy Institute.

3.3. Catalytic Tests

Acetic acid steam reforming reactions were performed at 600 °C on a MICROACTIVITY-PRO unit (PID Eng. & Tech. S.L., Alcobendas, Madrid, Spain) as described in previous works [7,38,85,87]. The reactor consists in a fixed-bed tubular reactor in stainless steel 316 (i.d. = 9.2 mm, L = 300 mm). The reactor is located inside an electric oven of low thermal, where temperature in the catalytic bed was measured by means of a K-thermocouple. All the components inside the hot box were maintained at 200 °C to prevent condensation in the pipes and to preheat the reactants. A schematic diagram is displayed in Figure 14.

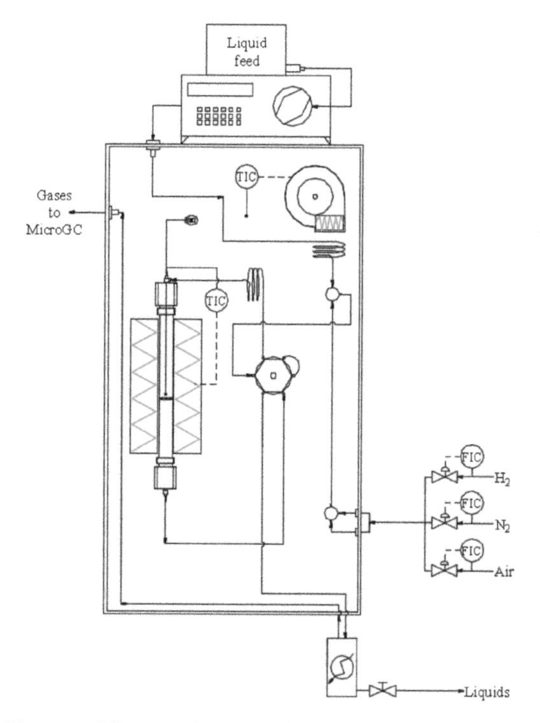

Figure 14. Schematic diagram of the catalytic testing setup [38].

The reactions were carried out isothermally at atmospheric pressure. Before tests, all catalysts were reduced under pure hydrogen (30 mL/min) up to 700 °C with a heating rate of 2 °C/min. Temperature was maintained for 30 min. Reaction feed was a mixture of acetic acid and water using a steam to carbon molar ratio of 2, using N_2 as carrier and internal standard (GHSV = 11,000 h^{-1}). The composition of the outlet gas was measured online with an 490 Micro-GC (Agilent, Santa Clara, CA, USA) equipped with a thermal conductivity detector (TCD), a PoraPlot U column (10 m) and a Molecular Sieve 5A column (20 m) using He and Ar as carrier gas, respectively. Condensable vapors were trapped in the condenser at 4 °C and analyzed in a Varian (Palo Alto, CA, USA) CP-3900 chromatograph equipped with a CP-WAX 52 CB (30 m × 0.25 mm, DF = 0.25) column and flame ionization detector (FID).

4. Conclusions

The incorporation of a second metal like Cu, Ag, Ce or Cr into Co/SBA-15 sample catalyst resulted in bimetallic catalysts with very different properties and catalytic behavior in acetic acid steam reforming. Co-Ag/SBA-15 presented some pore blockage of the SBA-15 structure due to the presence of isolated silver oxide particles. Cu and Ag addition to Co/SBA-15 led to a significant decrease in the reduction temperature, as shown in H_2-TPR profiles. Cu addition to Co/SBA-15 favors Co oxide reducibility, while maintaining almost unaltered Co^0 crystallites size. In contrast, Co-Ag/SBA-15 showed also lower reduction temperatures but larger Co^0 crystallites than Co/SBA-15. However, Ce addition does not affect significantly neither reducibility nor Co^0 crystallite size. Finally, Cr addition to Co/SBA-15 strongly decreases Co crystallites size, induced by the presence of chromium oxides, improving metal dispersion with a slight decrease in the reduction temperature.

Regarding acetic acid steam reforming, Co-Cu/SBA-15 and Co-Ag/SBA-15 gave lower hydrogen selectivity than unmodified Co/SBA-15 catalyst. However, Cr addition improved the catalytic behavior reaching the highest hydrogen selectivity next to the thermodynamic equilibrium. After the steam reforming tests, cobalt crystallites sizes in the used catalysts were very similar to those in fresh samples, indicating that coke deposition and not sintering is the cause of catalysts deactivation. Besides, the amount of coke formed on Co-Cr/SBA-15 was much lower than on the rest of the catalysts after 5 h of time on stream. Another difference resided in the nature of coke deposited because disordered aromatic structures such as amorphous or defective filamentous carbon were formed in a higher extent on Co-Cr/SBA-15 (I_D/I_G = 0.80) while the contribution of condensed graphitic aromatic structures increased in Co-Cu/SBA-15 (I_D/I_G = 0.48). Thus, Cr addition to Co/SBA-15 resulted in the best catalytic performance on acetic acid steam reforming, but Cr toxicity opens the way to the search for other metals providing similar catalytic properties.

Author Contributions: Conceptualization, A.C. and J.A.C.; methodology, A.J.V.; validation, A.J.V.; formal analysis, A.C.; investigation, P.J.M.; writing—original draft preparation, P.J.M.; writing—review and editing, A.C., J.A.C. and A.J.V.; supervision, A.C., J.A.C. and A.J.V.; project administration, A.C. and J.A.C.; funding acquisition, A.C. and J.A.C.

Funding: This research was funded by the Spanish Ministry of Economy and Competiveness (project ENE2017-83696-R) and the Regional Government of Madrid (project S2018/EMT-4344).

Acknowledgments: The authors acknowledge the IMDEA Energy Institute and the Complutense University of Madrid for the Raman and TEM analyses, respectively.

Conflicts of Interest: The authors declare no conflict of interest.

References

1. Dobosz, J.; Małecka, M.; Zawadzki, M. Hydrogen generation via ethanol steam reforming over Co/HAp catalysts. *J. Energy Inst.* **2018**, *91*, 411–423. [CrossRef]
2. Vizcaíno, A.J.; Carrero, A.; Calles, J.A. *Hydrogen Production from Bioethanol*; Nova Science Publishers: New York, NY, USA, 2012.
3. Agency, I.E. *Hydrogen Production and Storage: R&D Priorities and Gaps*; IEA Publications: Paris, France, 2006.
4. Ruocco, C.; Palma, V.; Ricca, A. Kinetics of Oxidative Steam Reforming of Ethanol Over Bimetallic Catalysts Supported on CeO_2–SiO_2: A Comparative Study. *Top. Catal.* **2019**, *62*, 467–478. [CrossRef]
5. He, L.; Parra, J.M.S.; Blekkan, E.A.; Chen, D. Towards efficient hydrogen production from glycerol by sorption enhanced steam reforming. *Energy Environ. Sci.* **2010**, *3*, 1046–1056. [CrossRef]
6. Wang, Y.; Wang, C.; Chen, M.; Tang, Z.; Yang, Z.; Hu, J.; Zhang, H. Hydrogen production from steam reforming ethanol over Ni/attapulgite catalysts - Part I: Effect of nickel content. *Fuel Process. Technol.* **2019**, *192*, 227–238. [CrossRef]
7. Carrero, A.; Vizcaíno, A.J.; Calles, J.A.; García-Moreno, L. Hydrogen production through glycerol steam reforming using Co catalysts supported on SBA-15 doped with Zr, Ce and La. *J. Energy Chem.* **2017**, *26*, 42–48. [CrossRef]

8. Shayan, E.; Zare, V.; Mirzaee, I. Hydrogen production from biomass gasification; a theoretical comparison of using different gasification agents. *Energy Convers. Manag.* **2018**, *159*, 30–41. [CrossRef]
9. Turner, J.; Sverdrup, G.; Mann, M.K.; Maness, P.C.; Kroposki, B.; Ghirardi, M.; Blake, D. Renewable hydrogen production. *Int. J. Energy Res.* **2008**, *32*, 379–407. [CrossRef]
10. Zheng, J.-L.; Zhu, Y.-H.; Zhu, M.-Q.; Kang, K.; Sun, R.-C. A review of gasification of bio-oil for gas production. *Sustain. Energy Fuels* **2019**, *3*, 1600–1622. [CrossRef]
11. López Barreiro, D.; Prins, W.; Ronsse, F.; Brilman, W. Hydrothermal liquefaction (HTL) of microalgae for biofuel production: State of the art review and future prospects. *Biomass Bioenergy* **2013**, *53*, 113–127. [CrossRef]
12. Chen, W.-H.; Lin, B.-J.; Huang, M.-Y.; Chang, J.-S. Thermochemical conversion of microalgal biomass into biofuels: A review. *Bioresour. Technol.* **2015**, *184*, 314–327. [CrossRef]
13. Chiaramonti, D.; Prussi, M.; Buffi, M.; Rizzo, A.M.; Pari, L. Review and experimental study on pyrolysis and hydrothermal liquefaction of microalgae for biofuel production. *Appl. Energy* **2017**, *185*, 963–972. [CrossRef]
14. Guo, Y.; Yeh, T.; Song, W.; Xu, D.; Wang, S. A review of bio-oil production from hydrothermal liquefaction of algae. *Renew. Sustain. Energy Rev.* **2015**, *48*, 776–790. [CrossRef]
15. Jacobson, K.; Maheria, K.C.; Kumar Dalai, A. Bio-oil valorization: A review. *Renew. Sustain. Energy Rev.* **2013**, *23*, 91–106. [CrossRef]
16. Remón, J.; Broust, F.; Volle, G.; García, L.; Arauzo, J. Hydrogen production from pine and poplar bio-oils by catalytic steam reforming. Influence of the bio-oil composition on the process. *Int. J. Hydrog. Energy* **2015**, *40*, 5593–5608. [CrossRef]
17. Zhou, D.; Zhang, L.; Zhang, S.; Fu, H.; Chen, J. Hydrothermal Liquefaction of Macroalgae Enteromorpha prolifera to Bio-oil. *Energy Fuels* **2010**, *24*, 4054–4061. [CrossRef]
18. Yang, C.; Jia, L.; Chen, C.; Liu, G.; Fang, W. Bio-oil from hydro-liquefaction of Dunaliella salina over Ni/REHY catalyst. *Bioresour. Technol.* **2011**, *102*, 4580–4584. [CrossRef]
19. Jena, U.; Das, K.C. Comparative Evaluation of Thermochemical Liquefaction and Pyrolysis for Bio-Oil Production from Microalgae. *Energy Fuels* **2011**, *25*, 5472–5482. [CrossRef]
20. Maddi, B.; Panisko, E.; Wietsma, T.; Lemmon, T.; Swita, M.; Albrecht, K.; Howe, D. Quantitative characterization of the aqueous fraction from hydrothermal liquefaction of algae. *Biomass Bioenergy* **2016**, *93*, 122–130. [CrossRef]
21. Silva, J.M.; Soria, M.A.; Madeira, L.M. Challenges and strategies for optimization of glycerol steam reforming process. *Renew. Sustain. Energy Rev.* **2015**, *42*, 1187–1213. [CrossRef]
22. Hu, X.; Lu, G. Comparative study of alumina-supported transition metal catalysts for hydrogen generation by steam reforming of acetic acid. *Appl. Catal. B Environ.* **2010**, *99*, 289–297. [CrossRef]
23. Banach, B.; Machocki, A.; Rybak, P.; Denis, A.; Grzegorczyk, W.; Gac, W. Selective production of hydrogen by steam reforming of bio-ethanol. *Catal. Today* **2011**, *176*, 28–35. [CrossRef]
24. Ishihara, A.; Andou, A.; Hashimoto, T.; Nasu, H. Steam reforming of ethanol using novel carbon-oxide composite-supported Ni, Co and Fe catalysts. *Fuel Process. Technol.* **2020**, *197*, 106203. [CrossRef]
25. Khodakov, A.Y.; Chu, W.; Fongarland, P. Advances in the Development of Novel Cobalt Fischer−Tropsch Catalysts for Synthesis of Long-Chain Hydrocarbons and Clean Fuels. *Chem. Rev.* **2007**, *107*, 1692–1744. [CrossRef] [PubMed]
26. Llorca, J.; Dalmon, J.-A.; Ramírez de la Piscina, P.; Homs, N.S. In situ magnetic characterisation of supported cobalt catalysts under steam-reforming of ethanol. *Appl. Catal. A Gen.* **2003**, *243*, 261–269. [CrossRef]
27. Tsoncheva, T.; Ivanova, L.; Minchev, C.; Fröba, M. Cobalt-modified mesoporous MgO, ZrO_2, and CeO_2 oxides as catalysts for methanol decomposition. *J. Colloid Interface Sci.* **2009**, *333*, 277–284. [CrossRef]
28. Calles, J.A.; Carrero, A.; Vizcaíno, A.J. Ce and La modification of mesoporous Cu–Ni/SBA-15 catalysts for hydrogen production through ethanol steam reforming. *Microporous Mesoporous Mater.* **2009**, *119*, 200–207. [CrossRef]
29. Pereira, E.B.; Homs, N.; Martí, S.; Fierro, J.L.G.; Ramírez de la Piscina, P. Oxidative steam-reforming of ethanol over Co/SiO2, Co–Rh/SiO2 and Co–Ru/SiO2 catalysts: Catalytic behavior and deactivation/regeneration processes. *J. Catal.* **2008**, *257*, 206–214. [CrossRef]
30. Chen, G.; Tao, J.; Liu, C.; Yan, B.; Li, W.; Li, X. Hydrogen production via acetic acid steam reforming: A critical review on catalysts. *Renew. Sustain. Energy Rev.* **2017**, *79*, 1091–1098. [CrossRef]

31. Wang, Y.; Chen, M.; Yang, Z.; Liang, T.; Liu, S.; Zhou, Z.; Li, X. Bimetallic Ni-M (M = Co, Cu and Zn) supported on attapulgite as catalysts for hydrogen production from glycerol steam reforming. *Appl. Catal. A Gen.* **2018**, *550*, 214–227. [CrossRef]
32. Eschemann, T.O.; Oenema, J.; de Jong, K.P. Effects of noble metal promotion for Co/TiO_2 Fischer-Tropsch catalysts. *Catal. Today* **2016**, *261*, 60–66. [CrossRef]
33. Jermwongratanachai, T.; Jacobs, G.; Ma, W.; Shafer, W.D.; Gnanamani, M.K.; Gao, P.; Kitiyanan, B.; Davis, B.H.; Klettlinger, J.L.S.; Yen, C.H.; et al. Fischer–Tropsch synthesis: Comparisons between Pt and Ag promoted Co/Al_2O_3 catalysts for reducibility, local atomic structure, catalytic activity, and oxidation–reduction (OR) cycles. *Appl. Catal. A Gen.* **2013**, *464–465*, 165–180. [CrossRef]
34. Harun, N.; Abidin, S.Z.; Osazuwa, O.U.; Taufiq-Yap, Y.H.; Azizan, M.T. Hydrogen production from glycerol dry reforming over Ag-promoted Ni/Al_2O_3. *Int. J. Hydrog. Energy* **2018**. [CrossRef]
35. Konsolakis, M.; Sgourakis, M.; Carabineiro, S.A.C. Surface and redox properties of cobalt–ceria binary oxides: On the effect of Co content and pretreatment conditions. *Appl. Surf. Sci.* **2015**, *341*, 48–54. [CrossRef]
36. Trimm, D.L. Coke formation and minimisation during steam reforming reactions. *Catal. Today* **1997**, *37*, 233–238. [CrossRef]
37. Cerdá-Moreno, C.; Da Costa-Serra, J.F.; Chica, A. Co and La supported on Zn-Hydrotalcite-derived material as efficient catalyst for ethanol steam reforming. *Int. J. Hydrog. Energy* **2019**, *44*, 12685–12692. [CrossRef]
38. Calles, J.A.; Carrero, A.; Vizcaíno, A.J.; García-Moreno, L.; Megía, P.J. Steam Reforming of Model Bio-Oil Aqueous Fraction Using Ni-(Cu, Co, Cr)/SBA-15 Catalysts. *Int. J. Mol. Sci.* **2019**, *20*, 512. [CrossRef]
39. Casanovas, A.; Roig, M.; de Leitenburg, C.; Trovarelli, A.; Llorca, J. Ethanol steam reforming and water gas shift over Co/ZnO catalytic honeycombs doped with Fe, Ni, Cu, Cr and Na. *Int. J. Hydrog. Energy* **2010**, *35*, 7690–7698. [CrossRef]
40. Casanovas, A.; de Leitenburg, C.; Trovarelli, A.; Llorca, J. Catalytic monoliths for ethanol steam reforming. *Catal. Today* **2008**, *138*, 187–192. [CrossRef]
41. Li, Z.; Si, M.; Li, X.; Lv, J. Effects of titanium silicalite and TiO_2 nanocomposites on supported Co-based catalysts for Fischer–Tropsch synthesis. *Appl. Organomet. Chem.* **2019**, *33*, e4640. [CrossRef]
42. Tang, Y.; Yang, M.; Dong, W.; Tan, L.; Zhang, X.; Zhao, P.; Peng, C.; Wang, G. Temperature difference effect induced self-assembly method for Ag/SBA-15 nanostructures and their catalytic properties for epoxidation of styrene. *Microporous Mesoporous Mater.* **2015**, *215*, 199–205. [CrossRef]
43. Carrero, A.; Calles, J.A.; García-Moreno, L.; Vizcaíno, A.J. Production of Renewable Hydrogen from Glycerol Steam Reforming over Bimetallic Ni-(Cu,Co,Cr) Catalysts Supported on SBA-15 Silica. *Catalysts* **2017**, *7*, 55. [CrossRef]
44. Vizcaíno, A.J.; Carrero, A.; Calles, J.A. Hydrogen production by ethanol steam reforming over Cu–Ni supported catalysts. *Int. J. Hydrog. Energy* **2007**, *32*, 1450–1461. [CrossRef]
45. Sun, X.; Sun, L.; Wang, J.; Yan, Y.; Wang, M.; Xu, R. Confination of Ag nanostructures within SBA-15 by a "two solvents" reduction technique. *J. Taiwan Inst. Chem. Eng.* **2015**, *57*, 139–142. [CrossRef]
46. Martínez, A.; López, C.; Márquez, F.; Díaz, I. Fischer–Tropsch synthesis of hydrocarbons over mesoporous Co/SBA-15 catalysts: The influence of metal loading, cobalt precursor, and promoters. *J. Catal.* **2003**, *220*, 486–499. [CrossRef]
47. Fierro, G.; Lo Jacono, M.; Inversi, M.; Dragone, R.; Porta, P. TPR and XPS study of cobalt–copper mixed oxide catalysts: Evidence of a strong Co–Cu interaction. *Top. Catal.* **2000**, *10*, 39–48. [CrossRef]
48. Aspromonte, S.G.; Miró, E.E.; Boix, A.V. FTIR studies of butane, toluene and nitric oxide adsorption on Ag exchanged NaMordenite. *Adsorption* **2012**, *18*, 1–12. [CrossRef]
49. Lin, S.S.Y.; Kim, D.H.; Ha, S.Y. Metallic phases of cobalt-based catalysts in ethanol steam reforming: The effect of cerium oxide. *Appl. Catal. A Gen.* **2009**, *355*, 69–77. [CrossRef]
50. Yun, D.; Baek, J.; Choi, Y.; Kim, W.; Jong Lee, H.; Yi, J. Promotional Effect of Ni on a CrO_x Catalyst Supported on Silica in the Oxidative Dehydrogenation of Propane with CO_2. *ChemCatChem* **2012**, 4. [CrossRef]
51. Chen, J.; Zhang, X.; Arandiyan, H.; Peng, Y.; Chang, H.; Li, J. Low temperature complete combustion of methane over cobalt chromium oxides catalysts. *Catal. Today* **2013**, *201*, 12–18. [CrossRef]
52. Zoican Loebick, C.; Lee, S.; Derrouiche, S.; Schwab, M.; Chen, Y.; Haller, G.L.; Pfefferle, L. A novel synthesis route for bimetallic CoCr–MCM-41 catalysts with higher metal loadings. Their application in the high yield, selective synthesis of Single-Wall Carbon Nanotubes. *J. Catal.* **2010**, *271*, 358–369. [CrossRef]

53. Scheffe, J.R.; Steinfeld, A. Thermodynamic Analysis of Cerium-Based Oxides for Solar Thermochemical Fuel Production. *Energy Fuels* **2012**, *26*, 1928–1936. [CrossRef]
54. Amin, N.A.S.; Tan, E.F.; Manan, Z.A. SCR of NO_x by C_3H_6: Comparison between Cu/Cr/CeO_2 and Cu/Ag/CeO_2 catalysts. *J. Catal.* **2004**, *222*, 100–106. [CrossRef]
55. Cheng, W.-H.; Chen, I.; Liou, J.-S.; Lin, S.-S. Supported Cu Catalysts with Yttria-Doped Ceria for Steam Reforming of Methanol. *Top. Catal.* **2003**, *22*, 225–233. [CrossRef]
56. Huang, X.; Ma, L.; Wainwright, M.S. The influence of Cr, Zn and Co additives on the performance of skeletal copper catalysts for methanol synthesis and related reactions. *Appl. Catal. A Gen.* **2004**, *257*, 235–243. [CrossRef]
57. Wang, Z.; Xi, J.; Wang, W.; Lu, G. Selective production of hydrogen by partial oxidation of methanol over Cu/Cr catalysts. *J. Mol. Catal. A Chem.* **2003**, *191*, 123–134. [CrossRef]
58. Xu, L.; Duan, L.E.; Tang, M.; Liu, P.; Ma, X.; Zhang, Y.; Harris, H.G.; Fan, M. Catalytic CO_2 reforming of CH_4 over Cr-promoted Ni/char for H_2 production. *Int. J. Hydrog. Energy* **2014**, *39*, 10141–10153. [CrossRef]
59. da Silva, A.L.M.; den Breejen, J.P.; Mattos, L.V.; Bitter, J.H.; de Jong, K.P.; Noronha, F.B. Cobalt particle size effects on catalytic performance for ethanol steam reforming—Smaller is better. *J. Catal.* **2014**, *318*, 67–74. [CrossRef]
60. Ma, H.; Zeng, L.; Tian, H.; Li, D.; Wang, X.; Li, X.; Gong, J. Efficient hydrogen production from ethanol steam reforming over La-modified ordered mesoporous Ni-based catalysts. *Appl. Catal. B Environ.* **2016**, *181*, 321–331. [CrossRef]
61. Hu, X.; Dong, D.; Shao, X.; Zhang, L.; Lu, G. Steam reforming of acetic acid over cobalt catalysts: Effects of Zr, Mg and K addition. *Int. J. Hydrog. Energy* **2017**, *42*, 4793–4803. [CrossRef]
62. Biswas, P.; Kunzru, D. Steam reforming of ethanol on Ni–CeO_2–ZrO_2 catalysts: Effect of doping with copper, cobalt and calcium. *Catal. Lett.* **2007**, *118*, 36–49. [CrossRef]
63. Thaicharoensutcharittham, S.; Meeyoo, V.; Kitiyanan, B.; Rangsunvigit, P.; Rirksomboon, T. Hydrogen production by steam reforming of acetic acid over Ni-based catalysts. *Catal. Today* **2011**, *164*, 257–261. [CrossRef]
64. Wang, Y.; Chen, M.; Liang, T.; Yang, Z.; Yang, J.; Liu, S. Hydrogen Generation from Catalytic Steam Reforming of Acetic Acid by Ni/Attapulgite Catalysts. *Catalysts* **2016**, *6*, 172. [CrossRef]
65. Nogueira, F.G.E.; Assaf, P.G.M.; Carvalho, H.W.P.; Assaf, E.M. Catalytic steam reforming of acetic acid as a model compound of bio-oil. *Appl. Catal. B Environ.* **2014**, *160–161*, 188–199. [CrossRef]
66. Valle, B.; Aramburu, B.; Benito, P.L.; Bilbao, J.; Gayubo, A.G. Biomass to hydrogen-rich gas via steam reforming of raw bio-oil over Ni/La_2O_3-αAl_2O_3 catalyst: Effect of space-time and steam-to-carbon ratio. *Fuel* **2018**, *216*, 445–455. [CrossRef]
67. Chen, J.; Yang, X.; Li, Y. Investigation on the structure and the oxidation activity of the solid carbon produced from catalytic decomposition of methane. *Fuel* **2010**, *89*, 943–948. [CrossRef]
68. Nagasawa, S.; Yudasaka, M.; Hirahara, K.; Ichihashi, T.; Iijima, S. Effect of oxidation on single-wall carbon nanotubes. *Chem. Phys. Lett.* **2000**, *328*, 374–380. [CrossRef]
69. Choong, C.K.S.; Zhong, Z.; Huang, L.; Wang, Z.; Ang, T.P.; Borgna, A.; Lin, J.; Hong, L.; Chen, L. Effect of calcium addition on catalytic ethanol steam reforming of Ni/Al_2O_3: I. Catalytic stability, electronic properties and coking mechanism. *Appl. Catal. A Gen.* **2011**, *407*, 145–154. [CrossRef]
70. Galetti, A.E.; Gomez, M.F.; Arrúa, L.A.; Abello, M.C. Hydrogen production by ethanol reforming over NiZnAl catalysts: Influence of Ce addition on carbon deposition. *Appl. Catal. A Gen.* **2008**, *348*, 94–102. [CrossRef]
71. Natesakhawat, S.; Watson, R.B.; Wang, X.; Ozkan, U.S. Deactivation characteristics of lanthanide-promoted sol–gel Ni/Al_2O_3 catalysts in propane steam reforming. *J. Catal.* **2005**, *234*, 496–508. [CrossRef]
72. Qi, W.; Chen, S.; Wu, Y.; Xie, K. A chromium oxide coated nickel/yttria stabilized zirconia electrode with a heterojunction interface for use in electrochemical methane reforming. *RSC Adv.* **2015**, *5*, 47599–47608. [CrossRef]
73. Garcia, L.A.; French, R.; Czernik, S.; Chornet, E. Catalytic steam reforming of bio-oils for the production of hydrogen: Effects of catalyst composition. *Appl. Catal. A Gen.* **2000**, *201*, 225–239. [CrossRef]
74. Helveg, S.; Sehested, J.; Rostrup-Nielsen, J.R. Whisker carbon in perspective. *Catal. Today* **2011**, *178*, 42–46. [CrossRef]

75. Natesakhawat, S.; Wang, X.; Zhang, L.; Ozkan, U.S. Development of chromium-free iron-based catalysts for high-temperature water-gas shift reaction. *J. Mol. Catal. A Chem.* **2006**, *260*, 82–94. [CrossRef]
76. Sierra Gallego, G.; Mondragón, F.; Tatibouët, J.-M.; Barrault, J.; Batiot-Dupeyrat, C. Carbon dioxide reforming of methane over La_2NiO_4 as catalyst precursor—Characterization of carbon deposition. *Catal. Today* **2008**, *133–135*, 200–209. [CrossRef]
77. Carrero, A.; Calles, J.A.; Vizcaíno, A.J. Effect of Mg and Ca addition on coke deposition over Cu–Ni/SiO_2 catalysts for ethanol steam reforming. *Chem. Eng. J.* **2010**, *163*, 395–402. [CrossRef]
78. Montero, C.; Ochoa, A.; Castaño, P.; Bilbao, J.; Gayubo, A.G. Monitoring Ni0 and coke evolution during the deactivation of a Ni/La_2O_3–αAl_2O_3 catalyst in ethanol steam reforming in a fluidized bed. *J. Catal.* **2015**, *331*, 181–192. [CrossRef]
79. Osorio-Vargas, P.; Flores-González, N.A.; Navarro, R.M.; Fierro, J.L.G.; Campos, C.H.; Reyes, P. Improved stability of Ni/Al_2O_3 catalysts by effect of promoters (La_2O_3, CeO_2) for ethanol steam-reforming reaction. *Catal. Today* **2016**, *259*, 27–38. [CrossRef]
80. Charisiou, N.D.; Siakavelas, G.; Papageridis, K.N.; Baklavaridis, A.; Tzounis, L.; Polychronopoulou, K.; Goula, M.A. Hydrogen production via the glycerol steam reforming reaction over nickel supported on alumina and lanthana-alumina catalysts. *Int. J. Hydrog. Energy* **2017**, *42*, 13039–13060. [CrossRef]
81. Silva, K.C.; Corio, P.; Santos, J.J. Characterization of the chemical interaction between single-walled carbon nanotubes and titanium dioxide nanoparticles by thermogravimetric analyses and resonance Raman spectroscopy. *Vib. Spectrosc.* **2016**, *86*, 103–108. [CrossRef]
82. Tzounis, L.; Kirsten, M.; Simon, F.; Mäder, E.; Stamm, M. The interphase microstructure and electrical properties of glass fibers covalently and non-covalently bonded with multiwall carbon nanotubes. *Carbon* **2014**, *73*, 310–324. [CrossRef]
83. Ferencz, Z.; Varga, E.; Puskás, R.; Kónya, Z.; Baán, K.; Oszkó, A.; Erdőhelyi, A. Reforming of ethanol on Co/Al_2O_3 catalysts reduced at different temperatures. *J. Catal.* **2018**, *358*, 118–130. [CrossRef]
84. Zhao, D.; Feng, J.; Huo, Q.; Melosh, N.; Fredrickson, G.H.; Chmelka, B.F.; Stucky, G.D. Triblock copolymer syntheses of mesoporous silica with periodic 50 to 300 angstrom pores. *Science* **1998**, *279*, 548–552. [CrossRef] [PubMed]
85. Vizcaíno, A.J.; Carrero, A.; Calles, J.A. Ethanol steam reforming on Mg- and Ca-modified Cu–Ni/SBA-15 catalysts. *Catal. Today* **2009**, *146*, 63–70. [CrossRef]
86. Carrero, A.; Calles, J.A.; Vizcaíno, A.J. Hydrogen production by ethanol steam reforming over Cu-Ni/SBA-15 supported catalysts prepared by direct synthesis and impregnation. *Appl. Catal. A Gen.* **2007**, *327*, 82–94. [CrossRef]
87. Vizcaíno, A.J.; Carrero, A.; Calles, J.A. Comparison of ethanol steam reforming using Co and Ni catalysts supported on SBA-15 modified by Ca and Mg. *Fuel Process. Technol.* **2016**, *146*, 99–109. [CrossRef]

Article

Single and Dual Metal Oxides as Promising Supports for Carbon Monoxide Removal from an Actual Syngas: The Crucial Role of Support on the Selectivity of the Au–Cu System

Bernay Cifuentes [1], **Felipe Bustamante** [2] and **Martha Cobo** [1,*]

1. Energy, Materials, and Environment Laboratory, Chemical Engineering Department, Universidad de La Sabana, Campus Universitario Puente del Común, Km. 7 Autopista Norte, Bogotá 250001, Colombia; bernay.cifuentes1@unisabana.edu.co
2. Environmental Catalysis Research Group, Chemical Engineering Department, Universidad de Antioquia UdeA, Calle 70 No. 52-21, Medellín 050010, Colombia; felipe.bustamante@udea.edu.co

* Correspondence: martha.cobo@unisabana.edu.co; Tel.: +571-8615555 (ext. 25207); Fax: +571-8615555

Received: 9 September 2019; Accepted: 10 October 2019; Published: 13 October 2019

Abstract: A catalytic screening was performed to determine the effect of the support on the performance of an Au–Cu based system for the removal of CO from an actual syngas. First, a syngas was obtained from reforming of ethanol. Then, the reformer outlet was connected to a second reactor, where Au–Cu catalysts supported on several single and dual metal oxides (i.e., CeO_2, SiO_2, ZrO_2, Al_2O_3, La_2O_3, Fe_2O_3, CeO_2-SiO_2, CeO_2-ZrO_2, and CeO_2-Al_2O_3) were evaluated. AuCu/CeO_2 was the most active catalyst due to an elevated oxygen mobility over the surface, promoting CO_2 formation from adsorption of C–O* and OH^- intermediates on Au^0 and CuO species. However, its lower capacity to release the surface oxygen contributes to the generation of stable carbon deposits, which lead to its rapid deactivation. On the other hand, AuCu/CeO_2-SiO_2 was more stable due to its high surface area and lower formation of formate and carbonate intermediates, mitigating carbon deposits. Therefore, use of dual supports could be a promising strategy to overcome the low stability of AuCu/CeO_2. The results of this research are a contribution to integrated production and purification of H_2 in a compact system.

Keywords: CO-PROX; CO-SMET; CO_2 methanation; hydrogen purification; process integration

---

1. Introduction

Synthesis gas (syngas) is used as a chemical building block in the synthesis of commodity chemicals and for energy applications. Specifically, syngas can be used in combustion processes [1], gas turbines [2], or hydrogen fuel cells (H_2-FC) [3] to produce energy. The H_2-FC are promising systems to provide sustainable energy for households, industry, transportation, and small devices. Likewise, the use of H_2-FC has been proposed as an alternative to supply energy in places that are not connected to the electrical network and for remote installations [4].

The syngas composition varies depending on the production source, but mostly contains H_2, carbon monoxide (CO), and light hydrocarbons. Bioethanol reforming is one of the most used pathways to produce syngas due to its high yield to H_2 [5]. In a previous study [6], we obtained a syngas containing H_2, CO, CO_2, CH_4, and H_2O from ethanol steam reforming (ESR) using a RhPt/CeO_2-SiO_2 catalyst. Syngas production remained stable for 72 h of continuous operation and on/off cycles. This syngas could be used for sustainable energy production in H_2-FC. However, CO must be removed from the syngas because of its harmful effect on fuel cell electrodes [7].

One of the most used strategies of CO removal from syngas is via chemical pathways, which includes preferential oxidation of CO (CO-PROX) [8,9], water gas shift reaction (WGSR) [10], and selective CO methanation (CO-SMET) [10]. Traditionally, the objective of the CO cleanup step is to ensure CO concentrations below 10 ppm, which requires several catalytic reactors in series [11] and presents a high operating cost. However, recent research studies have allowed the development of H_2-FC systems that tolerate CO concentrations above 100 ppm [12–14]. These contributions facilitate the use of less complex systems for syngas purification, which could lead to the development of more compact and economic H_2 technology.

Anticipating the commercialization of a new generation of more CO-tolerant H_2-FC, it has been proposed to redesign the CO removal stage to reduce the number of process units in syngas purification. The new approach seeks to carry out CO removal using a single catalytic reactor, where several reactions occur simultaneously (i.e., CO-PROX, WGSR, and CO-SMET). Kugai et al. [15] studied Pt–Cu and Pd–Cu bimetallic catalysts supported on CeO_2 for oxygen-enhanced water gas shift (OWGS), where WGSR and CO-PROX occur concurrently, reporting higher CO removal from a model reformate gas (synthetic syngas) in the OWGS compared to the WGSR carried out individually. Similarly, Xu and Zhang [16] reported that the presence of CO-SMET during CO-PROX on a commercial Ru/Al_2O_3 catalyst allows for wider temperature windows that ensure the CO removal of a synthetic syngas. Despite these valuable contributions, the CO removal from syngas in a compact system is still at laboratory scale. Among the limitations for evaluation at the pilot scale is the lack of consensus regarding the catalyst and the most appropriate operating conditions to carry out the syngas purification.

Au is recognized as a promising catalyst in the three cleaning reactions of syngas (i.e., CO-PROX, WGSR, and CO-SMET) [17,18]. Reina et al. [19] evaluated bimetallic catalysts of Au–M (M = La, Ni, Cu, Fe, Cr, Y), reporting that CO oxidation is favored by the Au–Cu combination because Cu interacts strongly with the support, favoring the oxygen mobility in the catalyst. Also, in a previous study [20], we evaluated Au–Cu bimetallic catalysts supported on CeO_2 for CO removal from a syngas obtained from ESR. It was possible to reduce the CO concentration below 100 ppm, but the catalyst showed rapid deactivation after 40 h. Deactivation was related to structural changes in the support and to the accumulation of carbonaceous compounds during continuous operation. Thus, this study illustrated that the support plays a key role in CO removal from an actual syngas, and led us to evaluate different supports for CO removal from a syngas in the search for a stable material.

Figure 1 shows the supports most used in the CO removal processes (i.e., WGSR, CO-PROX, CO-SMET, or their combinations). CeO_2, Fe_2O_3, ZrO_2, TiO_2, and Al_2O_3 are the most commonly used single supports in CO removal from synthetic syngas. However, there is a growing interest in mixed supports (dual metal oxides), because they may have characteristics not observed in individual supports [21]. Most combinations of dual metal oxides include CeO_2 in the matrix, usually combined with supports that provide larger surface area, such as Al_2O_3 [22] and SiO_2 [23], or with basic oxides, such as ZrO_2, to generate new active sites [24]. TiO_2 is mainly used in CO removal by photocatalytic processes [25] and was not considered in this study. On the other hand, although La_2O_3 is not among the most used supports in CO removal, it was recently reported that La_2O_3 is effective for avoiding carbon deposits during CO-SMET [26].

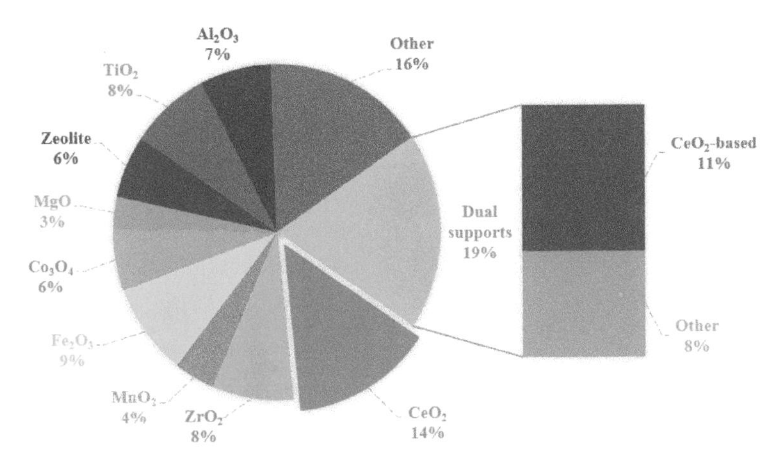

Figure 1. Supports used in CO removal from syngas streams using CO-PROX, WGSR or CO-SMET reactions.

Although several supports for syngas cleanup have been proposed, each investigation was carried out under different experimental conditions and using synthetic syngas, which makes it difficult to select the most suitable support for the CO removal. Therefore, the objective of this work was to study the CO removal from an actual syngas using bimetallic catalysts of AuCu-supported on single and dual metal oxides. Specifically, CeO_2, ZrO_2, La_2O_3, Fe_2O_3, Al_2O_3, and SiO_2 were selected as single metal oxides, and CeO_2-SiO_2, CeO_2-ZrO_2, and CeO_2-Al_2O_3 as dual metal oxides. The catalytic performance of the supports with and without active metals (i.e., Au and Cu) was evaluated. Then, the activity, selectivity, and stability were established as criteria for selecting the most suitable support for the CO elimination. In addition, characterization tests were conducted, such as temperature programmed reduction (TPR), surface area tests using the Brunauer-Emmett-Teller (BET) method, oxygen storage capacity (OSC) tests, thermogravimetric analysis (TGA), and in situ diffuse reflectance infrared Fourier transform spectroscopy (DRIFTS).

2. Results and Discussion

2.1. Activity, Selectivity, and Stability

Figure 2 shows the CO conversion in the cleanup reactor on the bare supports (i.e., without Au and Cu) and Au–Cu-supported catalysts. CeO_2 and ZrO_2 display the larger CO conversion between single metal oxides (Figure 2a). Indeed, the presence of oxygen vacancies on the surface of an oxide could favor a support showing high activity in the CO oxidation, despite the absence of active metals [27]; on the other hand, supports with low OSC, such as Al_2O_3 [28], present lower activity. The use of dual metal oxides has been proposed as a strategy to overcome the deficiencies of single supports [21]. Figure 2b shows that CeO_2-SiO_2 increases the CO conversion compared to SiO_2, which could be associated with the interaction between the two oxides. However, no significant improvement in the CO conversion with CeO_2-Al_2O_3 was observed, and even for CeO_2-ZrO_2, the combination of the two metal oxides leads to a less active material. Furthermore, below 260 °C the dual metal oxides showed less activity that single CeO_2, suggesting that the combination of several metal oxides does not always lead to more active materials in the syngas cleaning.

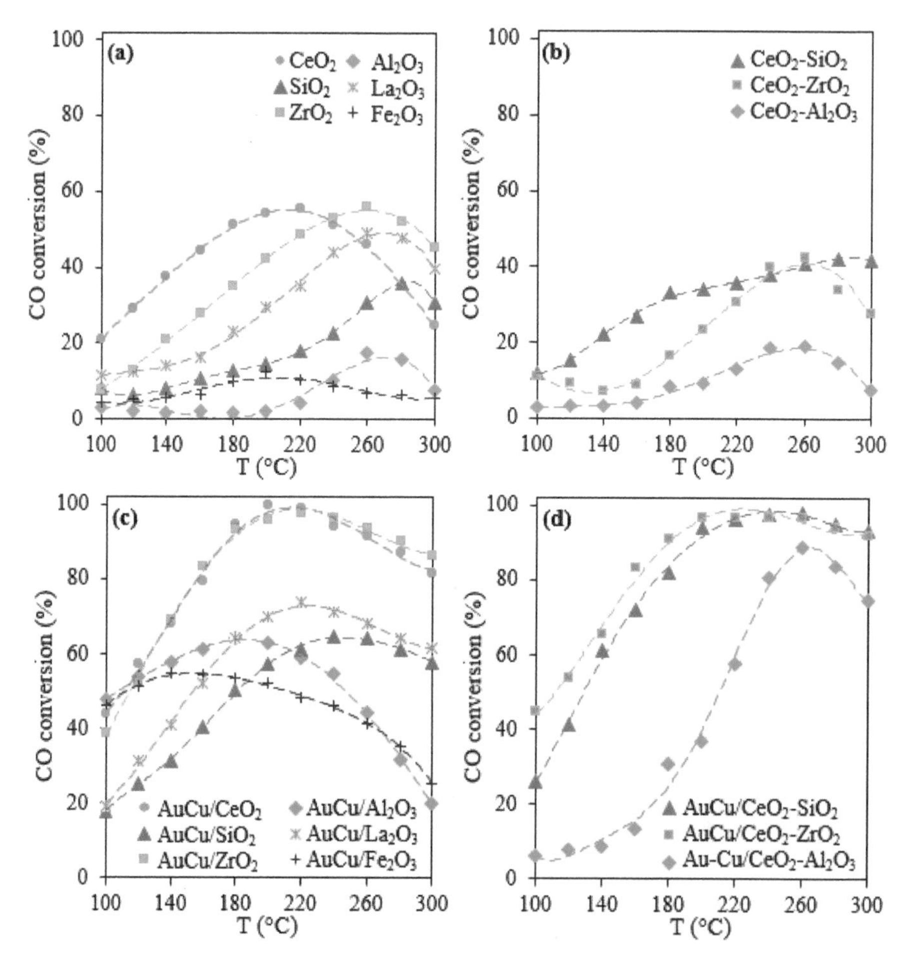

Figure 2. CO conversion obtained in the Cleanup reactor with supports (**a**,**b**) and supported 1 wt% Au–1 wt% Cu catalysts (**c**,**d**). Syngas feed: 7.8% H_2, 2.0% CO, 0.5% CO_2, 0.3% CH_4, 1.4% H_2O, 1.8% O_2, 6.8% N_2, and 79.4% Ar. Reaction conditions: Space velocity (SV) = 6.5 ± 0.2 L/g_{cat}*min and 0.3 g of the catalytic bed.

On the other hand, catalytic systems based on Au, Cu and Au–Cu have been studied extensively for the CO oxidation, CO-PROX, WGSR, and CO-SMET. In-depth descriptions for Cu/CeO_2 [9], AuCu/CeO_2 [29], AuCu/SiO_2 [30], AuCu/Al_2O_3 [31], Au/Fe_2O_3 [32], Au/La_2O_3/Al_2O_3 [33], and Au/CeO_2-ZrO_2 [34,35] are available in the literature. In general, Au favors the CO conversion through a mechanism that involves Au–CO and Au–OOH species [36], where the formation of C–O* intermediates determines the selectivity of the process [20], while CuO acts through a redox mechanism [8], promoting oxygen mobility in the oxide lattice [37] and facilitating the CO oxidation. A synergistic Au–Cu effect has also been proposed [19,29,38]. Therefore, the inclusion of 1 wt% Au and 1 wt% Cu in the single and dual metal oxides promotes greater CO conversion (Figure 2). Despite having the same active metals (i.e., Au and Cu), the catalysts showed maximum CO conversion at different temperatures, indicating that the properties of the support have a key role in the syngas cleaning. Table 1 shows that only AuCu/CeO_2 reached CO concentrations below 100 ppm in the actual syngas at 210 °C, whereas minimum CO concentrations of the other catalysts were above 500 ppm.

Table 1. Minimum concentration of CO obtained in syngas, apparent active metal dispersion (H/M ratio), surface area, OSC, and OSCC of Au–Cu catalysts supported on single and dual supports.

Catalyst[a]	Minimum CO Concentration in Outlet Gas (ppm)[b]	H/M Index	BET Surface Area (m^2/g_{cat})		OSC in AC Samples ($\mu mol\ O_2/g_{cat}$)		OSCC at 300 °C ($\mu mol\ O_2/g_{cat}$)	
			AC	Spent	100 °C	300 °C	Fresh	Spent
$AuCu/CeO_2$	75 at 210 °C	0.9	60	58 (U) 50 (S)	41	91	230	121 (U) 93 (S)
$AuCu/SiO_2$	8320 at 240 °C	0.7	364	277 (U)	21	37	45	41 (U)
$AuCu/ZrO_2$	507 at 225 °C	0.8	58	47 (U)	39	76	185	84 (U)
$AuCu/Al_2O_3$	745 at 180 °C	0.8	90	65 (U)	31	35	75	41 (U)
$AuCu/La_2O_3$	5365 at 225 °C	0.4	19	18 (U)	21	41	90	24 (U)
$AuCu/Fe_2O_3$	9416 at 140 °C	0.4	16	5 (U)	NR	NR	NR	NR
$AuCu/CeO_2$-SiO_2	861 at 230 °C	1.6	110	75 (U) 74 (S)	34	78	146	121 (U) 126 (S)
$AuCu/CeO_2$-ZrO_2	941 at 210 °C	0.9	42	30 (U)	42	94	210	162 (U)
$AuCu/CeO_2$-Al_2O_3	1521 at 260 °C	1.2	65	56 (U)	32	79	155	121 (U)

[a] Nominal metal loadings: 1 wt% Au and 1 wt% Cu. [b] Value includes the carrier gas. AC: activated catalyst, which were reduced with H_2 and stabilized in air before activity tests. U: sample used to obtain light-off curves. S: sample evaluated in the stability test. Note: NR = Not reported; OSC = oxygen storage capacity; OSCC = oxygen storage complete capacity; BET: Brunauer-Emmett-Teller test.

On the other hand, the selectivity in the CO removal has been attributed to the support rather than the active metal [39], being the consumption of H_2 an important criterion in catalyst selection [39]. Figure 3 shows that H_2 consumption increases with temperature, particularly in the supports and catalysts based on ZrO_2. The deficiency of ZrO_2 to adsorb/desorb bidentate carbonates above 150 °C has been associated with a promotion of the H_2 combustion over the CO oxidation [40]. Likewise, H_2 loss increases in the majority of the supported Au–Cu catalysts (Figure 3c,d) compared to their respective bare support (Figure 3a,b), possibly due to affinity of the Au–Cu system to form intermediates in the H_2 oxidation (e.g., hydroxyl groups [29,41]) and methane formation (e.g., C–O* species [18,20,42]). Also, the most active catalysts in the CO removal (i.e., $AuCu/CeO_2$, $AuCu/ZrO_2$, $AuCu/CeO_2$-SiO_2, and $AuCu/CeO_2$-ZrO_2) promote higher H_2 consumption. That is, an active catalyst in the CO conversion possibly has an inherent tendency to consume H_2. The high H_2 consumption, which in some cases exceeds 20%, could be associated with the syngas composition [20,37], specifically with the H_2/CO ratio. Table 2 shows the results obtained in the CO removal with catalytic systems based on Au–Cu. High H_2/CO ratios (>>10 [43]) are used in CO-PROX with synthetic syngas to favor CO oxidation [44] and reduce the H_2 consumption. To achieve such high H_2/CO ratios before CO-PROX several WGSR reactors are required, however [11]. Thence, aiming at reducing the number of units used in the traditional process, it has been proposed to carry out CO removal reactions in a single reactor using the syngas that comes directly from the reformer [15,20,35]. Nevertheless, the syngas obtained directly from the ESR contains larger amounts of CO. H_2/CO ratios around 4 have been reported for syngas obtained from ESR using Ir/CeO_2 [45] and $RhPd/CeO_2$ [46] catalysts. Thus, the low H_2/CO ratio in the actual syngas (e.g., the syngas used in this work has an H_2/CO = 4) could conduce to a high H_2 loss in the cleanup reactor. Simultaneous production of CO_2 and CH_4 was observed in all catalysts evaluated (Figures A1 and A2 in Appendix A), suggesting that CO-SMET and CO_2 methanation occur together with CO-PROX and WGSR. Then, H_2 oxidation and carbon hydrogenation would be the main causes of H_2 loss during the CO removal from an actual syngas.

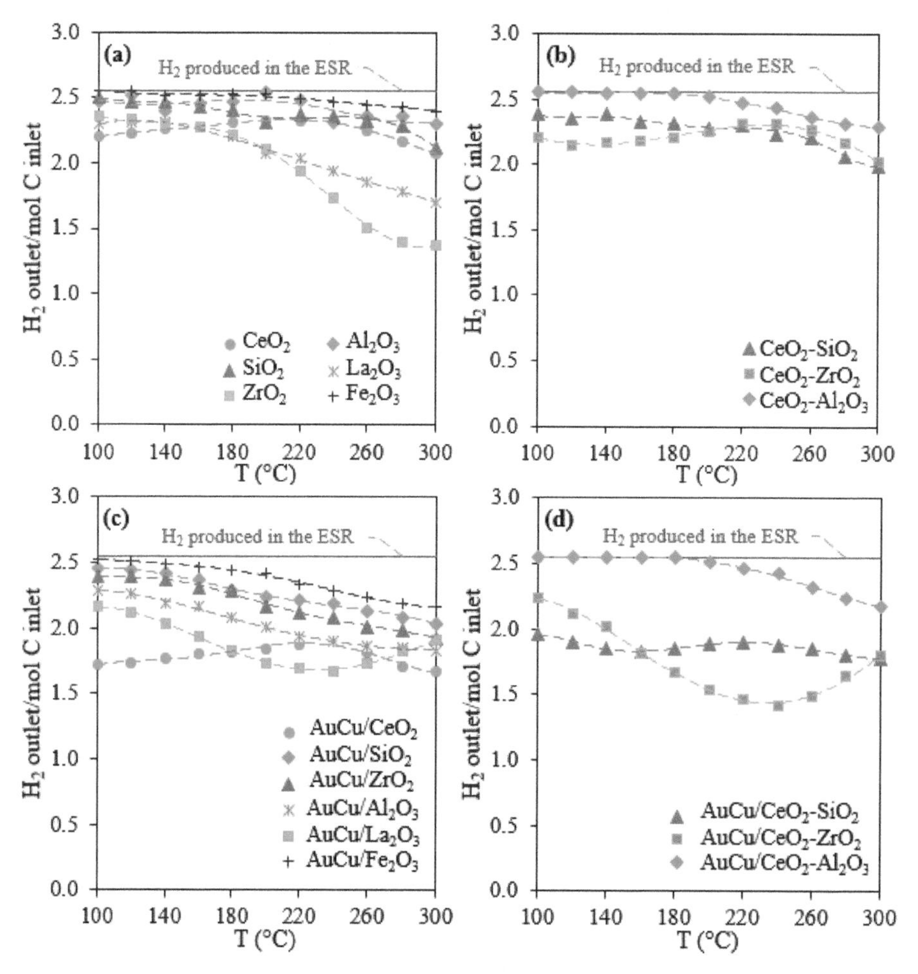

Figure 3. H_2 yield obtained from a system that integrate the ethanol steam reforming (ESR) reactor and the cleanup reactor, where the CO removal is performed with bare supports (**a**,**b**) and supported 1 wt% Au–1 wt% Cu catalysts (**c**,**d**). Reaction conditions: SV = 6.5 ± 0.2 L/g_{cat}*min and 0.3 g of the catalytic bed in both reactors.

Table 2. Comparison of various catalytic systems for the CO removal using Au–Cu catalysts.

Catalyst	Syngas Type	H_2/CO	T (°C)	CO Conversion (%)	H_2 Loss (%)	Ref.
AuCu/CeO_2	Synthetic	30	220	90	2	[29]
AuCu/SBA-15	Synthetic	>50	25	100	5[a]	[47]
Au/CuO-CeO_2/Al_2O_3	Synthetic	4.5	350	75	NR	[17]
Au/CeO_2-CuO_2/Al_2O_3	Synthetic	50	110	95	3[a]	[19]
Au/Al_2O_3	Synthetic	>50	80	99	2[a]	[36]
Au/CeO_2-ZrO_2	Actual	30	100	99	2[a]	[35]
AuCu/CeO_2	Actual	4	210	99	17	This work
AuCu/CeO_2-SiO_2	Actual	4	230	97	19	This work

[a] Calculated by O_2 mass balance. NR: Not reported.

Although CH_4 formation implicitly involves an undesirable H_2 consumption, it has been reported that a combination of CO-PROX and methanation improves CO removal compared to the CO-PROX alone, because of favoritism in the activation of adsorbed CO [16]. Then, C and H mass balances were carried out to determine the effect of CH_4 production on H_2 consumption and CO conversion. Figure 4 shows the H_2 and CO converted with respect to the CH_4 formed in the cleanup reactor. CH_4 formation appears to be directly proportional to H_2 loss (Figure 4a), but the amount of H_2 consumed is larger than the amount of H_2 contained in the formed CH_4 (yellow line); moreover, in most catalysts, H_2 loss is larger than the H_2 required by CO_2 methanation (green line). Hence, the remnant of H_2 loss may be associated with the production of water or hydrogenated compounds not detected by GC, indicating that methanation would have a secondary role in the H_2 loss during the syngas cleanup. On the other hand, CO conversion grows faster compared to the contribution of methanation (Figure 4b). Xu et al. [16] studied a Rh/Al_2O_3 catalyst and proposed that at temperatures above 150 °C, the methanation of CO_2 formed during the CO-PROX facilitates the CO oxidation caused by changes in the C–O* and H* adsorbed species. This possible beneficial effect of CO-PROX and subsequent CO_2 methanation seems to be stronger in some catalysts (e.g., $AuCu/CeO_2$-Al_2O_3 and $AuCu/CeO_2$-SiO_2), which would explain their higher activity at high temperatures (Figure 2d), where most CH_4 was produced (Figure A2 in Appendix A). $AuCu/CeO_2$ and $AuCu/La_2O_3$ show an atypical trend (Figure 4b), where the CO conversion decreases with the CH_4 formation, which could depend on the intermediates of C–O* formed on theses catalysts, as will be discussed later. Therefore, these results would confirm the beneficial effect of CO_2 methanation during the CO-PROX proposed in [16], but it was also identified that this effect depends on the support and composition of the syngas.

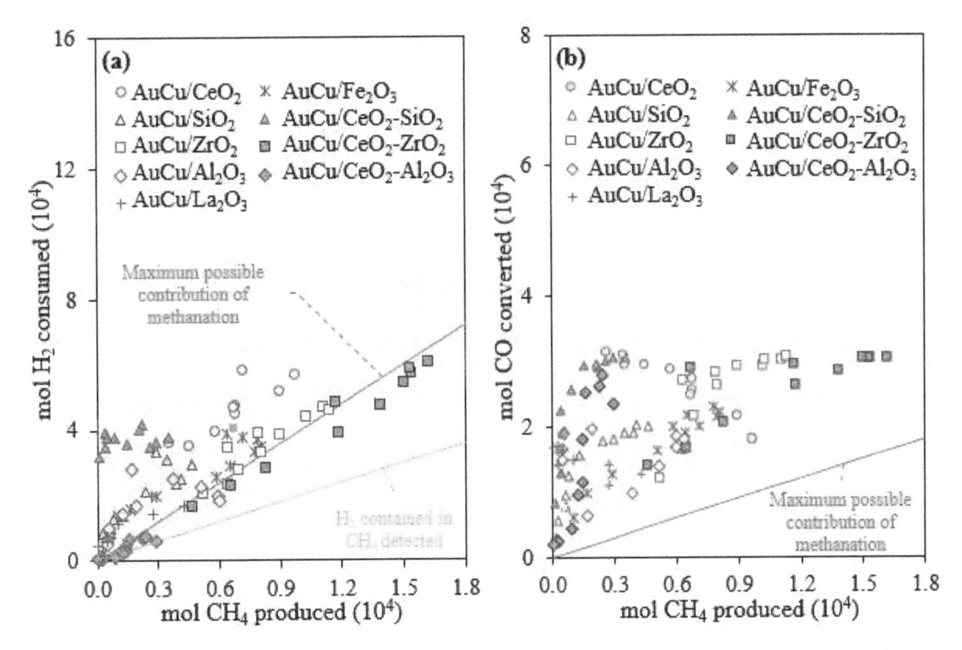

Figure 4. Contribution of methanation in (**a**) the H_2 consumption and (**b**) CO conversion during the CO removal from an actual syngas. The shaded area conveys the trend of the experimental data.

Although the main objective in the cleaning of the syngas is the CO removal, differences in the activity and selectivity could lead to changes in product distribution over prolonged periods of operation. Therefore, the stability of Au–Cu catalysts loaded on the best single (CeO_2) and dual support (CeO_2-SiO_2) was evaluated. Figure 5 shows the product distribution over time obtained from a system consisting of ESR and cleanup reactors, the latter of which is packed with either

$AuCu/CeO_2$ or $AuCu/CeO_2$-SiO_2. In both cases, a H_2-rich stream is obtained. However, $AuCu/CeO_2$ shows more variability in product distribution, and after around 42 h of operation deactivation was observed, at which point the test was stopped. In contrast, the $AuCu/CeO_2$-SiO_2 catalyst ensures a stable operation for longer periods of time (at least 30% more time-on-stream, Figure 5b) with CO concentration of about 1000 ppm. The results of the stability test show that the use of dual metal oxides leads to less active (i.e., CO concentration of 1000 ppm versus 75 ppm) but more stable materials, which could be more interesting in extended processes.

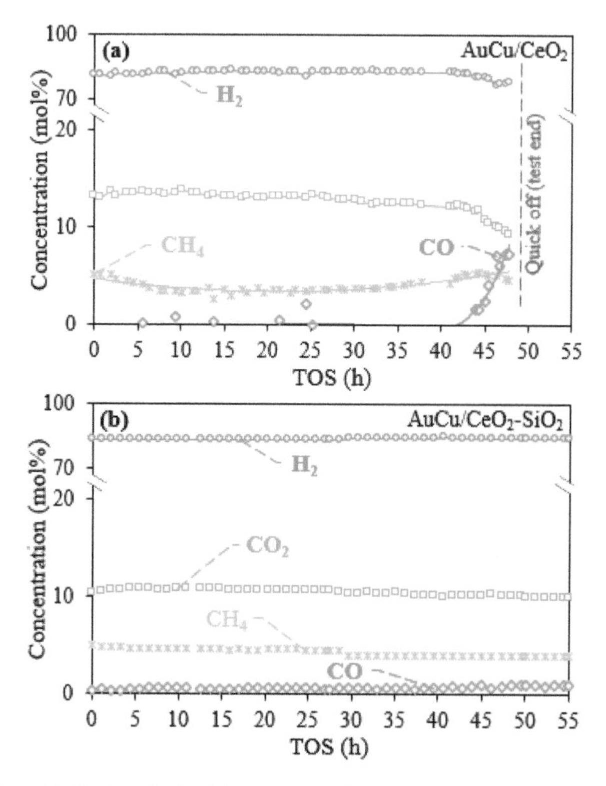

Figure 5. Products distribution obtained from a system that integrate the ESR reactor and the cleanup reactor, where the CO removal is performed with (**a**) $AuCu/CeO_2$ and (**b**) $AuCu/CeO_2$-SiO_2 catalysts. Syngas feed: 7.8% H_2, 2.0% CO, 0.5% CO_2, 0.3% CH_4, 1.4% H_2O, 1.8% O_2, 6.8% N_2, and 79.4% Ar. Reaction conditions: The space velocity (SV) = 6.5 ± 0.2 L/g_{cat}*min and 0.3 g of the catalytic bed. Note: TOS = Time-on-stream.

Activity, H_2 consumption, and stability were used as criteria for comparison among the Au–Cu-supported catalysts for the CO removal from an actual syngas. Now, catalytic properties, such as reducibility, surface area, OSC, carbon deposit formation, and the CO-support interactions, will be related to the activity, selectivity, and stability of the Au–Cu catalysts supported in single and dual metal oxides.

2.2. Catalysts Characterization

2.2.1. TPR

The redox properties of catalysts have a significant effect on CO oxidation and metal-support interactions [17]. Figure 6 shows the H_2-TPR profiles for the Au–Cu catalysts supported on single and dual metal oxides. Deconvolution peaks are presented to identify possible individual contributions in each reduction zone, but they are not intended to be exact. Contrary to bare supports (Figure A3 in Appendix A), discrepancies are observed between supported Au–Cu catalysts. The specific reduction temperatures for Au and Cu are very diverse in the literature, possibly because the reduction of metals strongly depends on the interaction with other species [48]. In this study, a first zone (<130 °C) observed was attributed to the reduction of Au^{3+} and Au^{+} nanoparticles [41]. The second zone (130 to 430 °C) was associated with the reduction of Cu, where at least three species [49] can be identified: (α) easily reducible CuO nano particles, (β) particles of CuO dispersed that interact moderately with the support, and (γ) isolated particles of Cu [50]. In the last zone (>430 °C), the reduction of surface layers and bulk of the support is likely happening [34]. The α and β species promote the formation of oxygen vacancies [51], contributing to the CO oxidation. Thus, preferential formation of CuO species in single and dual metal supports would explain the increase in CO_2 production over Au–Cu-supported catalysts compared to bare supports (Figures A1 and A2 in Appendix A).

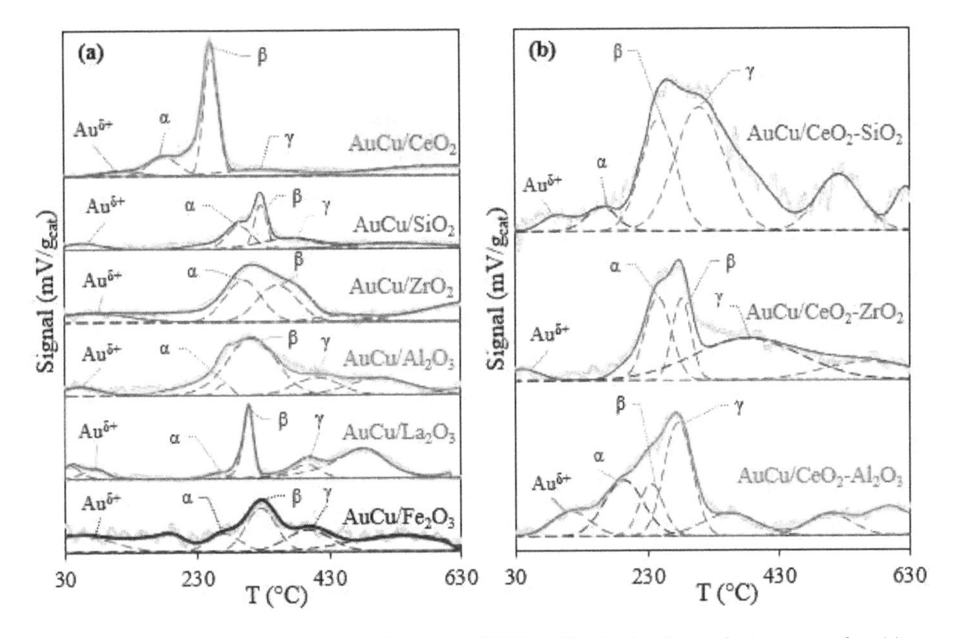

Figure 6. H_2-temperature programmed reduction (TPR) profiles for Au–Cu catalysts supported on (**a**) single and (**b**) dual metal oxides.

On the other hand, the displacement of the reduction peaks to lower temperatures has been associated with changes in metal-support interactions [48]. CeO_2 shows an exceptional ability to facilitate the reduction of Cu and the formation of (mostly) β species. This effect has been previously studied [9,52], correlating a stronger CuO-CeO_2 interaction with high activity during CO-PROX. However, the increase in the contribution of γ-species and a slight shift of reduction peaks to higher temperatures could indicate a variation of the CuO-CeO_2 interaction in the Au–Cu catalysts supported in dual oxides. Thus, a change in the redox properties of the support caused by the presence of a second metal oxide could explain why the Au–Cu catalysts supported on dual oxides (i.e., AuCu/CeO_2-SiO_2, AuCu/CeO_2-ZrO_2, and AuCu/CeO_2-Al_2O_3) showed less activity compared to AuCu/CeO_2 (Figure 2). However, an exceedingly strong CuO-support interaction could also mitigate the formation of selective Au–Cu alloys [31]. In fact, AuCu/ZrO_2 and AuCu/CeO_2-ZrO_2 show a significant contribution of α species, which could be related to the high H_2 loss observed in these catalysts (Figure 3c,d). On the contrary, the combination of inert metal oxides such as Al_2O_3 with CeO_2 could facilitate the migration of CuO towards Au particles [31], leading to lower H_2 loss compared to single CeO_2 support. Then, the change in redox properties of CeO_2 by the presence of inert metal oxides (e.g., SiO_2) could lead to less active but more selective materials during CO removal.

Table 1 shows the H/M index, which has been associated with apparent active metal dispersion [53]. The H/M index in AuCu/Fe_2O_3 and AuCu/La_2O_3 is particularly low, indicating that these catalysts are not as effective for dispersing active metals [53]. In the other catalysts, the H/M index was close to or larger than 1.0 (i.e., complete reduction of Au and Cu), which could be associated with a higher dispersion of Au and Cu on the catalytic surface. However, a high H/M value could also indicate an additional effect of superficial reduction of the supports by the interaction between metal oxides and active metals [54]. Au–Cu catalysts supported on dual metal oxides showed higher H/M index compared to their respective single supports, which could be associated with a favoring in the reduction of both active metals and support due to the interaction between metal oxides. If so, then the redox properties of Au–Cu catalysts supported on dual oxides would depend on several interactions: (i) active metal-active metal; (ii) active metal-support, and (iii) oxide I-oxide II. The variation of these interactions influences catalytic performance during CO removal.

2.2.2. BET Area

The surface area of the catalysts is key to the availability of the active sites and catalytic performance [55]. The BET area (Table 1) of the catalysts supported on basic oxides (i.e., CeO_2, ZrO_2 and La_2O_3 [56]) is larger than their respective bare supports (Table A1 in Appendix B), which has been previously associated with the formation of high disperse β species [20,57]. On the other hand, AuCu/Fe_2O_3 and AuCu/La_2O_3 show low surface areas, which match to the low capacity of these metal oxides to disperse active metals (low H/M index, Table 1). The synthesis method of the catalysts could influence the surface area of the support, overcoming some drawbacks of metal oxides such as Fe_2O_3 by using alternative synthesis methods [32,58]. In contrast, the higher surface area of AuCu/CeO_2-Al_2O_3 and AuCu/CeO_2-SiO_2 could favor the dispersion of Au and Cu, which is reflected by a larger H/M index (Table 1).

Although an increase in the surface area could contribute to improving the catalytic activity [55], the trend for surface area of the catalysts does not match their activity (Figure 2), indicating that the supports have other features that could be more relevant during the CO removal. Figure 7 shows the conversion rate of CO normalized by the surface area of catalysts. AuCu/La_2O_3 has a high normalized activity, possibly because the basic supports promote the formation of Au nanoparticles [33] and formation of β species [20,57], which are active in the CO conversion. In fact, basic oxides, such as CeO_2 and ZrO_2, also have higher normalized activity compared to less basic supports, such as Al_2O_3 and SiO_2. Also, the normalized activity of the AuCu/CeO_2-ZrO_2 catalyst increases compared to their respective single supports. Recently, it was reported that the replacement of Zr^{4+} ions in the lattice of the CaO-CeO_2 system leads to the formation of highly basic sites [59]. Then, we speculate that the high interaction in CeO_2-ZrO_2 observed by TPR could lead to the formation of sites with greater basicity. However, the low surface area of basic supports is a well-known limitation that affects their activity [6]. So, because of the possible role of basic sites in CO removal, the design of catalysts for CO removal should include a support with both a high surface area and elevated basicity. Modifications in the morphology of metal oxides have been proposed as a successful strategy to achieve this objective in other catalytic processes [60]. Then, preparation of Au–Cu catalysts supported on single and dual metal oxides can be optimized to improve their catalytic properties during CO removal.

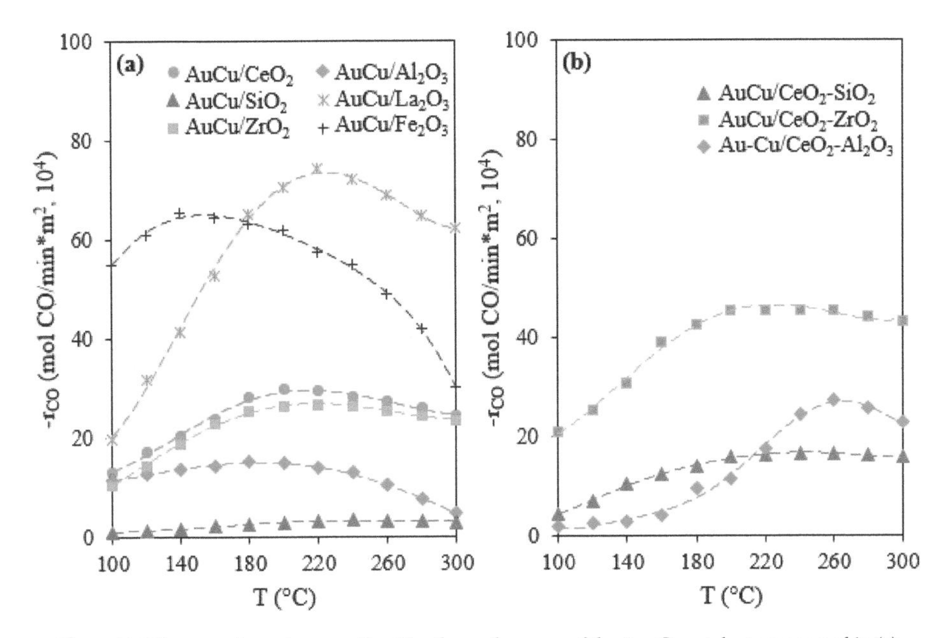

Figure 7. CO conversion rate normalized by the surface area of the Au–Cu catalysts supported in (**a**) single and (**b**) dual metal oxides.

2.2.3. OSC Measurements

The OSC of the support plays a central role in the oxidation of CO adsorbed on active sites [27]. Table 1 shows the OSC of Au–Cu catalysts supported on single and dual metal oxides. In general, the OSC of supported Au–Cu catalysts is higher than that of the bare supports (Table A1 in Appendix B), indicating that the presence of Au and Cu favors greater oxygen mobility in the catalyst. Also, the presence of α and β species has been associated with the formation of oxygen vacancies on the catalytic surface [52]. Catalysts that have a higher OSC at 300 °C (i.e., AuCu/CeO_2-ZrO_2, AuCu/CeO_2, AuCu/CeO_2-SiO_2, and AuCu/ZrO_2) were the most active (Figure 2), but also those that showed the highest consumption of H_2 (Figure 3). However, the OSC depends strongly on the temperature: at 100 °C, all catalysts except AuCu/Al_2O_3 showed an OSC up to 60% lower compared to 300 °C, which could be related to the lower activity of catalysts at low temperatures (Figure 2).

Likewise, the CO_2 formation depends on the availability of surface oxygen [15]. The first CO pulse (OSC) in AuCu/CeO_2 only corresponds to 39% of its oxygen storage complete capacity (OSCC), indicating that oxygen adsorbed on CeO_2 may not be easily released. The possible deficiency of CeO_2 to release the oxygen absorbed on its surface could limit the oxidation of carbon intermediates, which could, in turn, be related to the atypical trend observed in Figure 4b. The OSC in supports with larger surface area (i.e., AuCu/SiO_2, AuCu/Al_2O_3, AuCu/CeO_2-SiO_2, and AuCu/CeO_2-Al_2O_3), on the other hand, corresponds to more than 50% of their OSCC. A higher availability of surface oxygen (> OSC/OSCC) could be associated with the strong effect of CO_2 methanation on the CO removal for AuCu/CeO_2-SiO_2 and AuCu/CeO_2-Al_2O_3, as previously discussed. If so, then the beneficial effect of methanation during the CO-PROX proposed by [16] could be enhanced in catalysts that combine a high OSC and readiness to release their adsorbed oxygen (i.e., high OSC/OSCC ratio), which would require a high surface area.

On the other hand, the OSCC of the catalysts used decreases with respect to the fresh, activated ones (AC samples), reaching up to 73% reduction with AuCu/La_2O_3. This reduction could be associated with progressive oxidation of the catalyst surface by the presence of oxidants in the gas stream and deposits on the catalytic surface [20], conducive of a progressive deactivation. To clarify this, a TGA study was conducted.

2.2.4. TGA

Table 3 shows the weight loss of Au–Cu catalysts supported on single and dual metal oxides. Most AC samples show a weight loss of less than 1% that could correspond to a remnant of the precursors of the active metals. However, AuCu/Fe_2O_3 and AuCu/CeO_2-SiO_2 show an increase in weight that can be associated with an oxygen adsorption; specifically, the CeO_2-SiO_2 system can form a $Ce_{9.33}(SiO_4)\cdot 6O_2$ phase that is susceptible to consume oxygen above 600 °C [6]. The used catalysts have a higher weight loss than the fresh, activated ones (AC samples), indicating the presence of compounds deposited on the catalytic surface during the reaction. To determine the nature of the deposits, the TGA results were analyzed by weight loss in terms of rate of carbon equivalent formed in each temperature interval (Table 3). In the first interval (40–250 °C), light compounds, such as water, and adsorbed OH^- and gases are released [28]; in this interval, AuCu/SiO_2 and AuCu/Al_2O_3 showed the highest weight loss, which could be related to their high surface area, which favors moisture adsorption. In the second interval (250–600 °C), light hydrocarbons are oxidized [26]; AuCu/La_2O_3 and AuCu/Fe_2O_3 had the highest rate of carbon formation in this interval, which would explain the strong decrease in the OSCC and surface area, respectively, observed in these samples (Table 1). In the last interval (600–1000 °C), heavy hydrocarbons are oxidized, which are the type of deposits that could favor a faster deactivation of the catalyst [56]; in this zone, AuCu/CeO_2 showed a higher rate of carbon formation. Thus, rapid deactivation observed in AuCu/CeO_2 (Figure 5) could be associated with the decrease in surface area (17%, Table 1) and OSCC (59%, Table 1) promoted by the accumulation of deposits on the catalytic surface (Table 3). The formation of stable deposits could be associated with the formation of intermediates during the CO removal [28]; therefore, in situ DRIFTS was carried out to identify how the interaction between CO and support affects the performance of the supported Au–Cu catalysts.

Table 3. Weight loss of Au–Cu catalysts supported on single and dual supports evaluated in CO removal from an actual syngas.

Catalyst	Total Weight Loss (%)		Weight Loss of Spent Catalyst Samples by Temperature Intervals (mg of C/g_{cat}*h)		
	AC	Spent	40–250 °C	250–600 °C	600–1000 °C
AuCu/CeO_2	0.7	3.8 (U)	17.1 (U)	6.8 (U)	11.8 (U)
		5.6 (S)	14.5 (S)	15.1 (U)	18.1 (U)
AuCu/SiO_2	0.3	3.7 (U)	35.5 (U)	3.4 (U)	3.9 (U)
AuCu/ZrO_2	0.9	1.6 (U)	9.2 (U)	2.1 (U)	4.7 (U)
AuCu/Al_2O_3	0.5	3.7 (U)	28.9 (U)	2.1 (U)	9.4 (U)
AuCu/La_2O_3	0.6	2.1 (U)	9.2 (U)	7.5 (U)	2.4 (U)
AuCu/Fe_2O_3	−0.3	2.5 (U)	18.4 (U)	8.9 (U)	NR
AuCu/CeO_2-SiO_2	−0.9	0.3 (U)	3.9 (U)	4.8 (U)	NR
		1.3 (S)	15.8 (S)	4.1 (U)	NR
AuCu/CeO_2-ZrO_2	0.5	1.7 (U)	9.2 (U)	6.2 (U)	0.8 (U)
AuCu/CeO_2-Al_2O_3	0.6	2.6 (U)	17.1 (U)	2.7 (U)	7.1 (U)

Note: AC = activated catalyst, which were reduced with H_2 and stabilized in air before activity tests; U = sample used to obtain light-off curves; S = sample evaluated in the stability test; NR = Not reported.

2.2.5. In Situ DRIFTS

Figure 8 shows the DRIFTS spectra of CO adsorption on bare supports and supported Au–Cu catalysts. CeO_2 and ZrO_2 show higher intensity in the area associated with hydroxyl groups (~3500 cm^{-1}) that contributes to the CO conversion [36], which would explain their high activity among single metal oxides (Figure 2). Although the CO pulses were free of H_2 or water, hydroxyl groups may be formed from the interaction of H_2 with the surface of the support [61], which could occur during the H_2 reduction that was performed on the AC samples. In fact, Zhou et al. [62] studied the CO adsorption on bare ZrO_2 by DRIFTS and Fourier Transform Infrared Spectroscopy (FTIR), identifying up to three families of hydroxyl groups in the zone from 3675 to 3772 cm^{-1}, which are activated by the adsorption of CO, even at room temperature, and have an active role in the formation of surface intermediates. CeO_2 favors the formation of hydroxyl groups even with the first pulse of CO, which could be decisive in ensuring a syngas with a lower CO concentration. In the C–O* zone (1200 to 1700 cm^{-1} [63]), the formation of bidentate carbonates (1600 cm^{-1}) and formates (1300 and 1500 cm^{-1}) are observed, which are also intermediates in the CO conversion [20,51,63]. The formation of hydroxyl groups and C–O* species were lower than dual supports when compared to CeO_2; specifically, CeO_2-Al_2O_3 shows a significant reduction in the formation of C–O* intermediates, which would correspond to its lower activity among the dual supports (Figure 2).

The inclusion of Au–Cu in the single oxides (Figure 8c) favors the presence of hydroxyls and the formation of C–O* intermediates, possibly due to the ability of Au to form Au–CO and Au–OOH species [36]. In fact, most catalysts show an increase in CO adsorbed (2100 cm^{-1}), which is associated with CO–Au^0 species [64], indicating that Au could be present mostly as Au^0 on the catalytic surface, as previously reported for systems such as Au/CeO_2 [20] and Au/La_2O_3/Al_2O_3 [33], evaluated by XPS. However, in AuCu/CeO_2-SiO_2 and AuCu/CeO_2-ZrO_2, a weak peak of CO adsorption between 2075 and 2050 cm^{-1} is also observed, which has been associated with the formation of CO–$Au^{\delta-}$ species [65]. In the case of AuCu/CeO_2-ZrO_2, the formation of these species only occurs after several CO pulses. The presence of $Au^{\delta-}$ has been related to a stronger support-metal interaction, which could be ascribed to the high stability of AuCu/CeO_2-SiO_2 (Figure 5).

The formation of C–O* intermediates may occur on different active sites, including Au^0, Au^{δ}, and CuO, but the formation of carbonate species at approximately 1470 cm^{-1} occurs preferably on Cu^+ species [66], which are very active in CO-PROX [67]. The peak associated with Cu^+ is well defined in AuCu/CeO_2. Furthermore, the formation of active Cu^+ species due to the high affinity in CuO-CeO_2 has been extensively studied by XPS and DRIFTS [25,68]. Thus, a smaller amount of Cu^+ species on the other catalysts could explain their inability to ensure CO concentrations below 100 ppm (Table 1). Besides, the peaks associated with formate species, which are related to CH_4 formation, are better defined on CeO_2. It is accepted that CH_4 formation is promoted on several oxides (e.g., Al_2O_3, ZrO_2, Y_2O_3, MgO, and CeO_2 [69]), but the special ability to adsorb and activate carbon species makes CeO_2 an adequate support in CO_2 methanation and CO-SMET [70]. Nevertheless, during the CO removal the Boudouard reaction and the CH_4 decomposition could contribute to the production of carbon deposits [26], favoring the catalyst deactivation. Then, the ability of CeO_2 to form C–O* intermediates (Figure 8) assisted by Cu^+ species and its lower capacity to release the surface oxygen (low OSC/OSCC) could contribute to the generation of stable carbon deposits, as was observed by TGA, leading to its rapid deactivation (Figure 5). Besides, the deficiency of AuCu/CeO_2 to mitigate carbon deposition due to the excessive formation of C–O* intermediates could be also related to the atypical behavior of CH_4 formation (Figure 4b). However, the less active materials show low formation of intermediates (e.g., AuCu/La_2O_3, AuCu/Fe_2O_3, and AuCu/SiO_2). Thus, the selection of the support for the CO removal from a syngas must consider the balance between activity and stability.

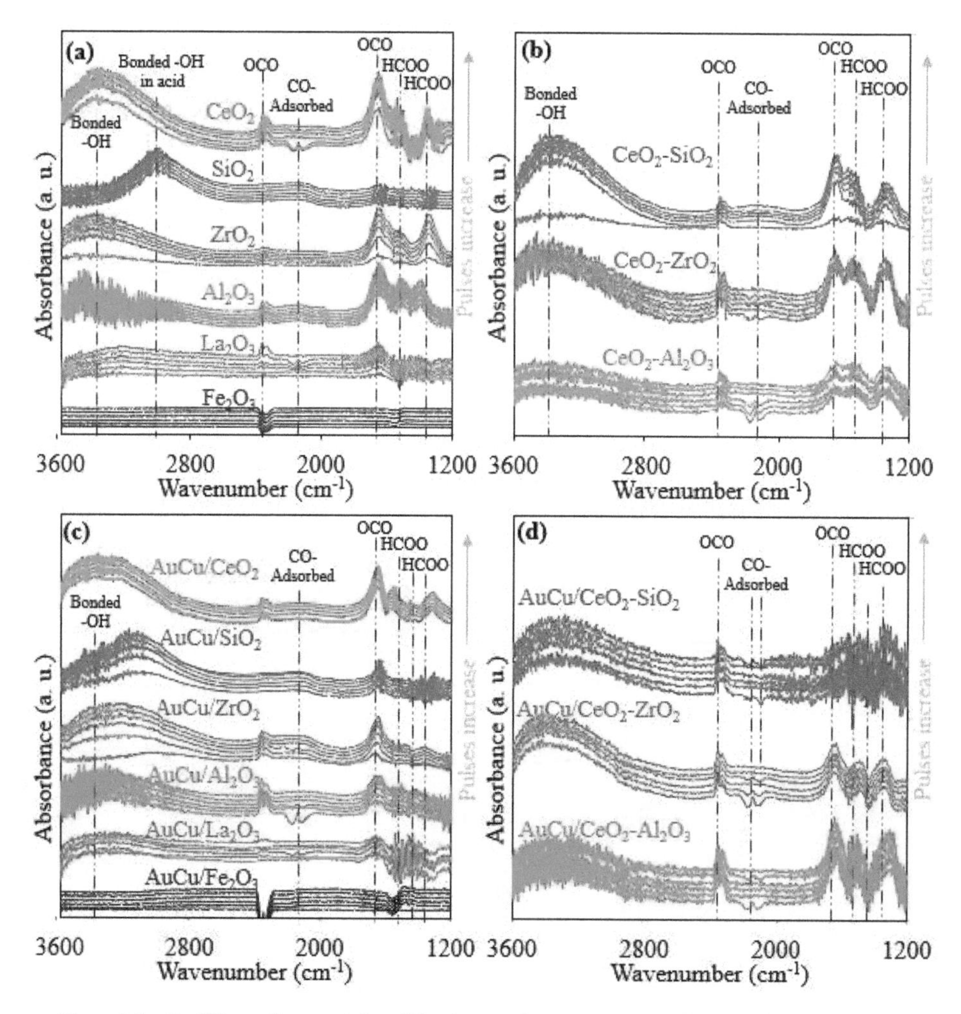

Figure 8. In situ diffuse reflectance infrared Fourier transform spectroscopy (DRIFTS) of CO adsorption of (**a**,**b**) bare supports and (**c**,**d**) supported Au–Cu catalysts.

The results of DRIFTS support the notion that the use of dual metal oxides favors less active but more stable catalysts. Therefore, in this study, CeO_2 is presented as the most promising support for developing a compact system to carry out the CO removal from an actual syngas. However, the selectivity and stability of CeO_2 require improvements. Furthermore, it was shown that the use of dual supports, specifically CeO_2-SiO_2 and CeO_2-ZrO_2, could be a promising strategy to overcome the deficiencies presented by CeO_2.

3. Materials and Methods

3.1. Support Selection

The supports evaluated in this work were selected according to a literature review, and are summarized in Figure 1. Scientific articles published between 2012 and 2019 that included at least one of the following reactions were reviewed: CO-PROX, WGSR, and CO-SMET. The detailed list of reviewed articles can be consulted in Table A2 (see Appendix B).

3.2. Catalyst Synthesis

The single supports of CeO_2, ZrO_2, and Fe_2O_3 were obtained by calcination at 500 °C for 2 h of $Ce(NO_3)_3{\cdot}6H_2O$ (CAS: 10294-41-4, Sigma Aldrich, Saint Louis, MO, USA), $ZrO(NO_3)_2{\cdot}xH_2O$ (CAS: 14985-18-3, Sigma Aldrich, Saint Louis, MO, USA), and $Fe(NO_3)_3{\cdot}9H_2O$ (CAS: 7782-61-8, Merck, Darmstadt, HE, Germany), respectively. Also, commercial oxides of La_2O_3 (CAS: 1312-81-8, Sigma Aldrich, Saint Louis, MO, USA), Al_2O_3 (CAS: 1344-28-1, Sigma Aldrich, Saint Louis, MO, USA), and SiO_2 (CAS: 60676-86-0, Merck, Darmstadt, HE, Germany) were used, which were also calcined at 500 °C in a muffle for 2 h.

Dual supports of CeO_2-ZrO_2, CeO_2-Al_2O_3, and CeO_2-SiO_2 were obtained from aqueous solutions of Ce $Ce(NO_3)_3{\cdot}6H_2O$ (CAS: 10294-41-4, Sigma Aldrich, Saint Louis, MO, USA) with $ZrO(NO_3)_2{\cdot}H_2O$ (CAS: 14985-18-3, Sigma Aldrich, Saint Louis, MO, USA), Al_2O_3 (CAS: 1344-28-1, Sigma Aldrich, Saint Louis, MO, USA), and SiO_2 (CAS: 60676-86-0, Merck, Darmstadt, HE, Germany), respectively, ensuring a molar ratio of Ce/M = 1 (M = Si, Zr and Al). Each solution was dried at 80 °C for 24 h and calcined at 500 °C in a muffle for 4 h. All supports (i.e., single and dual metal oxides) were screened with a 140-mesh sieve.

Bimetallic Au–Cu catalysts supported on each single and dual metal oxide were prepared according to the procedure described in [20], ensuring active metal loads of Au (1 wt%) and Cu (1 wt%). Au was first impregnated on each support by the precipitation-deposition method at pH 6 and 80 °C, using a solution of $HAuCl_4{\cdot}3H_2O$ (CAS: 16961-25-4 Sigma Aldrich, MO, USA). The filtered solid was dried at 80 °C for 24 h. Subsequently, Cu was included in the Au catalysts by the incipient wetness impregnation method, using a solution of $Cu(NO_3)_2{\cdot}3H_2O$ (CAS: 10031-43-3, Sigma Aldrich, Saint Louis, MO, USA). The catalyst obtained was dried at 80 °C for 24 h, calcined at 500 °C in a muffle for 2 h, and screened with a 140-mesh sieve.

The RhPt/CeO_2-SiO_2 catalyst for ESR was prepared according to the methodology described in [6]. Briefly, CeO_2-SiO_2 support was obtained from $Ce(NO_3)_3{\cdot}6H_2O$ (CAS: 10294-41-4, Sigma Aldrich, Saint Louis, MO, USA) and SiO_2 (CAS: 60676-86-0, Merck, Darmstadt, HE, Germany) solutions, ensuring a molar ratio of Ce/Si = 2. Rh and Pt were deposited on the CeO_2-SiO_2 support by the incipient wetness co-impregnation method, using $RhCl_3{\cdot}H_2O$ (CAS: 20765-98-4, Sigma Aldrich, Saint Louis, MO, USA) and $H_2PtCl_6{\cdot}6H_2O$ (CAS: 10025-65-7, Sigma Aldrich, Saint Louis, MO, USA) solutions. The catalyst obtained was dried at 80 °C for 24 h, calcined at 700 °C for 2 h, and screened with a 140-mesh sieve.

3.3. Obtaining Syngas

The syngas was obtained from ESR with a RhPt/CeO_2-SiO_2 catalyst at 700 °C in the first reactor (ESR reactor). The plug flow conditions in the ESR reactor were maintained ensuring L/Dp >50 ratios (i.e., catalytic bed height (L) and catalyst particle size (Dp)) and D/Dp >60 (i.e., diameter internal to the reactor (D)), as recommended in [71]. The catalyst bed consisted of 0.050 g of RhPt/CeO_2-SiO_2 and 0.250 g of inert quartz. The reactor feed consisted of 0.3 L/min of a mixture of ethanol (1.8 mol%), water (5.4 mol%), and Ar as carrier gas. The space velocity (SV) was set at 6.4 ± 0.2 L/g_{cat}*min. The syngas obtained in the ESR reactor, containing H_2 (8.4 mol%), CO (2.2 mol%), H_2O (1.6 mol%), CO_2 (0.6 mol%), CH_4 (0.3 mol%), and Ar (86.9 mol%), remained stable, with a variation <6.8%.

3.4. Catalytic Test

The supports and Au–Cu catalysts for the CO removal from the syngas were evaluated in a second reactor (cleanup reactor) between 100 and 300 °C. For this, the ESR reactor outlet was mixed with dry air, ensuring an excess oxygen factor (λ) of 1.8 ± 0.05 [20], and connected to the cleanup reactor inlet. The plug flow conditions in the cleanup reactor were maintained as previously described for the ESR reactor. The catalyst bed consisted of 0.050 g of catalyst (i.e., supports or Au–Cu catalysts) and 0.250 g of inert quartz. The SV in the cleanup reactor was set at 6.5 ± 0.3 L/g_{cat}^*min. Before the reaction, the supports and Au–Cu catalysts were pretreated in situ at 300 °C with streams of 8% H_2/Ar for 1 h, followed by Ar for 0.5 h, and finally 10% air/Ar for 0.5 h. These samples were labeled as "activated catalyst" (AC). Also, the samples used to obtain the light-off curves were labeled "U", while those used in the stability test were labeled "S".

The species at the outlet of each reactor (i.e., ESR reactor and cleanup reactor) were quantified by gas chromatography (GC) in a Clarus 580 (Perkin Elmer, Waltham, MA, USA), equipped with a Carboxen 1010 plot column (30 m, 0.53 mm ID, Restek, Bellefonte, PA, USA) connected to a thermal conductivity detector (TCD). Ar was used as carrier gas and N_2 as internal reference. The reaction conditions and GC data processed in Excel can be consulted in detail and downloaded from [72].

The conversion of CO (x_{CO}), the production of the main products (Y_{CO_2} and Y_{CH_4}), and the H_2 obtained (Y_{H_2}) from the integrated system were obtained considering the molar flows (F_i) to the output of each reactor (i.e., ESR reactor and cleanup reactor), according to Equations (1) and (3). Production of CO, CH_4, and H_2 were normalized with the amount of carbon entering the system ($F_{C,\ inlet\ to\ the\ system}$), which remained constant at $5.2*10^{-4}$ mol/min of C.

$$x_{CO} = \frac{F_{CO,\ ESR-reactor} - F_{CO,\ cleanup-reactor}}{F_{CO,\ ESR-reactor}} \tag{1}$$

$$Y_{CH_4;CO_2} = \frac{F_{CH_4;CO_2,\ cleanup-reactor} - F_{CH_4;CO_2,\ ESR-reactor}}{F_{C,\ inlet\ to\ the\ system}} \tag{2}$$

$$Y_{H_2} = \frac{F_{H_2,\ cleanup-reactor}}{F_{C,\ inlet\ to\ the\ system}} \tag{3}$$

3.5. Characterization Tests

The reducibility of supports and Au–Cu catalysts was determined by TPR. The experiments were carried out in a ChemBET Pulsar unit (Quantachrome Instruments, Boynton Beach, FL, USA) equipped with a TCD. Prior to the reduction, 0.07 ± 0.01 g of AC samples was pretreated with N_2 (0.02 L / min) at 120 °C for 1 h and then cooled to room temperature. Subsequently, 5 % H_2/N_2 was passed, and the temperature was increased to 700 °C (5 °C/min). The H_2 uptake was calculated by integrating the peaks associated with the reduction of active metals (i.e., Au and Cu). The apparent active metal dispersion (H/M ratio) was also determined [53], assuming that the adsorption stoichiometry is one hydrogen atom for one active metal atom (Au + Cu).

The surface area of the samples was determined by standard physisorption of N_2 in a ChemBET Pulsar unit (Quantachrome Instruments, Boynton Beach, FL, USA). For this, 0.06 ± 0.01 g of sample was pretreated with N_2 (0.02 L/min) at 100 °C for 1 h and then cooled to room temperature for 0.5 h. Subsequently, the sample was immersed in a liquid N_2 bath. The BET area was measured with a single point, using 30% N_2/He (0.02 L/min). The measurements were repeated until deviations lower than 5% were obtained.

The OSC values of the samples were measured in a ChemBET Pulsar unit (Quantachrome Instruments, Boynton Beach, FL, USA), according to the procedure described in [41]. Briefly, 0.06 ± 0.01 g of sample was degassed in Ar (0.02 L/min) at 300 °C for 1 h. OSC was measured at 300 and 100 °C with independent samples. For this, 10 pulses of pure O_2 (0.25 mL) were injected to oxidize the sample, followed by a 20 min purge with Ar. Then, pulses of a 5 % CO/Ar mixture (0.25 mL) were injected until a constant signal was obtained. The OSC value was calculated by the CO consumed in the first pulse, and the OSCC value was determined by the total CO consumed.

The weight loss, associated with the presence of impurities, moisture, and carbon deposition in samples, was measured by TGA. The change in mass was determined using a thermogravimetric analyzer (Mettler Toledo, Columbus, OH, USA). For this, 0.02 ± 0.01 g of sample was pretreated with a N_2 (0.1 L/min) at 100 °C for 1 h and then cooled to 40 °C for 0.5 h. Subsequently, the sample was heated to 1000 °C (5 °C/min) in a dry air stream (0.1 L/min). Then, the rate of carbon formation was calculated according to Equation (4).

$$\text{Rate of carbon formation} = \frac{Weight\ loss\ in\ term\ of\ C\ (mg)}{mass\ of\ catalyst\ (g) * TGA\ test\ time\ (h)} \quad (4)$$

The CO adsorption on supports and catalysts was studied by in situ DRIFTS in a Nicolet iS10 spectrum device (Thermo Scientific, Waltham, MA, USA) equipped with a diffuse reflection attachment DRK-3 Praying Mantis (Harrick Scientific Products, New York, NY, USA). Spectra were taken between 400 and 4000 cm^{-1}, with 64 scans per minute and a resolution of 4 cm^{-1}. The sample holder was sealed with an airtight hood with ZeSn windows. In addition, the airtight hood was isolated with an Ar stream to avoid interference from the environment. Approximately 0.02 g of AC samples were degassed in Ar (15 mL/min) at 50 °C for 30 min. Then, 10 pulses of 30 μL of CO, obtained from a certified 5% CO/Ar mixture, were injected into the cell; between each pulse, Ar (15 mL/min) was passed for 10 min.

Raw and processed Excel data for characterization tests can be downloaded from [72].

4. Conclusions

Several single and dual metal oxides were investigated as supports in a catalytic system based on Au–Cu for the CO removal from an actual syngas. The use of a syngas obtained directly from the ESR affects the effectiveness in the CO removal; specifically, a low H_2/CO ratio could favor greater H_2 loss. AuCu/CeO_2 was identified as the most active catalyst in the CO removal, but it also contributes to a higher H_2 consumption. H_2 is lost mainly by the formation of water and CH_4, where the occurrence of CO_2 methanation affected the CO removal differently. Over CeO_2-Al_2O_3 and CeO_2-SiO_2, methanation seems to improve CO removal because the CO-PROX product, CO_2, is constantly consumed to produce CH_4. On the contrary, methanation has a negative effect on CeO_2 and La_2O_3 because the formed CH_4 favors carbon deposition.

Differences among the catalysts were evaluated by several characterization techniques. DRIFTS spectra of CO adsorption showed that CeO_2 has a superior activity because it favors the formation of C–O* and OH^- intermediates, but it promotes the formation of carbon deposits that lead to its deactivation. Similarly, TPR showed that ZrO_2 has a high interaction with active metals (Au–Cu), which makes it active but less selective, favoring a high H_2 oxidation. In addition, the low OSC of Al_2O_3 and SiO_2, and the lower surface area of Fe_2O_3 and La_2O_3 make these metal oxides less active. Regarding dual supports, the inclusion of a second metal oxide weakens the interaction of CeO_2 with the active metals, reducing activity. However, dual metal oxides are more selective and stable than single CeO_2 because they mitigate the excess of C–O* species, as was observed by DRIFTS; specifically, CeO_2-SiO_2 mitigates the formation of stable carbon deposits that deactivate the catalyst. Thus, AuCu/CeO_2 was identified as a promising catalyst for carrying out the CO removal from a syngas using just one catalytic reactor, but improvements in CeO_2 stability are still required. Therefore, the use of dual supports (e.g., CeO_2-SiO_2) could be a strategy to overcome single CeO_2 deficiencies. Thus, the development of more compact systems for the purification of H_2 suitable for FC implicitly promotes greater H_2 consumption. The results of this work aim to contribute to the development and establishment of sustainable energies based on H_2.

Author Contributions: All the authors contributed to the development, writing, and review of this work.

Funding: This research was funded by Colciencias (Francisco Jose de Caldas Fund) and Universidad de La Sabana through the projects with code ING-163, ING-166, and ING-221 (Colciencias contracts 0608-2013, 174-2016 and 548-2019).

Acknowledgments: B. Cifuentes acknowledges Colciencias for the doctoral scholarship (grant number 727-2015). F. Bustamante acknowledges the support of the Environmental Catalysis Research Group of Universidad de Antioquia.

Conflicts of Interest: There are no conflicts of interest to declare.

Appendix A

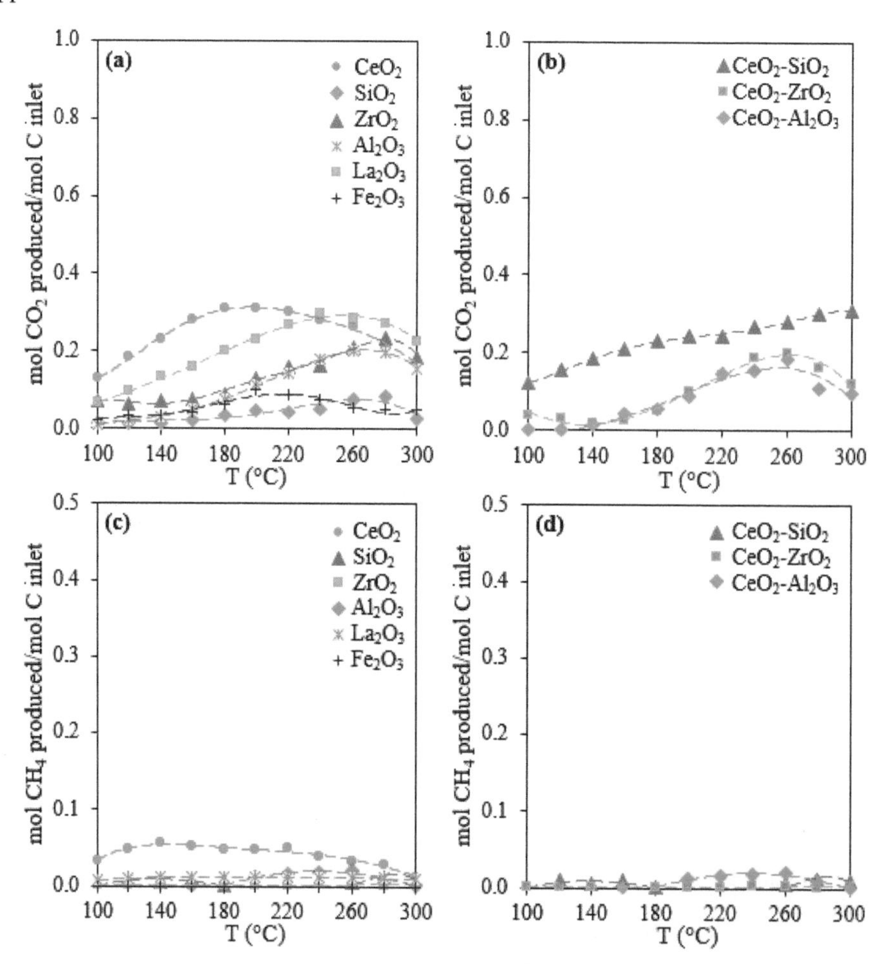

Figure A1. (**a**,**b**) CO_2 and CH_4 (**c**,**d**) production in the CO cleanup reactor with simple and dual supports. Syngas composition: 8.4% H_2, 2.2% CO, 0.6% CO_2, 0.3% CH_4, 1.6% H_2O, and Ar. $\lambda = 1.8$. Reaction conditions: SV = 6.5 ± 0.2 L/g_{cat}*min; 0.050 g of catalyst and 0.250 g of inert quartz.

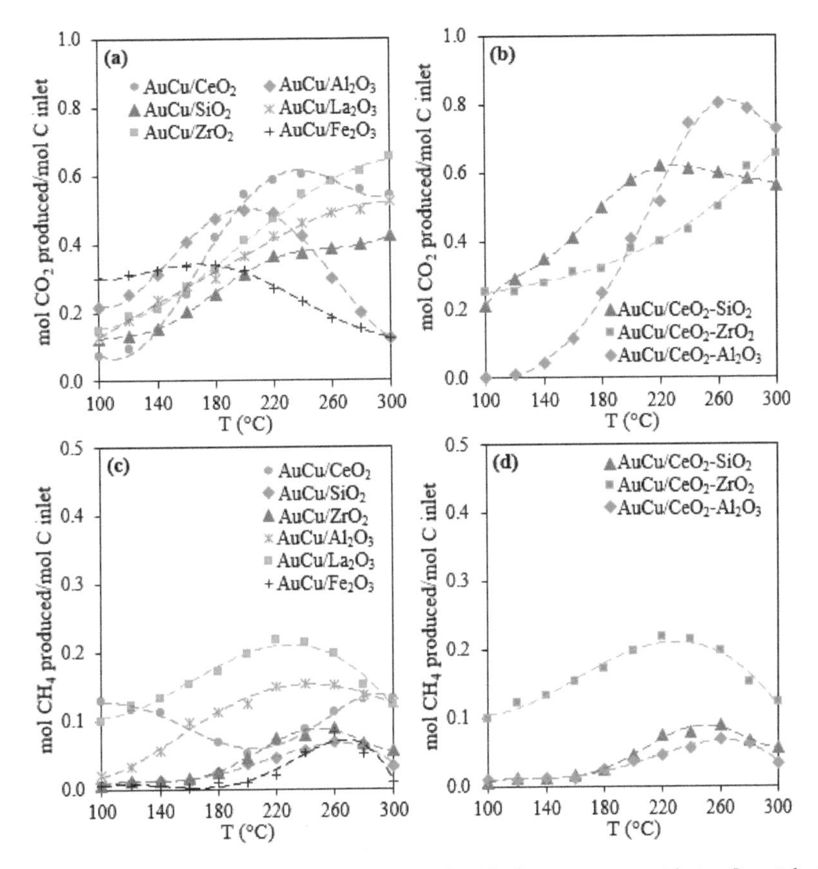

Figure A2. (**a**,**b**) CO_2 and CH_4 (**c**,**d**) production in the CO cleanup reactor with Au–Cu catalysts supported on simple and dual supports. Syngas composition: 8.4% H_2, 2.2% CO, 0.6% CO_2, 0.3% CH_4, 1.6% H_2O, and Ar. λ=1.8. Reaction conditions: SV = 6.5 ± 0.2 L/g_{cat}*min; 0.050 g of catalyst and 0.250 g of inert quartz.

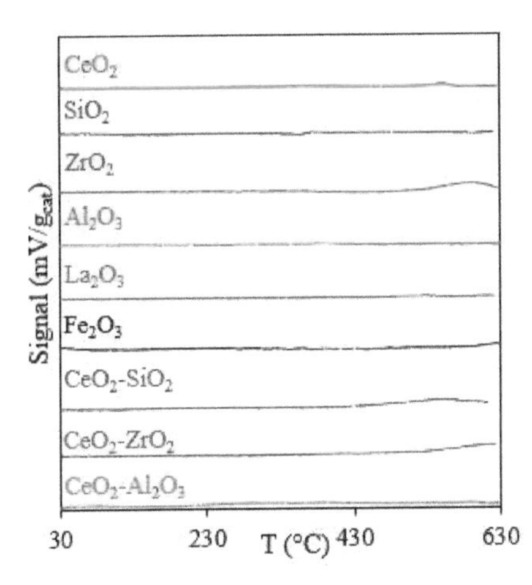

Figure A3. H_2-TPR profiles of bare supports evaluated in the CO removal from an actual syngas.

Appendix B

Table A1. Surface area, OSC, OSCC, and thermogravimetric analysis (TGA) results of single and dual supports.

Support	BET Surface Area (m^2/gcat)		OSC in Fresh Samples at 300 °C (μmol O_2/gcat)	OSCC in Fresh Samples at 300 °C (μmol O_2/gcat)	Weight Loss (%)	
	Fresh	Used			Fresh	Used
CeO_2	52.4	55.4	61	135	1.1	0.6
SiO_2	466.5	410.6	41	49	1.4	1.1
ZrO_2	51.6	44.9	55	99	1.6	0.6
Al_2O_3	96	68.6	36	55	1.7	0.2
La_2O_3	14.1	15.3	21	68	5.5	2.1
Fe_2O_3	38.1	36.7	5	16	0.7	0.8
CeO_2-SiO_2	163.2	155.2	54	105	−2.7	1.8
CeO_2-ZrO_2	44.3	40.5	46	110	0.2	2.0
CeO_2-Al_2O_3	72.7	69.1	41	120	2.0	1.0

Table A2. Reviewed articles for the selection of supports evaluated in the CO removal.

Date	Active Metals	Metal Oxide I	Metal Oxide II	Journal	Digital Object Identifier (DOI)
2012	CuO	Fe_2O_3	-	Chemical Engineering Journal	10.1016/j.cej.2012.01.017
2012	Pt	Other	-	Electrochimica Acta	10.1016/j.electacta.2012.04.150
2012	-	Fe_2O_3	-	Applied Surface Science	10.1016/j.apsusc.2011.10.092
2012	-	NiO_2	-	Journal of Molecular Catalysis A: Chemical	10.1016/j.molcata.2012.05.001
2013	Ni, Co	Co_3O_4	-	Journal of Alloys and Compounds	10.1016/j.jallcom.2013.04.053
2013	CuO	TiO_2	Al_2O_3	Surface and Coatings Technology	10.1016/j.surfcoat.2012.10.031
2013	Co	Fe2O3	-	Chemical Engineering Journal	10.1016/j.ces.2013.02.002
2014	Co	MgO	-	Process Safety and Environmental Protection	10.1016/j.psep.2013.12.003
2014	Pt	CeO_2	-	Chemical Engineering Journal	10.1016/j.cej.2014.06.058
2014	Pd	Fe_2O_3	-	Journal of Catalysis	10.1016/j.jcat.2014.06.019
2014	Ag	Zeolite	-	Fuel	10.1016/j.fuel.2014.07.011
2014	Au	NiO_2	-	Applied Catalysis A: General	10.1016/j.apcata.2014.02.003
2014	CuO	SiO_2	CeO_2	Journal of Environmental Chemical Engineering	10.1016/j.jece.2014.03.021
2015	Co, Fe, Cr	CeO_2	-	International Journal of Hydrogen Energy	10.1016/j.ijhydene.2015.03.044
2015	-	Co_3O_4	-	Applied Catalysis A: General	10.1016/j.apcata.2014.10.024
2015	CuO	Fe_2O_3	-	Chinese Journal of Catalysis	10.1016/S1872-2067(15)60922-6
2015	Au	Zeolite	-	Catalysis Communications	10.1016/j.catcom.2015.06.018
2015	Pt	CeO_2	-	Catalysis Today	10.1016/j.cattod.2014.12.038
2015	Au, Cu	CeO_2	ZrO_2	Catalysis Today	10.1016/j.cattod.2014.08.035
2015	-	PtO2	-	Applied Surface Science	10.1016/j.apsusc.2015.03.108
2015	CuO	CeO_2	ZrO_2	Journal of Industrial and Engineering Chemistry	10.1016/j.jiec.2015.06.038
2015	-	MnO_2	CeO_2	Applied Catalysis B: Environmental	10.1016/j.apcatb.2014.06.038
2016	Pd	Fe_2O_3	-	Journal of Environmental Chemical Engineering	10.1016/j.jece.2016.10.019
2016	-	CeO_2	ZrO_2	Applied Catalysis B: Environmental	10.1016/j.apcatb.2016.02.023
2016	Au	Zn_2SnO_4	-	Chinese Journal of Catalysis	10.1016/S1872-2067(16)62468-3
2016	-	Co_3O_4	-	Catalysis Communications	10.1016/j.catcom.2016.08.020
2016	Au	CeO_2	-	Applied Catalysis B: Environmental	10.1016/j.apcatb.2016.02.025
2016	Pd	CeO_2	-	Journal of Molecular Catalysis A: Chemical	10.1016/j.molcata.2016.08.035
2016	-	SiO_2	Al_2O_3	Journal of Molecular Graphics and Modelling	10.1016/j.jmgm.2016.08.005
2016	Ag	SiO_2	-	Catalysis Today	10.1016/j.cattod.2016.05.033
2016	-	PdO	-	Surface Science	10.1016/j.susc.2015.08.043
2016	-	Co_3O_4	-	Applied Catalysis A: General	10.1016/j.apcata.2016.03.027
2016	CuO	TiO_2	-	Catalysis Communications	10.1016/j.catcom.2016.02.001
2016	Pt	CeO_2	-	Applied Catalysis B: Environmental	10.1016/j.apcatb.2016.01.056
2016	CuO	MnO_2	-	Journal of Molecular Catalysis A: Chemical	10.1016/j.molcata.2016.08.024
2016	CuO	Peroskita	-	Applied Clay Science	10.1016/j.clay.2015.08.034
2016	Pd	ZnO	-	Catalysis Today	10.1016/j.cattod.2015.05.021
2016	-	Fe_2O_3	-	Chemical Engineering Journal	10.1016/j.cej.2016.04.136
2016	Au	TiO_2	-	Catalysis Today	10.1016/j.cattod.2015.09.040
2016	Au	Fe_2O_3	CeO_2	Catalysis Today	10.1016/j.cattod.2016.05.059
2016	-	Co_3O_4	-	Materials Letters	10.1016/j.matlet.2016.06.108
2016	-	Co_3O_4	-	Chinese Journal of Catalysis	10.1016/S1872-2067(15)60969-X
2016	Au	TiO_2	-	Applied Surface Science	10.1016/j.apsusc.2016.01.285
2016	-	Fe_2O_3	-	Journal of Molecular Catalysis A: Chemical	10.1016/j.molcata.2016.01.003
2016	Au	Other	-	Journal of Colloid and Interface Science	10.1016/j.jcis.2016.06.072
2016	Au	$LaPO_4$	-	Journal of the Taiwan Institute of Chemical Engineers	10.1016/j.jtice.2016.01.016
2016	Pt	Al_2O_3	-	International Journal of Hydrogen Energy	10.1016/j.ijhydene.2016.08.170
2016	Pt	Other	-	Surface Science	10.1016/j.susc.2015.08.024

Table A2. *Cont.*

Date	Active Metals	Metal Oxide I	Metal Oxide II	Journal	Digital Object Identifier (DOI)
2017	CuO	Nb_2O_5	-	Catalysis Communications	10.1016/j.catcom.2017.04.008
2017	Zn, Pt	CeO_2	-	Applied Catalysis B: Environmental	10.1016/j.apcatb.2017.04.044
2017	Pt, Fe	Fe_2O_3	Co_3O_4	Chinese Journal of Catalysis	10.1016/S1872-2067(17)62838-9
2017	CuO	MnO_2	CeO_2	Catalysis Communications	10.1016/j.catcom.2017.05.016
2017	Pt	MnO_2	-	Journal of Electroanalytical Chemistry	10.1016/j.jelechem.2016.09.031
2017	Au	$LaPO_4$	-	Chinese Journal of Chemical Engineering	10.1016/j.cjche.2017.08.008
2017	Fe, Mn	CeO_2	-	Catalysis Today	10.1016/j.cattod.2016.11.046
2017	Mn	Co_3O_4	-	Solid State Sciences	10.1016/j.solidstatesciences.2017.07.006
2017	Mn	Co_3O_4	-	Fuel	10.1016/j.fuel.2017.04.140
2017	Au	CeO_2	-	Applied Surface Science	10.1016/j.apsusc.2017.04.158
2017	-	MgO	-	Applied Catalysis B: Environmental	10.1016/j.apcatb.2016.11.043
2017	CuO	CeO_2	Zeolite	Microporous and Mesoporous Materials	10.1016/j.micromeso.2017.02.016
2017	-	Zeolite	-	Applied Catalysis B: Environmental	10.1016/j.apcatb.2017.06.083
2017	Co	ZnO	-	Ceramics International	10.1016/j.ceramint.2017.06.157
2017	Pd	TiO_2	SnO_2	Applied Catalysis B: Environmental	10.1016/j.apcatb.2017.02.017
2017	Pd	Fe_2O_3	-	Fuel Processing Technology	10.1016/j.fuproc.2017.02.037
2017	CuO	CeO_2	-	Journal of Power Sources	10.1016/j.jpowsour.2017.01.127
2017	Mn	CeO_2	-	Applied Catalysis B: Environmental	10.1016/j.apcatb.2017.03.049
2017	Co	Co_3O_4	-	Chemical Physics Letters	10.1016/j.cplett.2017.02.085
2017	Au	TiO_2	-	Catalysis Today	10.1016/j.cattod.2016.05.056
2017	CuO	CeO_2	-	International Journal of Hydrogen Energy	10.1016/j.ijhydene.2017.02.088
2017	CuO	CeO_2	-	Journal of Rare Earths	10.1016/j.jre.2017.05.015
2017	Pd	Al_2O_3	-	Applied Catalysis B: Environmental	10.1016/j.apcatb.2017.02.038
2017	Pt	TiO_2	-	Molecular Catalysis	10.1016/j.mcat.2017.01.014
2017	-	CeO_2	Other	Catalysis Today	10.1016/j.cattod.2017.06.017
2017	-	Al_2O_3	SnO_2	Applied Surface Science	10.1016/j.apsusc.2017.01.058
2017	Ag	Zeolite	-	Fuel	10.1016/j.fuel.2016.10.037
2017	Au	TiO_2	-	Applied Surface Science	10.1016/j.apsusc.2016.10.076
2017	-	Carbon	-	Molecular Catalysis	10.1016/j.molcata.2016.12.007
2017	Ag	SiO_2	-	Microporous and Mesoporous Materials	10.1016/j.micromeso.2017.01.016
2017	Pd, Rh	Al_2O_3	-	Catalysis Today	10.1016/j.cattod.2016.10.010
2017	Au, Cu	SiO_2	-	Catalysis Today	10.1016/j.cattod.2016.08.003
2017	Pd	CeO_2	MnO	Applied Catalysis B: Environmental	10.1016/j.apcatb.2017.01.020
2017	-	CeO_2	-	Catalysis Today	10.1016/j.cattod.2016.04.016
2017	Pd	Co_3O_4	-	Applied Catalysis A: General	10.1016/j.apcata.2016.12.021
2017	Pt	CeO_2	-	Applied Catalysis A: General	10.1016/j.apcata.2017.08.012
2017	Mn	Co_3O_4	-	Solid State Sciences	10.1016/j.solidstatesciences.2017.07.006
2017	Ni	ZrO_2	-	Applied Catalysis A: General	10.1016/j.apcata.2017.02.001
2018	-	SiO_2	Co_3O_4	Microporous and Mesoporous Materials	10.1016/j.micromeso.2017.07.016
2018	Pt	Fe_2O_3	-	Applied Catalysis A: General	10.1016/j.apcata.2018.09.014
2018	Pd	SiO_2	Al_2O_3	Applied Catalysis B: Environmental	10.1016/j.apcatb.2018.06.059
2018	Cu	CeO_2	-	Catalysis Today	10.1016/j.cattod.2018.10.037
2018	Cu -Ni	CeO_2	Al_2O_3	International Journal of Hydrogen Energy	10.1016/j.ijhydene.2018.12.127
2018	Ru	TiO_2	ZrO_2	International Journal of Hydrogen Energy	10.1016/j.ijhydene.2018.10.061
2018	Ni	ZrO_2	-	International Journal of Hydrogen Energy	10.1016/j.ijhydene.2018.06.173
2018	-	ZrO_2	-	Applied Catalysis B: Environmental	10.1016/j.apcatb.2018.03.001
2018	Ni	ZrO_2	-	Applied Catalysis B: Environmental	10.1016/j.apcatb.2018.06.045
2019	Au	TiO_2	-	International Journal of Hydrogen Energy	10.1016/j.ijhydene.2018.11.050
2019	Cu	Co_3O_4	-	Molecular Catalysis	10.1016/j.mcat.2019.01.020
2019	-	Other	-	Applied Catalysis B: Environmental	10.1016/j.apcatb.2018.12.022
2019	Pt	Zeolite	-	Applied Catalysis A: General	10.1016/j.apcata.2018.12.034
2019	Ni	ZrO_2	-	Applied Catalysis B: Environmental	10.1016/j.apcatb.2018.11.024

References

1. Zhang, K.; Jiang, X. An investigation of fuel variability effect on bio-syngas combustion using uncertainty quantification. *Fuel* **2018**, *220*, 283–295. [CrossRef]
2. Renzi, M.; Patuzzi, F.; Baratieri, M. Syngas feed of micro gas turbines with steam injection: Effects on performance, combustion and pollutants formation. *Appl. Energy* **2017**, *206*, 697–707. [CrossRef]
3. Rosa, A. Chapter 10—Hydrogen Production. In *Fundamentals of Renewable Energy Processes*, 3rd ed.; Academic Press: Cambridge, MA, USA, 2013; pp. 371–428.
4. Mohammed, H.; Al-Othman, A.; Nancarrow, P.; Tawalbeh, M.; El Haj Assad, M. Direct hydrocarbon fuel cells: A promising technology for improving energy efficiency. *Energy* **2019**, *172*, 207–219. [CrossRef]
5. Sengodan, S.; Lan, R.; Humphreys, J.; Du, D.; Xu, W.; Wang, H.; Tao, S. Advances in reforming and partial oxidation of hydrocarbons for hydrogen production and fuel cell applications. *Renew. Sustain. Energy Rev.* **2018**, *82*, 761–780. [CrossRef]
6. Cifuentes, B.; Hernández, M.; Monsalve, S.; Cobo, M. Hydrogen production by steam reforming of ethanol on a RhPt/CeO_2/SiO_2 catalyst: Synergistic effect of the Si:Ce ratio on the catalyst performance. *Appl. Catal. A Gen.* **2016**, *523*, 283–293. [CrossRef]

7. Xie, Y.; Wu, J.; Jing, G.; Zhang, H.; Zeng, S.; Tian, X.; Zou, X.; Wen, J.; Su, H.; Zhong, C.J.; et al. Structural origin of high catalytic activity for preferential CO oxidation over CuO/CeO_2 nanocatalysts with different shapes. *Appl. Catal. B Environ.* **2018**, *239*, 665–676. [CrossRef]
8. Kou, T.; Si, C.; Gao, Y.; Frenzel, J.; Wang, H.; Yan, X.; Bai, Q.; Eggeler, G.; Zhang, Z. Large-scale synthesis and catalytic activity of nanoporous Cu-O system towards CO oxidation. *RSC Adv.* **2014**, *4*, 65004–65011. [CrossRef]
9. Kou, T.; Si, C.; Pinto, J.; Ma, C.; Zhang, Z. Dealloying assisted high-yield growth of surfactant-free ⟨110⟩ highly active Cu-doped CeO_2 nanowires for low-temperature CO oxidation. *Nanoscale* **2017**, *9*, 8007–8014. [CrossRef]
10. Ashraf, M.A.; Ercolino, G.; Specchia, S.; Specchia, V. Final step for CO syngas clean-up: Comparison between CO-PROX and CO-SMET processes. *Int. J. Hydrogen Energy* **2014**, *39*, 18109–18119. [CrossRef]
11. Rossetti, I.; Compagnoni, M.; Torli, M. Process simulation and optimization of H2 production from ethanol steam reforming and its use in fuel cells. 2. Process analysis and optimization. *Chem. Eng. J.* **2015**, *281*, 1036–1044. [CrossRef]
12. Gubán, D.; Tompos, A.; Bakos, I.; Vass, Á.; Pászti, Z.; Szabó, E.G.; Sajó, I.E.; Borbáth, I. Preparation of CO-tolerant anode electrocatalysts for polymer electrolyte membrane fuel cells. *Int. J. Hydrogen Energy* **2017**, *42*, 13741–13753. [CrossRef]
13. Narayanan, H.; Basu, S. Regeneration of CO poisoned Pt black anode catalyst in PEMFC using break-in procedure and KMnO4 solution. *Int. J. Hydrogen Energy* **2017**, *42*, 23814–23820. [CrossRef]
14. Devrim, Y.; Albostan, A.; Devrim, H. Experimental investigation of CO tolerance in high temperature PEM fuel cells. *Int. J. Hydrogen Energy* **2018**, *43*, 18672–18681. [CrossRef]
15. Kugai, J.; Fox, E.B.; Song, C. Kinetic characteristics of oxygen-enhanced water gas shift on CeO2-supported Pt–Cu and Pd–Cu bimetallic catalysts. *Appl. Catal. A Gen.* **2015**, *497*, 31–41. [CrossRef]
16. Xu, G.; Zhang, Z.G. Preferential CO oxidation on Ru/Al2O3 catalyst: An investigation by considering the simultaneously involved methanation. *J. Power Sources* **2006**, *157*, 64–77. [CrossRef]
17. Reina, T.R.; Ivanova, S.; Laguna, O.H.; Centeno, M.A.; Odriozola, J.A. WGS and CO-PrOx reactions using gold promoted copper-ceria catalysts: "bulk $CuO–CeO_2$ vs. $CuO–CeO_2/Al_2O_3$ with low mixed oxide content". *Appl. Catal. B Environ.* **2016**, *197*, 67–72. [CrossRef]
18. Wang, F.; Zhang, J.-C.; Li, W.-Z.; Chen, B.-H. Coke-resistant Au–Ni/$MgAl_2O_4$ catalyst for direct methanation of syngas. *J. Energy Chem.* **2019**, *39*, 198–207. [CrossRef]
19. Reina, T.R.R.; Ivanova, S.; Centeno, M.A.A.; Odriozola, J.A.A. Catalytic screening of Au/CeO_2-MOx/Al_2O_3 (M = La, Ni, Cu, Fe, Cr, Y) in the CO-PrOx reaction. *Int. J. Hydrogen Energy* **2015**, *40*, 1782–1788. [CrossRef]
20. Cifuentes, B.; Bustamante, F.; Conesa, J.A.; Córdoba, L.F.; Cobo, M. Fuel-cell grade hydrogen production by coupling steam reforming of ethanol and carbon monoxide removal. *Int. J. Hydrogen Energy* **2018**, *43*, 17216–17229. [CrossRef]
21. Yao, X.; Gao, F.; Dong, L. The application of incorporation model in γ–Al_2O_3-supported single and dual metal oxide catalysts: A review. *Cuihua Xuebao/Chin. J. Catal.* **2013**, *34*, 1975–1985. [CrossRef]
22. MacIel, C.G.; Profeti, L.P.R.; Assaf, E.M.; Assaf, J.M. Hydrogen purification for fuel cell using $CuO/CeO_2–Al_2O_3$ catalyst. *J. Power Sources* **2011**, *196*, 747–753. [CrossRef]
23. Águila, G.; Gracia, F.; Araya, P. CuO and CeO_2 catalysts supported on Al_2O_3, ZrO_2, and SiO_2 in the oxidation of CO at low temperature. *Appl. Catal. A Gen.* **2008**, *343*, 16–24. [CrossRef]
24. Zhao, Z.; Lin, X.; Jin, R.; Dai, Y.; Wang, G. High catalytic activity in CO PROX reaction of low cobalt-oxide loading catalysts supported on nano-particulate $CeO_2–ZrO_2$ oxides. *Catal. Commun.* **2011**, *12*, 1448–1451. [CrossRef]
25. Moretti, E.; Rodríguez-Aguado, E.; Molina, A.I.; Rodríguez-Castellón, E.; Talon, A.; Storaro, L. Sustainable photo-assisted CO oxidation in H_2-rich stream by simulated solar light response of Au nanoparticles supported on TiO_2. *Catal. Today* **2018**, *304*, 135–142. [CrossRef]
26. Li, S.; Tang, H.; Gong, D.; Ma, Z.; Liu, Y. Loading Ni/La_2O_3 on SiO_2 for CO methanation from syngas. *Catal. Today* **2017**, *297*, 298–307. [CrossRef]
27. Liu, S.; Wu, X.; Weng, D.; Ran, R. Ceria-based catalysts for soot oxidation: A review. *J. Rare Earths* **2015**, *33*, 567–590. [CrossRef]
28. Martínez, L.M.; Laguna, O.H.; López-Cartes, C.; Centeno, M.A. Synthesis and characterization of Rh/MnO_2-CeO_2/Al_2O_3 catalysts for CO-PrOx reaction. *Mol. Catal.* **2017**, *440*, 9–18. [CrossRef]

29. Gamboa-Rosales, N.K.; Ayastuy, J.L.; González-Marcos, M.P.; Gutiérrez-Ortiz, M.A. Effect of Au promoter in CuO/CeO_2 catalysts for the oxygen-assisted WGS reaction. *Catal. Today* **2011**, *176*, 63–71. [CrossRef]
30. Destro, P.; Marras, S.; Manna, L.; Colombo, M.; Zanchet, D. AuCu alloy nanoparticles supported on SiO_2: Impact of redox pretreatments in the catalyst performance in CO oxidation. *Catal. Today* **2017**, *282*, 105–110. [CrossRef]
31. Destro, P.; Kokumai, T.M.; Scarpellini, A.; Pasquale, L.; Manna, L.; Colombo, M.; Zanchet, D. The Crucial Role of the Support in the Transformations of Bimetallic Nanoparticles and Catalytic Performance. *ACS Catal.* **2017**, *8*, 1031–1037. [CrossRef]
32. Ayastuy, J.L.; Gurbani, A.; Gutiérrez-Ortiz, M.A. Effect of calcination temperature on catalytic properties of Au/Fe_2O_3 catalysts in CO-PROX. *Int. J. Hydrogen Energy* **2016**, *41*, 19546–19555. [CrossRef]
33. Lakshmanan, P.; Park, E. Preferential CO Oxidation in H_2 over $Au/La_2O_3/Al_2O_3$ Catalysts: The Effect of the Catalyst Reduction Method. *Catalysts* **2018**, *8*, 183–193. [CrossRef]
34. Córdoba, L.F.; Martínez-Hernández, A. Preferential oxidation of CO in excess of hydrogen over $Au/CeO_2–ZrO_2$ catalysts. *Int. J. Hydrogen Energy* **2015**, *40*, 16192–16201. [CrossRef]
35. Arévalo, J.D.; Martinez-Hernández, Á.; Vargas, J.C.; Córdoba, L.F. Hydrogen Production and Purification by Bioethanol Steam Reforming and Preferential Oxidation of CO. *Tecciencia* **2018**, *13*, 55–64.
36. Chen, Z.; Pursell, C.J.; Chandler, B.D.; Saavedra, J.; Whittaker, T.; Rioux, R.M. Controlling activity and selectivity using water in the Au–catalysed preferential oxidation of CO in H_2. *Nat. Chem.* **2016**, *8*, 584–589.
37. Dasireddy, V.D.B.C.; Valand, J.; Likozar, B. PROX reaction of CO in $H_2/H_2O/CO_2$ Water–Gas Shift (WGS) feedstocks over $Cu–Mn/Al_2O_3$ and $Cu–Ni/Al_2O_3$ catalysts for fuel cell applications. *Renew. Energy* **2018**, *116*, 75–87. [CrossRef]
38. Zhan, W.; Wang, J.; Wang, H.; Zhang, J.; Liu, X.; Zhang, P.; Chi, M.; Guo, Y.; Guo, Y.; Lu, G.; et al. Crystal Structural Effect of AuCu Alloy Nanoparticles on Catalytic CO Oxidation. *J. Am. Chem. Soc.* **2017**, *139*, 8846–8854. [CrossRef]
39. Reina, T.R.; Megías-Sayago, C.; Florez, A.P.; Ivanova, S.; Centeno, M.Á.; Odriozola, J.A. H_2 oxidation as criterion for PrOx catalyst selection: Examples based on Au–CoOx-supported systems. *J. Catal.* **2015**, *326*, 161–171. [CrossRef]
40. Yung, M.M.; Zhao, Z.; Woods, M.P.; Ozkan, U.S. Preferential oxidation of carbon monoxide on $CoOx/ZrO_2$. *J. Mol. Catal. A Chem.* **2008**, *279*, 1–9. [CrossRef]
41. Fonseca, J.D.; Ferreira, H.S.; Bion, N.; Pirault-Roy, L.; do Carmo Rangel, M.; Duprez, D.; Epron, F. Cooperative effect between copper and gold on ceria for CO-PROX reaction. *Catal. Today* **2012**, *180*, 34–41. [CrossRef]
42. Proaño, L.; Tello, E.; Arellano-Trevino, M.A.; Wang, S.; Farrauto, R.J.; Cobo, M. In-Situ DRIFTS study of two-step CO_2 capture and catalytic methanation over Ru,"Na_2O"/Al_2O_3 Dual Functional Material. *Appl. Surf. Sci.* **2019**, *479*, 25–30. [CrossRef]
43. Jhalani, A.; Schmidt, L.D. Preferential CO oxidation in the presence of H_2, H_2O and CO_2 at short contact-times. *Catal. Lett.* **2005**, *104*, 103–110. [CrossRef]
44. Tanaka, K.I.; He, H.; Shou, M.; Shi, X. Mechanism of highly selective low temperature PROX reaction of CO in H_2: Oxidation of CO via HCOO with OH. *Catal. Today* **2011**, *175*, 467–470. [CrossRef]
45. Wang, F.; Zhang, L.; Zhu, J.; Han, B.; Zhao, L.; Yu, H.; Deng, Z.; Shi, W. Study on different CeO2 structure stability during ethanol steam reforming reaction over Ir/CeO_2 nanocatalysts. *Appl. Catal. A Gen.* **2018**, *564*, 226–233. [CrossRef]
46. Divins, N.J.; Casanovas, A.; Xu, W.; Senanayake, S.D.; Wiater, D.; Trovarelli, A.; Llorca, J. The influence of nano-architectured CeOx supports in $RhPd/CeO_2$ for the catalytic ethanol steam reforming reaction. *Catal. Today* **2015**, *253*, 99–105. [CrossRef]
47. Li, X.; Fang, S.S.; Teo, J.; Foo, Y.L.; Borgna, A.; Lin, M.; Zhong, Z. Activation and deactivation of Au–Cu/SBA-15 catalyst for preferential oxidation of CO in H_2-Rich Gas. *ACS Catal.* **2012**, *2*, 360–369. [CrossRef]
48. Laguna, O.H.; Hernández, W.Y.; Arzamendi, G.; Gandía, L.M.; Centeno, M.A.; Odriozola, J.A. Gold supported on $CuOx/CeO_2$ catalyst for the purification of hydrogen by the CO preferential oxidation reaction (PROX). *Fuel* **2014**, *118*, 176–185. [CrossRef]
49. Redina, E.A.; Greish, A.A.; Mishin, I.V.; Kapustin, G.I.; Tkachenko, O.P.; Kirichenko, O.A.; Kustov, L.M. Selective oxidation of ethanol to acetaldehyde over Au–Cu catalysts prepared by a redox method. *Catal. Today* **2015**, *241*, 246–254. [CrossRef]

50. Wang, J.; Zhong, L.; Lu, J.; Chen, R.; Lei, Y.; Chen, K.; Han, C.; He, S.; Wan, G.; Luo, Y. A solvent-free method to rapidly synthesize CuO–CeO_2 catalysts to enhance their CO preferential oxidation: Effects of Cu loading and calcination temperature. *Mol. Catal.* **2017**, *443*, 241–252. [CrossRef]
51. Wang, X.; Rodriguez, J.A.; Hanson, J.C.; Gamarra, D.; Martínez-Arias, A.; Fernández-García, M. In Situ Studies of the Active Sites for the Water Gas Shift Reaction over Cu–CeO_2 Catalysts: Complex Interaction between Metallic Copper and Oxygen Vacancies of Ceria. *J. Phys. Chem. B* **2006**, *110*, 428–434. [CrossRef]
52. Wang, F.; Büchel, R.; Savitsky, A.; Zalibera, M.; Widmann, D.; Pratsinis, S.E.; Lubitz, W.; Schüth, F. In Situ EPR Study of the Redox Properties of CuO–CeO_2 Catalysts for Preferential CO Oxidation (PROX). *ACS Catal.* **2016**, *6*, 3520–3530. [CrossRef]
53. Mallát, T.; Szabó, S.; Petró, J.; Mendioroz, S.; Folgado, M.A. Real and apparent dispersion of carbon supported palladium-cobalt catalysts. *Appl. Catal.* **1989**, *53*, 29–40. [CrossRef]
54. Barbato, P.S.; Colussi, S.; Di Benedetto, A.; Landi, G.; Lisi, L.; Llorca, J.; Trovarelli, A. Origin of High Activity and Selectivity of CuO/CeO_2 Catalysts Prepared by Solution Combustion Synthesis in CO-PROX Reaction. *J. Phys. Chem. C* **2016**, *120*, 13039–13048. [CrossRef]
55. Schüth, F.; Ward, M.D.; Buriak, J.M. Common Pitfalls of Catalysis Manuscripts Submitted to Chemistry of Materials. *Chem. Mater.* **2018**, *30*, 3599–3600. [CrossRef]
56. Cifuentes, B.; Valero, M.; Conesa, J.; Cobo, M. Hydrogen Production by Steam Reforming of Ethanol on Rh-Pt Catalysts: Influence of CeO_2, ZrO_2, and La_2O_3 as Supports. *Catalysts* **2015**, *5*, 1872–1896. [CrossRef]
57. Nagaraja, B.M.; Padmasri, A.H.; Raju, B.D.; Rao, K.S.R. Vapor phase selective hydrogenation of furfural to furfuryl alcohol over Cu-MgO coprecipitated catalysts. *J. Mol. Catal. A Chem.* **2007**, *265*, 90–97. [CrossRef]
58. Li, J.; Zhan, Y.; Lin, X.; Zheng, Q. Influence of Calcination Temperature on Properties of Au/Fe_2O_3 Catalysts for Low Temperature Water Gas Shift Reaction. *Acta Phys.-Chim. Sin.* **2008**, *24*, 932–938. [CrossRef]
59. Liu, F.; Xiao, Y.; Sun, X.; Qin, G.; Song, X.; Liu, Y. Synergistic catalysis over hollow CeO_2–CaO–ZrO_2 nanostructure for polycarbonate methanolysis with methanol. *Chem. Eng. J.* **2019**, *369*, 205–214. [CrossRef]
60. Chen, W.; Ran, R.; Weng, D.; Wu, X.; Zhong, J.; Han, S. Influence of morphology on basicity of CeO_2 and its use in 2-chloroethyl ethyl sulfide degradation. *J. Rare Earths* **2017**, *35*, 970–976. [CrossRef]
61. Malik, A.S.; Zaman, S.F.; Al-Zahrani, A.A.; Daous, M.A.; Driss, H.; Petrov, L.A. Selective hydrogenation of CO_2 to CH_3OH and in-depth DRIFT analysis for PdZn/ZrO_2 and CaPdZn/ZrO_2 catalysts. *Catal. Today* **2019**, in press. [CrossRef]
62. Zhou, W.; Ma, Z.; Guo, S.; Wang, M.; Wang, J.; Xia, M.; Jia, L.; Hou, B.; Li, D.; Zhao, Y. Comparative study of CO adsorption on zirconia polymorphs with DRIFT and transmission FT-IR spectroscopy. *Appl. Surf. Sci.* **2018**, *427*, 867–873. [CrossRef]
63. Agarwal, S.; Mojet, B.L.; Lefferts, L.; Datye, A.K. Ceria Nanoshapes-Structural and Catalytic Properties. In *Catalysis by Materials with Well-Defined Structures*; Elsevier: Amsterdam, The Netherlands, 2015; pp. 31–70.
64. Leba, A.; Davran-Candan, T.; Önsan, Z.I.; Yildirim, R. DRIFTS study of selective CO oxidation over Au/γAl_2O_3 catalyst. *Catal. Commun.* **2012**, *29*, 6–10. [CrossRef]
65. Fernández-García, S.; Collins, S.E.; Tinoco, M.; Hungría, A.B.; Calvino, J.J.; Cauqui, M.A.; Chen, X. Influence of {111} nanofaceting on the dynamics of CO adsorption and oxidation over Au supported on CeO_2 nanocubes: An operando DRIFT insight. *Catal. Today* **2019**, *336*, 90–98. [CrossRef]
66. Gamarra, D.; Martínez-Arias, A. Preferential oxidation of CO in rich H_2 over CuO/CeO_2: Operando-DRIFTS analysis of deactivating effect of CO_2 and H2O. *J. Catal.* **2009**, *263*, 189–195. [CrossRef]
67. Moretti, E.; Lenarda, M.; Storaro, L.; Talon, A.; Frattini, R.; Polizzi, S.; Rodríguez-Castellón, E.; Jiménez-López, A. Catalytic purification of hydrogen streams by PROX on Cu supported on an organized mesoporous ceria-modified alumina. *Appl. Catal. B Environ.* **2007**, *72*, 149–156. [CrossRef]
68. Polster, C.S.; Nair, H.; Baertsch, C.D. Study of active sites and mechanism responsible for highly selective CO oxidation in H_2 rich atmospheres on a mixed Cu and Ce oxide catalyst. *J. Catal.* **2009**, *266*, 308–319. [CrossRef]
69. Suzuki, Y.; Hayakawa, K.; Fukuhara, C.; Watanabe, R.; Kawasaki, W. A novel nickel-based structured catalyst for CO_2 methanation: A honeycomb-type Ni/CeO_2 catalyst to transform greenhouse gas into useful resources. *Appl. Catal. A Gen.* **2016**, *532*, 12–18.
70. Le, T.A.; Kim, M.S.; Lee, S.H.; Kim, T.W.; Park, E.D. CO and CO_2 methanation over supported Ni catalysts. *Catal. Today* **2017**, *293–294*, 89–96. [CrossRef]